Lecture Notes in Control and Information Sciences

Edited by A. V. Balakrishnan and M. Thoma

Vol. 1: Distributed Parameter Systems: Modelling and Identification
Proceedings of the IFIP Working Conference, Rome, Italy, June 21–26, 1976
Edited by A. Ruberti
V, 458 pages. 1978

Vol. 2: New Trends in Systems Analysis
International Symposium, Versailles, December 13–17, 1976
Edited by A. Bensoussan and J. L. Lions
VII, 759 pages. 1977

Vol. 3: Differential Games and Applications
Proceedings of a Workshop, Enschede, Netherlands, March 16–25, 1977
Edited by P. Hagedorn, H. W. Knobloch, and G. J. Olsder
XII, 236 pages. 1977

Vol. 4: M. A. Crane, A. J. Lemoine
An Introduction to the Regenerative Method for Simulation Analysis
VII, 111 pages. 1977

Vol. 5: David J. Clements, Brian D. O. Anderson
Singular Optimal Control: The Linear Quadratic Problem
V, 93 pages. 1978

Vol. 6: Optimization Techniques
Proceedings of the 8th IFIP Conference on Optimization Techniques, Würzburg, September 5–9, 1977
Part 1
Edited by J. Stoer
XIII, 528 pages. 1978

Vol. 7: Optimization Techniques
Proceedings of the 8th IFIP Conference on Optimization Techniques, Würzburg, September 5–9, 1977
Part 2
Edited by J. Stoer
XIII, 512 pages. 1978

Vol. 8: R. F. Curtain, A. J. Pritchard
Infinite Dimensional Linear Systems Theory
VII, 298 pages. 1978

Vol. 9: Y. M. El-Fattah, C. Foulard
Learning Systems:
Decision, Simulation, and Control
VII, 119 pages. 1978

Vol. 10: J. M. Maciejowski
The Modelling of Systems with Small Observation Sets
VII, 241 pages. 1978

Vol. 11: Y. Sawaragi, T. Soeda, S. Omatu
Modelling, Estimation, and Their Applications for Distributed Parameter Systems
VI, 269 pages. 1978

Vol. 12: I. Postlethwaite, A. G. J. McFarlane
A Complex Variable Approach to the Analysis of Linear Multivariable Feedback Systems
IV, 177 pages. 1979

Vol. 13: E. D. Sontag
Polynomial Response Maps
VIII, 168 pages. 1979

Vol. 14: International Symposium on Systems Optimization and Analysis
Rocquentcourt, December 11–13, 1978;
IRIA LABORIA
Edited by A. Bensoussan and J. Lions
VIII, 332 pages. 1979

Vol. 15: Semi-Infinite Programming
Proceedings of a Workshop, Bad Honnef, August 30 – September 1, 1978
V, 180 pages. 1979

Vol. 16: Stochastic Control Theory and Stochastic Differential Systems
Proceedings of a Workshop of the „Sonderforschungsbereich 72 der Deutschen Forschungsgemeinschaft an der Universität Bonn"
which took place in January 1979 at Bad Honnef
VIII, 615 pages. 1979

Vol. 17: O. I. Franksen, P. Falster, F. J. Evans
Qualitative Aspects of Large Scale Systems
Developing Design Rules Using APL
XII, 119 pages. 1979

Vol. 18: Modelling and Optimization of Complex Systems
Proceedings of the IFIP-TC 7 Working Conference
Novosibirsk, USSR, 3–9 July, 1978
Edited by G. I. Marchuk
VI, 293 pages. 1979

Vol. 19: Global and Large Scale System Models
Proceedings of the Center for Advanced Studies (CAS)
International Summer Seminar
Dubrovnik, Yugoslavia, August 21–26, 1978
Edited by B. Lazarević
VIII, 232 pages. 1979

Vol. 20: B. Egardt
Stability of Adaptive Controllers
V, 158 pages. 1979

Vol. 21: Martin B. Zarrop
Optimal Experiment Design for Dynamic System Identification
X, 197 pages. 1979

For further listing of published volumes please turn over to inside of back cover.

Lecture Notes in Control and Information Sciences

Edited by A.V. Balakrishnan and M.Thoma

INRIA

60

Modelling and Performance Evaluation Methodology

Proceedings of the International Seminar
Paris, France, January 24–26, 1983

Edited by F. Baccelli and G. Fayolle

Springer-Verlag Berlin Heidelberg GmbH 1984

Editors
F. Baccelli
G. Fayolle

INRIA
Institut National de Recherche en Informatique et en Automatique
Domaine de Voluceau, Rocquencourt, B. P. 105
F-78153 Le Chesnay/Cedex
France

ISBN 978-3-540-13288-2 ISBN 978-3-540-38838-8 (eBook)
DOI 10.1007/978-3-540-38838-8

Library of Congress Cataloging in Publication Data
Main entry under title:
Modelling and performance evaluation methodology.
(Lecture notes in control and information sciences; 60)
1. Computer networks -- Congresses. 2. Queuing theory -- Congresses.
3. Computer network protocols -- Congresses.
I. Bacelli, F. (Francois)
II. Fayolle, G. (Guy)
III. Series.
TK5105.5.M63 1984 384 84-1423

Originally published by Springer-Verlag Berlin Heidelberg New York in 1984.

2061/3020-543210

Preface

The purpose of this international seminar was to bring together promising probabilistic tools and advanced research in computer system performance evaluation and, more generaly, to contribute to the enlargment of the interface between computer science and probability.

The computer science oriented papers collected in this volume cover a wide range of applications including communication systems, architecture, data structures and algorithms.

Salient features of these systems such as throuput, response time or stability condition can be formulated in term of probabilistic problems of specific nature.

These problems, in turn, often reduce to the analysis of some steady state properties of basic continuous (or point) processes related to queueing models.

The main mathematical methods -which rely on Markovian features, functional equations and ergodic theory- are presented in more theoretical or survey papers.

F. Baccelli G. Fayolle

Preface

L'objectif de ce séminaire international était de mettre en contact des outils probabilistes prometteurs et les recherches sur l'évaluation de performances des systèmes informatiques.

Les articles "informatiques" continus dans ce volume concernent de nombreux domaines d'applications. On trouvera notamment des analyses de systèmes de communication, d'architectures, de structures de données ou d'algorithmes.

On peut voir dans le détail dans quelle mesure la détermination de caractéristiques essentielles de ces systèmes -tels que le débit, le temps de réponse ou la condition de stabilité- peut se formuler comme un problème probabiliste.

A leur tour, ces problèmes se réduisent le plus souvent à l'analyse de certaines propriétés stationnaires de processus continus ou ponctuels spécifiques liés à des modèles de files d'attente.

Les outils probabilistes les plus utilisés, qui reposent sur l'analyse markovienne, les méthodes d'équations fonctionnelles et la théorie ergodique, sont aussi présentés dans des articles de nature plus théorique.

F. BACCELLI G. FAYOLLE

TABLE OF CONTENTS - TABLE DES MATIERES

I

QUEUES AND NETWORKS 1

FILES D'ATTENTE ET RESEAUX 1

AN ASYMPTOTIC ANALYSIS
OF BLOCKING

F.P. Kelly

Statistical Laboratory,
University of Cambridge,
16 Mill Lane,
Cambridge CB2 1SB
England

1. INTRODUCTION

Product form queueing networks have proved valuable in modelling a variety of computer and communication systems, and have been flexible enough to represent adequately many of the features arising in such applications ([2], [3], [6]). They have not, however, been able to provide much insight into the phenomenon of blocking, a phenomenon which appears particularly unyielding to any general form of exact analysis. In this paper we outline the progress made with alternative, asymptotic approaches to blocking.

The approach to be discussed is based on a model which can be described as follows. Messages are transmitted through a series of nodes linked by communication channels. The lengths of successive messages are independent identically distributed random variables, and the time taken to transmit a message through a channel is equal to its length. Each node has a finite buffer, and when the number of messages at a node reaches the buffer size transmission from the preceding node is interrupted. The most basic measure of the performance of such a system is the maximum rate at which it can accept messages, which we term the throughput. A system's throughput is in general difficult to calculate exactly, but there are available fairly tractable bounds. For a given system these bounds are not especially tight, but they do make possible a number of qualitative insights into the phenomenon of blocking. In particular, we discuss the rate of decay of throughput as series length increases, and the rate at which buffer sizes should grow to ensure that throughput does not decline to zero. We also consider the effect of reordering the sequence of messages, and the effect of segregating long messages from the rest.

The reader is referred to [4] and [5] for further discussion of the topics of this paper and for detailed proofs of the results quoted here, and to [1] and [7] for reviews of previous work on queueing systems with blocking.

2. THROUGHPUT AND SERIES LENGTH

An infinite sequence of messages is to be transmitted through a series of n nodes linked by communication channels (Figure 1). A message is not available for transmission from node i $(i=2,3,\ldots,n)$ until its transmission from node $i-1$ has been completed, and each node transmits messages in the order of their arrival. Each node has a buffer able to hold up to B messages. If node i contains B messages then transmission from node $i-1$ is blocked and must wait until node i has completed transmission of a message. Input of messages to node 1 is instantaneous, so that this node always contains B messages, and transmission from node n is never blocked. The time taken by a node to transmit a message is equal to the length of the message, and is thus the same at each of the n nodes. The length of the uth message is X_u , where $X_1, X_2, \ldots$ are a sequence of independent positive random variables with common distribution F.

Suppose that at time $t=0$ the system begins operation with nodes $2,3,\ldots,n$ empty and the first B messages of the input sequence present at node 1. Let N_t be the number of messages which have completed transmission from node n by time t. Then we define the throughput of the system to be

$$\lambda(n,B,F) = \lim_{t\to\infty} \frac{EN_t}{t}$$

$$= \lim_{t\to\infty} \frac{N_t}{t} \qquad \text{a.s.}$$

where the existence and equality of the limits can be demonstrated. It will be convenient to use the symbol $\lambda(n,B,F)$ as a label for the system itself as well as for the numerical value of the system's throughput.

Although it is in general difficult to calculate $\lambda(n,B,F)$, bounds can be obtained quite easily. For example, consider the system $\lambda(n,1,F)$. The throughput of this system can only be improved if node i is allowed to begin, but not complete, transmission of a message when node $i+1$ is full. This corresponds to the rule common in models of manufacturing job-shops where the server at node i can process a job even though node $i+1$ is full, but the job cannot move on and release the server at node i until there is a space available at node $i+1$. Now the time taken to input a message to the amended system is simply the maximum of the previous n message lengths. Hence the throughput of the amended system is $M(n,F)^{-1}$, where

$$M(n,F) = E \max\{X_1, X_2, \ldots, X_n\}$$

$$= \int_0^\infty [1 - F(x)^n] dx \ .$$

We thus have the upper bound

$$\lambda(n,1,F) \le M(n,F)^{-1} \ . \tag{2.1}$$

A lower bound can be found by a related argument. Suppose that operation of the system $\lambda(2n,1,F)$ is restricted as follows: no message can complete transmission from node i $(i=1,2,\ldots,2n-2)$ until node $i+2$ (in addition to node $i+1$) is empty. For the restricted system the time that elapses between message $u-1$ and message u $(u>n)$ leaving node 1 is $\max\{X_{u-1},X_{u-2},\ldots,X_{u-n}\}$ $+ \max\{X_u,X_{u-1},\ldots,X_{u-n+1}\}$ and so the throughput of the restricted system is $[2M(n,F)]^{-1}$. Thus

$$\lambda(2n,1,F) \ge \tfrac{1}{2}M(n,F)^{-1} \tag{2.2}$$

To illustrate these bounds suppose that message lengths are exponentially distributed, with

$$F(x) = 1 - e^{-x} \qquad x \ge 0 \ .$$

Then

$$M(n,F) = \sum_{i=1}^{n} \frac{1}{i}$$

and so

$$\left(2\sum_{i=1}^{n} \frac{1}{i}\right)^{-1} \le \lambda(2n,1,F) \le \left(\sum_{i=1}^{2n} \frac{1}{i}\right)^{-1} \ . \tag{2.3}$$

Thus

$$\lambda(n,1,F) = O((\log n)^{-1}) \tag{2.4}$$

where the notation $g(n) = O(h(n))$ indicates that

$$0 < \liminf_{n\to\infty} \frac{g(n)}{h(n)} \le \limsup_{n\to\infty} \frac{g(n)}{h(n)} < \infty$$

Although simpler to state, the order relationship (2.4) gives a good deal less information than the bounds (2.3). It is worth noting that when in this paper we give order relationships such as (2.4) they are obtained from explicit bounds such as (2.3) whose calculation depends on the precise form of the distribution F.

The order relationship (2.4) can be generalized to values of B other than one, and to arbitrary distributions: the result is as follows.

Theorem 2.1

$$\lambda(n,B,F) = O(\zeta(n,F)^{-1})$$

where

$$\zeta(n,F) = \inf\{K : K^{-1}\int_K^\infty x\,dF(x) \le \frac{1}{n}\}$$

The theorem puts no conditions on the distribution F, but care is needed in interpreting certain extreme cases. It is clear that if F has infinite mean then $\lambda(n,B,F) = 0$ and $\zeta(n,F) = \infty$, while if there is an upper bound M on message lengths, so that $F(M) = 1$ and $F(x) < 1$ for $x<M$, then as n tends to infinity $\lambda(n,B,F)$ decreases to a limit value not less than $\frac{1}{2}M^{-1}$ (or M^{-1} if $B\ge 2$).

The infimum is necessary in the definition of $\zeta(n,F)$ since the distribution F may have atoms: when the function F is continuous $\zeta = \zeta(n,F)$ satisfies

$$\int_{\zeta}^{\infty} x dF(x) = \frac{\zeta}{n} .$$

Suppose, for example,

$$1-F(x) = \frac{2}{\sqrt{2\pi}} \int_{x}^{\infty} e^{-z^2/2} dz , \tag{2.5}$$

so that X_1 is distributed as $|Y|$ where Y has a standard normal distribution. Then

$$\int_{K}^{\infty} x dF(x) = \frac{2}{\sqrt{2\pi}} e^{-K^2/2}$$

and so

$$\zeta^2 = 2 \log \left(\frac{n}{\zeta} \sqrt{\frac{2}{\pi}} \right)$$

Thus

$$\zeta(n,F) = 0(\sqrt{\log n})$$

and

$$\lambda(n,B,F) = 0((\sqrt{\log n})^{-1})$$

The following Corollaries deal with two important classes of distribution. Corollary 2.3 includes the exponential and normal examples discussed above.

Corollary 2.2 If

$$1 - F(x) = 0(x^{-\rho})$$

where $\rho>1$ then

$$\lambda(n,B,F) = O(n^{-1/\rho})$$

<u>Corollary 2.3</u> If

$$-\log[1 - F(x)] = O(x^{\rho})$$

then

$$\lambda(n,B,F) = O((\log n)^{-1/\rho})$$

3. THROUGHPUT AND BUFFER SIZE

To assess the effect of increasing the buffer size B it is necessary to obtain a further set of bounds on the throughput $\lambda(n,B,F)$. In this Section we indicate the arguments needed.

We begin by considering a system $\lambda(n,B,G)$ in which the distribution G concentrates probability one on the interval $[q,(B-1)q]$. Thus the longest possible message is no more than $(B-1)$ times the length of the shortest possible message. For this system the buffer at node n can never be full - even when $X_u = (B-1)q$ and $X_{u+1} = X_{u+2} = \ldots = X_{u+B-1} = q$ node n must be empty when message u arrives, and a time $(B-1)q$ later it just completes transmission of message u as its buffer receives message $u+B-1$. Thus transmission from node $n-1$ is never blocked, and an inductive argument shows that transmission is never blocked from any node. The throughput of this system is therefore maximal:

$$\lambda(n,B,G) = \left[\int_q^{(B-1)q} x\,dF(x)\right]^{-1} .$$

For an arbitrary distribution F the throughput $\lambda(n,B,F)$ can be bounded by a comparison with a system of the above form. Message lengths generated by F which are less than q are simply increased to q ; and, to deal with messages of length greater than

$(B-1)q$, after transmission of such a long message from node 1 no more messages are transmitted from that node until the long message has cleared the system. This comparison leads to the bound

$$\lambda(n,B,F) \geq \left[qF(q) + \int_q^\infty xdF(x) + (n-1)\int_{(B-1)q}^\infty xdF(x)\right]^{-1} \qquad (3.1)$$

for any $q \in (0,\infty)$.

By comparing the system $\lambda(n,B,F)$ with one in which messages of length less than K are replaced by messages of length zero it is possible to show that

$$\lambda(n,B,F) \leq BK^{-1} + [n\int_K^\infty xdF(x)]^{-1} \qquad (3.2)$$

for any $K \in (0,\infty)$. It is interesting to note the appearance of the tail integrals in the bounds (3.1) and (3.2). In both cases they arise from a calculation of the time taken for a long message to traverse the system.

If we are given the exact form of the distribution F it may be possible to refine the bounds (3.1) and (3.2), and the form of F will certainly affect the best choice of the constants q and K . The rather crude choice $q = c^{-1}$ and $K = Bc^{-1}$ gives the following result.

<u>Theorem 3.1</u> Let F have unit mean and let

$$\eta(n,F) = \inf\{K : \int_K^\infty xdF(x) \leq \frac{1}{n}\} .$$

Then for any $c>0$

$$B \geq c\eta(cn,F) + 1 \Rightarrow \lambda(n,B,F) \geq c/(c+2)$$

$$B < c\eta(cn,F) \Rightarrow \lambda(n,B,F) < 2c \ .$$

If F is continuous then $\eta = \eta(n,F)$ satisfies

$$\int_{\eta}^{\infty} x dF(x) = \frac{1}{n} \ .$$

For example, if F has the normal form (2.5) then

$$\eta(n,F) = \sqrt{2 \log n\sqrt{\frac{2}{\pi}}} \ .$$

We know that if message lengths are not bounded above $\lambda(n,B,F)$ tends to zero as n grows. This tendency can clearly be offset by sufficiently increasing the buffer size B, and Theorem 3.1 determines the required rate of increase. Examples are given in the following Corollaries.

<u>Corollary 3.2</u> Suppose that

$$1 - F(x) = O(x^{-\rho})$$

where $\rho>1$. Then

$$B(n) \geq O(n^{1/(\rho-1)}) \Leftrightarrow \lambda(n,B(n),F) = O(1)$$

<u>Corollary 3.3</u> Suppose that

$$-\log[1 - F(x)] = O(x^{\rho}) \ .$$

Then

$$B(n) \geq O((\log n)^{1/\rho}) \iff \lambda(n,B(n),F) = O(1) .$$

4. REORDERING THE INPUT

Theorems 2.1 and 3.1 relate the throughput $\lambda(n,B,F)$ to the tail behaviour of the distribution F, as measured by the functions $\zeta(n,F)$ and $\eta(n,F)$. This is intuitively reasonable: since messages cannot overtake one another we expect throughput to be greatly influenced by the occasional long message blocking the passage of a large number of subsequent, relatively short, messages, and the tail of F provides information on the frequency and length of long messages. If the input sequence could be reordered so that consecutive messages were closer in length we should expect this blocking effect to be reduced, but how much reordering is necessary?

To answer this question suppose that the input sequence $X_1,X_2,\ldots$ to the system $\lambda(n,1,F)$ is reordered to form an input sequence $Y_1,Y_2,\ldots$ as follows. Messages are reordered in blocks of size J so that messages $Y_{rJ+1},Y_{rJ+2},\ldots,Y_{(r+1)J}$ are the messages $X_{rJ+1},X_{rJ+2},\ldots,X_{(r+1)J}$ arranged in increasing order of length for r odd and decreasing order of length for r even. Let $\beta(J,n,F)$ be the throughput of the resulting system. The following results indicate how J should grow as a function of n in order to bound the throughput $\beta(J,n,F)$ away from zero.

<u>Proposition 4.1</u> Suppose that

$$1 - F(x) = O(x^{-\rho})$$

where $\rho>1$. Then

$$J(n) \geq O(n^{\rho/(\rho-1)}) \iff \beta(J(n),n,F) = O(1)$$

Proposition 4.2 Suppose that

$$-\log[1 - F(x)] = 0(x^{\rho}) .$$

Then

$$J(n) \geq 0(n(\log n)^{1/\rho}) \iff \beta(J(n),n,F) = 0(1) .$$

The reordering of the input sequence could be achieved by having a sorting buffer of size $J(n)$ in front of the system. Comparing Proposition 4.1 and 4.2 with Corollaries 3.2 and 3.3 we see that for the two classes of distribution function concerned the required rate of growth of $J(n)$ is of the same order as that of the total buffer capacity $nB(n)$ when no reordering is allowed.

5. SEGREGATING THE INPUT

The deleterious effect of long messages on throughput might be mitigated if they could be segregated from the rest. In this Section we indicate that by using two series in parallel, one dealing with long messages and the other dealing with the rest, throughput can be substantially improved.

In the system to be considered messages first pass through a segregating node S . A message leaving S goes to the lower or upper series, depending on whether its length is greater than K or not (Figure 2). It takes no time to transfer a message from S to the first node in the relevant series, but transference may be delayed if the buffer of the receiving node is full. Thus the only effect of the segregating node S is to ensure that messages enter the system in the order of the input sequence. Let

$$F^K(x) = \frac{F(x)}{F(K)} \qquad 0 \le x \le K$$

$$F_K(x) = \frac{F(x)-F(K)}{1-F(K)} \qquad K \le x < \infty$$

Messages directed to the upper series of nodes have length distribution F^K, and so the upper series behaves as does the system $\lambda(n,1,F^K)$, except that input to it is interrupted whenever S contains a long message and the buffer of the first node in the lower series is full. Similarly, apart from interruptions to the input, the lower series behaves as does the system $\lambda(n,1,F_K)$. Let $\alpha(n,F,K)$ be the throughput of the combined system.

Bounds on $\alpha(n,F,K)$ in terms of $\lambda(n,1,F^K)$ and $\lambda(n,1,F_K)$ can be obtained as follows. Consider first the effect of reducing to zero transmission times at each node in the lower series. The throughput of the upper series will be $\lambda(n,1,F^K)$, but for each message accepted by the upper series on average $F(K)^{-1}-1$ are accepted instantaneously by the lower series, and so

$$\alpha(n,F,K) \le F(K)^{-1}\lambda(n,1,F^K) \ . \tag{5.1}$$

Similarly by considering the effect of making transmission times zero in the upper series we obtain the bound

$$\alpha(n,F,K) \le [1 - F(K)]^{-1}\lambda(n,1,F_K) \ . \tag{5.2}$$

To obtain a lower bound suppose that operation of the system $\alpha(n,F,K)$ is restricted in the following way: transmission from any node in the upper (respectively, lower) series is interrupted during the periods when the message in S is of length greater than (respectively, not greater than) K . Thus each series functions only when the next message is to be allocated to it. Calculation of the throughput of the restricted system establishes that

$$\alpha(n,F,K) \geq \{F(K)\lambda(n,1,F^K)^{-1} + [1 - F(K)]\lambda(n,1,F_K)^{-1}\}^{-1} \quad (5.3)$$

The best choice of K will depend upon both n and F, and the question of interest is whether

$$\alpha(n,F) = \sup_K\{\alpha(n,F,K)\}$$

declines less quickly with n than $\lambda(n,B,F)$. In general it does, as we shall prove in the case where message lengths are exponentially distributed. The result holds more generally, but the memoryless property of the exponential distribution considerably simplifies the required calculation of expected maxima.

Suppose, then, that

$$F(x) = 1 - e^{-x} \qquad x \geq 0 .$$

For this distribution

$$M(n,F_K) = K + M(n,F)$$

$$= K + \sum_{i=1}^{n} \frac{1}{i}$$

$$\leq K + 1 + \log n .$$

Also

$$M(n,F^K) \leq K .$$

Thus, from the bounds (2.2) and (5.3),

$$\alpha(n,F,K) \geq \tfrac{1}{2}[K + e^{-K}(K + 1 + \log n)]^{-1} .$$

But with the choice

$$K(n) = \log\log n$$

we obtain

$$\alpha(n,F,K(n)) \geq \tfrac{1}{2}[\log\log n + (\log n)^{-1}(\log\log n + 1 + \log n)]^{-1}$$

$$\geq O((\log\log n)^{-1}) ,$$

and so

$$\alpha(n,F) \geq O((\log\log n)^{-1}) .$$

Using the bounds (2.1), (5.1) and (5.2) it is possible to obtain an upper bound of the same order. This suggests that perhaps more generally

$$\alpha(n,F) \doteq O(\zeta(\zeta(n,F),F)^{-1}) .$$

For the classes of distribution considered earlier this is indeed true. The following results should be compared with Corollaries 2.2 and 2.3.

Proposition 5.1 If

$$1 - F(x) = O(x^{-\rho})$$

where $\rho>1$ then

$$\alpha(n,F) = O(n^{-1/\rho^2})$$

Proposition 5.2 If

$$-\log[1 - F(x)] = O(x^{\rho})$$

then

$$\alpha(n,F) = O((\log\log n)^{-1/\rho}) .$$

REFERENCES

[1] Boxma, O.J. and Konheim, A.G. (1981) Approximate analysis of exponential queueing systems with blocking. Acta Informatica, 15, 19-66.

[2] Gelenbe, E. and Mitrani, I. (1980) Analysis and Synthesis of Computer Systems. Academic Press, London.

[3] Kelly, F.P. (1979) Reversibility and Stochastic Networks. Wiley, Chichester and New York.

[4] Kelly, F.P. (1982) The throughput of a series of buffers. Adv. Appl. Prob. 14, 633-653.

[5] Kelly, F.P. (1982) Segregating the input to a series of buffers.

[6] Kleinrock, L. (1976) Queueing Systems, Vol. II: Computer

Applications. Wiley, New York.

[7] Neuts, M.F. (1981) Matrix-geometric Solutions in Stochastic Models. John Hopkins University Press, Baltimore.

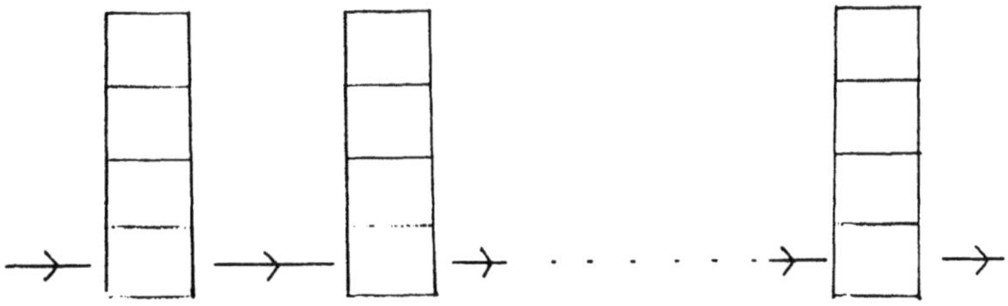

Figure 1. The system $\lambda(n,B,F)$

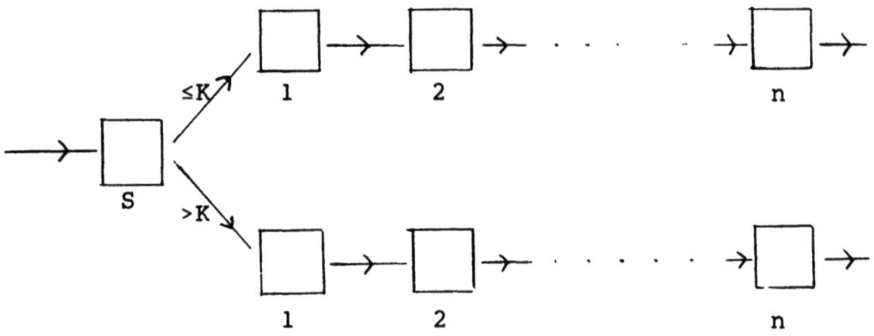

Figure 2. The system $\alpha(n,F,K)$

A CONVERSE TO THE OUTPUT THEOREM OF QUEUEING THEORY: A MARTINGALE APPROACH

Armand M. Makowski
Electrical Engineering Department
University of Maryland
College Park, Maryland 20742

ABSTRACT

A converse to Burke's celebrated output theorem is established by the martingale calculus for point processes. It is shown that when the departure process of an M|M|1 queue is Poisson, then its rate is necessarily equal to the rate of the input process and the system is stable and stationary. Extensions to more complex queueing systems can be obtained by similar methods coupled with filtering results.

I. INTRODUCTION

In 1956, Burke initiated the study of output processes for queueing systems as he proved the following result [2]: The output process of a stable stationary M|M|1 queue is Poisson and has the same rate as the input process. Over the years, this result - widely known as the Output Theorem of Queueing Theory - was refined and given several generalizations with the help of a wide variety of techniques ranging from reversibility arguments to methods from martingale theory. The interested reader is referred to [1], [3] and [4] for more details.

In this note, one explores the following related question: Assume the departure process of a given queueing system to be Poisson. What are then the operating characteristics which are imposed on the system by this statistical structure of the output process? Results in that direction have been discussed in the literature [4] [5] and can be viewed as converses to the Output Theorem, in one form or another. The approach adopted here seems novel in the treatment of these questions and draws on the stochastic calculus associated with the martingale theory for point processes [1]; it also seems to provide additional evidence of the usefulness of martingale-theoretic techniques for studying statistical properties of flows in queueing systems [1] [7].

The simple situation of an M|M|1 queue is discussed and a somewhat ad-hoc line of argumentation is provided; it is nevertheless felt that the proposed methodology could be suitably modified so as to encompass more complex situations.

II. MODEL AND MAIN RESULT

The formulation of an $M|M|1$ queueing system adopted here is borrowed from [6]. Let $(\Omega, \mathcal{F}, P)$ be a complete probability triple that carries a pair of independent Poisson processes $\{A(t), t>0\}$ and $\{Y(t), t>0\}$ with rate λ and μ, respectively, and independent of a given N-valued random variable Z.

On $(\Omega, \mathcal{F}, P)$, construct the N-valued process $\{Q(t), t>0\}$ as the unique process with right-continuous piecewise constant paths solving the equation

$$Q(t) = Z + A(t) - D(t) \tag{2.1}$$

for all $t>0$, where the N-valued process $\{D(t), t>0\}$ is given by

$$D(t) := \int_0^t 1(Q(s-))dY(s) \tag{2.2}$$

Here, the expression $1(n)$ is defined for all n in N, to be $1(n) = 0$ for $n = 0$ and $1(n) = 1$ for $n \neq 0$. A moment of reflection should convince the reader that $\{Q(t), t>0\}$ behaves statistically as the queue size process of an $M|M|1$ queue with initial queue size given by Z and arrivals modelled by the Poisson process $\{A(t), t>0\}$. In this context, $\{D(t), t>0\}$ can then be interpreted as the output or departure process for this queueing system.

To fix the notation, let $(\mathcal{F}_t, t>0)$ denote the (P-completion of the) filtration generated on Ω by the random variable Z and by the processes $\{A(t), t>0\}$ and $\{Y(t), t>0\}$; the corresponding filtration associated with the output process $\{D(t), t>0\}$ is denoted by $(\mathcal{D}_t, t>0)$. Clearly, the inclusion

$$\mathcal{D}_t \subseteq \mathcal{F}_t \tag{2.3}$$

holds for all $t>0$ in view of (2.2).

The result proved in this note can now be stated as follows.

THEOREM. *If* $\{D(t), t>0\}$ *is a Poisson process with rate* σ, *then under the foregoing assumptions, the following holds:*

(i): *For all* $t>0$, *the random variable* $Q(t)$ *is independent of the* σ*-field* $\mathcal{D}_t$.

(ii): *The queue described by* (2.1)-(2.2) *is stable in the sense that* $\lambda<\mu$ *and stationary, with*

$$P[Q(t) = n] = (1 - \frac{\lambda}{\mu})(\frac{\lambda}{\mu})^n \tag{2.4}$$

for all n in N and all $t>0$.

(iii): *In addition, the equality*

$$\sigma = \lambda \tag{2.5}$$

holds.

III. DISCUSSION

The proof of the main result of this note is given in this section; the discussion proceeds in two stages, the first step being taken in the following auxiliary proposition.

PROPOSITION. *For all* z *in* (0,1), *set*

$$\beta(z) := \frac{(1 - z)(\lambda z - \mu)}{z} \tag{3.1}$$

If $\{D(t), t>0\}$ *is a Poisson process with rate* σ, *then under the foregoing assumptions*,

$$E[z^{Q(t)} | \mathcal{D}_t] = \frac{\sigma - \mu}{\lambda z - \mu} \tag{3.2}$$

for all $t>0$, *whenever* $\beta(z)<0$.

PROOF. For all $t>0$ and z in (0,1), set

$$G(t,z) := z^{Q(t)} \exp[\beta(z)t] \tag{3.3}$$

and observe by a simple path analysis, as given in [6], that

$$G(t+h,z) - G(t,z)$$

$$= \int_t^{t+h} \beta(z)G(s,z)ds + (z - 1) \int_t^{t+h} G(s-,z)dA(s) + (\frac{1}{z} - 1) \int_t^{t+h} G(s-,z)dD(s). \tag{3.4}$$

for every $h>0$. Now, by direct inspection of (2.2) and (3.3), one readily concludes that

$$\int_t^{t+h} G(s-,z)dD(s)$$

$$= \int_t^{t+h} G(s-,z)dY(s) - [\int_t^{t+h} \exp[\beta(z)s]dY(s) - \int_t^{t+h} \exp[\beta(z)s]dD(s)]. \tag{3.5}$$

Recall [1] that under the assumptions made,

$$A(t) - \lambda t = (P,\mathcal{F}_t) - \text{martingale} \tag{3.6}$$

$$Y(t) - \mu t = (P,\mathcal{F}_t) - \text{martingale} \tag{3.7}$$

and

$$D(t) - \sigma t = (P,\mathcal{D}_t) - \text{martingale}. \tag{3.8}$$

Hence, by invoking standard results on stochastic integration [1, chapter II], one obtains that

$$E\left[\int_t^{t+h} G(s-,z)dA(s)\,|\,\mathcal{F}_t\right] = \lambda E\left[\int_t^{t+h} G(s-,z)ds\,|\,\mathcal{F}_t\right], \tag{3.9}$$

$$E\left[\int_t^{t+h} G(s-,z)dY(s)\,|\,\mathcal{F}_t\right] = \mu E\left[\int_t^{t+h} G(s-,z)ds\,|\,\mathcal{F}_t\right] \tag{3.10}$$

and, after integration,

$$E\left[\int_t^{t+h} \exp[\beta(z)s]dY(s) - \int_t^{t+h} \exp[\beta(z)s]dD(s)\,|\mathcal{D}_t\right]$$

$$= (\mu - \sigma)\left[\frac{e^{\beta(z)h} - 1}{\beta(z)}\right] e^{\beta(z)t} . \tag{3.11}$$

One has repeatedly used the inclusion (2.3) as well as the ${}_t$-predictability of $\{G(t-,z),t>0\}$ and the $\mathcal{D}_t$-predictability of $\{\exp[\beta(z)t],\ t>0\}$.

In the next step, substitute (3.5) into (3.4) and condition both sides of the resulting equality with respect to $\mathcal{D}_t$; making use of (3.9)-(3.11), one gets from easy calculations that

$$E[G(t+h,z) - G(t,z)|\mathcal{D}_t] = \left[\frac{\sigma - \mu}{\lambda z - \mu}\right]\left[e^{\beta(z)h} - 1\right] e^{\beta(z)t} , \tag{3.12}$$

or equivalently, that

$$E[z^{Q(t+h)}\exp[\beta(z)h]\,|\mathcal{D}_t] - E[z^{Q(t)}\,|\mathcal{D}_t] = \left[\frac{\sigma - \mu}{\lambda z - \mu}\right]\left[e^{\beta(z)h} - 1\right]. \tag{3.13}$$

For every z in (0,1), $|z^{Q(t+h)}|<1$ and therefore

$$\lim_{h\uparrow+\infty} |z^{Q(t+h)}\exp[\beta(z)h]| = 0 \tag{3.14}$$

<u>whenever</u> $\beta(z)<0$. The conclusion (3.2) is now immediate from (3.13) and (3.14) after an easy application of the Bounded Convergence Theorem for conditional expectations.

One is now ready to give a proof of the theorem stated in Section II.

PROOF OF THEOREM. First one shows that $\mu<\lambda$ is necessarily <u>impossible</u> under the assumptions made; this is done by an argument ab absurdo.

If $\lambda=\mu$, then $\beta(z) = -\lambda(1-z)^2/z$ and by the auxiliary result (3.2), one readily concludes that

$$E[z^{Q(t)}] = \frac{\sigma-\lambda}{\lambda(z-1)} \tag{3.15}$$

for all z in (0,1).

Now, in the event that $\lambda \neq \sigma$, one easily gets that

$$0 < \frac{|\lambda - \sigma|}{\lambda(1-z)} = |E[z^{Q(t)}]| < 1 \tag{3.16}$$

for <u>all</u> z in (0,1). Letting $z \uparrow 1$ in (3.16), one obtains a <u>contradiction.</u> If, on the contrary, it is assumed that $\lambda=\sigma$, then

$$E[z^{Q(t)}] = 0 \tag{3.17}$$

for all z in (0,1). But this is clearly impossible since

$$\lim_{z \uparrow 1} E[z^{Q(t)}] = 1 \tag{3.18}$$

from basic principles and a <u>contradiction</u> again arises. Hence $\lambda \neq \mu$.

If one assumes that $\mu<\lambda$, it is natural to introduce z^* in (0,1) defined by $z^* := \mu/\lambda$. Observe that for $0<z<z^*$, $\beta(z)<0$ and (3.2) immediately implies that

$$E[z^{Q(t)}] = \frac{\sigma-\mu}{\lambda z-\mu} \tag{3.19}$$

for these values of z. If by chance $\sigma=\mu$, then $E[z^{Q(t)}] = 0$ for $0<z<z^*$ and therefore

$$P[Q(t) = n] = 0 \tag{3.20}$$

for all n in N, an obvious impossibility! If, on the other hand, $\sigma\neq\mu$, then let $z\uparrow z^*$ in (3.19) and observe that a contradiction again arises since $E[z^{Q(t)}]<1$ and $\lim_{z\uparrow z^*} \left|\frac{\sigma-\mu}{\lambda z-\mu}\right| = +\infty$. Hence $\lambda<\mu$ necessarily in view of the first part of the discussion.

Since $\lambda<\mu$, then $\beta(z)<0$ for <u>all</u> z in (0,1) and (3.2) yields

$$E[z^{Q(t)}|\mathcal{D}_t] = \frac{\sigma-\mu}{\lambda z-\mu} = E[z^{Q(t)}] \tag{3.21}$$

for every $t>0$. Part (i) of the theorem is now immediate. From basic principles, one gets that

$$1 = \lim_{z\uparrow 1} E[z^{Q(t)}] = \frac{\sigma-\mu}{\lambda-\mu} . \tag{3.22}$$

and the relation (2.5) follows. Part (ii) is now straightforward as one substitute (2.5) into (3.21).

The approach adopted here in obtaining the simple characterization result of Section II could be used to analyze more complex situations such as the ones described by Markovian queues. The analysis becomes substantially more involved and seemingly requires use of results in filtering theory as in the work of Bremaud [1].

REFERENCES

[1] P. Bremaud, Point Processes and Queues: Martingale Dynamics, Springer-Verlag, New York, 1981.

[2] P.J. Burke, The output of a queueing system, Opns. Res. 4(1956), pp. 699-704.

[3] P.J. Burke, Output processes and tandem queues, in Proceedings of the Symposium on Computer-Communications Networks and Teletraffic (Ed. J. Fox), New York, April 1972.

[4] D.J. Daley, Queueing output processes, Adv. Appl. Prob. 8 (1976), pp. 395-415.

[5] R.L. Disney, R.L. Farrell and P.R. DeMorais, A characterization of M|G|1|N queues with renewal departure processes, Management Sc. 19 (1973), pp. 1222-1228.

[6] A.F. Martins-Netto and E. Wong, A martingale approach to queues, in Mathematical Programming Study 6 (1976) (Ed. R. J-B. Wets), pp. 97-110.

[7] J. Walrand and P. Varayia, Flows in queueing networks: a martingale approach, Mathematics of Operations Research 6(3) (1981), pp. 387-404.

II

STATIONARITY AND ERGODICITY 1

STATIONNARITE ET ERGODICITE 1

CONSTRUCTION DE FILES D'ATTENTE STATIONNAIRES

J. NEVEU

FILES A UN SERVEUR

1.1.- Soit $N(\omega,.) = \sum_{n\in\mathbb{Z}} \sigma_n(\omega)\, \varepsilon_{T_n(\omega)}$ la mesure aléatoire sur R définie sur l'espace de probabilité $(\Omega,\mathcal{A},P)$ qui représente les demandes de services $\sigma_n : \Omega \to R_+^*$ de clients arrivant aux instants $T_n (\ldots T_{-1} < T_o \leq 0 < T_1 < \ldots$; $\lim_{n\to\pm\infty} T_n = \pm\infty)$. Cette mesure aléatoire est supposée stationnaire par rapport au flot mesurable $(\theta_t,\ t\in R)$ qui préserve P et est ergodique ; cela signifie que $N(\theta_t\omega,.) = \tau_t\, N(\omega,.)$ $(\omega\in\Omega,\ t\in R)$ où le second membre désigne la mesure translatée $\sum_{\mathbb{Z}} \sigma_n\, \varepsilon_{T_n - t}$. Il existe alors une constante $i>0$ que nous supposerons finie telle que $E[N(]s,t)] = i(t-s)$ pour tous $s<t$ dans R ; c'est l'intensité de N qui représente la demande de service moyenne par unité de temps.

Soit en outre $(D(t),\ t\in R)$ une fonction aléatoire réelle positive stationnaire par le flot θ $(D(t)\circ\theta_s = D(t+s)$ pour tous $s,t\in R)$. Si elle ne prend que des valeurs entières, cette f.a. pourra représenter le nombre de serveurs disponibles à chaque instant lorsque le travail est constamment divisé entre ces serveurs ; le cas le plus simple est celui où $D\equiv 1$! La charge du système est alors donnée en régime stationnaire pour la f.a. W de la proposition suivante ; cette f.a. permet ensuite de construire les autres caractéristiques du système notamment à partir des instants où W s'annule.

<u>Proposition 1</u> : <u>Si</u> $i<E[D(.)]$, <u>il existe sur</u> $(\Omega,\mathcal{A},P)$ <u>une f.a. cadlag, stationnaire pour</u> θ, <u>unique telle que</u>

$$(1) \qquad dW(t) + D(t)\, 1_{\{W(t)>0\}}dt = dN(t) \quad \underline{\text{sur}}\ R$$

<u>(i.e. telle que</u> $W(t) - W(s) + \int_s^t D(u)\, 1_{\{W(u)>0\}}du = N]s,t])$ <u>pour tous</u> $s<t$).

<u>En outre</u> $E[D(t)\ 1_{W(t)>0}] = i$ $(t\in R)$.

<u>Si</u> $i=E[D(.)]$, <u>une solution stationnaire de</u> (1) <u>peut encore exister mais elle est nécessairement strictement positive et doit donc vérifier l'équation de cocycle</u> $dW(t) = dN(t) - D(t)dt$; <u>les f.a. positives</u> $W(.)+c^{te}$ = <u>sont alors toutes les solutions. Nous considérerons ce cas de saturation du serveur comme exceptionnel</u>.

<u>Si</u> $i>E[D(.)]$, <u>l'équation</u> (1) <u>n'a pas de solution stationnaire</u>.

Une démonstration consiste à construire les f.a. W_u nulles sur $]-\infty,u]$ et solutions de (1) sur $]u,+\infty]$ et à remarquer que W_u croit lorsque $u\searrow-\infty$; la limite $W_\infty(t) = \lim\limits_{u\searrow-\infty} \nearrow W_u(t)$ est alors la plus petite solution stationnaire de (1) à valeurs dans $\overline{R}_+$ et par l'ergodicité de θ elle est p.s. finie ou p.s. infinie. La croissance des W_u lorsque $u\searrow-\infty$ entraîne que $E[D(t)1_{\{W_\infty(t)>0\}}] \leq i$ et donc que $W_\infty(t) <\infty$ si $E[D(t)] > i$. Inversement si W est une solution p.s. finie stationnaire de (1), $E[D(t)\ 1_{\{W(t)>0\}}] = i$ et donc $E[D(t)] \geq i$; de plus $W \geq W_\infty$ et comme $D(t)\ 1_{\{W(t)>0\}}$ et $D(t)\ 1_{\{W_\infty(t)>0\}}$ ont la même espérance, ces v.a. sont p.s. égales ce qui entraîne que $dW(t) = dW_\infty(t)$. On en déduit facilement l'unicité de W lorsque $E[D(t)]>i$.

La proposition précédente permet par exemple de construire les files stationnaires d'un système à priorité. Si N^p et N^o sont deux mesures aléatoires stationnaires sur $(\Omega,\mathcal{A},P;\theta)$ représentant les demandes prioritaires et ordinaires respectivement, la solution W^p de (1) pour $D \equiv 1$, $N=N^p$ donne à chaque instant la charge en demandes prioritaires tandis que la solution W^o de (1) pour $D=1_{\{W^p=0\}}$, $N=N^o$ donne la charge en demandes ordinaires. La condition $i^p+i^o<1$ sur les intensités assure l'existence de N^p et de N^o (car en ce qui concerne N^o, $E[D(t)]= 1-i^p$ d'après la proposition) ; cette condition est aussi nécessaire si on exclut les cas exceptionnels.

1.2.- Soit $\hat{\Omega}=\{T_o=0\}$ l'espace de Palm de N, soient σ et τ les restrictions de σ_o et T_1 à $\hat{\Omega}$, soit $(\hat{\theta}^n, n\in\mathbb{Z})$ le groupe discret formé par les restrictions à $\hat{\Omega}$ des applications $\theta_{T_n} : \Omega \to \hat{\Omega}(n\in\mathbb{Z})$ et soit $\hat{P}$ la mesure de Palm de N, mesure positive sur Ω portée par $\hat{\Omega}$, de masse totale i et invariante par $\hat{\theta}$. L'espace $\hat{\Omega}$ muni de $\hat{P}$ et de $\hat{\theta}$ permet avec σ et τ de reconstruire la mesure aléatoire stationnaire N sur $(\Omega,\mathcal{A},P;\theta)$.

La v.a. réelle $\hat{W}=W(0-)$ sur $\hat{\Omega}$ associée à une solution W de (1) vérifie l'équation classique des files d'attente

$(\hat{1})$ $\hat{W}\circ\hat{\theta} = (W+\sigma-\tau)^+$ sur $\hat{\Omega}$

et réciproquement toute solution de $(\hat{1})$ permet de construire une solution de (1) par la formule explicite $W(t)=[\hat{W}\circ\theta_{T_n}+\sigma_n-(t-T_n)]^+$ si $T_n \leq t < T_{n+1}$. Nous nous restreignons ici au cas où $D\equiv 1$ dans (1) faute de quoi τ devrait être remplacée par $\int_0^\tau D(t)dt$. Mais pour les applications il est important de considérer une solution de (1) aux instants d'un processus ponctuel $N^* = \sum_{\mathbb{Z}} \varepsilon_{T_n^*}$ éventuellement distinct de $\sum_{\mathbb{Z}} \varepsilon_{T_n}$.

La variable aléatoire $W^*=W(0-)$ restreinte à l'espace de Palm Ω^* de N^* vérifie cette fois l'équation

(1^*) $W^*\circ\theta^* = (W^*+\sigma^*-\tau^*) \vee \rho^*$ sur Ω^*

dans laquelle τ^* et θ^* désignent les restrictions de T_1^* et $\theta_{T_1^*}$ à Ω^*, où $\sigma^*=N([0,\tau^*[)$ représente sur Ω^* la somme des demandes σ_n formulées entre les instants $T_0^*=0$ et $T_1^*=\tau^*$ et où ρ^* est la partie de σ^* qui ne pourrait être effectuée par un serveur libre à l'instant 0 pendant le temps τ^* soit parce que τ^* est trop court, soit parce que les demandes σ_n sont formulées à des instants T_n trop tardifs). Lorsque $T_n^* \equiv T_n (n\in\mathbb{Z})$, on a $\sigma^*=\sigma$ et $\rho^* = (\sigma-\tau)^+$ de sorte que (1^*) se réduit à $(\hat{1})$.

Les équations $(\hat{1})$ et (1^*) se résolvent sur $\hat{\Omega}$ et Ω^* resp. par approximations successives comme (1) sur Ω pourvu que σ,τ resp. σ^*,τ^* et $\rho^*\ \theta^*-\rho^*$ soient intégrables ; elles admettent une solution unique si $\int\sigma\, d\hat{P} < \int\tau\, d\hat{P}$ resp. $\int\sigma^* < \int\tau^*\, dP^*$ et plus précisément une proposition $(\hat{1})$ resp. (1^*) analogue à la proposition ci-dessus peut être énoncée. D'autre part (1^*) peut aussi se ramener à $(\hat{1})$ par le changement de v.a. $w=W-\rho^*\circ(\theta^*)^{-1}$ car cette v.a. vérifie l'égalité $w\circ\theta = (w+\sigma^*+\rho^*(\theta\ )^{-1}-\rho^*-\tau^*)^+$ équivalente à (1^*).

1.3.- La construction suivante de mesures aléatoires stationnaires images intervient constamment dans l'étude des files d'attente, notamment pour étudier les processus des débuts de service ou des sorties.

Lemme 2 : Etant donnée une mesure aléatoire stationnaire N sur $(\Omega,\mathcal{A},P;\theta)$ d'espace de Palm $(\hat{\Omega},\hat{\mathcal{A}},\hat{P};\hat{\theta})$, pour toute v.a. réelle S définie sur $\hat{\Omega}$ telle que $S\circ\hat{\theta}-S$ soit $\hat{P}$-intégrable l'image $N_S(\omega,.)$ de $N(\omega,.)$ par

$t \to t+S(\theta_t(\omega)$ (<u>où</u> S <u>est prolongée de manière arbitraire à</u> Ω) <u>soit</u>

$$N_S(\omega,.) = \sum_{n\in\mathbb{Z}} \sigma_n(\omega)\varepsilon_{T_n+S\theta_{T_n}(\omega)} \quad (\omega\in\Omega)$$

<u>est une mesure aléatoire stationnaire par le flot</u> θ, <u>de même intensité</u> i <u>que</u> N, <u>d'espace et de mesure de Palm</u> $\theta_S(\hat{\Omega})$ <u>et</u> $\theta_S(\hat{P})$ <u>respectivement</u>.

La démonstration de ce lemme est facile. L'hypothèse d'intégrabilité de $S\hat{\theta}-S$ assure que $\lim_{n\to\pm\infty} T_n+S\theta_{T_n} = \pm\infty$ p.s. sur Ω mais la suite $(T_n+S\theta_{T_n}, n\in\mathbb{Z})$ n'est croissante que si $\tau+S\hat{\theta} \geq S$ sur $\hat{\Omega}$; cette condition est vérifiée par exemple par la solution $\hat{W}$ de $(\hat{1})$. On pourra noter aussi que l'image de $N(\omega,.)$ par $t\to S(t,\omega)$ ne peut être stationnaire en général que si $S(t,\omega)$ est de la forme précédente $S(t,\omega) = t+S(\theta_t\omega)$.

Par exemple si des clients arrivant à des instants T_n demandent des services σ_n^1 et σ_n^2 respectivement à deux serveurs en série suivant la discipline premier arrivé, premier servi et si $N^i = \sum_{\mathbb{Z}} \sigma_n^i \varepsilon_{T_n}$ sont des mesures aléatoires stationnaires sur $(\Omega,\mathcal{A},P;\theta)$, en régime stationnaire la charge $\dot{W}^i$ des deux serveurs sont les solutions stationnaires de (1) respectivement pour $N=N^1$ et pour $N=(N^2)_S$ où $S=\hat{W}^1+\sigma^1$ sur $\hat{\Omega} = \{T_o=0\}$. Elles existent et sont uniques dès que les intensités i^1 et i^2 de N^1 et N^2 sont <1 puisque N^2 et N_S^2 ont la même intensité, cette condition étant aussi nécessaire aux cas exceptionnels près.

2. FILES D'ATTENTE A k SERVEURS

Soient σ et τ deux v.a. réelles strictement positives définies sur l'espace de probabilité $(\hat{\Omega},\hat{\mathcal{A}},\hat{P})$ muni d'une transformation inversible bi-mesurable $\hat{\theta}$, préservant $\hat{P}$ et ergodique. La mesure aléatoire $N = \sum_{\mathbb{Z}} \sigma\hat{\theta}^n \varepsilon_{T_n}$ où les T_n sont définis par $T_o=0$, $T_{n+1}-T_n = \tau\circ\hat{\theta}^n (n\in\mathbb{Z})$ constitue alors la description de Palm de la mesure aléatoire stationnaire du §1, la mesure de Palm ayant été normalisée.

Le vecteur des charges $W=(W^i ; 1\leq i\leq k)$ à l'instant 0- de k serveurs assurant la demande N doit vérifier en régime stationnaire les équations

$$(2) \qquad W^i\circ\hat{\theta} = (W^i+\sigma 1_{A^i}-\tau)^+ \quad (i=1,\dots k)$$

où A désigne la partition de $\hat{\Omega}$ formée par les événements $A^i = \{$le $i^{\text{ème}}$ serveur sert le client 0$\}$ $(1\leq i\leq k)$. Nous nous intéresserons ici à la discipline où chaque client se dirige vers le (ou l'un des) serveur le moins chargé à son instant d'arrivée, ce qui revient à exiger que

(2') $W^i = \min_j W^j$ sur A^i $(i=1,\dots k)$.

Le système (2.2') a-t-il des solutions ?

Le vecteur réordonné des charges, soit $V=r\circ W$ où le réordonnement $r:R^k \to R^k$ est défini par $r[x_1,\dots x_k)] = (\min_i x_i,\dots,\max_i x_i)$, vérifie en conséquence de (2.2') l'équation

(3) $V\circ\hat\theta = r[(V^i+\sigma\delta_{i,1}-\tau)^+ \; ; \; 1 \leq i \leq k]$

sur $\hat\Omega$. La v.a. V^1 représente l'attente du client 0 ; l'équation la plus simple que l'on puisse écrire entre V^1 et $V^1\circ\hat\theta$ fait nécessairement intervenir $V^2,\dots V^k$ et est précisément (3).

<u>Proposition 3 - [Loynes] [5] : Si</u> $\int\sigma\, d\hat P < k\int\tau\, d\hat P$, <u>l'équation (3) possède une solution finie p.s. finie minimale. De plus</u>

$$\hat P(V^1=0) \neq 0 \quad \underline{\text{et}} \quad \hat P(V^k<V^1+\sigma) \neq 0.$$

<u>Démonstration</u> : La suite des vecteurs $(V_n,\, n\in N)$ définis par approximation successive par $V_o=0$, $V_{n+1}\,\hat\theta = r[(V_n+\sigma\delta_1-\tau)^+]$ est croissante car r préserve l'ordre de R^k. La limite $V_\infty = \lim_n\uparrow V_n$ est alors la solution minimale de (3) à valeurs dans $\bar R^k_+$.

Les événements $\{V^1_\infty<\infty\}$ et $\{V^k_\infty<\infty\}$ sont invariants et donc négligeables ou de complémentaires négligeables. De plus il est impossible que $V^1_\infty<\infty = V^k_\infty$ p.s. car sinon l'équation satisfaite par les V^k_n, soit

$$V^k_{n+1}\,\hat\theta = [V^k_n \vee (V^1_n+\sigma)-\tau]^+$$

qui entraîne avec l'inégalité $V^k_n \leq V^k_{n+1}$ que

$$\int\tau\, d\hat P \leq \int (V^1_n - V^k_n + \sigma)\vee(\tau - V^k_n)\vee 0\, d\hat P$$

conduirait à la contradiction $\int\tau\, d\hat P \leq 0$ en faisant tendre $n\to\infty$. Si $V^1_\infty<\infty$ p.s., on voit de plus sur l'inégalité limite que $\hat P[V^k<V^1+\sigma] \neq 0$.

Enfin comme

$$\sum_j V^j_{n+1}\circ\hat\theta = \sum_j V^j_n + \sigma - \sum_j (V^j_n+\sigma\delta_{j1})\wedge\tau,$$

la croissance des V_n entraîne encore que

$$\int \sum_j (V_n^j + \sigma\delta_{j1}) \wedge \tau \, d\hat{P} \leq \int \sigma \, d\hat{P}$$

pour tout $n \in N$ et à la limite pour $n=\infty$ de sorte que si $V_\infty^1 = \infty$ p.s., nécessairement $k\int \tau \, d\hat{P} \leq \int \sigma \, d\hat{P}$. Cette dernière conclusion est même valable dès que $V_\infty^j + \sigma\delta_{j1} > \tau$ pour tout j, c'est-à-dire si $V_\infty^j \, \hat{\theta} > 0$ pour tout j. On en conclut bien que si $\int \sigma \, d\hat{P} < k \int \tau \, d\hat{P}$, la v.a. V_∞^1 est p.s. finie et nulle avec une probabilité positive. □

Si (W,A) est une solution de (2.2') et si V désigne le réordonnement de W, il existe au moins une permutation aléatoire $\phi : \hat{\Omega} \to S_k$ (k^e groupe symétrique des permutations de $1,\dots k$) telle que

$$W^i(\omega) = V^{\phi(\omega)i}(\omega) \quad (\omega \in \hat{\Omega} \; ; \; 1 \leq i \leq k) \tag{4}$$

et la permutation aléatoire ρ définie sur $\hat{\Omega}$ par $\phi\hat{\theta} = \rho\hat{\theta}.\phi$ est alors telle que

$$V^{\rho(\hat{\theta}\omega)i}(\hat{\theta}\omega) = [V^i(\omega) + \sigma(\omega)\,\delta_{i1} - \tau(\omega)]^+ . \tag{5}$$

Inversement si V est une solution de (3), la solution minimale de Loynes par exemple, il n'existe pas nécessairement de permutation aléatoire ρ vérifiant (5) qui soit un cocycle (c'est-à-dire pour laquelle il existe une permutation ϕ telle que $\phi\hat{\theta} = \rho\hat{\theta}.\phi$) ; dans ce cas il ne peut exister de W sur $\hat{\Omega}$ et il faut s'appuyer alors sur le lemme 4 ci-dessous, bien connu des ergodiciens, pour construire W sur une extension de $\hat{\Omega}$.

<u>Remarque</u> : Si pour une solution V de (3), l'événement $B=\{V^1=V^k\}$ n'est pas négligeable, par exemple si $\hat{P}(V^k=0) \neq 0$; il existe au moins une solution (W,A) de (2.2') telle que $r(W)=V$ et que $\{A^iB, \; 1 \leq i \leq k\}$ soit une partition de B donnée à l'avance. En effet définissons une application $\phi : B \to S_k$ pour que $(B \cap \{\phi i=1\}, \; 1 \leq i \leq k)$ soit cette partition , choisissons une permutation ρ vérifiant (5) et prolongeons ρ à $\hat{\Omega}$ pour que $\phi\hat{\Omega} = \rho\hat{\theta}\,\phi$ hors de $\hat{\theta}^{-1}B$ (en posant $\phi(\hat{\theta}^n\hat{\omega}) = \rho(\hat{\theta}^n\hat{\omega})\dots\rho(\hat{\theta}\hat{\omega})\,\phi(\omega)$ si $\omega \in B$ et $0 \leq n < \nu_B(\omega)$, modifions enfin ρ sur $\theta^{-1}B$ pour avoir la relation de cocycle partout, ce qui est possible puisque (5) laisse ρ arbitraire sur $\hat{\theta}^{-1}B$. Malheureusement l'hypothèse $\hat{P}(B) \neq 0$ est très forte ! Elle a été introduite par Borovkov dans [1].

<u>Lemme 4</u> : <u>Pour toute permutation aléatoire</u> $\rho : \hat{\Omega} \to S_k$, <u>il existe un sous-groupe</u> H <u>de</u> S_k <u>(unique à une conjugaison près) et des permutations aléatoires</u> $\phi : \hat{\Omega} \to S_k$ et $\rho' : \hat{\Omega} \to H$ <u>telles que</u>

a). $\phi\hat{\theta}.\rho'\hat{\theta} = \rho\hat{\theta}.\phi$ sur $\hat{\Omega}$

b). $\Theta(\omega,h) = (\hat{\theta}\omega, \rho'(\hat{\theta}\omega)h)$ soit une transformation sur $\hat{\Omega} \times H$ qui soit inversible, préserve $\hat{P} \otimes \lambda_H$ (λ_H : probabilité uniforme sur H) et soit ergodique.

Ce lemme se démontre facilement en étudiant la décomposition ergodique de la transformation $(\omega,g) \to (\hat{\theta}\omega, \rho(\hat{\theta}\omega)g)$ de $\Omega \times S_k$. Il permet en posant sur $\hat{\Omega} \times H$

$$W^i(\omega,h) = V^{\phi(\omega)hi}(\omega)$$

(6)

$$A^i = \{(\omega,h) : \phi(\omega)\ hi=1\} \quad (i=1\ldots k)$$

de définir un vecteur aléatoire W et une partition A qui sont solution de (2.2') sur l'extension $\hat{\Omega}\times H$ [avec $\sigma(\omega,h) = \sigma(\omega)$, $\tau(\omega,h) = \tau(\omega)$]. On peut montrer de plus que toute solution de (2.2') sur une extension de $\hat{\Omega}$ se projète sur une solution du type précédent.

On notera aussi que la transformation $(\omega,h) \to (\omega, h\,h_o)$ de $\Omega\times H$ associée à h_o H, commute avec Θ et envoie W^i sur W^j si $j=h_o i$. Les charges W^i de serveurs dont les indices appartiennent à une même classe de transitivité de H sont donc semblables. De plus si J est une telle classe de transitivité, $A^J = \sum_J A^i$ est un événement de $\hat{\Omega}$ et le service $\sigma 1_{A^J}$ est effectué par les serveurs d'indices $\in J$.

Les résultats précédents permettent de montrer que la non-unicité des solutions de (3) dont Loynes a donné un exemple, correspond à un phénomène de saturation d'un sous-ensemble de serveurs. En effet d'abord, grâce au lemme suivant, il existe une permutation aléatoire ρ sur $\hat{\Omega}$ vérifiant (6) pour toutes les solutions V de (3).

Lemme 5 : Si V est une solution de (3) sur $\hat{\Omega}$ distincte de la solution minimale de Loynes V_∞, il existe une permutation aléatoire ρ sur $\hat{\Omega}$ vérifiant (5) à la fois pour V et $\dot{V}_\infty$ et de plus $V^i_\infty + \sigma\delta_{i1} \geq \tau$ sur $\{V^i > V^i_\infty\}$.

Démonstration : Le réordonnement de R^k contracte la distance euclidienne et plus précisément $|r(v)-r(u)| < |v-u|$ sauf si r(v) et r(u) peuvent s'obtenir à partir de v et u par une même permutation de coordonnées. D'autre part pour tout couple de réels a,b, on a $|b^+-a^+| \leq |b-a|$ avec l'inégalité stricte sauf si a=b ou si a,b sont positifs.

Cette double remarque entraîne que

$$|V-V_\infty|\circ\hat{\theta} = |r(V+\sigma\delta_1-\tau)^+ - r(V_\infty+\sigma\delta_1-\tau)^+|$$

$$\leq |(V+\sigma\delta_1-\tau)^+ - (V_\infty+\sigma\delta_1-\tau)^+$$

$$\leq |V-V_\infty|$$

et comme $\hat{\theta}$ préserve la mesure, de strictes inégalités sont ici p.s. impossibles ce qui permet de conclure d'après le début. □

Par conséquent les divers vecteurs W définis sur $\hat{\Omega}\times H$ par (6) à partir des diverses solutions V de (3) et d'une même permutation ρ vérifient tous les équations (2) avec la même partition A. D'après la proposition $(\hat{1})$ cela n'est possible pour un vecteur W et le vecteur W_∞ correspondant à V_∞ que si $W^i_\infty = W^i + C^i$ pour une constante positive C^i qui ne dépend d'ailleurs d'après (6) que de la classe de transitivité J de i. Pour que la constante C^i ne soit pas nulle, il faut que $\sigma 1_{A^i} - \tau = W^i_\infty\circ\hat{\theta} - W^i$ sur $\hat{\Omega}\times H$ et donc que

$$\sigma 1_{A^J} - |J|\tau = W^J_\infty\circ\hat{\theta} = W^J_\infty \quad \text{sur } \hat{\Omega} \tag{7}$$

pour la v.a. positive définie sur $\hat{\Omega}$ par $W^J_\infty(\omega) = \sum_J W^i_\infty(\omega,h) = \sum_J V^{\phi(\omega)j}(\omega)$ indépendamment de h ; en particulier $\int_{A^J}\sigma d\hat{P} = |J|\int\tau\, d\hat{P}$ et la classe de serveurs J travaille donc à saturation. Remarquons encore que si H est transitif sur $\{1,\ldots k\}$, l'équation (3) a nécessairement une solution unique puisque nous avons supposé que $\int\sigma\, d\hat{P} < k\int\tau\, d\hat{P}$.

3. SYSTEMES A REJET

Etant donné $(\hat{\Omega},\hat{\mathcal{A}},\hat{P};\theta)$ et σ,τ avec les mêmes hypothèses qu'au paragraphe 2, désignons par ν la v.a. entière > 0

$$\nu(\omega) = \min\{n\ ;\ n\in N^*,\ \sum_{0}^{n-1}\tau(\hat{\theta}^j\omega) \geq \sigma(\omega)\}$$

donnant pour le système à un serveur fonctionnant suivant la discipline du rejet, l'indice du premier client accepté après le client 0 lorsque ce dernier a été lui-même accepté. La construction du régime stationnaire de ce système se ramène à celle d'une partie aléatoire $M(\omega)$ de $\mathbb{Z}$, l'ensemble des indices des clients acceptés, telle que

$$\begin{aligned} &M(\hat{\theta}\omega) = M(\omega) - 1 \quad \text{(stationnarité)},\\ &\nu(\omega) = \min[n\ ;\ n\in N^*\cap M(\omega)] \quad \text{si } 0\in M(\omega). \end{aligned} \tag{8}$$

On demande en outre que $\hat{P}(M\neq 0) \neq 0$ auquel cas M est nécessairement p.s.

infini. Si $A=\{0\in M(.)\}$, le problème de la construction de M est équivalent à celui de la construction d'un événement non-négligeable A de $\hat{\Omega}$ tel que

$$(8') \qquad \nu = \nu_A \quad \text{sur } A,$$

si ν_A désigne le temps de retour à A des orbites de $\hat{\theta}$ (soit $\nu_A(\omega) = \min(n : n\in N^*, \hat{\theta}^n \omega\in A)$.

Ce problème est résolu par la proposition suivante. Cette proposition est apparue aussi dans [6] sous une forme un peu **différente.**

Proposition 6 : Si $\int \nu\, d\hat{P}<\infty$, l'espace $\Omega^* = \lim_{j\uparrow\infty} \downarrow (\theta^\nu)^j \hat{\Omega}$ n'est pas vide et $c = \int_{\Omega^*} \nu d\hat{P}$ est une transformation p.s. inversible de Ω^*, son inverse étant de la forme $\theta^{-\nu^*}$ pour une v.a. $\nu^* : \Omega^* \to N^*$, qui préserve en outre la restriction P^* de $\hat{P}$ à Ω^*. La tribu invariante de θ^ν sur Ω^* est finie. Soit Ω^*_o un de ses atomes.

Considérons alors la "tour" $\overline{\Omega} = \Omega^*_o \times [0,\nu[$ qui est le sous-ensemble $(\omega^*,n) : 0\leq n < \nu(\omega^*)\}$ de $\Omega^*_o \times \mathbb{Z}$, munie de la mesure $\overline{P} = 1_{\overline{\Omega}}(P \otimes \lambda_{\mathbb{Z}})$ où $\lambda_{\mathbb{Z}}$ désigne la mesure de comptage de $\mathbb{Z}$ et de la transformation $\overline{\theta}(\omega^*,n)=(\omega^*,n+1)$ ou si $n+1 = \nu(\omega^*)$, $(\theta^\nu \omega^*,0)$; cette transformation est ergodique p.s. inversible et préserve $\overline{P}$. L'application $\phi : \overline{\Omega} \to \Omega$ définie par $\phi(\omega^*,n) = \theta^n \omega^*$ est alors un homomorphisme $(c_o,1)$ de $(\overline{\Omega},\overline{P},\overline{\theta})$ sur $(\Omega,c\hat{P},\hat{\theta})$ et l'ensemble aléatoire $\overline{M}$ défini sur $\overline{\Omega}$ par

$$\overline{M}(\overline{\omega}) = \{n \; ; \; n\in\mathbb{Z}, \; \overline{\theta}^n \overline{\omega}\in\Omega^* \times \{0\}\}$$

est solution de (8) sur $\overline{\Omega}$ relativement à $\overline{\theta}$ et $\nu\circ\phi$ (c_o désigne l'entier $\int_{\Omega^*_o} \nu d\hat{P}$).

Démonstration : Les sous-espaces $\Omega_j=(\theta^\nu)^j \; \Omega (j\in N)$ décroissent avec j et θ^ν applique Ω_j sur Ω_{j+1}. Par suite $\nu\geq\nu_{\Omega_j}$ sur Ω_j et d'après **la** formule de Kac

$$\int_{\Omega_j} \nu \, d\hat{P} \geq \int_{\Omega_j} \nu_{\Omega_j} \, d\hat{P}=1 \quad (j\in N).$$

A la limite $\int_{\Omega^*} \nu \, d\hat{P}\geq 1$ puisque ν est intégrable et Ω^* ne peut être négligeable.

Pour tout $\omega\in\Omega$, le sous-ensemble $\xi_j(\omega) = \{n : \theta^{-n} \omega \in \Omega_j, \nu(\theta^{-n}) = n\}$ de N^* décroit avec j, est non vide si et seulement si $\omega \in \Omega_{j+1}$ et est p.s. fini car par l'invariance de $\hat{P}$

$$\int \text{Card } \xi_j(\omega)\, d\hat{P}(\omega) = \sum_{N^*} \hat{P}[\Omega_j \cap (\nu=n)] = \hat{P}(\Omega_j) \leq 1.$$

Donc $\xi^*(\omega) = \lim_j \downarrow \xi_j(\omega) = \{n : \theta^{-n} \omega\in\Omega^*,\ \nu(\theta^{-n}\omega) = n\}$ est par compacité non vide si et seulement si $\omega\in\Omega^*$ tandis que

$$\int \text{Card } \xi^*(\omega)\, d\hat{P}(\omega) = P(\Omega^*).$$

Ainsi pour presque tout $\omega\in\Omega^*$, $\xi^*(\omega)$ se réduit à un seul point, soit $\nu^*(\omega)$, et nous avons donc $\{\nu^*=n\} = \{\nu\, \theta^{-n}=n\}$ dans Ω^* pour tout $n\in N^*$. Il s'en suit facilement que $\theta^{-\nu^*}$ inverse θ^ν sur Ω^* et que θ^ν préserve P.

Ces propriétés de θ^ν sur Ω^* entraînent comme il est bien connu que $\overline{\theta}$ définit une transformation inversible de $\overline{\Omega}$ préservant $\overline{P}$. D'autre part la fonction $\eta : \Omega \to N$ défini par

$$\eta(\omega) = \text{Card } \phi^{-1}(\{\omega\}) = \sum_N 1_{(\hat{\theta}^{-n} \omega\in\Omega^*,\ \nu(\hat{\theta}^{-n} \omega) > n\}}$$

est p.s. invariante par $\hat{\theta}$ car

$$\eta-\eta\ \hat{\theta}^{-1} = 1_{\Omega^*} - \text{Card } \xi^*(\omega) = 0 \quad \text{p.s.} ;$$

elle est donc égale p.s. à un entier positif c qui n'est pas nul puisque $\eta\geq 1_{\Omega^*}$. De plus pour toute fonction mesurable $f : \hat{\Omega} \to R_+$

$$\int_{\overline{\Omega}} f\circ\phi\, d\overline{P} = \int_{\Omega^*} \sum_{o}^{\nu-1} f\hat{\theta}^{-n}\, d\hat{P} = \int_{\hat{\Omega}} f\eta\, d\hat{P} = c\int_{\hat{\Omega}} f\, d\hat{P}.$$

Ces propriétés de η restent valables si on remplace Ω^* par un de ses sous-ensembles θ^ν-invariant ; il s'en suit facilement que la décomposition ergodique de Ω^* pour θ^ν ne peut contenir plus de c atomes. La fin de la démonstration est alors immédiate.

De plus on peut démontrer que toute extension ergodique de $(\hat{\Omega},\hat{\mathcal{A}},\hat{P};\hat{\theta})$ portant une solution de (8) est nécessairement une extension d'un des espaces $\overline{\Omega}$ associés aux parties ergodiques Ω^*_o de Ω^* construits ci-dessus.

<u>Remarque</u> : Si l'événement $C = \bigcap_{N^*} (\nu\, \theta^{-n} \leq n)$ n'est pas négligeable, la fonction η de la démonstration précédente est égale à 1 sur C et donc sur Ω ; aucune extension de Ω n'est donc nécessaire dans ce cas pour construire une solution de (8). D'ailleurs dans ce cas il est immédiat de voir que $A = \bigcup_N (\theta^\nu)^j C$ vérifie (8') comme l'a montré Borokov. [1].

Lisek a également dans [3] considéré une méthode proche de l'extension pour obtenir sinon $\overline{M}$, du moins sa loi.

Je remercie MM. Flipo, Jaïbi, Puka et Robert pour leur cordiale collaboration.

BIBLIOGRAPHIE

[1] A.A. BOROVKOV : Random processes in queuing theory. Springer-Verlag, (1976).

[2] P. FRANKEN, D. KÖNIG, U. ARNDT et V. SCHMIDT : Queues and Point processes. Akademie Verlag, Berlin (1981).

[3] J.C. KIEFFER et J.G. DUNHAM : On the long-run behavior of the sequence of outputs of a sequential machine. Preprint.

[4] B. LISEK : A method for solving a class of recursive stochastic equations. Z.f. Wahrsch, 60 (1982), 151-162.

[5] R.M. LOYNES : The stability of queues with non-independent inter-arrival and service times. Proc. Cambridge Phil. Soc. 58 (1962), 497-520.

[6] J. NEVEU : Processus ponctuels. Lect. Notes Math. 598, Springer-Verlag (1977), 249-447.

COMPARISON THEOREMS FOR QUEUES WITH DEPENDENT INTER-ARRIVAL TIMES

Tomasz Rolski
(Wroclaw University, Mathematics Institute,
pl.Grunwaldzki 2/4, 50 384 Wroclaw, Poland)

Summary. We consider a G/GI/1 queue under a "first in first out" discipline and a G/M/s loss system (s=1,2). Customers may arrive at the system according to two classes of arrival processes. The first class consists of $\underline{\Phi} \circ \underline{\Lambda}$ point processes where $\underline{\Phi}$ is a point process and $\underline{\Lambda}$ is a random measure (if $\underline{\Phi}$ is a stationary Poisson process then $\underline{\Phi} \circ \underline{\Lambda}$ is a familiar doubly stochastic Poisson process). The second class includes Markov renewal point processes. The notion of an ergodically stable sequence of random variables introduced by Rolski (1981a), (1981b) is utilized to define the mean stationary delay and the stationary probability of loss. Denoting the mean waiting time in two G/GI/1 queues by $c(\Sigma_1)$ and $c(\Sigma_2)$ respectively we look for conditions on arrival processes in these queues for

$$c(\Sigma_1) \leq c(\Sigma_2) .$$

Similar results are stated for the stationary probability of loss in a G/M/s loss system. One of a proven result contains as a special case a Rolski (1981b) solution of Ross's (1978) conjecture.

I. PRELIMINARIES

§. 1. Introduction

In this paper we consider a G/GI/1 queue under a "first in first out" discipline and a G/M/s loss system (s=1,2). In the first system a performance measure for the system is the mean stationary delay while in the second the stationary probability of loss. Supposing the same service process in two G/GI/1 (or G/M/s) systems Σ_1 and Σ_2 we are looking for conditions on arrival processes for

$$c(\Sigma_1) \leq c(\Sigma_2) ,$$

where $c(\Sigma)$ denotes the respective performance measure for a system Σ . Such results we call comparison theorems. Theorems where either Σ_1 or Σ_2 is such that the characteristic c in this system is computable are especially interesting. This is in the case where Σ is M/GI/1 queue or M/M/s loss system respectively.

There is a large body of literature consisting of comparison theorems of queues with arrivals according to renewal processes. For a recent review of results in this field, see the monograph by Stoyan (1977).

If the input is not a renewal process the difficulties increase rapidly. According to my knowledge there exist a few papers on comparison theorems in such queues. Among them are: Stoyan (1973), Fleischmann (1976), König and Schmidt (1980), Rolski (1981b). In this paper we give further results in this direction. There also exist papers on stochastic ordering of queues under stochastic ordering type assumptions; see e.g. O'Brien (1975). Compared with conditions assumed in this paper, the notion of stochastic ordering is stronger than conditions supposed here. The most important feature of the conditions required in the paper is that the asymptotic arrival rate (called the stationary arrival rate) in both compared queues is the same.

Two types of arrivals are considered in this paper. These are $\underline{\Phi} \circ \underline{\Lambda}$ point processes defined later in Chapter 3.1 and Markov renewal point processes. Notice that the class of $\underline{\Phi} \circ \underline{\Lambda}$ point processes includes as a special case the class of doubly stochastic Poisson processes (DSPP). Queues with double stochastic Poisson arrivals were considered by Ross (1978) and subsequently by Fond and Ross (1978) and Niu (1980). Actually in these papers the arrivals, although assumed to constitute a DSPP, are of the renewal type as the results are obtained for a very special subclass of DSPP's. Ross (1978) set forth two conjectures that his results are valid in general cases too. He conjectured some bounds for the mean stationary delay in G/GI/1 queues and the probability of loss in G/M/s loss systems under double stochastic Poisson arrivals. The problem was resolved in full generality by Rolski (1981b) in a case of G/GI/1 queues. In this paper both the conjectures follow from more general results. Conjectured bounds are obtained via comparison theorems. Similar results are given for the second type of arrivals when they form a Markov renewal point process.

The main tools employed in the paper are the Palm theory of point processes, ergodic theory and n-dimensional version of conditional Jensen's inequality. We follow here the stability theory from Rolski (1981a), developed subsequently by Rolski (1981b), Rolski and Szekli (1982). Some new notions and results on stability are given in the paper. For ergodic theory we refer the reader to Breiman (1968), Brown (1976), Doob (1953).

The paper consists of three chapters. After the first introductory chapter the second deals with the stability theory which is useful for stating main results of the paper. These are comparison theorems which are given in Chapter 3.

§. 2. Some notations and definitions

Let R_+ be the non-negative reals, Z_+ the non-negative integers, Z the integers, N the positive integers and E a Polish space. For denoting the Borel σ-field of subsets of a space, we write B before the symbol denoting the space. The set of all possible mappings of A into B is denoted by B^A. We write simply $B^s = B \times \ldots \times B$ (s-times). By I we denote Z_+ or Z. If $\underline{e}_i = \{e_{ij}, j \in I\}$ $(i=1,2)$ then $(\underline{e}_1,\underline{e}_2) = \{(e_{1j},e_{2j}), j \in I\}$. We define the shift γ on E^I by

$$\gamma\{e_i, i \in I\} = \{e_{i+1}, i \in I\} .$$

The k-fold superposition of γ we denote by γ^k. The shift $\gamma: E^I \to E^I$ is a measurable mapping.

All random elements are defined on a probability space (Ω,F,Pr). The distribution of a random element $\underline{X}$ we denote by $P_{\underline{X}}$. We write $\underline{X} \overset{d}{=} \underline{Y}$ if $P_{\underline{X}} = P_{\underline{Y}}$.

We now recall definitions of random measures and point processes on R_+. Similar ones exist for random measures and point processes on R.

A random measure $\underline{\Lambda}$ on R_+ or R is defined as a mapping

$$\underline{\Lambda}: (\Omega,F,Pr) \to (M,BM)$$

where M is a space of locally finite measures on R_+ or R respectively. We state now a few definitions for random measures on R_+; similar ones exist for random measures on R. By $\mu(t) = \mu[0,t]$ $(t \in R_+)$ we denote the distribution function of a measure $\mu \in M$. A measure μ on R_+ is called diffuse if its distribution function $\mu(t)$ is continuous for $t \in R_+$ and $\mu(0) = 0$. A random measure $\underline{\Lambda}$ is diffuse if its trajectories are diffuse almost surely (a.s.). Let N be the set of non-negative integer-valued measures from M. A random measure $\underline{\Phi}$ such that $Pr(\underline{\Phi} \in N) = 1$ is called a point process. We endow the set M with vague topology. Then M is a Polish space and N is a measurable subset of M.

A measure $\nu \in N$ is called simple if $\nu(\{t\}) \leq 1$ $(t \in R_+)$. A random measure $\underline{\Phi}$ is called simple if trajectories of $\underline{\Phi}$ are simple a.s.. For details about random measures and point processes we refer to Kallenberg's book (1976). The family of shifts $\{\tau^t, \ t \in R_+\}$ on M is defined by

$$\tau^t\mu(B) = \mu(B + t) \qquad (t \in R_+, \quad B \in BM) .$$

The mapping

$$M \times R_+ \ni (\mu,s) \to \tau^s\mu \in M$$

is measurable.

Any simple measure $\nu \in N$ has the representation $\nu = \sum_i 1_{\{\zeta_i\}}$ where $\zeta_i = \zeta_i(\nu)$ can be chosen in a measurable way and such that

$$\ldots < \zeta_{-1} < \zeta_0 \leq 0 < \zeta_1 < \ldots \quad .$$

If ν is a measure on R_+ we set

$$t_0 = \zeta_1 ,$$

$$t_i = \zeta_{i+1} - \zeta_i \qquad (i \in N) ,$$

otherwise if ν is a measure on R

$$t_i = \zeta_{i+1} - \zeta_i \qquad (i \in Z) .$$

If $\underline{\Phi}$ is a point process then $T_i = t_i(\underline{\Phi})$ and $Z_i = \zeta_i(\underline{\Phi})$. We adopt the convention that a symbol attached to a letter which denotes a point process (random measure) is also used with letters denoting characteristics of this point process (random measures) e.g. $T_i^* = t_i(\underline{\Phi}^*)$.

In the case of N we define another mapping $\sigma\colon N \to N$ by

$$\sigma\nu = \tau^{\zeta_1(\nu)}(\nu) .$$

II. STABILITY THEOREMS

§. 1. Stability notions

We now recall a concept of stable random elements. We follow here the terminology introduced by Rolski (1981a),(1981b) . Gray and Keiffer (1980) also investigated this notion for sequences of random elements and called them asymptotically mean stationary. We state definitions and properties for random sequences indexed by Z_+ and random measures (point processes) on R_+ . Similar definitions exist on Z or R respectively.

Definition 2.1: The sequence of random elements $\underline{X} = \{X_i, \ i \in Z_+\} \in E^{Z_+}$ is said to be ergodically stable if

$$\lim_{n\to\infty} \frac{1}{n} \sum_{i=0}^{n-1} Pr(\gamma^i \underline{X} \in D) = P_{\underline{X}^0}(D) \qquad (D \in \mathcal{B}E^{Z_+}) \tag{2.2}$$

where $P_{\underline{X}^0}$ is a probability measure on E^{Z_+} and the shift γ is measure preserving ergodic on $(E^{Z_+}, \mathcal{B}E^{Z_+}, P_{\underline{X}^0})$. Any sequence of random elements $\underline{X}^0$ having the distribution $P_{\underline{X}^0}$ defined in (2.2) is called a stationary version of $\underline{X}$ and the distribution $P_{\underline{X}^0}$ is called the stationary distribution of $\underline{X}$.

Following the general definition of stable random elements from Rolski (1981a) we define now stable random measures.

Definition 2.3: A random measure $\underline{\Lambda}$ is said to be ergodically stable if

$$\lim_{t\to\infty} \frac{1}{t} \int_0^t Pr(\tau^s \underline{\Lambda} \in D)ds = P_{\underline{\Lambda}}*(D) \qquad (D \in \mathcal{B}M) \tag{2.4}$$

where $P_{\underline{\Lambda}*}$ is a probability measure on M and $\{\tau^s, \ s \in R_+\}$ is measure preserving ergodic on $(M,\mathcal{B}M,P_{\underline{\Lambda}*})$. Any random measure $\underline{\Lambda}^*$ having the distribution $P_{\underline{\Lambda}*}$ defined in (2.4) is called a stationary version of $\underline{\Lambda}$ and the distribution $P_{\underline{\Lambda}*}$ is called the stationary distribution of $\underline{\Lambda}$.

We denote $\underline{X} \overset{s}{\equiv} \underline{Y}$ if random elements $\underline{X}$, $\underline{Y}$ (assuming values at E^{Z_+} or M respectively) have the same stationary distribution.

The second notion of stability is defined for point processes only. It is defined by the mapping σ .

Definition 2.5: A point process $\underline{\Phi}$ is σ-ergodically stable if

$$\lim_{n\to\infty} \frac{1}{n} \sum_{i=0}^{n-1} Pr(\sigma^i \underline{\Phi} \in D) = P_{\underline{\Phi}^0}(D) \qquad (D \in \mathcal{B}N)$$

where $P_{\underline{\Phi}^0}$ is a probability measure on N and σ is measure preserving ergodic on $(N^0 = \{\nu \in N,\ \nu(\{0\}) = 1\},\ BN^0,\ P_{\underline{\Phi}^0})$. Any point process $\underline{\Phi}^0$ having the distribution $P_{\underline{\Phi}^0}$ is called a Palm $\underline{\Phi}^0$ version of $\underline{\Phi}$ and the distribution $P_{\underline{\Phi}^0}$ is called the $\underline{\Phi}^0$ Palm distribution of $\underline{\Phi}$.

There is a close relationship between σ-stability and γ-stability. Let $\underline{\Phi}$ be a simple point process with the inter-point distances sequence $\underline{T} = \{T_i,\ i \in Z_+\}$. By noting the fact that $\sigma\underline{\Phi} = \gamma\underline{T}$ we have that $\underline{\Phi}$ is σ-ergodically stable if and only if $\underline{T}$ is ergodically stable. An example of an ergodically stable sequence of random elements is a stationary ergodic sequence of random elements. The stationary sequence of independent random elements is known to be ergodic (in the case of random variables see e.g. Breiman (1968)). We consequently append the star * to a stationary version of a random measure. Similarly we append a little circle o to a stationary version of a sequence of random elements or to a Palm version of a point process. An equivalent condition for a sequence of random elements (random measures) to be ergodically stable is given in the following theorem. The theorem also gives the reader some idea of what stable random elements look like.

Theorem 2.6:

(i) A sequence of random elements $\underline{X} = \{X_i,\ i \in Z_+\}$ is ergodically stable if and only if for any measurable function $f\colon E^{Z_+} \to R_+$ we have

$$\lim_{n\to\infty} \frac{1}{n} \sum_{i=1}^{n} f(\gamma^i \underline{X}) = Ef(X^0) \qquad \text{a.s..}$$

(ii) A random measure $\underline{\Lambda}$ is ergodically stable if and only if for any measurable function $f\colon M \to R_+$ we have

$$\lim_{t\to\infty} \frac{1}{t} \int_0^t f(\tau^s \underline{\Lambda})\,ds = Ef(\underline{\Lambda}^*) \qquad \text{a.s..}$$

(iii) A point process $\underline{\Phi}$ is σ-ergodically stable if and only if for any measurable function $f\colon N \to R_+$ we have

$$\lim_{n\to\infty} \frac{1}{n} \sum_{i=0}^{n-1} f(\sigma^i \underline{\Phi}) = Ef(\underline{\Phi}^0)\ . \qquad \text{a.s..}$$

Proof: Part (i) is proved in Rolski (1981b), Proposition 2. Part (ii) and (iii) can be proved similarly.

It can be proved for an ergodically stable random measure that

$$\lim_{t\to\infty} \frac{\Lambda[0,t]}{t} = \lambda_{\underline{\Lambda}} \qquad \text{a.s.} \tag{2.7}$$

where $\lambda_{\underline{\Lambda}}$ is a constant to be called the stationary intensity of the random measure $\underline{\Lambda}$. The proof of (2.7) is based on the fact that the sequence of random variables $\{\Lambda[i,i+1),\ i \in Z\}$ is ergodically stable; see Rolski (1981a, Lemma

3.3). Since

$$\lim_{t\to\infty} \frac{\Lambda[0,t]}{t} = \lim_{t\to\infty} \frac{\Lambda^*[0,t]}{t} \qquad \text{a.s.}$$

and $\underline{\Lambda}^*$ is stationary we have

$$\lambda_{\underline{\Lambda}} = \lambda_{\underline{\Lambda}^*} = E\Lambda^*[0,1] \ .$$

It is known that the assumption of ergodicity of two independent sequences $\underline{X}_1$ and $\underline{X}_2$ does not suffice for $(\underline{X}_1,\underline{X}_2)$ to be ergodic. A stronger assumption is needed that one of the components of $(\underline{X}_1,\underline{X}_2)$ is weakly mixing. We state here the definition of a weakly mixing random measure refering for the definition in the case of sequences to Brown (1976). A stationary random measure $\underline{\Lambda}$ is called weakly mixing if

$$\lim_{t\to\infty} \frac{1}{t}\int_0^t |\Pr(\underline{\Lambda}\in A, \tau^s \Lambda \in B) - \Pr(\underline{\Lambda}\in A)\Pr(\underline{\Lambda}\in B)|\,ds = 0 \ , \qquad (A,B \in \mathcal{BM}).$$

An example of the weakly mixing sequence of random variables is a stationary independent sequence of random elements. From Westcott (1972) it follows that the stationary Poisson process is weakly mixing.

Proposition 2.8:

(i) Let $\underline{X}_1$ and $\underline{X}_2$ be two independent ergodically stable sequences and a stationary representation $\underline{X}_1^0$ of $\underline{X}_1$ is weakly mixing. Then $(\underline{X}_1,\underline{X}_2)$ is ergodically stable.

(ii) Let $\underline{\Lambda}_1$ and $\underline{\Lambda}_2$ be two independent ergodically stable random measures and a stationary representation $\underline{\Lambda}_1^0$ of $\underline{\Lambda}_1$ is weakly mixing. Then $(\underline{\Lambda}_1,\underline{\Lambda}_2)$ is ergodically stable.

Proof: We omit giving details of the proof which follows from Proposition 1.6 and Exercise 1.19 from Brown (1976).

§. 2. Stable G/GI/1 queues

Customers arrive at a queueing system according to instants determined by a point process $\underline{\Phi}$. Following notations introduced in the foregoing section, Z_i $(i \in N)$ are arrival instants on $(0,\infty)$ and T_i $(i \in N)$ are inter-arrival times; $T_0 = Z_1$. If $\underline{\Phi}$ has a point at zero then we set $Z_0 = 0$ and $\underline{Z} = \{Z_i, i \in Z_+\}$ forms a sequence of arrival instants. The service time of the i-th customer is S_i $(i \in Z_+)$. In the paper we assume that $\underline{S} = \{S_i, i \in Z_+\}$ consists of independent random variables and $\{S_i, i \in N\}$ are identically distributed. Moreover $\underline{\Phi}$ and $\underline{S}$ are independent. The delay sequence $\underline{W} = \{W_i, i \in Z_+\}$ in a single server queue operating according to a first in first out discipline, fulfills the recurrence relationship

$$W_{i+1} = (W_i + S_i - T_i)_+ \qquad (i \in Z_+) \ .$$

We assume in this section that the sequence $\underline{T} = \{T_i, i \in Z_+\}$ is ergodically stable. Hence by Proposition 2.8 the sequence $(\underline{T},\underline{S})$ is ergodically stable. We assume that a stationary representation $(\underline{T}^o,\underline{S}^o)$ of $(\underline{T},\underline{S})$ is a double-ended sequence. This can be done because the simple-ended version of $(\underline{T}^o,\underline{S}^o)$ is stationary (see Breiman (1968)). There exists (see e.g. Loynes (1962), Borovkov (1972)) a stationary ergodic sequence $\underline{W}^o = \{W_i^o, i \in Z\}$ such that

$$W_{i+1}^o = (W_i^o + S_i^o - T_i^o)_+ \qquad (i \in Z) .$$

The condition

$$\rho = \frac{ES_o^o}{ET_o^o} < 1$$

makes the random variables W_i^o $(i \in Z)$ finite almost surely. We can express W_o^o by

$$W_o^o = \sup(0, S_{-1}^o - T_{-1}^o, S_{-1}^o + S_{-2}^o - T_{-1}^o - T_{-2}^o, \ldots) .$$

The following theorem was proved by Rolski (1981b) .

Theorem 2.9: If $\rho < 1$ then $\underline{W}^o \overset{s}{\equiv} \underline{W}$.

From Theorem 2.9 it follows that

$$\lim_{n\to\infty} \frac{1}{n} \sum_{i=0}^{n-1} W_i = EW_o^o \qquad \text{a.s.} \tag{2.10}$$

Definition 2.11: The mean stationary delay in a queue generated by the sequence $(\underline{T},\underline{S})$ is

$$d(\underline{T},\underline{S}) = \lim_{n\to\infty} \frac{1}{n} \sum_{i=0}^{n-1} W_i ,$$

provided the limit in (2.10) exists a.s.. A random variable $D(\underline{T},\underline{S})$ distributed identically to W_o^o is said to be the stationary delay in a queue generated by the sequence $(\underline{T},\underline{S})$.

Note the following convention that in $d(\cdot,\cdot)$ $(D(\cdot,\cdot))$ we identify the input process and service time sequence respectively. For example if customers arrive according to instants of a point process $\underline{\Phi}$ and service times form a stationary independent sequence, each service time is exponentially distributed with intensity μ then we denote the mean stationary delay in the queue by $d(\underline{\Phi},\underline{\mu})$. The following lemma will be useful later on.

Lemma 2.12: For any non-decreasing convex function $f: R_+ \to R_+$ there exist non-increasing convex functions $g_n : R_+^n \to R_+$ such that

$$Ef(D(\underline{T},\underline{S})) = \lim_{n\to\infty} E\, g_n\, (T_{-1}^o, \ldots, T_{-n}^o) .$$

Proof: By the monotone convergence theorem

$$Ef(D(\underline{T},\underline{S})) = Ef(\sup(0,S^o_{-1} - T^o_{-1},\ldots)) =$$

$$= \lim_{n\to\infty} Ef(\sup(0,S^o_{-1} - T^o_{-1},\ldots,S^o_{-1} + \ldots + S^o_{-n} - T^o_{-1} - \ldots - T^o_{-n})) .$$

Let $P_{(\underline{T}^o)_n}$ be the distribution of $(T^o_{-1},\ldots,T^o_{-n})$ and $P_{(\underline{S}^o)_n}$ be the distribution of $(S^o_{-1},\ldots,S^o_{-n})$, Since $\underline{T}^o$ and $\underline{S}^o$ are independent, by Fubini's theorem

$$A_n = Ef(\sup(0,S^o_{-1} - T^o_{-1},\ldots,S^o_{-1} + \ldots + S^o_{-n} - T^o_{-1} - \ldots - T^o_{-n}))$$

$$= \int_{R^{2n}} f(\sup(0,s_{-1}-t_{-1},\ldots,s_{-1}+\ldots+s_{-n}-t_{-1}-\ldots-t_{-n}))P_{(\underline{S}^o)_n}(d\underline{s})P_{(\underline{T}^o)_n}(d\underline{t})$$

$$= \int_{R^n_+} (\int_{R^n_+} f(\sup(0,s_{-1}-t_{-1},\ldots,s_{-1}+\ldots+s_{-n}-t_{-1}-\ldots-t_{-n}))P_{(\underline{S}^o)_n}(d\underline{s}))P_{(\underline{T}^o)_n}(d\underline{t}) .$$

The function

$$g_n(t_{-1},\ldots,t_{-n}) = \int_{R^n_+} f(\sup(0,s_{-1}-t_{-1},\ldots,s_{-1}+\ldots+s_{-n}-t_{-1}-\ldots-t_{-n}))P_{(\underline{S}^o)_n}(d\underline{s})$$

is non-increasing convex. Thus the proof is completed because $A_n = Eg_n(T^o_{-1},\ldots,T^o_{-n})$.

§. 3. Stable G/M/s loss systems

In this section we prove that a G/M/s loss system is stable if an arrival process constitutes an ergodically stable point process $\underline{\Phi}$. As before $\underline{T} = \{T_i,\ i \in Z_+\}$ is the sequence of inter-arrival times. The service times $\underline{S} = \{S_i,\ i \in Z_+\}$ are independent exponentially distributed with intensity μ . The input is independent of the service. There are s servers at the system and queueing is not allowed. The characteristic under investigation is the number of customers in the system just prior to arrival instants L_i $(i \in Z_+)$. We are going to define a convenient (for further consideration) version of the process $\underline{L} = \{L_i,\ i \in Z_\pm\}$. The family of random processes

$$\underline{\delta}_i = (\{\delta(j,t),\quad t \in \mathbf{R}_+\},\quad j=1,\ldots,s) \quad (i \in Z_+)$$

consists of independent death processes, with the death rate μ and $\delta_{ij}(0) = j$ $(i \in Z_+)$. We assume that any $\{\delta_{ij}(t)\ ,\ t \geq 0\}$ has right continuous non-increasing trajectories.

Denote $\underline{\Delta} = \{\underline{\delta}_i,\ i \in Z_+\}$. The following is the recurrence relationship defining the sequence $\underline{L}$,

$$L_1 = \begin{cases} (\min(L_0+1,s)-\delta_0(\min(L_0+1,s),T_0))_+ , & \text{if there is an arrival at zero} \\ (L_0 - \delta_0(L_0,T_0))_+ , & \end{cases}$$

and

$$L_{i+1} = (\min(L_i+1,s) - \delta_i(\min(L_i+1,s),T_i) \qquad (i \in N) .$$

The proof of the following theorem is given in the Appendix.

Theorem 2.13: The sequence $\underline{L}$ is ergodically stable with the common stationary distribution $P_{\underline{L}0}$ regardless of the initial condition $L_0 = k$ $(k=1,\dots,s)$.

Remark 2.14: The existence of a stationary queue size process in the system at arrival instants was demonstrated by Lisek (1979); see also Franken et al (1981).

From Theorem 2.13 it follows that

$$\lim_{n\to\infty} \frac{1}{n} \sum_{i=0}^{n-1} 1_{\{s\}}(L_i) = p(\underline{T},\mu) \qquad \text{a.s.} .$$

Moreover, by the bounded convergence theorem we have

$$\lim_{n\to\infty} \frac{1}{n} \sum_{i=0}^{n-1} Pr(L_i = s) = p(\underline{T},\mu) . \tag{2.15}$$

Definition 2.16: The limit in (2.15) is said to be the stationary probability of loss.

The following lemma is of importance for further considerations. There exists a proof of the lemma in the case of $s=1$ (see Stoyan (1973)) and $s=2$ (see Fleischmann (1976)).

Lemma 2.17: For the single or two server loss system there exists a sequence of non-increasing convex functions $f_n: R_+^n \to R_+$ $(n \in N)$ such that

$$\sum_{i=0}^{n-1} Pr(L_i = s) = Ef_n(T_0,\dots,T_{n-1}) \qquad (n \in N) .$$

III. COMPARISON THEOREMS

§. 1. Ross's conjectures

Ross (1978) conjectured bounds for the mean stationary delay in a G/GI/1 queue and for the stationary probability of loss in a G/M/s loss system with doubly stochastic Poisson arrivals. In this section we state theorems from which both Ross's conjectures follow. We assume that the input process belongs to a class of point processes which included DSPP's. Point processes in this new class have the structure of DSPP's. We recall first a definition of DSPP and then we give a generalisation of the DSPP's. Consider a stationary Poisson process $\underline{\Pi}$ on R_+ with intensity 1 and a diffuse random measure $\underline{\Lambda}$ on R_+ independent on $\underline{\Pi}$. The following representation of $\underline{\Pi}$ is useful. Let $\underline{M} = \{M_i, \quad i \in Z_+\}$ be a stationary independent sequence with a common distribution function

$$Pr(M_i \leq x) = 1 - e^{-x} \qquad (x \in R_+) \quad .$$

Define $\underline{Y} = \{Y_j, \quad j \in N\}$ by

$$Y_j = \begin{cases} \sum_{i=0}^{j-1} M_i & j=1,2,\dots \\ 0 & j=0 \\ \sum_{i=-1}^{j} M_i & j=-1,-2,\dots \, . \end{cases}$$

Then $\underline{\Pi} = \sum_{j=1}^{\infty} 1_{\{Y_j\}}$ is a stationary Poisson process on R_+ with intensity 1.

As usual we set $\Lambda(t) = \Lambda[0,t]$ and $\Pi(t) = \Pi[0,t] \quad (t \in R_+)$.

Definition 3.1: A point process $\underline{\Pi} \circ \underline{\Lambda}$ defined by the random process $\{\Pi \circ \Lambda(t), t \in R_+\}$ is called a doubly stochastic Poisson process (DSPP) .

The random measure $\underline{\Lambda}^{-1}$ defined by the random process

$$\Lambda^{-1}(t) = \inf\{s, \Lambda(s) \geq t\} \qquad (t \in R_+)$$

is called the inverse random measure to $\underline{\Lambda}$.

Note that

$$\underline{\Pi} \circ \underline{\Lambda} = \sum_{i=1}^{\infty} 1_{\{\Lambda^{-1}(Y_j)\}} \quad .$$

A class of input processes including DSPP's can be defined in the following way. Let $\underline{\Phi}$ (a basic point process) be a point process, and $\underline{\Lambda}$ be a diffuse random measure. The random measure $\underline{\Lambda}$ will define the change of time for the point process $\underline{\Phi}$. We always suppose that the basic point process and the time changing random measure are independent.

Definition 3.2: A point process $\underline{\Phi} \circ \underline{\Lambda}$ is a point process generated by the random process $\{\Phi \circ \Lambda(t),\ t \in R_+\}$.

Note that setting $\underline{\Phi} = \underline{\Pi}$ we obtain a DSPP.

In this section we establish some conditions for a basic point process $\underline{\Phi}$ and the time changing random measure $\underline{\Lambda}$.

Assumption 3.3: The time changing random measure $\underline{\Lambda}$ is ergodically stable. A stationary version $\underline{\Lambda}^*$ is such that both the $\underline{\Lambda}^*$ and $\underline{\Lambda}^{*-1}$ are diffuse, and

$$\lambda_{\underline{\Lambda}^*} = E\Lambda^*[0,1] = E\Lambda^{*-1}[0,1] = 1 .$$

The basic point process $\underline{\Phi}$ is supposed to be ergodically stable. Moreover a stationary version $\underline{\Phi}^*$ is simple, weakly-mixing.

The requirement that $\underline{\Lambda}^*$ and $\underline{\Lambda}^{*-1}$ are both diffuse random measures yields that $\underline{\Lambda}^{o-1}$ is a stationary random measure where $\underline{\Lambda}^o$ is a Palm version of $\underline{\Lambda}^*$; see for definition Geman and Horowitz (1973). Recall that $\underline{\Lambda}^o$ is a Palm version of $\underline{\Lambda}^*$ if

$$P_{\underline{\Lambda}^o}(B) = E\int_0^1 1_B(\tau^s\underline{\Lambda}^*)\Lambda^*(ds) \qquad (B \in \mathcal{BM}) .$$

In Theorem 3.4 we shall utilize the mentioned characterization of $\underline{\Lambda}^{o-1}$. However it is plausible that results of the main Theorem 3.5 continue to hold for every random measure $\underline{\Lambda}$ such that $\underline{\Lambda}^*$ is diffuse.

It can be easily proved that for every invariant set $B \in \mathcal{BM}$ we have $P_{\underline{\Lambda}^o}(B) = 0$ if and only if $P_{\underline{\Lambda}^*}(B) = 0$. Hence by a result analogous to that of Proposition 1 of Rolski (1981b) we obtain that $\underline{\Lambda}^o \overset{s}{\equiv} \underline{\Lambda}^*$.

Theorem 3.4: The point process $\underline{\Phi} \circ \underline{\Lambda}$ is ergodically stable with a stationary representation $\underline{\Phi}^* \circ \underline{\Lambda}$ and Palm version $\underline{\Phi}^o \circ \underline{\Lambda}^o$. Moreover

$$\lambda_{\underline{\Phi}^* \circ \underline{\Lambda}^*} = \lambda_{\underline{\Phi}^*} .$$

Proof: We demonstrate first that $\underline{\Phi} \circ \underline{\Lambda}$ is ergodically stable. The proof of this fact consists of four parts.

(a) The random element $\underline{\Lambda}^{-1}$ is ergodically stable with a stationary representation $\underline{\Lambda}^{o-1}$. To demonstrate this fact we remark first that for every measurable function $g\colon M \to R_+$ the sequence

$$\underline{U} = \{\int_i^{i+1} g(\tau^s\underline{\Lambda})\Lambda(ds) , \quad i \in Z_+\}$$

is ergodically stable and it has a stationary representation

$$\underline{U}^* = \{\int_i^{i+1} g(\tau^s\underline{\Lambda}^*)\Lambda^*(ds), \quad i \in Z_+\} .$$

This can be proved by Lemma 3.3 of Rolski (1981a). Now let $i(\lambda) = \lambda^{-1}$ $(\lambda \in M)$ and $[t] = \max\{i,\ i < t\}$. By Serfozo (1977) the mapping i is measurable. Clearly we have

$$\frac{1}{t}\int_0^{[t]} g(\tau^s\underline{\Lambda})\Lambda(ds) \le \frac{1}{t}\int_0^{t} g(\tau^s\underline{\Lambda})\Lambda(ds) \le \frac{1}{t}\int_0^{[t]+1} g(\tau^s\underline{\Lambda})\Lambda(ds)\ .$$

Bearing in mind that

$$\tau^s\underline{\Lambda}^{-1} = i \circ (\tau^{\Lambda^{-1}(s)}\underline{\Lambda}) \qquad \text{a.s.,}$$

$$\lim_{t\to\infty} \frac{\Lambda^{-1}(t)}{t} = 1 \qquad \text{a.s.,}$$

$$\lim_{t\to\infty} \frac{[t]}{t} = 1\ ,$$

and that $\underline{U}$ is ergodically stable we have almost surely that

$$\lim_{t\to\infty} \frac{1}{t}\int_0^t g(\tau^s\underline{\Lambda}^{-1})ds = \lim_{t\to\infty}\frac{1}{t}\int_0^t g \circ i(\tau^{\Lambda^{-1}(s)}\underline{\Lambda})ds$$

$$= \lim_{t\to\infty}\frac{1}{t}\int_0^{\Lambda^{-1}(t)} g \circ i(\tau^s\underline{\Lambda})\Lambda(ds) = E\int_0^1 g \circ i\,(\tau^s\underline{\Lambda}^*)\Lambda^*(ds) = Eg(\underline{\Lambda}^{o-1})\ .$$

(b) The random element $(\underline{\Lambda}^{-1},\underline{\Phi})$ is ergodically stable and $(\underline{\Lambda}^{o-1},\underline{\Phi}^*)$ is its stationary representation. This follows, by (a) and Proposition 2.8.

(c) For any measurable mapping $g\colon M \to M$ the sequence of random variables

$$\{\int_j^{j+1} g(\tau^s \circ (\underline{\Lambda}^{-1},\underline{\Phi}))\Lambda^{-1}(ds)\ , \qquad j \in Z_+\}$$

is ergodically stable and its stationary representation is

$$\{\int_j^{j+1} g(\tau^s \circ (\underline{\Lambda}^{o-1},\underline{\Phi}^*))\underline{\Lambda}^{o-1}(ds), \quad j \in Z_+\}\ .$$

(d) The mappings $M \times N \ni (\lambda,\nu) \to \nu \circ \lambda \in N$ and $M \ni \lambda \to \lambda^{-1} \in M$ are measurable, see Serfozo (1977). Thus we have

$$\nu \circ \lambda = f(\lambda^{-1},\nu)$$

where $f\colon M \times N \to N$ is a measurable mapping. Note that

$$\tau^s \circ (\underline{\Phi} \circ \underline{\Lambda}) = f(\tau^{\Lambda(s)}\underline{\Lambda}^{-1},\ \tau^{\Lambda(s)}\underline{\Phi}) \qquad \text{a.s.} \quad (s \in R_+)\ .$$

Hence for every measurable function $g\colon N \to R_+$

$$\lim_{t\to\infty} \frac{1}{t}\int_0^t g(\tau^s \circ (\underline{\Phi}\circ\underline{\Lambda})ds = \lim_{t\to\infty}\frac{1}{t}\int_0^t g\circ f(\tau^{\Lambda(s)}\underline{\Lambda}^{-1},\tau^{\Lambda(s)}\underline{\Phi})ds$$

$$= \lim_{t\to\infty}\frac{1}{t}\int_0^{\Lambda(t)} g\circ f(\tau^s\underline{\Lambda}^{-1},\tau^s\underline{\Phi})\Lambda^{-1}(ds) \qquad \text{a.s.}$$

and by (c)

$$\lim_{t\to\infty}\frac{1}{t}\int_0^t g(\tau^s\circ(\underline{\Phi}\circ\underline{\Lambda}))ds = E\int_0^1 g\circ f(\tau^s\underline{\Lambda}^{0-1},\ \tau^s\underline{\Phi}^*)\Lambda^{0-1}(ds) \qquad \text{a.s.}\ .$$

Hence

$$\lim_{t\to\infty}\frac{1}{t}\int_0^t g(\tau^s\circ(\underline{\Phi}\circ\underline{\Lambda}))ds =$$

$$= \lim_{t\to\infty}\frac{\Lambda(t)}{t}\frac{1}{\Lambda(t)}\int_0^{\Lambda(t)} g\circ f(\tau^s\underline{\Lambda}^{-1},\tau^s\underline{\Phi})\Lambda^{-1}(ds) =$$

$$= E\int_0^1 g\circ f(\tau^s\underline{\Lambda}^{0-1}),\tau^s\underline{\Phi}^*)\Lambda^{0-1}(ds) \qquad \text{a.s.}\ .$$

Since g is arbitrary we have by Theorem 2.6 (ii) that $\underline{\Phi}\circ\underline{\Lambda}$ is ergodically stable.

In the next part of the proof we demonstrate that

(e) $\underline{\Phi}^*\circ\ \underline{\Lambda}^*$ is stationary,

(f) $\underline{\Phi}^0\circ\underline{\Lambda}^0$ is a Palm version.

This would yield that $(\underline{\Phi}\ {}^*\circ\underline{\Lambda}^*)^0 = \underline{\Phi}^0\circ\underline{\Lambda}^0$ because the stationary distribution of $\underline{\Phi}\circ\underline{\Lambda}$ depends on the distribution of $(\underline{\Lambda}^{0-1},\underline{\Phi}^*)$ which in turn depends on the distribution of $(\underline{\Lambda}^*,\underline{\Phi}^*)$; see Geman and Horowitz (1973).

The proof of (e) is as follows. We have for every $s\in R_+$

$$\Pr(\tau^s\circ(\underline{\Phi}^*\circ\underline{\Lambda}^*)\in B) = \Pr((\tau^{\Lambda^*(s)}\underline{\Phi}^*)\circ\ (\tau^s\underline{\Lambda}^*)\in B)$$

$$= E\Pr((\tau^{\Lambda^*(s)}\underline{\Phi}^*)\circ(\tau^s\underline{\Lambda}^*)\in B\,|\,\Lambda^*(s),\tau^s\underline{\Lambda}^*)$$

$$= E\Pr(\underline{\Phi}^*\circ\tau^s\underline{\Lambda}^*\in B\,|\,\Lambda^*(s),\tau^s\underline{\Lambda}^*)$$

$$= E\Pr(\underline{\Phi}^*\circ\tau^s\underline{\Lambda}^*\in B) = E\Pr(\underline{\Phi}^*\circ\tau^s\underline{\Lambda}^*\in B\,|\,\underline{\Phi}^*)$$

$$= E\Pr(\underline{\Phi}^*\circ\underline{\Lambda}^*\in B) = \Pr(\underline{\Phi}^*\circ\underline{\Lambda}^*\in B)\ .$$

We now prove (f). The coordinate of the i-th point of $\underline{\Phi}^0$ on R_+ is

$$X_i^0 = \zeta_i(\underline{\Phi}^0) \qquad (i\in Z_+)\ .$$

Since $\underline{\Phi}^o$ is a Palm version the sequence $T_i^o = X_{i+1}^o - X_i^o \quad (i \in Z_+)$ is stationary. Bearing in mind that $\underline{\Lambda}^{o-1}$ is a stationary random measure we have for $a_i \in R_+ \quad (i \in N)$

$$\Pr(\Lambda^{o-1}(X_2^o)-\Lambda^{o-1}(X_1^o)\geq a_1, \quad \Lambda^{o-1}(X_3^o)-\Lambda^{o-1}(X_2^o)\geq a_2,\ldots)$$

$$= \int_{R_+} \Pr(\Lambda^{o-1}(X_2^o)-\Lambda^{o-1}(X_1^o)\geq a_1,\Lambda^{o-1}(X_3^o)-\Lambda^{o-1}(X_2)\geq a_2,\ldots|X_1^o=x)P_{X_1^o}(dx)$$

$$= \int_{R_+} \Pr(\Lambda^{o-1}(T_1^o+x)-\Lambda^{o-1}(x)\geq a_1,\Lambda^{o-1}(T_1^o+T_2^o+x)-\Lambda^{o-1}(T_1^o+x)\geq a_2,\ldots|X_1^o=x)P_{X_1^o}(dx)$$

$$= \int_{R_+} \Pr(\Lambda^{o-1}(T_1^o)-\Lambda^{o-1}(0)\geq a_1,\Lambda^{o-1}(T_1^o+T_2^o)-\Lambda^{o-1}(T_1^o)\geq a_2,\ldots|X_1^o=x)\ P_{X_1^o}(dx)$$

$$= \Pr(\Lambda^{o-1}(T_1^o)-\Lambda^{o-1}(0)\geq a_1,\Lambda^{o-1}(T_1^o+T_2^o)-\Lambda^{o-1}(T_1^o)\geq a_2,\ldots)$$

$$= \Pr(\Lambda^{o-1}(X_1^o)-\Lambda^{o-1}(0)\geq a_1,\Lambda^{o-1}(X_2^o)-\Lambda^{o-1}(X_1^o)\geq a_2,\ldots)\ .$$

This demonstrates that the sequence $\{\Lambda^{o-1}(X_{i+1}^o) - \Lambda^{o-1}(X_i^o),\ i \in Z_+\}$ of inter-point distances in $\underline{\Phi}^o \circ \underline{\Lambda}^o$ is stationary. Thus the point process $\underline{\Phi}^o \circ \underline{\Lambda}^o$ is a Palm version. This completes the proof of (f) .

We now demonstrate that $\lambda_{\underline{\Phi}^* \circ \underline{\Lambda}^*} = \lambda_{\underline{\Phi}^*}$. Notice first that if

$$\psi_k(t) = \Pr(\Phi^*(t) = k) \qquad (k \in Z_+, \quad t \in R_+)$$

then

$$\Pr(\Phi^* \circ \Lambda^*[0,1] = k) = E\ \psi_k(\Lambda^*(1))\ .$$

Hence

$$\lambda_{\underline{\Phi}^* \circ \underline{\Lambda}^*} = E\ \Phi^* \circ \Lambda^*(1) = \sum_{k=0}^{\infty} k\ E\psi_k(\Lambda^*(1))$$

$$= \int_{R_+} (\sum_{k=0}^{\infty} k\ E\psi_k(t))P_{\Lambda^*(1)}(dt) = \lambda_{\underline{\Phi}^*} \int_{R_+} tP_{\Lambda^*(1)}(dt) = \lambda_{\underline{\Phi}^*}$$

which completes the proof of Theorem 3.4.

Having proved Theorem 3.4 we worked out a result from which both Ross's conjectures follow. The input process at the considered system is $\underline{\Phi} \circ \underline{\Lambda}$ and the service process $\underline{S} = \{S_i,\ i \in Z_+\}$ is stationary independent and independent on $(\underline{\Phi},\underline{\Lambda})$.

Theorem 3.5: (i). In G/GI/1 queues if $\rho = \lambda_{\underline{\Phi}^*} E\ S_0 < 1$ then

$$d(\underline{\Phi} \circ \underline{\Lambda},S) \geq d(\underline{\Phi},\underline{S}) \quad .$$

(ii) In G/M/s loss systems (s=1,2)

$$p(\underline{\Phi} \circ \underline{\Lambda},\mu) \geq p(\underline{\Phi},\mu) \quad .$$

Proof: We use notations of Theorem 3.4 . We extend the sequence of inter-point distances in $\underline{\Phi}^o \circ \underline{\Lambda}^o$ to a double ended one $\underline{D}^o = \{D_i^o\ ,\ i \in Z\}$. Since $\underline{\Phi}^o \circ \underline{\Lambda}^o$ is a σ-ergodic Palm version we have that $\underline{D}^o$ is an ergodic sequence. Similarly we extend the stationary random measure $\underline{\Lambda}^{o-1}$ from R_+ to R . Remark that

$$E\Lambda^{o-1}(t) = t \qquad (t \in R) . \tag{3.6}$$

By lemma 2.12 we have

$$d(\underline{\Phi}^o \circ \underline{\Lambda}^o, S) = \lim_{n\to\infty} E\, g_n(D^o_{-1},\ldots,D^o_{-n}) ,$$

where g_n $(n \in N)$ are convex functions.

Using the n-dimensional version of conditional Jensen's inequality (see Pfanzagl (1974)), recalling (3.6)

$$\begin{aligned} &Eg_n(D^o_{-1},\ldots,D^o_{-n}) \\ &= Eg_n(\Lambda^{o-1}(X^o_o)-\Lambda^{o-1}(X^o_{-1}),\ldots,\Lambda^{o-1}(X^o_{-n+1})-\Lambda^o(X^o_{-n})) \\ &= E(E(g_n(\Lambda^{o-1}(X^o_o)-\Lambda^{o-1}(X^o_{-1}),\ldots,\Lambda^{o-1}(X^o_{-n+1})-\Lambda^{o-1}(X^o_{-n})\,|\,\underline{X})) \\ &\geq Eg_n(E(\Lambda^{o-1}(X^o_o)-\Lambda^{o-1}(X_{-1})\,|\,\underline{X},\ldots,E(\Lambda^{o-1}(X_{-n+1})-\Lambda^{o-1}(X^o_{-n})\,|\,\underline{X})) \\ &= Eg_n(X^o_o-X^o_{-1}\ ,\ldots,X^o_{-n+1}-X^o_{-n}) = Eg_n(T^o_{-1},\ldots,T^o_{-n}) \qquad (n \in N) . \end{aligned}$$

Since

$$d(\underline{\Phi}^o \circ \underline{\Lambda}^o, S) = \lim_{n\to\infty} Eg_n(D^o_{-1},\ldots,D^o_{-n})$$

and

$$d(\underline{\Phi}^o, S) = \lim_{n\to\infty} Eg_n(T^o_{-1},\ldots,T^o_{-n})$$

we have

$$d(\underline{\Phi}^o \circ \underline{\Lambda}^o, \underline{S})) \geq d(\underline{\Phi}^o, \underline{S}) .$$

To complete the proof it suffices to prove

$$d(\underline{\Phi}^o \circ \underline{\Lambda}^o, \underline{S}) = d(\underline{\Phi} \circ \underline{\Lambda}, \underline{S}) \tag{3.7}$$

and

$$d(\underline{\Phi}^o, \underline{S}) = d(\underline{\Phi}, \underline{S}) . \tag{3.8}$$

The equality in (3.7) follows from Theorem 2.9 because by Theorem 3.4, $\underline{\Phi} \circ \underline{\Lambda} \overset{S}{\equiv} \underline{\Phi}^o \circ \underline{\Lambda}^o$. Similarly we obtain (3.8) . The proof of (i) is completed.

(ii) This part can be proved similarly to (i) utilizing (2.15) and Lemma 2.17. The proof of Theorem 3.5 is completed.

Solutions of Ross's conjectures are given in the following corollary.

Corollary 3.9: Let the input process $\underline{\Phi}\circ\underline{\Lambda}$ be a DSPP ($\underline{\Phi} \overset{d}{\equiv} \underline{\Pi}$) .

(i) In G/GI/1 queues

(3.10) $$d(\underline{\Pi}\circ\underline{\Lambda},\underline{S}) \geq \frac{\lambda_{\underline{\Pi}}\, ES_0^2}{2(1-\lambda_{\underline{\Pi}}\, ES_0)}$$

(ii) In G/M/s loss systems (s=1,2)

(3.11) $$p(\underline{\Pi}\circ\underline{\Lambda},S) \geq \frac{\rho^s}{s!} \Big/ \sum_{k=0}^{s} \frac{\rho^k}{k!}$$

where $\rho = \lambda_{\underline{\Pi}}\, E S_0$.

Proof: The proof of both the parts follows easily from Theorem 3.5 because on the right-hand sides we have M/GI/1 and M/M/s (with losses) systems respectively.

We now give a few remarks about Theorem 3.5 and the Corollary 3.8 after Theorem 3.5.

Remarks: (a) Corollary 3.8 generalizes Theorem 2 of Rolski (1981b) where the random measure $\underline{\Lambda}$ was assumed to be stationary ergodic.

(b) From the proof of Theorem 3.5 it follows that a stronger result is valid. Namely for each convex non-decreasing function $f: R_+ \to R_+$

$$E\, f(D(\underline{\Phi}\circ\underline{\Lambda},\underline{S})) \geq Ef(D(\underline{\Phi},\underline{S})) .$$

(c) Applying the Little formula (see Corollary 2 after Theorem 1 of Rolski (1981b)) which says that for the mean stationary queue size $1(\underline{\Pi}\circ\underline{\Lambda},\underline{S})$

$$1(\underline{\Pi}\circ\underline{\Lambda},\underline{S}) = \lambda_{\underline{\Pi}}(d(\underline{\Pi}\circ\underline{\Lambda},\underline{S}) + ES_0) ,$$

utilizing Corollary 3.8 we obtain

$$1(\underline{\Pi}\circ\underline{\Lambda},\underline{S}) \geq \frac{\lambda^2_{\underline{\Pi}^*}\, ES_0^2}{2(1-\rho)} + \rho.$$

(d) In the case of the stationary Poissonian input Lemoine (1981) derived a useful representation of $d(\underline{\Pi}\circ\underline{\Lambda}^*,\underline{S})$ using more straightforward arguments (without dealing with the Palm theory of stationary random measures).

(e) In the case of simple server loss systems (s=1) the inequality (3.11) was verified by Fond and Ross (1978) (see also Niu (1980)) for a related result when $\Lambda(t) = \int_0^t \lambda(s)ds$ $(t \in R_+)$ and $\{\lambda(t),\ t \in R_+\}$ is a two-state continuous time Markov chain with one state equal to zero). In the mentioned paper some modifications of G/M/1 loss systems were derived. The following model was considered by Kokotushkin (1974). He considered a G/M/s loss system $(s \in N)$ with arrivals according to a DSPP $\underline{\Pi}\circ\underline{\Lambda}$ where

$$\Lambda(t) = \int_0^t \lambda(s)\,ds \quad (t \in R_+) \quad \text{and} \quad \{\lambda(t),\ t \in R_+\}$$

is a two-state random process assuming values with one state equal to zero and the holding time at zero is exponential. Under such assumptions Kokotushkin established that (3.11) holds. Heyman (1981) found a counter-example that the exponential service time cannot be supressed. He disproved (3.11) for some classes of G/D/1 loss systems.

We close the section with a few examples of an ergodically stable random measure. We seek ergodically stable random measures in the class of random measures of the form

$$\Lambda(t) = \int_0^t \lambda(s)ds \qquad (t \in R_+) .$$

Here we suppose that the random process $\underline{\lambda} = \{\lambda(t),\ t \in R_+\}$ assumes values in $\mathcal{D}(R_+)$. If $\underline{\lambda}$ is an ergodically stable random process i.e. for every $D \in \mathcal{B}\mathcal{D}(R_+)$

$$\lim_{t\to\infty} \frac{1}{t} \int_0^t \Pr(\tau^s \underline{\lambda} \in D)ds = P_{\underline{\lambda}*}(D) ,$$

where $P_{\underline{\lambda}*}$ is a probability measure and $\{\theta^s,\ s \in R_+\}$ is a family of measure preserving ergodic transformations on $(\mathcal{D}(R_+), \mathcal{B}\mathcal{D}(R_+), P_{\underline{\lambda}*})$ then the random measure $\underline{\Lambda}$ is ergodically stable too. For the proof we note that

$$\underline{\Lambda} = f(\underline{\lambda})$$

and

$$\tau^s f(\lambda) = f(\theta^s \lambda) \qquad (\lambda \in \mathcal{D}(R_+)) ,$$

where $f: \mathcal{D}(R_+) \to M$ is a measurable mapping. Then, for every $g: M \to R_+$ we have

$$\lim_{t\to\infty} \frac{1}{t} \int_0^t g(\tau^s \underline{\Lambda})ds = \lim_{t\to\infty} \frac{1}{t} \int_0^t g \circ f(\tau^s \underline{\Lambda})ds = E\, g \circ f(\lambda^*) \qquad \text{a.s.}$$

which demonstrates that $\underline{\Lambda}$ is indeed ergodically stable.

The simplest example we obtain assuming that $\underline{\lambda}$ is a periodic deterministic function with the period equal for example to one. Then $\underline{\Lambda}$ is an ergodically stable measure because for every $g: M \to R_+$ the function $\{g(\tau^s \Lambda),\ s \in R_+\}$ is periodic; this yields that

$$\lim_{t\to\infty} \frac{1}{t} \int_0^t g(\tau^s \underline{\Lambda})ds = \int_0^1 g(\tau^s \Lambda)ds .$$

Thus a stationary representation of $\underline{\Lambda}$ is $\underline{\Lambda}^* = \tau^{\Theta} \underline{\Lambda}$ where Θ is a random variable independent of $\underline{\Lambda}$ uniformly distributed on [0,1].

We generalize slightly the foregoing example supposing that the intensity function $\underline{\lambda}$ is random. To define the random periodic intensity function $\underline{\lambda}$ consider an ergodically stable sequence $\{\underline{\lambda}_i,\ i \in Z_+\}$ where each $\underline{\lambda}_i$ is a random element assuming values at $\mathcal{D}[0,1)$. Then it can be proved that

$$\lambda(t) = \sum_{i=0}^{\infty} \lambda_i(t-1) 1_{[i,i+1)}(t) \qquad (t \in R_+)$$

is an ergodically stable random process.

§. 2. Markov renewal input

We now define a Markov renewal point process. A suitable definition is given to make the consideration of this section easier to follow. Consider a double array of independent, positive random variables $\underline{X} = \{X_{ij}, \ i \in Z_+\}$. For each $i \in Z_+$ the sequence $\underline{X}_i = \{X_{ij}, \ j \in Z_+\}$ is assumed to be stationary independent with a common distribution F_i . Denote $\underline{F} = \{F_i, \ i \in Z_+\}$. There is also given a Markov chain $\underline{N} = \{N_i, \ i \in Z_+\}$ with a state space Z_+ . Let

$$p_{ij} = Pr(N_{k+1} = j | N_k = i) \qquad (i,j \in Z_+)$$

be transition probabilities of the chain $\underline{N}$ and $\pi_i = Pr(N_0 = i)$ $(i \in Z_+)$ its initial probabilities. Set

$$\underline{p} = \{p_{ij}, \ i,j \in Z_+\} \ , \qquad \underline{\pi} = \{\pi_i, \ i \in Z_+\} \ .$$

We assume that $\underline{N}$ is an indecomposable, positive-recurrent Markov chain. This implies the existence of a unique invariant probability measure $\underline{\pi}^o$ on Z_+ , i.e.

$$\pi_j^o = \sum_{i=0}^{\infty} \pi_i^o p_{ij} \ . \qquad (j \in Z_+) \ .$$

Definition 3.12: A point process $\underline{\Psi}$ with inter-point distances sequence

$$\underline{T} = \{T_j = X_{N_j,j} \ , \quad j \in Z_+\}$$

is called a Markov renewal point process.

The distribution of $\underline{T}$ (or equivalently of Ψ) depends on $[\underline{\pi},\underline{p},\underline{F}]$.

Theorem 3.13: The sequence $\underline{T}$ is ergodically stable.

Proof: The sequence $\{(X_{N_j,j},N_j) \ , \ j \in Z_+\}$ is an indecomposable Markov chain. A stationary version of $\underline{N}^J$ is a Markov chain $\underline{N}^o$ with transition probabilities $\underline{p}$ and initial probabilities $\underline{\pi}^o$. By Theorem 7.16 from Breiman (1968) it can be demonstrated that the stationary Markov chain $(\underline{T}^o,\underline{N}^o) = \{(X_{N_j^o j},N_j^o), \ j \in Z_+\}$ is ergodic.

Hence $\underline{T}^o$ is ergodic too. Since

$$Pr(\underline{T} \in \cdot) = \sum_{k=0}^{\infty} \pi_k \, Pr(\underline{T}^o \in \cdot | N_0^o = k) \quad ,$$

we have $P_T(B) = 0$ if and only if $P_{T^o}(B) = 0$. Hence by Proposition 1 of Rolski (1981b) the sequence $\underline{T}$ is ergodically stable.

Consider now two G/GI/1 queues with Markov renewal inputs $\underline{\Psi}_1$ and $\underline{\Psi}_2$ generated by $[\underline{\pi}_1,\underline{p},\underline{F}_2]$ and $[\underline{\pi}_2,\underline{p}_2,\underline{F}_2]$ respectively. In both the queues the service time sequence $\underline{S}$ is stationary independent, also independent on $\underline{\Psi}_1$ and $\underline{\Psi}_2$. The stationary arrival rates are

$$\lambda_i = \sum_{j=0}^{\infty} \pi_{ij}^o \int_{R_+} x\, F_{ij}(dx)$$

and we assume the stability conditions to be fulfilled, i.e.

$$\rho_i = \lambda_i \, E\, S_o < 1 \qquad (i=1,2) .$$

From Theorems 2.9 and 3.13 it follows that

$$(3.14) \qquad d([\underline{\pi}^o, \underline{p}, \underline{F}_i], \underline{S}) = d([\underline{\pi}_i, \underline{p}, \underline{F}_i], \underline{S}) \qquad (i=1,2) .$$

Recall the notion of convex ordering of distribution functions. It is said that a distribution function F is earlier than G regarding the order $\leq_c$ if for every non-decreasing convex function $f: R \to R_+$

$$(3.15) \qquad \int_R f(x)F(dx) \leq \int_R f(x)G(dx) .$$

In such a case we write $F \leq_c G$. It is known that for distribution functions F and G with a common mean (i.e. $\int_R xF(dx) = \int_R xG(dx)$) $F \leq_c G$ if and only if for every convex function $f: R \to R_+$ the inequality (3.15) holds. If random variables X,Y have distribution functions, F,G respectively we write $X \leq_c Y$ whenever $F \leq_c G$.

Theorem 3.16: If

$$(3.17) \qquad F_{1i} \leq_c F_{2i}, \quad \int_R xF_{1i}(dx) = \int_R xF_{2i}(dx) \qquad (i \in Z_+)$$

then

$$D([\underline{\pi}_1, \underline{p}, \underline{F}_1], \underline{S}) \leq_c D([\underline{\pi}_2, \underline{p}, \underline{F}_2], \underline{S}) .$$

Proof: Let $\{X_{ij}^k, \ i,j \in Z_+\}$ be a double array of random variables which correspond to $\underline{F}_k$ $(k=1,2)$. By lemma 2.12 we have that for every non-decreasing function $f: R \to R_+$

$$(3.18) \qquad E\, f(D[\underline{\pi}_i^o, \underline{p}, \underline{F}_i], \underline{S}) = \lim_{n\to\infty} E\, g_n(X^i_{N^o_{-1},-1}, \ldots, X^i_{N^o_{-n},-n})$$

where $g_n: R_+^n \to R_+$ are convex. Here $\{N_i^o, \ i \in Z\}$ is stationary extension of $\{N_i^o, \ i \in Z_+\}$ to a double ended stationary sequence. By (3.17) we have for every $\ell_{-1}, \ldots, \ell_{-n} \in Z_+$

$$E\, g_n(X^1_{\ell_{-1},-1}, \ldots, X^1_{\ell_{-n},-n}) \leq E\, g_n(X^2_{\ell_{-1},-1}, \ldots, X^2_{\ell_{-n},-n}) .$$

Hence

$$E\, g_n(X^1_{N^o_{-1},-1}, \ldots, X^2_{N^o_{-n},-n}) \leq E\, g_n(X^2_{N^o_{-1},-1}, \ldots, X^2_{N_{-n},-n})$$

which by (3.18), (3.14) completes the proof.

Recall that a non-negative random variable X with the distribution function F and mean m_F is said to be of NBUE (NWUE) type if

$$\int_x^{\infty} (1-F(t))dt \leq (\geq)\ m_F(1-F(x)) \qquad (x \in R_+)\ .$$

It is known (see Marshall and Proschan (1972), Stoyan (1977)) that if X is of NBUE (NWUE) type then $F \leq_c M_{m_F}$ where M_{m_F} is the exponential distribution function with the mean m_F. Bearing the above fact in mind we easily prove the following corollary.

Corollary 3.19: If $F_{1i}(x) = 1 - e^{-x/m_i}$ $(x \in R_+,\ i \in Z_+)$, F_{2i} is of NBUE (NWUE) type and $\int_{R_+} xF_{2i}(dx) = m_i$ then

$$d([\underline{\pi}_1, \underline{p}, \underline{F}_1], \underline{S}) \leq (\geq)\ d([\underline{\pi}_2, \underline{p}, \underline{F}_2], \underline{S})\ .$$

Consider now a $G/GI/1$ queue with a Markov renewal input defined by $[\underline{\pi}, \underline{p}, \underline{F}]$. We obtain an interesting case setting

$$F_i(x) = 1 - e^{-x/m_i} \qquad (i \in Z_+)\ .$$

Characteristics of queues with such arrivals are often computable (see e.g. Neuts (1978a),(1978b)). In this case $\underline{T}$ has the following simple representation

$$T_i = m_{N_i} M_i \qquad (i \in Z_+)\ ,$$

where $\underline{M} = \{M_i,\ i \in Z_+\}$ is a stationary independent sequence with the common standard exponential distribution function (i.e. $Pr(M_i \leq x) = 1 - e^{-x}$ $(x \in R_+)$) and $\underline{M}$, $\underline{N}$ are mutually independent. Let

$$m = E\, m_{N_0^o}\ , \qquad \rho = m^{-1} E\, S_0\ .$$

Theorem 3.20: If $\rho < 1$ then

$$d(\{m_{N_i} M_i\ ,\ i \in Z_+\}, \underline{S}) \geq \frac{m^{-1}\, E\, S_0^2}{2(1-\rho)}\ . \tag{3.21}$$

Proof: Clearly by Theorem 2.9

$$d(\{m_{N_i} M_i,\ i \in Z_+\}, \underline{S}) = d(\{m_{N_i^o} M_i\ ,\ i \in Z_+\}, \underline{S})$$

and by Lemma 2.12

$$d(\{m_{N_i^o} M_i,\ i \in Z_+\}, \underline{S}) = \lim_{n\to\infty} Eg_n(m_{N_{-1}^o} M_{-1}, \ldots, m_{N_{-n}^o} M_{-n})\ (n \in N)\ ,$$

where g_n is convex $(n \in N)$. Using the n-dimensional version of conditional Jensen's inequality (see Pfanzagl (1974)) we have

$$Eg_n(m_{N_{-1}^o} M_{-1}, \ldots, m_{N_{-n}^o} M_{-n}) = E(E(g_n(m_{N_{-1}^o} M_{-1}, \ldots, m_{N_{-n}^o} M_{-n}) | \underline{M}))$$

$$\geq Eg_n(E(m_{N_{-1}^o} M_{-1} | \underline{M}), \ldots, E(m_{N_{-n}^o} M_{-n} | \underline{M})) = Eg_n(mM_{-1}, \ldots, mM_{-n})\ .$$

Since $\lim_{n\to\infty} Eg_n(mM_{-1},\ldots,mM_{-n})$ is the waiting time in an M/G/1 queue, by the Pollaczek-Khinchine formula the inequality (3.21) follows. This completes the proof.

Remark 3.22: The exponentially distributed random variables are of NBUE type. König and Schmidt (1980) proved that if all the distributions F_j are of m-MRLA type (i.e. $\int_x^\infty (1-F_j(t))dt/(1-F_j(x)) \le m$, $x \in R_+$) then the distribution of $D([\underline{\pi},\underline{p},\underline{F}],\underline{S})$ is stochastically less than the distribution of the waiting time in an M/GI/1 queue with the arrival rate m^{-1} and the service time sequence $\underline{S}$. Hence

$$d([\underline{\pi},\underline{p},\underline{F}]\ ,\ \underline{S}) \le \frac{m^{-1}ES_0^2}{2(1-\rho)} .$$

From the formula

$$m = \sum_{i=0}^{\infty} \pi_i^0 m_i$$

it follows that for some i_0 we have $m_{i_0} > m_i$ unless $m = m_i$ $(i \in Z_+)$. This is impossible because in that case (namely if F_{i_0} is of m-MRLA type) we have $m_{i_0} \le m$. Thus their inequality is valid in the case $m=m_i$ $(i \in Z_+)$.

Remark 3.23: Parallel results can be proved for G/M/s loss systems (s=1,2). Namely if (3.17) are fulfilled we have

$$p([\underline{\pi}_1,\underline{p}_1,\underline{F}_1],\mu) \le p([\underline{\pi}_2,\underline{p},\underline{F}_2],\mu) .$$

It also holds that

$$p(\{m_{N_i} M_i,\ i \in Z_+\},\mu) \ge \frac{\rho^s}{s!} \Big/ \sum_{k=0}^{s} \frac{\rho^s}{s!}$$

where $\rho = m^{-1}ES_0 = (m\mu)^{-1}$.

APPENDIX

We prove here Theorem 2.13. All notations and assumptions introduced in Section 2.3 are valid. We start with the following lemma.

Lemma A.1: The sequence of random vectors

$$\underline{\Delta} \circ \underline{T} = \{(\delta_j(1,T_j),\ldots,\delta_j(s,T_j)) , \qquad j \in Z_+\}$$

is ergodically stable.

Proof: By Proposition 2.8 (i) the sequence $(\underline{\Delta},\underline{T})$ is ergodically stable because $\underline{\Delta}$ is stationary independent and $\underline{T}$ is ergodically stable. A stationary version of $(\underline{\Delta},\underline{T})$ is $(\underline{\Delta},\underline{T}^0)$ where $\underline{T}^0$ is a stationary version of $\underline{T}$ which is independent of $\underline{\Delta}$. Denote by $D(R_+)$ the space of right continuous functions on R_+ having limits from the left. Let γ_1 denote the shift on $(\mathcal{D}^s(R_+))^{Z+}$ and γ_2 denote the shift on R_+^{Z+}. We have

$$\gamma_2 \circ(\underline{\Delta} \circ \underline{T}) = (\gamma_1 \circ \underline{\Delta}) \circ (\gamma_2 \circ \underline{T}) .$$

Hence, by Lemma 3.3 of Rolski (1981a) the sequence $\underline{\Delta} \circ \underline{T}$ is ergodically stable and $\underline{\Delta} \circ \underline{T}^0$ is a stationary version of $\underline{\Delta} \circ \underline{T}$. This completes the proof of Lemma A.1.

Proof of Theorem 2.13: Set $\underline{s} = (1,2,\ldots,s)$. We have that

$$Pr(\underline{\delta}_0 \circ T_0^0 = s) > 0$$

because

$$Pr(\delta_0(1,t) = 1,\ldots,\delta_0(s,t) = s) > 0 \qquad (t \in R_+)$$

Since the sequence $(\underline{\Delta},\underline{T}^0)$ is stationary we may assume that it is a double ended one. There exists a sequence of random elements $(\hat{\underline{\Delta}},\hat{\underline{T}}) = \{(\hat{\underline{\delta}}_i,\hat{\underline{T}}_i), \quad i \in Z\}$ with the distribution $Pr((\underline{\Delta},\underline{T}^0) \in \cdot\,|\,\underline{\delta}_{-1} \circ T^0_{-1} = s)$. Define the sequence $\hat{\underline{N}} = \{\hat{N}_i \ , \ i \in Z\}$

$$\hat{N}_i = 1_{\{\underline{s}\}}(\hat{\underline{\delta}}_{i-1} \circ \hat{T}_{i-1}) \qquad (i \in Z)$$

and the sequence $\{\hat{L}_i \ , \quad i \in Z_+\}$ by

$$\hat{L}_0 = 0 \ ,$$

$$\hat{L}_{i+1} = (\min(\hat{L}_i+1,s) - \delta_i(\min(\hat{L}_i+1,s),\hat{T}_i))_+ \qquad (i \in Z_+) .$$

Let $\hat{Y}_0 = \min\{i \in N, \hat{N}_i = 1\}$. We have

(A.2) $$E\,\hat{Y}_0 < \infty .$$

The notion of a discrete random process associated with a point process (r.p. & p.p.) was introduced by Rolski (1981a) . Consider the r.p. & p.p. $(\hat{L},\hat{N})$. From the definition of $(\hat{\underline{\Delta}},\hat{\underline{T}})$ and (A.2) we have

(A.3) $$\Pr(\gamma^{\hat{Y}_0}(\hat{\underline{L}},\hat{\underline{N}}) \in \cdot) = \Pr((\hat{\underline{L}},\hat{\underline{N}}) \in \cdot)$$

(If (A.3) holds then $(\hat{\underline{L}},\hat{\underline{N}})$ is said to be a Palm version). Moreover $(\hat{\underline{L}},\hat{\underline{N}})$ is a σ-ergodic Palm version (for the definition of a σ-ergodic Palm version see Rolski (1981a)). To demonstrate this fact consider the stationary r.p. & p.p. $((\underline{\Delta},\underline{T}^0),\underline{N}^0)$ where $\underline{N}^0$ is defined by

$$N_i^0 = 1_{\{s\}}(\underline{\delta}_{i-1} \circ T_{i-1}^0) \qquad (i \in Z) .$$

Since $(\underline{\Delta},T^0)$ is ergodic we have that $((\underline{\Delta},\underline{T}^0),\underline{N}^0)$ is ergodic and by Lemma 2.2 of Rolski (1981a) we obtain that $((\hat{\underline{\Delta}},\hat{\underline{T}}),\hat{\underline{N}})$ is σ-ergodic. There is a function ϕ such that $\hat{\underline{L}} = \phi(\hat{\underline{\Delta}},\hat{\underline{T}})$. Hence, using Proposition 1.2 of Rolski (1981a) it can be proved that $(\hat{L},\hat{N})$ is σ-ergodic. Now we apply Theorem 2.2 of Rolski (1981a) to obtain that $(\hat{\underline{L}},\hat{\underline{N}})$ is ergodically stable. Let $\underline{L}^0$ be a stationary version of $\hat{L}$. We are now going to prove that $\underline{L}^0$ is a stationary version of $\underline{L}$ independent of an initial condition $L_0 = k$. For a sequence

$$\underline{\psi} = \{\underline{\psi}_i = (\psi_i(1),\dots,\psi_i(s)), \quad i \in Z\} \in (\{0,\dots,s\}^s)^{Z_+}\}$$

of scalar vectors define a sequence $\underline{\ell} = \{\ell_i, \quad i \in Z_+\}$ by

$$\ell_0 = k ,$$

$$\ell_{i+1} = (\min(\ell_i+1,s) - \psi_i(\min(\ell_i+1,s))_+ \qquad (i \in Z_+) .$$

For a measurable function

$$f\colon \{0,\dots,s\}^{Z_+} \to R_+$$

set

$$A = \{\underline{g} \in \{0,\dots,s\}^{Z_+}, \quad \lim_{n\to\infty} \frac{1}{n} \sum_{i=0}^{n-1} f(\gamma^i \underline{g}) = Ef(\underline{L}^0)\} .$$

We can proceed similarly as in the proof of Theorem 1 of Rolski (1981b) to demonstrate that the set

$$B = (\underline{\ell}^{-1} \circ A) \cap \{\underline{\psi} \in (\{0,\dots,s\}^s)^{Z_+}, \ \underline{\lim}_{n\to\infty} \frac{1}{n} \sum_{i=0}^{n-1} 1_{\{\underline{s}\}}(\underline{\psi}_i) > 0\}$$

is invariant. Then applying arguments used to prove Theorem 1 of Rolski (1981b) we demonstrate that

$$P_{\underline{L}}(A) = P_{\underline{\Delta}}(B) = P_{\underline{\Delta}^0}(B)$$

and

$$1 = P_{\underline{\Delta}^0}(B) = P_{\underline{L}^0}(A) .$$

Hence $P_{\underline{L}}(A) = 1$ which means that for any function $f\colon R_+^{Z_+} \to R_+$

$$\lim_{n\to\infty} \frac{1}{n} \sum_{i=0}^{n-1} f(\gamma^i \underline{L}) = Ef(\underline{L}^0) \qquad \text{a.s. .}$$

This means that $\underline{L}$ is ergodically stable and $\underline{L}^0$ is a stationary version of $\underline{L}$. The proof of Theorem 2.13 is completed.

REFERENCES

Barlow, R.E. and Proschan, F. (1972) Classes of distributions applicable in replacement with renewal theory implications. Proc. 6th Berkeley Symp.Math.Statist. Probab. 1970, vol.1, 495-515, Berkeley-Los Angeles.

Borovkov, A.A. (1972) Stochastic Processes in Queueing Theory. (in Russian). Nauka, Moskva.

Breiman, L. (1968) Probability. Addison-Wesley, Reading Mass..

Brown, J.R. (1976) Ergodic Theory and Topological Dynamics. Academic Press, New York.

Doob, J.L. (1953) Stochastic Processes, John Wiley and Sons, New York.

Fleischmann, K. (1976) Optimal input for the loss systems G/M/2. Math.Operationsforsch. u. Statist. 7, 129-137.

Fond, S. and Ross, S.M. (1978) A heterogeneous arrival and service queueing loss model, Naval Res. Logist. Quart. 25, 483-488.

Franken, P., König, D., Arndt, U. and Schmidt, V. (1981) Queues and Point Processes, Akademie-Verlag, Berlin.

Geman, D. and Horowitz, J. (1973) Remarks on Palm measures, Ann. Inst. H. Poincaré 9, 215-232.

Gray, R.M. and Kieffer, J.C. (1980) Asymptotically mean stationary measures. Ann.Probability. 8, 962-973.

Heyman, D.P. (1982) On Ross's conjecture about queues with non-stationary Poisson arrivals. J. Appl. Probability, 19, 245-249.

Kallenberg, O. (1976) Random measures, Academic Press, London.

Kokotushkin, V.A. (1974) A generalization of the Palm-Khinchin theorem (in Russian) Teor. Verojatnost. i Primenen. 19, 622-625.

König, D. and Schmidt, V. (1980) Stochastic inequalities between customer-stationary and time-stationary characteristics of queueing systems with point processes. J.Appl. Probability, 17, 768-777.

Lemoine, A. (1981) On queues with periodic Poisson input. J.Appl.Probability. 18, 889-900.

Lisek, B. (1979) Construction of stationary distributions for loss systems, Math. Operationsforsch. u. Statist., Ser. Statist. 10, 561-581.

Loynes, R.M. (1962) The stability of a queue with non-independent inter-arrival and service times. Proc. Camb. Phil. Soc. 58, 497-520.

Neuts, M.F. (1978a) The M/M/1 queue with randomly varying arrival and service rates. Opsearch 15, 139-157.

Neuts, M.F. (1978b) Further results on the M/M/1 queue with randomly varying rates. Opsearch, 15, 158-168.

Niu, S.C. (1980) A single server queueing loss model with heterogeneous arrival and service. Operations Res. 28, 585-593.

O'Brien, G.L. (1975) Inequalities for queues with dependent interarrival and service times. J.Appl. Probability. 12, 653-656.

Pfanzagl, J. (1974) Convexity and conditional expectation. Ann.Probability. 2, 490-494.

Rolski, T. (1981a) Stationary Random Processes Associated with Point Processes. Lecture Notes in Statistics, Springer-Verlag, New York.

Rolski, T. (1981b) Queues with non-stationary input stream: Ross's conjecture. Advances in Appl. Probability. 13, 603-618.

Rolski, T. and Szekli, R. (1982) Networks of work-conserving normal queues. In Proceedings of the conference, Applied Probability-Computer Science, The Interface, Boca Raton 1981, USA, ed. Ralph Disney.

Ross, S.M. (1978) Average delay in queues with non-stationary Poisson arrivals. J.Appl. Probability. 15, 602-609.

Serfozo, D. (1977) Compositions, inverses and thinnings of random measures. Z.Wahrscheinlichkeitstheorie verw. Gebiete. 37, 253-265.

Stoyan, D. (1973) Monotonieeigenschaften Einliniger Bedienungs-systeme Mit Exponentiallen Bedienungszeiten. Apl. Mat. 18, 268-279.

Stoyan, D. (1977) Qualitative Eigenschaften und Abschätzungen stochastisher Modelle, Akademie-Verlag, Berlin.

Westcott, M. (1972) The probability generating functional. J.Australian Math. Soc., Ser. A. 14, 448-466.

III

ANALYSIS OF PROTOCOLS 1

ANALYSE DE PROTOCOLES 1

Stack Algorithms for Collision-Detecting Channels and Their Analysis: A Limited Survey

Micha HOFRI[1]
Faculty of Computer Science,
Technion, Haifa - ISRAEL

ABSTRACT

We consider a single-channel based local network of broadcasting processes constrained to operate independently and employ identical channel-access and transmission protocols. Around 1977 Capetanakis in the US and independently Mikhailov and Tsybakov in the USSR proposed what came to be known as Stack-Algorithm for the collision resolution part of the transmission protocol. The concept has undergone considerable evolution since; we describe some of the newer ways of looking at those algorithms, and discuss several attempts of analyzing their performance. The analysis turns out to be singularly hard, and is related to that of multitype branching processes. Some of the whys and wherefores of the difficulties will be discussed and the modest achievements that we are aware of will be presented. The known results nearly all concern the calculation of the capacity of the channel, with very little known about attendant variables, such as queue lengths and delays.

The research was supported by the Technion VPR Fund - Lawrence Deutch Research Fund.

1. Introduction

1-1 The intention of this paper was to present a family of algorithms used in communications technology, and the information that standard analysis techniques can yield about their behaviour and characterization. It became immediately apparent that the field is too wide for anything like a comprehensive coverage. Communications protocols seem to appeal to the algorithmic ingenuity of numerous people, and the profusion of methods and systems with their varieties and subspecies would daunt an entomologist. Thus, we rather arbitrarily selected a narrow field - one, naturally, with which we are relatively familiar - and explore what is known and done there. We suggest that this particular choice is a good one to illustrate the application of a variety of analytical techniques.

1-2 The reader is assumed to be familiar with the concepts that underlie computer communications via packet-switching in general, and the ALOHA network concepts in particular. These we shall briefly recall, sometimes specializing to the environment we intend to analyze:

– An infinite number of users (nodes, transmitters ...) wish to communicate.

– A single, error-free channel is available. Each user can receive any other on this channel. Alternatively, all the users may be viewed as transmitting to some central location. No other means exist for a user to know about the communications activity of others.

– Communications are performed by transmitting packets. The time axis is "slotted", a slot being the time needed to transmit one packet; propagation delay is negligible and may be taken as zero (whatever delay there is is considered part of the packet duration); thus we are really concerned with local networks only. All the users are always synchronized with respect to these slots and transmit packets at slot-start only.

– The users are algorithmically and statistically identical, follow the same communications protocols and are not ordered in any sense. They are adamant: in the face of channel over-commitment each user with a packet will keep it indefinitely, until it is successfully transmitted.

– At the end of each slot each user that monitors the channel can determine whether during the slot none, one or more users transmitted (ternary feedback). The corresponding channel responses are denoted by *lack*, *ack* (short for acknowledgement) and *nack*. So far nothing is beyond the usual assumptions that are used to describe the slotted ALOHA channel, except for one detail: we have not specified the protocols actually used by the nodes in accessing and using the channel.

1-3 Under the above circumstances the protocol used in the ALOHA network when a transmitting user detected a 'collision' - more than one message was attempted (and none is assumed to be correctly received) - is known to be unstable, in the sense that at any packet arrival rate there is a nonzero probability that as those users retransmit their messages, more and more users will join the fray leading to unbounded delays.

In 1977, Capetanakis [CAP1], and independently Tsybakov and Mikhailov [TM1][2] suggested similar Collision Resolution Algorithms (CTM - CRAs) that avoid this instability for low enough, CRA-specific packet generation rates. All of these algorithms have the flavour of "divide and conquer". This feature proves the decisive one in their robustness, and we shall largely disregard most of the attendant modifications (that do impact their performance materially). Such a CRA, coupled with an access algorithm that specifies when a user may broadcast a newly generated packet for the first time comprises a "channel protocol".

In the next Section we describe two basic varieties of those protocols in detail, and in later Sections outline analyses that those algorithms, and related ones, have been subjected to.

2. CRA via a Stack

2-1 We shall now describe the basic CTM - CRA in terms that are different from the original presentation, but are probably easier to visualize:

a) A user becomes active once it has a packet to broadcast, and may do so. The access protocol specifies when it may broadcast for the first time, and two "obvious" protocols are described later in this Section.

b) Each active user maintains a conceptual stack; at each slot-end it determines its position in the stack according to the following procedure (identical to all users, who are unable however to communicate their stack states):

b1) When a user becomes active and is allowed its first broadcast - it enters level 0 in its stack. It will transmit at the nearest slot, and will always do so when at stack level 0.

b2) At slot end, if it was not a collision slot, a user in stack level 0 (there can be at most one such user, system-wide) becomes inactive, and all other users decrease their stack level by 1.

b3) At slot end, if it was a collision slot, all users at stack level i, $i \geq 1$ change to level $i+1$. The users at level 0 are split into two groups, one remains at level 0, while the other pushes itself into level 1. We repeat that no such user is cognizant of how many users are at each level, including his own. This partition can be made on the basis of a random variable, (much like the flipping of a coin), on the basis of the time when the user became active [GAL] etc.[3] We assume, however that each user, independently of the other active users, has the same probability, p, of staying at level 0 (or $q=1-p$ of having to wait at level 1), and these common probabilities are not varied.

[2] These references contains several cognate CRAs, some entirely out of the scope of this paper as they are designed for a finite population of users (and have special properties). The papers contain elaborate discussions of the motivation and significance of various nuances and variations of these CRAs. Only a few of the variations are analyzed at all, though substantial ground is covered in assessing channel capacity and bounding the packet delay in the infinite population cases.

[3] In principle, the users could also incorporate a priority partitioning scheme into this step, but such an option is not handled by the analysis we present.

c) This algorithm is effected at each successive slot end, and is instantaneous in our time scale.

Elaborations of this algorithm are usually confined to step b3, such as using more than a two-way split ([TM1, MAT]), or making p depend in some way on the packet history ([GAL, ROS]). A modification of b2 is presented in subsection 2-3; more abound, cf.[GAL, LAM1, MAS, ROS, TM1, TM2].

2-2 The time it takes to dispose under this algorithm of a group of n colliding users is denoted by L_n. This includes the slot of the initial collision and subsequent slots, until all users that monitor the stacks depths via the channel responses know it reverted to the pre-collision state[4]. We use this seemingly indirect formulation, rather than saying that L_n is the time until the n-th successful transmission after the collision because n is an unobservable quantity *during* a CRI. The effect of this may best be shown through an example: consider the following sequence of the numbers of users transmitting at successive slots: 2, 0, 2, 2, 1, 1. All the colliding packets are out, but the same sequence of feedback levels would have been obtained from the following sequence: 3, 0, 3, 2, 1, 1 - and there is one more packet panting to get out. Indeed, the counter implementation suggested in the footnote indicates that the CRI is not over; the next slot may be silent, a successfull transmission, or yet another collision, if the actual multiplicities were say, 4, 0, 4, 2, 1, 1.

For uniformity we shall also define L_i for $i=0,1$ as the time required to dispose of the consequences of zero or one packet; these two are clearly of size 1.

2-3 The above algorithm does not utilize the ability of users to distinguish successful transmissions from idle slots. Making use of this ability the users can better utilize slots that under the above algorithm are "doomed" - certain to produce a collision. This transpires if the following sequence occurs: first, a collision is sensed; then all colliders elect level 1 in their stacks, and hence the next slot is idle. In the following slot they will all retransmit. The example sequences above were selected to display this. The sequence nack-lack is heard by all users, specifically by those that are now in level 1, and though they do not know how many they are, they know they number two or more. They can simply avoid the superfluous collision and immediately activate the level selection procedure b3. This holds whether the collision in the above sequence started a CRI or was an intermediate one. Following the silent slot of such a sequence, this modified algorithm requires the other users not to change their stack level, or the CRI-state counter described above. The modified CRA will be denoted by MCRA.

An interesting - and undesireable - phenomenon is possible under the MCRA if the channel is not error-free: if a silent slot following a collision is erroneously sensed as a collision, none of the colliding users will ever return to level 0 of their stacks, the CRA

[4] This can be implemented by requiring each user to maintain a counter, kept at zero when the system is quiescent, incremented by one for each collision and decremented by one for any silent slot or a successful transmission slot; when it is decremented to -1, a CRI just terminated, and the counter is reset to zero again.

will not terminate, resulting in a deadlock!

2-4 Two access protocols that could be used in conjunction with such CRAs suggest themselves:

a) Delayed Access (DA) protocol. All the users monitor the channel all the time. Thus when a user has a packet to transmit it knows whether the channel is quiescent or a CRA is in process. In the first case it immediately becomes active, while in the second it waits till the CRI is over and then enters level 0 of its stack.

b) Immediate Access (IA) protocol. Under IA an inactive user need not monitor the channel. When it obtains a packet it will become active at once and transmit at the nearest slot boundary. It may hit an otherwise idle slot (there are many such in the above CRAs) and thus succeed and exit; or it will join the CRA as if this is the collision that started a CRI. Besides saving the constant monitoring required by the DA protocol it also can disentangle the deadlock that the MCRA can experience, and furthermore it can be shown to increase channel capacity under both types of CRA (at the expense of increasing the variance of the delay).

2-5 The combination of two CRAs and two access protocols results in four realizable channel protocols.

3. Channel Capacity.

3-1 We shall describe analytical procedures that can be used to determine the channel capacity under the channel protocols discussed above. The results may be compared to those obtained in [FH] by an essentially numerical procedure.

Channel capacity under the DA protocol.

3-2 Let $\{N_i, i \geq 0\}$ be the process of collision multiplicities: N_i is the number of users transmitting at the first slot of the i–th CRI (which may be degenerate, when that number is 0 or 1). Let λ be the rate of new packet generation, then

$$E(N_{i+1}) = \lambda E(L_{N_i}) \tag{1}$$

The mean time between CRI completions would be finite iff the process $\{N_i\}$ is ergodic. This is also the condition for finite mean delay for individual packets. A standard result on the ergodicity of Markov chains over the natural numbers can be converted to the following

Requirement: The process $\{N_i, i \geq 0\}$ is ergodic iff

$$E(N_{i+1} - N_i \mid N_i = n) < 0 \tag{2}$$

uniformly in i for all n larger than some finite n_0.

To use (2), in light of (1), it suffices to evaluate $E(L_n)$ as a function of n. This is the main objective of the analysis we present, though higher moments can be computed in the same way.

3-3 Consider first the original CRA as described in 2-1; it can be used to relate L_n to its "descendants":

$$L_n = \begin{cases} 1 & n=0,1 \\ 1+L_I+L_{n-I} & n>1 \end{cases} \tag{3}$$

where the three components in the relation for $n>1$ are the initial collision slot, the time required to clear the packets remaining at level 0 (if any), and the time required to transmit successfully those that initially elected level 1 (if any), respectively. The variable I has the distribution $B(n,p)$. Equation (3) can be used to manufacture an intuitive argument to the effect of L_n converging in distribution to Normal variables. This was recently proved rigorously in [BAR] for the MCRA, but the same procedure holds for the CRI durations under all the CRA's we discuss.

Defining the probability generating functions (pgf's)

$$g_n(u) = \sum_{i\geq 1} P(L_n=i)u^i$$

and their generating function (gf)

$$h(u,z) = \sum_{n\geq 0} g_n(u)z^n/n!$$

we successively obtain

$$g_n(u) = \begin{cases} u & n=0,1 \\ u\sum_{i=0}^{n}\binom{n}{i}p^i q^{n-i} g_i(u) g_{n-i}(u) & n>1 \end{cases} \tag{4}$$

and

$$h(u,z) = u(1-u^2)(1+z)+uh(u,pz)h(u,qz). \tag{5}$$

The analytic properties of $h(\cdot,\cdot)$ are not easy to glean from (5). It is remarkably similar to equations arising in the investigations of multitype branching processes; what we designate as "collision multiplicity" would serve there as "descendant type". The objectives of the computations are however rather different, and so are the required treatments.

Note that equation (5) is really an equation in *one* variable, as u can be considered there a mere parameter.

3-4 Whereas for the resolution of (2) only the expected values $E(L_n)$ are required, we procedd as follows: Differentiating (5) with respect to u, at $u=1$, and using the notation

$$\alpha_n = E(L_n), \quad \alpha(z) = \sum_{n\geq 0} \alpha_n z^n/n!$$

we obtain

$$\alpha(z) = e^{z} - 2(1+z) + e^{qz}\alpha(pz) + e^{pz}\alpha(qz), \tag{6}$$

and introducing $\varphi(z) = e^{-z}\alpha(z)$,

$$\varphi(z) - \varphi(pz) - \varphi(qz) = 1 - 2(1+z)e^{-z}. \tag{7}$$

Equation (7) is a convenient starting point, though not necessarily the only one. Note that directly from (3) one can write the following relation between the expectations, which is equivalent to (7):

$$\alpha_n = 1 + \sum_{i=0}^{n} \binom{n}{i} p^i q^{n-i} (\alpha_i + \alpha_{n-i}). \tag{8}$$

And furthermore, equation (8) can be readily reduced to a recursion:

$$\alpha_n = \left[1 + p^n + q^n + \sum_{i=1}^{n-1} \binom{n}{i} p^i q^{n-i} (\alpha_i + \alpha_{n-i})\right] / (1 - p^n - q^n). \tag{9}$$

3-5 Equation (7) can be used to obtain a closed form expression for α_n: Let $\varphi(z) = \sum_{n\geq 0} \varphi_n z^n$; equating coefficients of powers of z in (7) we get

$$\varphi_n = \begin{cases} 1 & n=0 \\ \dfrac{2(-1)^n(n-1)}{n!(1-p^n-q^n)} & n>1 \end{cases}. \tag{10}$$

The same relation follows from taking the Mellin transform of (7) [DAV].

The relation $\alpha(z) = e^{z}\varphi(z)$, in terms of the generated sequences is

$$\alpha_n = n! \sum_{k=0}^{n} \frac{\varphi_k}{(n-k)!}$$

and thus

$$\alpha_n = 1 + \sum_{k=2}^{n} \binom{n}{k} \frac{2(-1)^k(k-1)}{1-p^k-q^k}. \qquad n \geq 0 \tag{11}$$

3-6 Differentiating equation (11) twice with respect to p, it is easy to see that the only solution of the equation $\alpha_n' = 0$ for $p \in (0,1)$ is $p = \frac{1}{2}$, and that the second derivative indeed is positive there, thus this value minimizes α_n, as intuition and symmetry might have led one to believe. It is remarkable that the *same* p minimizes *all* the α_n.

The optimum is very flat though, and typically α_n varies by less than one percent when p varies in the range (.44,.56).

For $p = \frac{1}{2}$ equation (11) can be converted to much more stable expressions, such as (cf.[FH]):

$$\begin{aligned} \alpha_n &= 3 + 2\sum_{k=2}^{n} \binom{n}{k} \frac{(k-1)(-1)^k}{2^{k-1}-1} \\ &= \alpha_{n-1} + 2(n-1)\sum_{k\geq 1} (1-2^{-k})^{n-2} / 2^k \qquad n \geq 3 \end{aligned} \tag{12}$$

3-7 The explicit relations in eqs. (8), (11) and (12), while of interest in their own right are not satisfactory for the analytical determination of the channel capacity through

(2) - although they do suffice for numerical search, [TM1 ,FH, MAS]. We need an explicit expression of the dependence of a_n on n, and in view of the requirement in subsection 3-2, an asymptotic evaluation would be expedient. It is not surprising that methods that have been found useful in dealing with the expected complexity of graph algorithms are found just as good in treating an algorithm that was originally presented as the "serial tree algorithm".

3-8 We now derive in detail the channel capacity, under the DA protocol, for both the original CTM-CRA and the modified version.

We begin with equation (11) which can be rewritten as

$$\begin{aligned} a_n &= 1 + 2\sum_{k=2}^{n}\binom{n}{k}(k-1)(-1)^k\sum_{i\geq 0}(p^k+q^k)^i \\ &= 1 + 2\sum_{i\geq 0}\sum_{r=0}^{i}\binom{i}{r}\sum_{k=2}^{n}\binom{n}{k}(k-1)\beta^k, \qquad \beta\equiv -p^r q^{i-r} \end{aligned} \tag{13}$$

Summing over k and shifting the n subscript one obtains

$$a_{n+1} = 1 + 2\sum_{i\geq 0}\sum_{r=0}^{i}\binom{i}{r}[(1+\beta)^n n\beta-(1+\beta)^n+1] \tag{14}$$

Writing $x=-n\beta$ we approximate, for large n, $(1+\beta)^n=(1-\frac{x}{n})^n$ by e^{-x}. In the Appendix we show that the approximated value, a_n differs from the original a_n by only $O(1)$. (Though for the purpose at hand an approximation to within $O(n^{1-\varepsilon})$, for some $\varepsilon>0$ would be adequate). We proceed then with

$$a_{n+1} = 1 + 2\sum_{i\geq 0}\sum_{r=0}^{i}\binom{i}{r}[1-xe^{-x}-e^{-x}]. \quad x=np^r q^{i-r} \tag{15}$$

A standard relation for the gamma function is [5]

$$e^{-x} = \int_{(c)}\Gamma(z)x^{-z}dz \qquad c>0 \tag{16}$$

and moving the contour of integration past the poles of $\Gamma(z)$ at $z=0, -1$ one gets

$$e^{-x}+x-1 = \int_{(c)}\Gamma(z)x^{-z}dz \qquad -2<c<-1 \tag{17}$$

which is precisely what we need in order to rewrite (15), taking, arbitrarily, the midpoints in the integration strips:

$$\begin{aligned} a_{n+1} &= 1-2\sum_{i\geq 0}\sum_{r}\binom{i}{r}[x(e^{-x}-1)+(e^{-x}+x-1)]. \\ &= 1-2\sum_{i\geq 0}\sum_{r}\binom{i}{r}[\, x\int_{(-\frac{1}{2})}\Gamma(z)x^{-z}dz + \int_{(-\frac{3}{2})}\Gamma(z)x^{-z}dz\,] \end{aligned} \tag{18}$$

The second integral of (18) is, by change of variable, equal to $\int_{(-\frac{1}{2})}\Gamma(z-1)x^{1-z}dz = \int_{(-\frac{1}{2})}\frac{\Gamma(z)}{z-1}x^{1-z}dz$, hence

[5] The notation $\int_{(c)}$ stands for $\frac{1}{2\pi i}\int_{c-i\infty}^{c+i\infty}$.

$$a_{n+1} = 1-2\sum_{i\geq 0}\int_{(-\frac{1}{2})}\Gamma(z)\frac{z}{z-1}\sum_{r=0}^{i}\binom{i}{r}x^{1-z}dz$$

$$= 1-2\sum_{i\geq 0}\int_{(-\frac{1}{2})}\Gamma(z)\frac{zn^{1-z}}{z-1}q^{i(1-z)}(1+(\frac{p}{q})^{1-z})^{i}dz$$

$$= 1-2\int_{(-\frac{1}{2})}\Gamma(z)\frac{z}{z-1}\;\frac{n^{1-z}}{1-p^{1-z}-q^{1-z}}dz \qquad (19)$$

Note that the last summation converges only if $\mathrm{Re}(z)<0$, hence the special representation chosen in equation (18). The merit of equation (19) over all preceding steps is that n is exposed here "in splendid isolation". The evaluation of the contour integral is routine: one goes from $(-\frac{1}{2}, -iN_1)$ to $(-\frac{1}{2}, iN_1)$ to (N_2, iN_1) to $(N_2, -iN_1)$ to $(-\frac{1}{2}, -iN_1)$, – a negative sense. The contribution of the horizontal parts of the contour decays when N_1 increases because $|\Gamma(t+iN)|=O(|t+iN|^{t-\frac{1}{2}}e^{-t-\pi N/2})$, and the n^{1-z} takes care of the receding vertical component. Hence the required integral is minus the sum of residues of the integrand to the right of $-\frac{1}{2}$. Determining these requires information about the roots of $1-p^{s}-q^{s}$. For a few isolated values of p these can be determined[6], but in general all one can say is: a) $\mathrm{Re}(s)\leq 1$ and for the most part $-1<\mathrm{Re}(s)$, with rather rare exceptions. b) The roots are well separated and thus easy to determine numerically.

The pole at $z=0$ is the first to consider; it is simple and the residue there contributes $\frac{-n}{plogp+qlogq}$; the rest we list implicitly, yielding:

$$a_n = 1+2n\left[\frac{-1}{plogp+qlogq} + 2\mathrm{Re}\left\{\sum_{\zeta}\frac{\zeta\Gamma(\zeta)}{\zeta-1}n^{-\zeta}\mathrm{Res}[1-p^{1-z}-q^{1-z}]^{-1}_{z=\zeta}\right\}\right] + O(1) \qquad (20)$$

where 'Res' denotes the residue of the quantity in brackets at the designated point, and ζ goes over all roots of $1-p^{1-z}-q^{1-z}$ that lie to the right of the contour. The sum in the braces turns out to be extremely small, due to cancellations which are quite hard to estimate analytically. It is typically four to seven orders of magnitude below the leading term.

3-9 A subsidiary result from the above, that may be unsettling at first is that while a_n is essentially linear in n, the quantity a_n/n does not approach a limit as $n\to\infty$, but oscillates with minute amplitude and ever increasing period around the value of the leading term above, that is fixed in n. The first to notice this, in the present context as far as we know was Hajek ([HAJ]). Veterans of combinatorial graph algorithms would not be too surprised, though.

3-10 The main information supplied then by equation (20) is that

$$r_n(p) \equiv E(L_n)/n = \frac{-2}{plogp+qlogq} + c_n(p) \qquad (21)$$

where $c_n(p)$ is extremely small for all n and p, with a maximal value over n of $\bar{c}(p)$. Example: for $p=\frac{1}{2}$ the leading term is $r^{*}=\frac{2}{log\,2}\approx 2.8854$, and $\bar{c}(\frac{1}{2})\approx 6\cdot 10^{-5}$.

---

[6] E.g. for $p=\frac{1}{2}$ the roots are $s=1+2\pi ik/log2$, k being any integer; for $p=2/(1+\sqrt{5})=\varphi^{-1}$, the golden ratio, $s=(-1)^{k+1}+k\pi i/log\,\varphi$.

Using this in eqs. (1) and (2) we obtain the constraint over λ as $\lambda r_n(\frac{1}{2})-1<0$ or

$$\lambda_{\max} = \frac{1}{r^{*}+\bar{c}(\frac{1}{2})} \approx 0.346569 \tag{22}$$

3-11 Consider now the CTM-MCRA. By its definition it satisfies a relation similar to (3), cf. [FH]:

$$L_n = 1 \qquad n=0,1$$

$$L_n = 1 + \begin{cases} L_n & I=0 \\ L_I + L_{n-I} & I>0 \end{cases} \qquad n>1. \tag{23}$$

The equation corresponding here to (7) is

$$\varphi(z)-\varphi(pz)-\varphi(qz) = 1-e^{-pz}-e^{-z}(1+qz+2pz), \tag{24}$$

which is obviously *not* symmetrical with respect to p and q, and leads to the analogue of (13):

$$a_n = 1+\sum_{k\ge 0}\sum_{r=0}^{k}\binom{k}{r}\sum_{j=2}^{n}\binom{n}{j}[j(1+p)-1-p^j]\beta^j \qquad \beta\equiv -p^r q^{k-r} \tag{25}$$

or

$$a_n = 1+\sum_{k\ge 0}\sum_{r=0}^{k}\binom{k}{r}[2+(1+p)n\beta(1+\beta)^{n-1}-(1+\beta)^n-(1+p\beta)^n]. \tag{26}$$

The analogue of equation (11) here, though for $p=\frac{1}{2}$ only, has already been obtained in [TB]. Again, the same exponential approximation is used, writing x for $-\beta n$ and replacing $(1-\frac{x}{n})^n$ by e^{-x}, leading to

$$a_n = 1+\sum_{k\ge 0}\sum_{r=0}^{k}\binom{k}{r}[2-(1+p)xe^{-x}-e^{-x}-e^{-xp}],$$

regrouping the terms in the brackets to identify with values of gamma function integrals:

$$= 1+\sum_{k\ge 0}\sum_{r=0}^{k}\binom{k}{r}[(e^{-px}-1+px)+px(e^{-x}-1)+(xe^{-x}+e^{-x}-1)]. \tag{27}$$

The last group in (27) is precisely the one appearing in (15), without the factor 2. The rest of the terms are handled similarly to produce

$$a_n = \frac{-n}{p\log p+q\log q} + \sum_{k\ge 0}\sum_{r}\binom{k}{r}\int_{(-\frac{1}{2})}\Gamma(z)x^{1-z}[p+\frac{p^{1-z}}{z-1}]dz \tag{28}$$

Performing the summations and representing the integral by the leading residue, again at z=0, a simple pole (we shall omit the term corresponding to $c_n(p)$, as it is of the same minute value here too), we obtain

$$r_n(p) \equiv \frac{E[L_n]}{n} \to r(p) = -\frac{1+p-p\log p}{p\log p+q\log q} \tag{29}$$

which is, as expected, always less than the RHS of equation (21).

3-12 Differentiating $r(p)$ twice with respect to p we find that the root of the equation

$$\log p \cdot \log q + (1+p)(\log p - \log q) = 0 \tag{30}$$

is indeed a minimum for $r(p)$. Numerical solution yields the single root $p^* = .41750778$, and r^* is then, with $c(p^*)$ allowed for, 2.6229. These values compare well with [FH], where the corresponding quantities found by direct search over the solution of (24) were .4175 and 2.626, respectively.

For the CTM-MCRA we find then

$$\lambda_{max} = \frac{1}{r^*} \approx .38126 \tag{31}$$

In distinction from the unmodified CRA, here p^* does *not* minimize all the α_n (strictly speaking, it minimizes none!). Thus α_2 is minimized by $p_2^* = \sqrt{2}-1$, and for increasing n p_n^* oscillates with decreasing amplitude around p^*: p_{20}^* coincides with p^* to five decimal positions.

Channel capacity under the IA protocol

3-13 Under IA newly arrived packets plunge right in. A new CRI then always begins with the packets that arrived during one slot, which was either a quiescent one or terminated a previous CRI with a successful transmission. While under DA ***every*** CRI with finite multiplicity terminates with probability 1, regardless of the input rate or distribution, this is not the case here. In a sense, the IA protocol seems more amenable for analysis, since all the CRIs are iid, but as we shall soon see this simplicity is illusory.

3-14 Consider the unmodified CTM-CRA. In analogy with equation (3) we now write

$$L_n = \begin{cases} 1 & n=0,1 \\ 1 + L_{I+X} + L_{n-I+Y} & n>1 \end{cases} \tag{32}$$

where the new variables X and Y are the numbers of arriving packets during the initial slot of the CRI and during the last slot of the CRI of multiplicity $I+X$ (which may be degenerate), respectively, and may be assumed iid. If X and Y vanish with probability 1 we get equation (3) back, and all the analysis below can be shown to reduce to what was presented above. In [TV] equation (32) was considered with no assumptions concerning the distribution of X, but for $p=\frac{1}{2}$ only; while for numerical investigations it is not important whether $p=\frac{1}{2}$ or not, for analytical purposes it turns out to be of the utmost significance. The same relation was also analyzed there for the MCRA. At the end of [TV] there is a remark to the effect that the case $p \neq \frac{1}{2}$ was considered too, (for Poisson input), but the numerical value given there for the channel capacity appears to be incorrect. Involved stability conditions were proven, but explicit results were not obtained. In [FH] numerical investigation of (32) (and its corresponding MCRA version) were carried out, for $p \neq \frac{1}{2}$ as well, but only for Poisson distribution of the input (under the assumption of infinite user population we see little point in considering any other one). Equation (32) was investigated analytically in [FFH], for Poisson input and resulted in the following:

3-15 Taking the expectation of both sides in (32) we obtain

$$\alpha_n = 1+\sum_{k\geq 0}\sum_{l\geq 0}\sum_{i=0}^{n}\binom{n}{i}p^i q^{n-i} e^{-2\lambda}\frac{\lambda^{k+l}}{k!\,l!}(\alpha_{i+k}+\alpha_{n-i+l}), \quad n>1 \tag{33}$$

Note that equation (33), unlike (8), is not a recursion!

The equation corresponding to (7), using the same notations, is here

$$\varphi(z)-\varphi(\lambda+pz)-\varphi(\lambda+qz) = 1-e^{-z}[2\varphi(\lambda)(1+z)+z\varphi'(\lambda)] \tag{34}$$

One of the unknown values in the RHS can be eliminated by evaluating (34) at suitably chosen values of z (at $z=-\lambda/p$, and $z=-\lambda/q$), leading to

$$\varphi(z)-\varphi(\lambda+pz)-\varphi(\lambda+qz) = 1-2\varphi(\lambda)e^{-z}(1+Kz) \tag{35}$$

where K should properly be written as $K(p,\lambda)$.

3-16 Equation (34) introduces us to a semigroup H generated by iteration of the two non-commutative linear transformations

$$\sigma_1(z) = \lambda+pz, \quad \sigma_2(z) = \lambda+qz. \tag{36}$$

In order to use H, equation (35) could not serve as is, but had to be differentiated twice, since then we obtain equation of the form

$$f(z)-uf(\sigma_1(z))-vf(\sigma_2(z)) = t(z) \quad |u+v|<1, \quad t(\cdot) \text{ entire} \tag{37}$$

where here u and v are p^2 and q^2. Such an equation can be shown to possess the *unique* analytic solution

$$f(z) = \sum_{\sigma\in H}(u,v)^{\sigma}t(\sigma(z)) \tag{38}$$

where σ is any - finite or infinite - iteration of the transformations σ_1 and σ_2. Denoting by $|\sigma|_i$ the number of times σ_i is used in σ, we use $(u,v)^{\sigma}$ for $u^{|\sigma|_1}v^{|\sigma|_2}$.

An observation found instrumental in the proof is that the points $\sigma(z)$ are constrained to be in (and on) the triangle with vertices (in the complex plane) z, λ/p and λ/q. Furthermore, assuming $p>q$, the interval $(\lambda/p, \lambda/q)$ is the locus of $\sigma(z)$ for the members of H that are infinitely long, for any z, and one can represent any point on that interval, for any z, as the transform of that z by a suitably chosen σ. H thus induces on the interval a distribution - which may be shown to be uniform over it.

3-17 The crux of the analysis lies in showing that an exponential approximation can be obtained for the sum in (38) in terms of the above distribution. However, while similar in principle to the one we used above it turns out to be much too cumbersome to be presented here; in particular, it required further investigation of H, derivation of intermediate approximations etc. We refer the interested reader to [FFH].

4. Packet Waiting Times

4-1 Waiting times characteristics are more intimately interwoven with the dynamical behaviour of the scheduling mechanisms than the processes that determine the channel capacity. Such is the case in virtually every service system, and the present ones conform. This is the underlying cause for the comparative paucity of results concerning packet delay in transmission systems. Since the CRAs discussed here have a relatively complex system-state descriptor – the union of the stacks of all active users, it may come as no surprise that very little indeed is known about the distribution of packet delays, and even for the mean values usually only bounds (albeit rather good ones, for low traffic levels) are available. For example, [TM1] contains a discussion of delay under the DA protocol, but the authors do little more than give the delay distribution in terms of other processes, which are not easy to analyse. They do manage to bound the expected delay, and the estimates have been considerably tightened in [MAS].

4-2 We shall demonstrate the reasons for this difficulty for the the unmodified CRA, under the DA access protocol. A user that begets a packet either has to wait out an ongoing CRI or becomes active at once. Correspondingly its delay is the sum of two dependent components each of which may vanish: $W = W_1 + W_2$,

W_1 = Time from packet-ready time to the end of a CRI (if any);

W_2 = Time from first attempted transmission until the beginning of the slot in which the packet is successfully transmitted (this component is also called the release delay).

The component W_1 is either zero or the "random modification" of a CRI duration. Its distribution can be thus naturally tied to the pgf's we defined in 3-3, but explicit results are not forthcoming. For the mean value we can write, with some unknown quantities interspersed that can be later approximated:

$$E(W_1) = \sum_{l \geq 1} \frac{l-1}{2} \Pr(\text{an arrival hits a CRI of duration } l)$$

$$= \sum_{l \geq 1} \frac{l-1}{2} \frac{lP(L=l)}{E(D)+E(L)} = \frac{E(L^2)-E(L)}{2(E(L)+E(D))} \tag{39}$$

where D is the duration between end of a CRI and the start of the next one, and L the duration of a CRI (averaged over all positive multiplicities according to their relative frequency). Because of the independence of arrivals over successive slots $E(D)=P(D>0)/(1-p_0)$, where p_0 is the probability of no arrivals in a slot, and $P(D>0)=1-g_L(p_0)$; $g_L(\cdot)$ being the pgf of L. The last expression in equation (39) differs from a corresponding one in [TM1], who apparently evaluate $E(W_1)$ given that it is nonzero.

4-3 While the component W_2 depends on W_1, for the evaluation of the mean delay the dependence is of no consequence.

There is an extremely interesting variation of the CTM-CRA that has an impact on the W_2 component of the waiting time [GAL]: in this variation each user when he becomes active knows the time the current CRI started, and the period from that earlier time till the beginning of "his" CRI is the resolution period. When resolving colliders that arrived during this period, the assignment of transmitters to levels 0 and 1 is done on the basis of the packet arrival time, with transmitters that "arrived" during the first p-portion of the interval staying at level 0, and the rest pushed to level 1; this is repeated in the case of further collisions, with reduced resolution intervals, and the result is that messages are transmitted (and also successfully so) in the order of their generation! A FIFO server is thus created. This however does not impact the mean waiting time, only its higher moments are effected.[7]

Given that a CRI is of multiplicity n, denote by R_n the release time of a packet during such a CRI, and the desired W_2 will be obtained by removing the conditioning on n. It is not hard to show that the mean release times, r_n, satisfy a recurrence relation similar to (8):

$$r_n = 1+\sum_{i=0}^{n}\binom{n}{i}p^i q^{n-i}\left[\frac{i}{n}r_i + \frac{n-i}{n}(l_i+r_{n-i})\right] \tag{40}$$

for $n \geq 2$ ($r_0=r_1=1$). The equation for the gf of these r_n is tedious, but writing $f_n=nr_n$, and $\psi(z) = \sum_{n\geq 2} f_n z^n/n!$ we obtain

$$\psi(z) - e^{qz}\psi(pz) - e^{pz}\psi(qz) = qze^{qz}\alpha(pz)+ze^{z}-qz-z \tag{41}$$

where α is the same as in equation (6). Clearly (41) can be also handled in precisely the manner of of equation (6).

4-4 The stumbling block for an accurate evaluation of $E(W)$ is the lack of a solution for equation (5) and its like; these would yield the distribution of the process $\{N_i\}$, defined in 3-3. In the absence of precise results one can bound the probabilities of higher multiplicities in terms of P(N=0,1) and get bounds through them on $E(W)$. Naturally these bounds can be only adequate for low λ, when higher multiplicities are rare. This has been done in [TM1] and considerably improved in [MAS] by Massey and his student N. Amati. There we find that for λ=0.1 the upper bound of $E(W)$ exceeds the lower one by 11%, whereas for λ=0.25 the corresponding figure is already 70%.

4-5 I am not aware of any published results concerning delays under the IA access protocol, except the crude bound (of the nature $E(W)<E(L)-1$) in [TV]. Some work is now being done, though. For the evaluation of waiting times, the independence between successive CRI's that was already alluded to in section 3, is of prime importance, and here IA is simpler than DA to analyze – in a very limited sense, though – Cf. [FFHJ].

4-6 This concludes our limited survey. Clearly the field is wide open for much work that will be interesting both from the point of view of mathematical content and

[7] Perhaps more important than the possibility of ordered service is the fact that this description spurred a number of investigators to introduce a richer parametrization of the interval partitioning approach, and by optimizing over the values of these parameters obtain channel protocols with much higher capacity, up to .4877! cf. [ROS, TM2], and further references at the first one.

possible application. We wish to make just two points:

(i) Concerning waiting time computation - the interest there in the second moment is nearly as great as in the first one, so that its robustness and some quantiles can be assessed.

(ii) The above CRAs can be coupled with a great many access protocols besides the ones we mentioned, including such that use reservations, carrier-sensing, priorities and even certain external controls, [HK, ROS].

APPENDIX: Proof of the Asymptotic Approximation

A-1 We wish to prove - see equations (14) and (15) - that

$$\delta(p,n) \equiv \alpha_{n+1} - a_{n+1} = O(1) \tag{A.1}$$

uniformly in p, asymptotically in n. This is done below directly, though a more general approach is currently being investigated. Disregarding a factor of 2,

$$\delta(p,n) = \sum_{i\geq 0}\sum_{r=0}^{i}\binom{i}{r}[xe^{-x}+e^{-x}+(1+\beta)^n n\beta-(1+\beta)^n]$$

$$= \sum_{i\geq 0}\sum_{r=0}^{i}\binom{i}{r}(x+1)[e^{-x}-(1-\frac{x}{n})^n] \quad x=np^r q^{i-r} \tag{A.2}$$

We shall only prove for the x component of the $(x+1)$ factor; perusing the proof one sees that this is the dominant term (i.e. - the same proof steps show the required for the second component too).

A-2 First note

$$e^{-x}-(1-\frac{x}{n})^n = e^{-x}-e^{-n\log(1-\frac{x}{n})}$$

developing the logarithm function as a power series to two terms and writing the remainder explicitly,

$$= e^{-x}(1-exp[-\frac{x^2/n}{2(1-\vartheta x/n)^2}]) \quad 0<\vartheta<1$$

$$< e^{-x}(1-e^{-x^2/n})$$

where the last inequality actually only holds for values of i of 2 or more, but as the first terms in the sum over i give contributions that are exponentially decreasing in n, it should be taken as "less than ... for our purposes"; for $i>2$, $2(1-\vartheta\frac{x}{n})^2>1$. Another overestimate we shall use is $1-e^{-c}<c$ for any positive c. To emphsize the role of n we shall write $x=na(i,r)$.

A-3 The proof consists of splitting the sum over i into three parts and showing that each is $O(n^{-s})$ for some nonnegative s. The partial-sum limits one should choose turn out to be $n_1\equiv(\log n-\log\log n)/\log(1/q)$ and $n_2\equiv\log n/\log(1/r_2)$ where $r_2=p^2+q^2$.

For $0\leq i<n_1$ we use the last overestimate to write

$$\delta_1(p,n) = \sum_{i=0}^{n_1-1}\sum_{r=0}^{i}\binom{i}{r}n^2a^3(i,r)e^{-na(i,r)} \tag{A.3}$$

For the above range of i $a_1\equiv q^{n_1}\leq a(i,r)$, leading to $e^{-na_1}\geq e^{-na(i,r)}$, and

$$\delta_1(p,n) < n^2e^{-na_1}\sum_{i=0}^{n_1-1}\sum_{r=0}^{i}\binom{i}{r}a^3(i,r)$$

$$= n^2 e^{-na_1} \sum_{i=0}^{n_1-1} (p^3+q^3)^i = O(e^{2\log n - na_1}) \qquad \text{(A.4)}$$

and this is indeed small as required, by our choice of n_1.

A-4 For the second part of the sum we use the same bound as in (A.3), in the range $n_1 \le i < n_2$. We use a form of the mean value theorem adapted to finite sums, to write a value that is asymptotically equal to δ_2:

$$\delta_2(p,n) = O(\ (n_2-n_1) j n^2 \binom{j}{k} a^3(j,k) e^{-na(j,k)}\ ) \qquad \text{(A.5)}$$

for some $n_1 \le j < n_2$, $0 \le k \le j$. For the claim to hold, (A.5) must be $O(n^{-s})$ with nonnegative s for *every* j and k in that range. This can be shown by replacing the value in (A.5) by a larger value

$$n^2 n_2 p^{3k} q^{3j-3k} j^{k+1} e^{-np^k q^{j-k}}, \qquad \text{(A.6)}$$

observing that for $p \ge q$ and $j \ge n_1$ this is an increasing function of k, and taking $k=j$, the resulting value is an increasing function of j, and taking the value at $j=n_2$ one finally obtains $exp[2\log n + (2+n_2)\log n_2 - 3n_2 \log(1/p) - np^{n_2}]$; with the above choice of n_2 and $p > \frac{1}{2}$[8] the exponent is indeed negative.

A-5 For the last part, $i \ge n_2$, note that at this range the natural bound $e^{-na}(1-e^{-na^2}) < a$ is finally effective, since we have

$$\delta_3(p,n) < \sum_{i \ge n_2} \sum_r \binom{i}{r} na^2 = \sum_{i \ge n_2} n(p^2+q^2)^i \qquad \text{(A.7)}$$

which is $O(nr_2^{n_2})$, and by our choice of n_2 the claim is satisfied.

BIBLIOGRAPHY

[BAR] Barzilay Zeev: The Distribution of Modified-Capetanakis Collision Resolution Intervals. Operations Research, Statistics and Economics Mimeograph Series No. 339, Faculty of Industrial and Management Engineering, Technion-IIT, January 1983.

[BKR] de Bruijn N. G., Knuth D. E., Rice S. O.: The Average Height of Planted Plane Trees. In "Graph Theory and Computing" R-C. Read (ed.), Academic Press New York 1972 pp.15-22.

[CAP1] Capetanakis J. I.: Tree Algorithms for Packet Broadcast Channels. IEEE Trans. Inf. Th. IT- **25** pp. 505-515 (1979)

[CAP2] ------: Generalized TDMA: The Multiaccessing Tree Protocol. IEEE Trans. Comm. COM- **27** pp. 1476-1487 (1979)

[DAV] Davies B.: *Integral Transforms and Their Applications*. Springer-Verlag New York, 1978.

[FFH] Fayolle Guy, Flajolet Philippe, Hofri Micha: On a Functional Equation Arising in the Analysis of a Protocol for a Multi-Access Broadcast Channel. INRIA

[8] For $p=\frac{1}{2}$, $\delta_2=0$, since then n_1 and n_2 coincide.

Rapport de Recherche, #131, April 1982. Revised version - In preparation.

[FFHJ] -----, -----, -----, Jacquet Philippe: On Packet Waiting Times in Collision Channels with Stack-Based Collision Resolution Algorithms. In preparation.

[FH] -----, Hofri Micha: On the Capacity of a Collision Channel Under Stack-Based Collision Resolution Algorithms. Technion, Haifa Israel TR #237. March 1982.

[FO] Flajolet P., Odlyzko A.: The Average Height of Binary Trees and Other Simple Trees. J. of Comp. & System Sc. **25** #2 pp. 171-213 (1982)

[GAL] Gallager R.G.: Conflict Resolution in Random Access Broadcast Networks. Proceedings AFOSR Workshop in Communications Theory and Applications, pp. 74-76, 17-20 September 1978, Provincetown, Mass.

[HAJ] Hajek B.: Expected number of slots needed for the Capetanakis Collision-Resolution Algorithm. Preliminary Report, CSL University of Illinois, Urbana IL, July 25, 1980.

[HK] Hofri M., Konheim A.: The Analysis of a Finite Quasi-Symmetric ALOHA Network with Reservation. Technion Faculty of Computer science TR #281, April 1983.

[KNU] Knuth D. E.: *The Art of Computer Programming*, Vol III; *Sorting and Searching*, Addison-Wesley Publ. Co. 1973 p. 131ff.

[LAM1] Lam Simon S.: Multiple Access Protocols. Chapter 4 in *Computer Communications* Vol.I, W. Chou (Editor), Prentice-Hall Englewood Cliffs, NJ 1982.

[LAM2] -----: Packet Broadcast Networks - A Performance Analysis of the R-ALOHA Protocol. IEEE Trans. Comp. C- **29** pp. 596-603 (1980)

[MAS] Massey James L.: Collision Resolution Algorithms and Random Access Communications. In *Multiuser Communications* (Ed. G. Longo) CISM Courses and Lectures No. 265, pp.73-137 Springer Verlag, 1981.

[MAT] Mathys Peter: A Comparison of Q-ary Tree Algorithms for Collision Resolution.. Memo of AFIP, ETH Zurich Switzerland, November 10, 1981.

[ROS] Rosberg Zvi: A Note on Multiple Access Protocols. Faculty of Computer Science TR #229, Technion, Haifa, Israel. December 1981.

[TB] Tsybakov B. S., Berkovskii M. A.: Multiple Access with Reservation. Prob. of Inf. Trans. (English Translation) **16** #1, pp. 35-54 (1980)

[TM1] -----, Mikhailov V. A.: Free Synchronous Packet Access in a Broadcast Channel with Feedback. Prob. of Inf. Trans.(English Translation) **14** #4 pp. 259-280 (1978)

[TM2] -----, -----: Random Multiple Packet Access: Part-and-Try Algorithm. Prob. of Inf. Trans.(English Tranlation) **16** #4, pp. 305-317 (1980)

[TMF] -----, -----, Fedortsev S.P.: Allowance for Packet Propagation Time in Random Multiple Access. Prob. of Inf. Trans.(English Tranlation) **17** #2, pp. 131-134 (1981)

[TV] -----, Vvedenskaya N. D.: Random Multiple-Access Stack Algorithm. Prob. of Inf. Trans.(English Tranlation) **16** #3 pp. 230-243 (1980)

PRIORITY QUEUEING SYSTEMS
WITH APPLICATIONS TO PACKET-RADIO NETWORKS

Moshe Sidi[1] and Adrian Segall[2]

Department of Electrical Engineering
Technion - Israel Institute of Technology
Haifa, Israel

ABSTRACT

In this paper we investigate a certain class of systems containing dependent discrete time queues. This class of systems consists of N nodes transmitting packets to each other or to the outside of the system over a common shared channel, and is characterized by the fact that access to the channel is assigned according to priorities that are preassigned to the nodes. To each node, a given probability distribution is attached, that indicates the probabilities that a packet transmitted by the node is forwarded to each of the other nodes or to the outside of the system.

Using the analyticity of the joint generating function of the queue lengths distribution, we obtain an expression for this joint generating function for N=3. The method we use can easily e extended for an arbitrary number of nodes. From the generating function any moment of the queue lengths as well as average time delays can be obtained.

The main motivation for investigating the class of systems of this paper is its applicability to several packet-radio networks. We give two examples: the first is a certain access scheme for a network where the nodes can hear each other and the second is a three-node tandem packet-radio network.

---

1. Currently on leave at the Lab. for Information and Decision Systems at M.I.T., Cambridge, MA 02139, U.S.A.

2. Currently on leave with I.B.M. Thomas J. Watson Research Center, Box 218, Yorktown Heights, N.Y. 10598, U.S.A.

1. Introduction

The purpose of the present paper is to analyze a certain class of systems containing dependent discrete-time queues. The system under consideration consists of N nodes transmitting messages to each other or to the outside of the system over a common shared channel. Fixed-length packets of data enter the system at all nodes and are buffered until the channel is made available to the node. The time is divided into slots of size corresponding to the transmission time of a packet and transmissions are started only at the beginning of a slot. The system under consideration is characterized by the fact that access to the channel is assigned according to priorities that are preassigned to the nodes. No two nodes have the same priority and a given node is allowed to transmit in a given slot only if those nodes with higher priority have empty queues. To each node we attach a given probability distribution that indicates the probabilities that a packet of data transmitted by the node is forwarded to each of the other nodes or to the outside of the network. All packets received by a node from outside of the system or from other nodes, are buffered in a common outgoing queue. In Section 2 we formulate the system we consider and in Section 3 we present the steady-state analysis of it for N=3 and obtain the condition for steady-state and the joint generating function of the queue lengths at the nodes. From this generating function any moment of the queue lengths at the nodes can be derived and also average time delays can be obtained by using Little's law [5]. As will be seen, the solution method we use, can be easily extended for a system having an arbitrary number of nodes.

The present model is an extension to the "loop-system" considered in [2] in two respects: in [2] nodes transmit only to the outside of the system and also the arrival processes are assumed to be independent. Using a different approach, we are able to analyze in the present paper systems with dependent inputs as well as those where nodes may transmit packets to each other.

Our motivation for investigating the class of systems of this paper is its applicability to several packet-radio networks. In Section 4 we indicate two such applications: the first is the head of the line protocol suggested in [3] for multiplexing a small number of fully connected buffered users over a common radio channel, and the second is a three-node tandem packet-radio network.

2. Formulation

In this section we formulate the class of discrete-time queueing systems that is considered in this paper. We assume that packets arrive randomly at the N nodes of the system from N different sources, and the arrival processes may in general be correlated. Let $A_i(t)$ i=1,2,...,N; t=0,1,2,... be the number of packets entering node i from its corresponding source in the time interval (t, t+1]. The input process $\{A_i(t)\}_{i=1}^{N}$ is assumed to be a sequence of independent and identically distributed random vectors with integer-valued elements. Let the corresponding probability distribution and generating function of the input process be

$$a(i_N,i_{N-1},...,i_1) = Prob\{A_N(t) = i_N,A_{N-1}(t) = i_{N-1},...,A_1(t) = i_1\} \tag{1}$$

$$F(\underline{z}) = E\left\{\prod_{i=1}^{N} z_i^{A_i(t)}\right\}, \tag{2}$$

where we use the notation $\underline{z} = (z_N,z_{N-1},...,z_1)$.

Next, we describe the departure processes from the nodes. We assume that no more than one packet may leave each node in any given time slot. Let node i have priority i, namely a packet leaves node i whenever the queues at nodes $1,2,...,(i-1)$ are empty and the one at node i is nonempty. In this case, node i transmits the packet at the head of its buffer to node j ($j=1,2,...,N$, $j \neq i$) with probability $\nu_i(j)$ or to the outside of the system with probability $\nu_i(0)$. Here we assume $\nu_i(i)=0$. The above implies that packets in the system are routed randomly through the nodes, until they eventually leave the system. It is assumed that packets indeed arrive at every node with nonzero probability and that the buffers at the nodes have infinite length. A schematic figure of a node i in the system appears in Fig. 1.

3. Steady-State Distribution

To describe the evolution of the system we need several definitions. Let $L_i(t)$ be the number of packets at node $i (i=1,2,...,N)$ at time t and let $U_i(L_i(t))$ $(i=1,2,...,N)$ be a binary-valued random variable that takes value 1 if $L_i(t)>0$ and 0 otherwise. Also let $D_i^j(t)$ $1 \le i \le N, 0 \le j \le N$ be a binary-valued random variable that takes value 1 if a packet is transmitted from i to j at time t, where $j=0$ stands for the outside of the system.

Using these definitions, it is easy to see that the system under consideration evolves according to the following equations:

For $i=1,2,\ldots,N$ and $t=0,1,2,\ldots$

$$L_i(t+1) = L_i(t)+A_i(t) + \sum_{m=1}^{N} D_m^i(t)-U_i(L_i(t))\prod_{m=1}^{i-1}[1-U_m(L_m(t))] \tag{3}$$

Consider now the steady-state joint generating function of the queue lengths distribution:

$$G(\underline{z}) = \lim_{t\to\infty} E\left\{\prod_{i=1}^{N} z_i^{L_i(t)}\right\} \tag{4}$$

Here we assume that the Markov chain $\left\{L_i(t)\right\}_{i=1}^{N}$ is ergodic, namely $G(\underline{z})|_{z_1=z_2=\cdots=z_N=0}>0$. For notational convenience let us define the "routing polynoms":

$$P_i(\underline{z}) = \sum_{m=1}^{N} \nu_i(m)z_m+\nu_i(0) \tag{5}$$

for $i=1,2,\ldots,N$.

To facilitate the presentation let us assume henceforth that $N=3$. As will be seen, the solution method that follows can easily be extended for an arbitrary number of nodes [7].

Using a standard technique we obtain from (3) and (4) that for $N=3$:

$$G(\underline{z}) = F(\underline{z}) \tag{6a}$$

$$\frac{[z_2^{-1}P_2(\underline{z})-z_1^{-1}P_1(\underline{z})]G^{(1)}(\underline{z})+[z_3^{-1}P_3(\underline{z})-z_2^{-1}P_2(\underline{z})]G^{(2)}(\underline{z})+[1-z_3^{-1}P_3(\underline{z})]G^{(3)}(\underline{z})}{1-F(\underline{z})z_1^{-1}P_1(\underline{z})}$$

where

$$\underline{z} = (z_3,z_2,z_1) \tag{6b}$$

$$G^{(1)}(\underline{Z}) = G(z_3,z_2,0) \tag{6c}$$

$$G^{(2)}(\underline{z}) = G(z_3,0,0) \tag{6d}$$

$$G^{(3)}(\underline{z}) = G(0,0,0) \tag{6e}$$

In (6) we see that $G(\underline{z})$ is expressed in terms of the boundary functions $G(z_3,z_2,0)$, $G(z_3,0,0)$ and the constant $G(0,0,0)$. In order to determine $G(\underline{z})$ uniquely we still have to determine these boundary terms.

By using the normalization condition $G(1,1,1)=1$ we find that $G(0,0,0)$ is given by:

$$G(0,0,0) = 1 - \sum_{l=1}^{3} \lambda_l \tag{7a}$$

where for $1 \le l \le 3$

$$\lambda_l = r_l + \sum_{j=1}^{3} \lambda_j \nu_j(l) \tag{7b}$$

$$r_l = \frac{\partial F(\underline{z})}{\partial z_l} \Big|_{z_1=z_2=z_3=1}. \tag{7c}$$

The proof of (7) appears in [7]. It is clear that the condition for steady-state is that $G(0,0,0)>0$.

Now we shall turn to determine the boundary functions $G(z_3,z_2,0)$ and $G(z_3,0,0)$. The process of determining these boundary functions uses the fact that the function $G(\underline{z})$ is analytic in the polydisk $|z_i|<1$, $1 \le i \le 3$.

Using Rouche's theorem [6], we can show that for given z_2,z_3 with $|z_2|<1$, $|z_3|<1$, the following equation in z_1

$$F(\underline{z})P_1(\underline{z}) = z_1 \tag{8}$$

has a unique solution in the unit circle $|z_1|<1$. Let $\hat{z}_1=\hat{z}_1(z_2,z_3)$ denote this unique solution and let $\underline{z}^{(1)}=(z_3,z_2,\hat{z}_1)$. Then from (6) we obtain:

$$G(z_3,z_2,0) = F(\underline{z}^{(1)}) \frac{[z_3^{-1}P_3(\underline{z}^{(1)})-z_2^{-1}P_2(\underline{z}^{(1)})]G(z_3,0,0)+[1-z_3^{-1}P_3(\underline{z}^{(1)})]G(0,0,0)}{1-F(\underline{z}^{(1)})z_2^{-1}P_2(\underline{z}^{(1)})} \tag{9}$$

This is true since $G(\underline{z})$ is an analytic function in the polydisk $|z_i|<1$, $1 \le i \le 3$. Then in this polydisk whenever the denominator of $G(\underline{z})$ vanishes, the numerator must also vanish. Since the denominator of $G(\underline{z})$ vanishes at $\hat{z}_1$, we obtain (9) from (6) and (8). Notice that in (9) the boundary function $G(z_3,z_2,0)$ is expressed in terms of $G(z_3,0,0)$ and $G(0,0,0)$.

Now exploiting the similarity between (6) and (9) we readily see that the following equation in z_2

$$F(\underline{z}^{(1)})P_2(\underline{z}^{(1)}) = z_2 \tag{10}$$

has a unique solution in the unit circle $|z_2|<1$ for given $|z_3|<1$. Let $\hat{z}_2=\hat{z}_2(z_3)$ denote this solution and let $\underline{z}^{(2)}=(z_3,\hat{z}_2(z_3),\hat{z}_1(\hat{z}_2(z_3),z_3))$. Then from (9) and (10) we obtain:

$$G(z_3,0,0) = F(\underline{z}^{(2)})\frac{[1-z_3^{-1}P_3(\underline{z}^{(2)})]G(0,0,0)}{1-F(\underline{z}^{(2)})z_3^{-1}P_3(\underline{z}^{(2)})}. \tag{11}$$

Since $G(0,0,0)$ has already been obtained in (9), we obtain $G(z_3,0,0)$ from (11) and then by substituting it into (9) we obtain $G(z_3,z_2,0)$. Thus all the required boundary terms have been obtained, and the joint generating function $G(\underline{z})$ has been uniquely determined. It is easily seen that this solution method can be extended for an arbitrary number of nodes [7]. From $G(\underline{z})$ any moment of the queues lengths at the nodes can be derived. Specifically if we denote $\overline{L}_i$ the average queue length at node i in steady-state we obtain

$$\overline{L}_i = \frac{\partial G(\underline{z})}{\partial z_i}\Big|_{z_1=z_2=z_3=1}. \tag{12}$$

If we assume that packets arrive at the nodes only at the end of a slot, then using Little's law, we may obtain also the average time delays at node i denoted by T_i as follows:

$$T_i = \frac{\overline{L}_i}{\lambda_i}. \tag{13}$$

where λ_i - the total arrival rate at node i - is defined in (7b).

The total average delay is given by

$$T = \frac{\overline{L}_1+\overline{L}_2+\overline{L}_3}{r_1+r_2+r_3}. \tag{14}$$

4. Examples and Applications to Packet-Radio Networks

In this section we shall give two examples how the results of Section 3 can be applied to packet-radio networks. Our first example would be a fully connected packet-radio network (all nodes can hear each other) that apply the channel access scheme suggested in [3]. This scheme actually allows each node in the network to know when the channel is made available to it. The second example is a three-node tandem packet-radio network where, as will be seen, the priority of the nodes is determined by the network's topology. We shall assume here that all nodes in the network share a common radio channel and that a node cannot transmit and receive a packet simultaneously.

Fully-Connected Packet-Radio Networks

By a fully-connected packet-radio network we mean that all nodes in the network can hear each other. The head-of-the-line protocol suggested in [3] for multiplexing a small number of buffered users over the common channel can be applied in such networks. Assume that priorities are preassigned to the nodes. Then, according to this protocol, each slot is divided into two parts. The first part is used to determine the node with the highest priority that has a packet ready for transmission in that slot, and the second part is used for the actual transmission of the packet at the head of the buffer at this node. In order to determine this node, the first part of the slot is divided into N equal parts (N-number of nodes in the network) which are called minislots. A node with priority i that has a packet ready for transmission, senses the channel during minislots 1,2,...,(i-1) and only if the channel is idle during this period, it starts to transmit a signal until the end of the first part of the slot. This protocol allows all nodes in the network to know, during each slot, which is the node with the highest priority that has a packet ready for transmission, and thus letting it transmit it successfully.

From the above discussion it is clear that a fully-connected network that applies the protocol suggested in [3] belongs to the class of systems analyzed in this paper.

Example

Let us consider the packet-radio network depicted in Fig. 2. Assume that in this network the three nodes can hear each other and they apply the protocol suggested in [3]. We shall assume that node i has priority i ($1 \le i \le 3$). Packets arrive to the nodes from their corresponding sources. Nodes 2 and 3 transmit their packets to node 1 and node 1 finally transmits all the packets to the outside of the system. Therefore we have here: $\vartheta_1(2) = \vartheta_1(3) = \vartheta_3(2) = \vartheta_2(3) = \vartheta_3(0) = \vartheta_2(0) = 0$; $\vartheta_1(0) = \vartheta_2(1) = \vartheta_3(1) = 1$. From (7) we obtain in this case:

$$G(z_3,z_2,z_1) = F(z_3,z_2,z_1) \tag{15}$$
$$\frac{(z_2^{-1}z_1-z_1^{-1})G(z_3,z_2,0)+(z_3^{-1}-z_2^{-1})z_1G(z_3,0,0)+(1-z_3^{-1}z_1)G(0,0,0)}{1-z_1^{-1}F(z_3,z_2,z_1)}$$

We can now apply the solution method presented in Section 3. To facilitate the presentation we shall assume that packets do not arrive at node 1 from its corresponding source and that the arrival processes into nodes 2 and 3 are independent Bernoulli processes, i.e.

$$F(z_3,z_2,z_1) = (z_3r_3+\bar{r}_3)(z_2r_2+\bar{r}_2) \tag{16}$$

where here $\bar{r}_i=1-r_i$ i=2,3 and r_i is the average arrival rate (in units of packets/slot) at node i. Now from (6)-(11) we obtain in this case that

$$\hat{z}_1(z_2,z_3) = (z_3r_3+\bar{r}_3)(z_2r_2+\bar{r}_2) \tag{17}$$

$$\hat{z}_2(z_3) = \frac{1-2\alpha r_2\bar{r}_2-\sqrt{1-4\alpha r_2\bar{r}_2}}{2\alpha r_2^2} \tag{18}$$

where $\alpha=(z_3r_3+\bar{r}_3)^2$ and

$$G(z_3,0,0) = -\frac{1-z_3^{-1}\hat{z}_1(\hat{z}_2(z_3),z_3)}{[z_3^{-1}-\hat{z}_2(z_3)]\hat{z}_1(\hat{z}_2(z_3),z_3)}G(0,0,0) \tag{19}$$

$$G(z_3,z_2,0) = \tag{20}$$
$$-\frac{[1-z_3^{-1}\hat{z}_1(z_2,z_3)]G(0,0,0)+(z_3^{-1}-z_2^{-1})\hat{z}_1(z_2,z_3)G(z_3,0,0)}{z_2^{-1}\hat{z}_1(z_2,z_3)-\hat{z}_1^{-1}(z_2,z_3)}$$

where

$$G(0,0,0) = 1-2(r_2 + r_3) \tag{21}$$

and the condition for steady-state is $r_2+r_3<1/2$. From (15)-(21) we obtain after tedious algebra:

$$\bar{L}_1 = r_2+r_3 \tag{22a}$$

$$\bar{L}_2 = r_2\left[1+\frac{r_2+r_3}{1-2r_2}\right] \tag{22b}$$

$$\bar{L}_3 = r_3 + \frac{2r_2r_3+A_2(r_2+r_3)-(\bar{L}_1+A_2-1)(1-r_2-r_3)}{1-2r_2-2r_3} - 1 \tag{22c}$$

where

$$A_2 = \bar{L}_2 - r_2 \tag{22d}$$

and

$$T_1 = 1 \tag{23a}$$

$$T_i = \frac{\bar{L}_i}{r_i} \quad 2\le i\le 3. \tag{23b}$$

From (22b) we notice that though node 3 has lower priority than node 2, it affects the delay at node 2 since both nodes 2 and 3 route their packets through node 1. In Fig. 3, T_2 and T_3, the average delays (in units of slots) at nodes 2 and 3 respectively, and the total average delay in the network - T are plotted versus the total arrival rate (in units of packets/slot) into the network - γ when $r_2=r_3=r$ (clearly $\gamma=2r$).

Tandem Packet-Radio Network

A tandem packet radio network with N nodes is depicted in Fig. 4. In this network all nodes share a common radio channel and are equipped with radio transmitting and receiving devices. We assume that every node can either transmit or receive but not simultaneously. Instantaneous feedback to the transmitter is assumed meaning that a node knows at the end of the slot if the transmitted packet has been successfully received. All nodes are assumed to have full access capability to the common channel. This means that each node always transmits a packet when its buffer is nonempty, while when it is empty, it does not transmit and is able to receive packets transmitted by other nodes. We also assume that the network topology is such that when node i ($2\leq i\leq N-1$) transmits only nodes i+1 and i-1 can hear the transmission. When nodes 1 or N transmit then only nodes 2 and N-1 can hear the transmission respectively. We finally assume that packets leave the network only when they are transmitted by node 1.

Tandem packet-radio networks with arbitrary number of nodes do not belong to the class of systems considered in this paper since in these networks nodes i and i+3 may succeed in their transmissions simultaneously. However, a three-node tandem packet radio network does belong to the class of systems considered in this paper. This is seen by noticing that in such a network, it is clear that transmissions from node 1 are always successful. Also, a packet cannot leave node 2 unless node 1 is empty, because node 1 always transmits when it is nonempty, and in such a case it cannot receive packets from node 2. Finally, a packet may not leave node 3 unless nodes 2 and 1 are both empty. The reason that node 2 should be empty is clear. Node 1 must also be empty, otherwise it transmits, and the transmissions of nodes 1 and 3 interfere, so that the packet transmitted by node 3 cannot be successfully received at node 2. Therefore, this network belongs to the systems considered in this paper. Also, from the network topology we find that $P_3(\underline{z})=z_2$; $P_2(\underline{z})=z_1$; $P_1(\underline{z})=1$.

Now we can apply the solution method presented in Section 3. In order to give numerical results we have done it when the arrival processes into nodes 1, 2 and 3 are independent Bernoulli processes with average rates r_1, r_2 and r_3 respectively, (in units of packets/slot) i.e.,

$$F(z_3,z_2,z_1) = (z_3 r_3+\bar{r}_3)(z_2 r_2+\bar{r}_2)(z_1 r_1+\bar{r}_1) \qquad (24)$$

where $\bar{r}_i=1-r_i$ for i=1,2,3. For this case we have obtained explicit expressions for the average number of packets at each node and then by using Little's law we have found T_1, T_2, T_3 the average time delays at nodes 1, 2 and 3 respectively (in units of slots) and also T - the average time in the system. The results are :

$$T_1 = \frac{1}{r_1+r_2+r_3}(r_1+A_1)$$

$$T_2 = \frac{1}{r_2+r_3}(r_2+A_2)$$

$$T_3 = \frac{1}{r_3}(r_3+$$

$$+\frac{(r_2+r_3)A_2+2r_2r_3-(A_1+A_2)(1-r_1-r_2-r_3)+r_2+2r_3+r_1r_2+r_1r_3)}{1-r_1-2r_2-3r_3})$$

$$T = T_1+\frac{r_2+r_3}{r_1+r_2+r_3}T_2 + \frac{r_3}{r_1+r_2+r_3}T_3$$

where

$$A_1 = \frac{r_2+r_3}{1-r_1}$$

$$A_2 = \frac{r_2(r_2+r_3)+(1-r_1)(r_3+r_1r_2)}{(1-r_1)(1-r_1-2r_2)}$$

and the condition for steady state is that

$1-r_1-2r_2-3r_3>0$.

In Fig. 5 the average time delays T_1, T_2, T_3 and T are plotted versus γ - the total throughput when $r_1=r_2=r_3=r$ ($\gamma=3r$).

References

[1] Sidi, M. and A. Segall, "Two Interfering Queues in Packet-Radio Networks", IEEE Transactions on Communications, Jan. 1983.

[2] Konheim, A.G. and B. Meister, "Service in a Loop System", *J. Ass. Comput. Mach.*, Vol. 19, Jan. 1972, pp. 92-106.

[3] Scholl, M., "Multiplexing Techniques for Data Transmission over Packet Switched Radio Systems", Ph.D. Thesis, Computer Science Dept., UCLA, 1976.

[4] Kobayashi, H. and A.G. Konheim, "Queueing Models for Computer Communications System Analysis", *IEEE Transactions on Communications*, Vol. COM-25, Jan. 1977, pp. 2-26.

[5] Little, J.D.C., "A Proof for the Queueing Formula $L = \lambda W$", *Op. Res.* Vol. 9, 1961, pp. 383-387.

[6] Copson, E.T., "Theory of Functions of a Complex Variable", London: Oxford University Press, 1946.

[7] Sidi, M. and A. Segall, "Structured Priority Queueing Systems with Applications to Packet-Radio Networks", EE Pub. No. 437, Technion - Israel Institute of Technology, Haifa, July 1982.

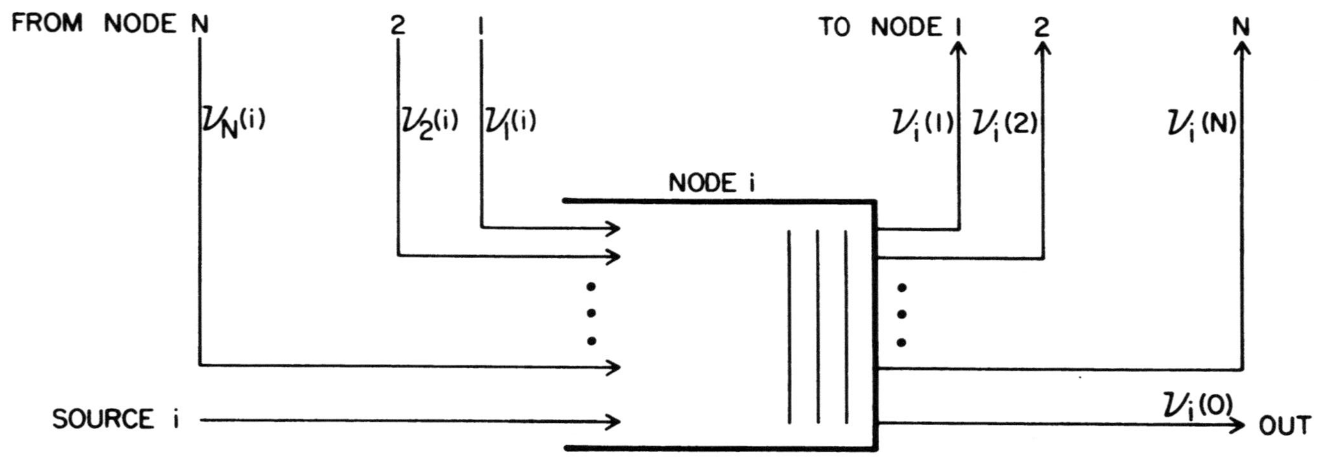

Fig. 1 - An example for a node i in the network.

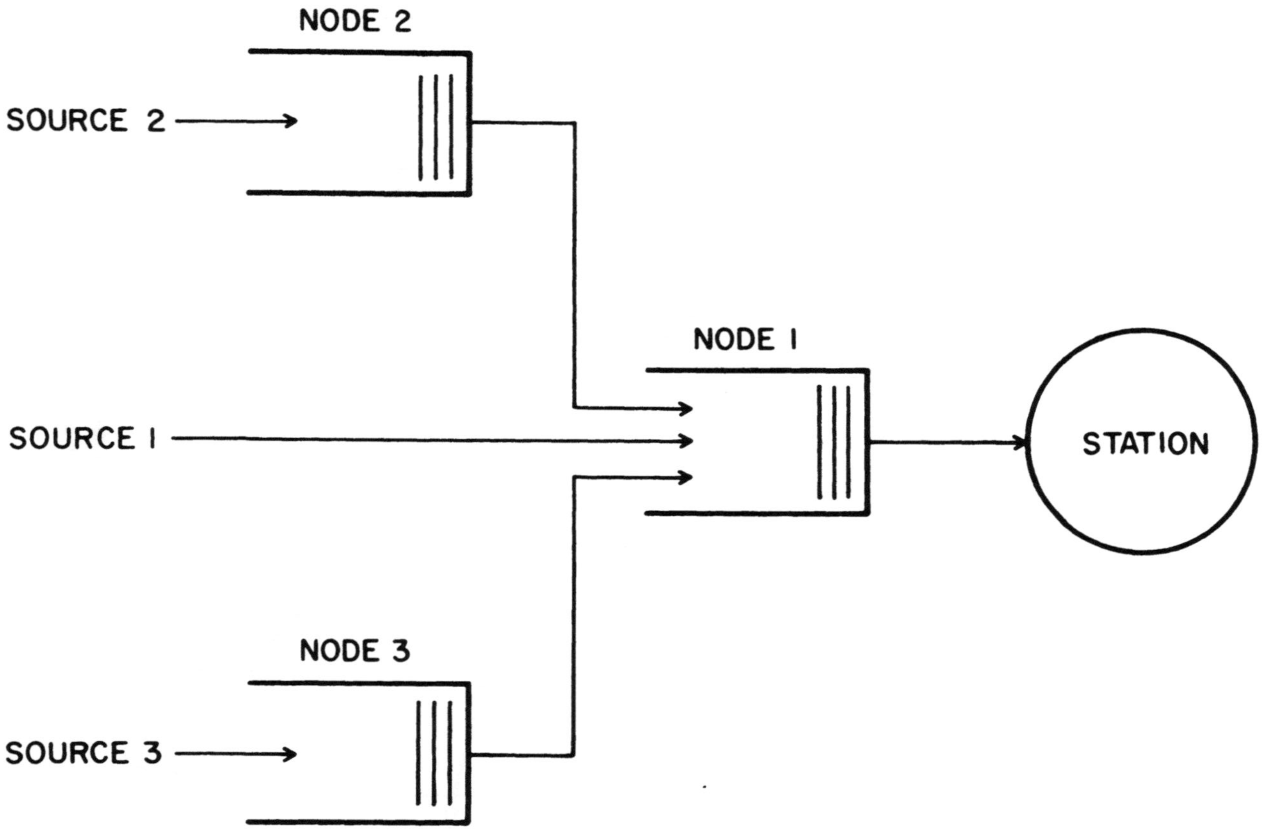

Fig. 2 - A three-node packet-radio network.

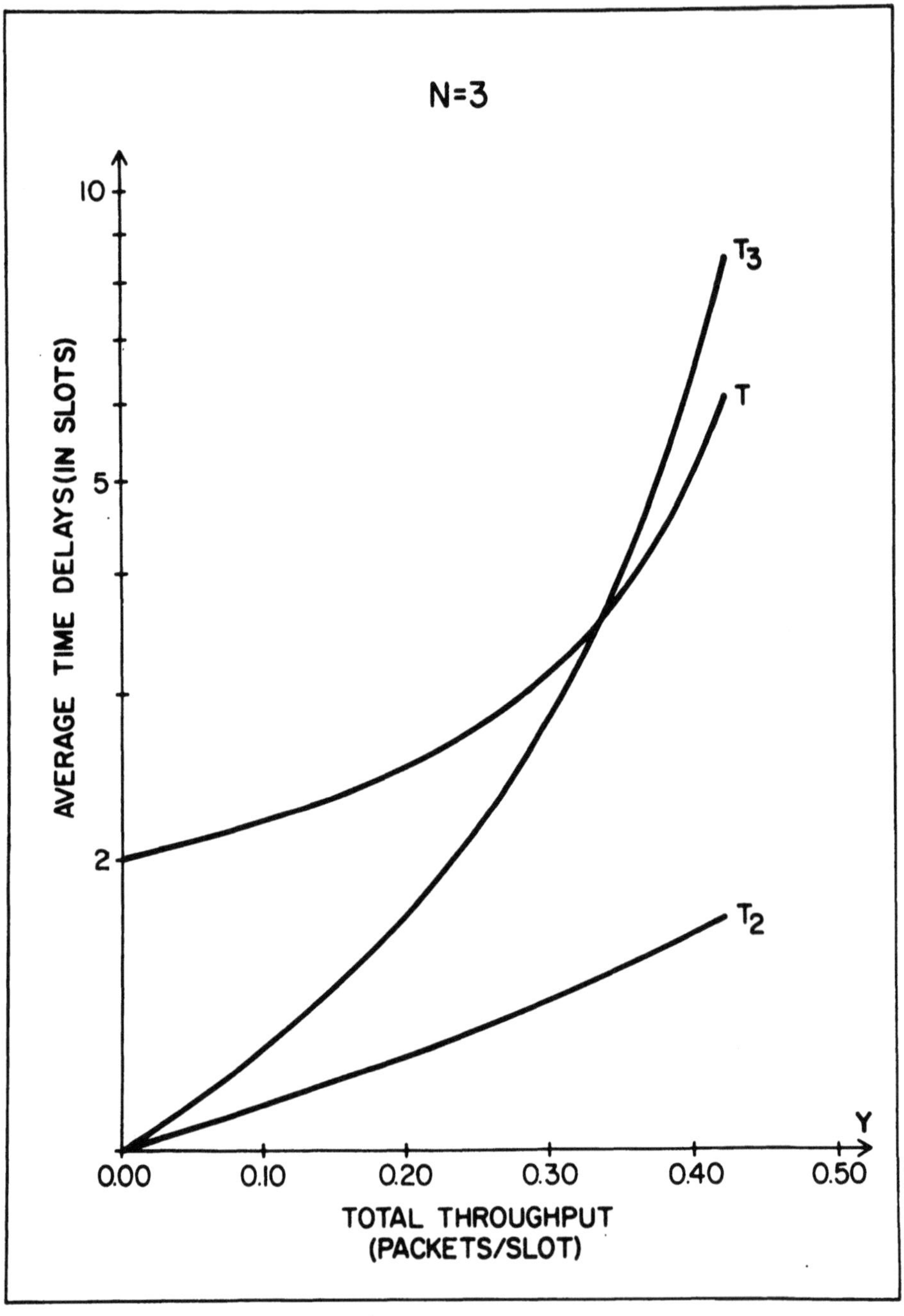

Fig. 3 - Time delays vs γ for the network of Fig. 2, with $r_1 = 0$, $r = r_2 = r_3$.

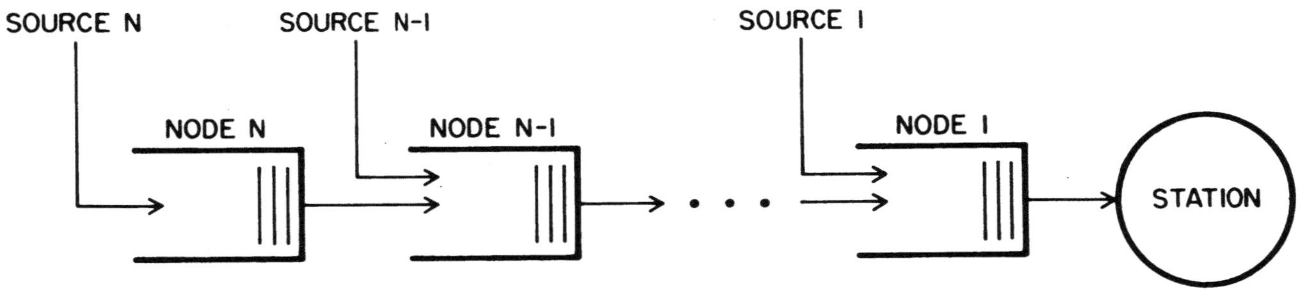

Fig. 4 - Tandem packet radio network with N nodes.

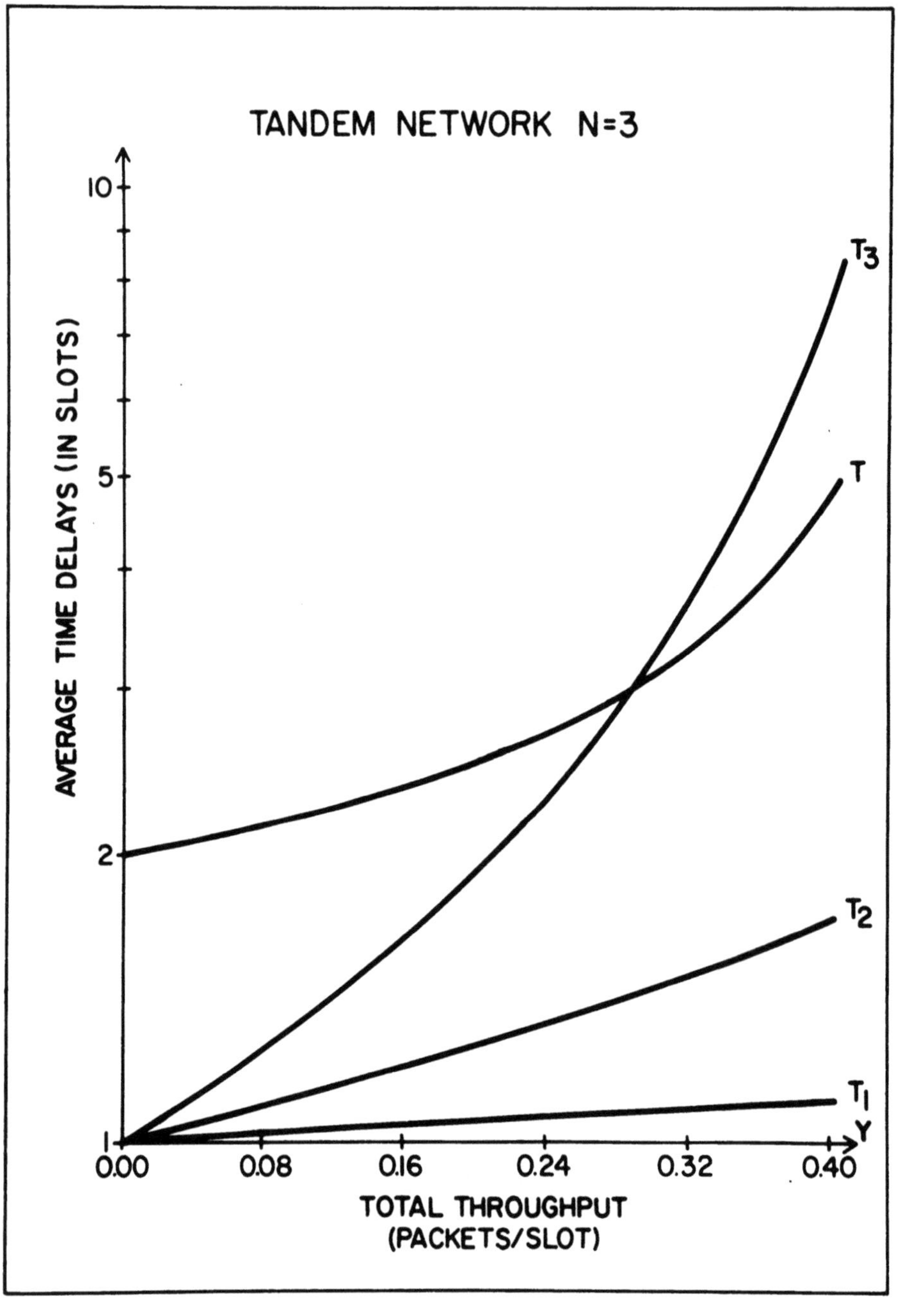

Fig. 5 - Three-node tandem packet-radio network: delays vs γ.

HISTORY DEPENDENT ACCESS SCHEMES IN A TWO-NODE PACKET-RADIO NETWORK

Moshe Sidi[1] and Adrian Segall[2]
Department of Electrical Engineering
Technion - Israel Institute of Technology
Haifa, Israel

ABSTRACT

A two-node packet-radio network with infinite buffers at the nodes is considered. The two nodes transmit data packets to a common station through a shared radio channel. We assume that one of the nodes is granted full rights in accessing the channel while the other node accesses the channel according to some prescribed probabilities, that depend on the number of slots the packet at the head of the queue is waiting in that position. Different probabilities define different access schemes. For these access schemes we derive the joint generating function of the queues contents in steady-state for general arrival processes and give several numerical results for independent Bernoulli arrival processes.

1. Currently on leave at the Lab. for Information and Decision Systems at M.I.T., Cambridge, MA 02139, U.S.A.

2. Currently on leave with I.B.M. Thomas J. Watson Research Center, Box 218, Yorktown Heights, N.Y. 10598, U.S.A.

1. Introduction

In the present paper we consider certain packet-radio networks consisting of two nodes that transmit digital information to a central station by using a distributed access algorithm on a common radio channel. A scheme of the network appears in Fig. 1. Because of simultaneous transmissions, any attempt to transmit may be unsuccessful and in this case the information must be retransmitted. The fact that the activity at one node affects the behavior of the queue at the other node, gives rise to statistical dependence between the queues at the two nodes. In general, the queue length dependence is quite complicated, and there is little hope to obtain explicit analytical results for general access schemes. The purpose of this paper is to present several analytic results for a certain class of distributed access schemes.

We assume throughout this paper that the buffers at the two nodes have *infinite length* and that all information units have equal length (an information unit is called a packet). The time is divided into slots corresponding to the transmission time of a packet and a node may start packet transmissions only at the beginning of a slot. Propagation delay is assumed to be negligible. Also, we neglect channel noise and assume no channel errors, so unsuccessful transmissions may occur only when both nodes transmit simultaneously.

The distributed channel access schemes that are considered in this paper are characterized by the fact that one of the nodes (node 2 in this paper) is granted full rights in accessing the channel. This means that node 2 always attempts to transmit whenever its buffer is nonempty. The other node (node 1) has only partial access capability to the channel and the exact schemes it uses will be described in Section 2.

For these channel access schemes we obtain the generating function of the steady-state joint probability distribution of the queue lengths at the two nodes. From this generating function, any moment may be derived, as well as average time delays. Before proceeding we may mention that the channel schemes considered in this paper are a generalization of the models considered in [1].

2. The Model

Packets arrive randomly at two nodes from the outside of the system, and in general the arrival processes may be correlated. Let $A_1(t)$ and $A_2(t)$ be the number of packets entering node 1 and node 2 from their corresponding sources in the time interval (t,t+1]. The input process $[A_1(t),A_2(t)]$ is assumed to be a sequence of independent and identically distributed random vectors with integer-valued elements. Let

$$a(i,j) = Prob\,(A_1(t) = i \ , A_2(t) = j); \tag{1}$$

and

$$F(x,y) = E[x^{A_1(t)}y^{A_2(t)}] = \sum_{i=0}^{\infty}\sum_{j=0}^{\infty} a(i,j)x^i y^j \tag{2}$$

We assume that $F(x,y)$ depends on both x and y, namely that packets arrive at the two nodes with nonzero probability and that the two nods have infinite buffers. The packets at the nodes must be transmitted to a main station over a common channel.

We now describe the channel access schemes used by the nodes. As mentioned earlier, in these schemes node 2 has full channel access capability. This means that, at each slot when its buffer is nonempty, node 2 tries to transmit a packet. Whether this transmission is successful or not depends on the activity of the other node during the same slot. Node 1 accesses the channel by tossing a coin just before every slot, provided its buffer is nonempty, and attempting transmission of the packet at the head of the queue in case of success. The schemes considered in this paper are characterized by the fact that the coin probability of success is a function of the number of slots the considered packet is waiting at the head of the queue. Let q_k $(k=0,1,2,...)$ be the coin probability of success if the packet at the head of the queue has been waiting exactly k slots in that position. We restrict our analysis to schemes for which $q_k = p = const.$ for $k \geq K$, where K is an arbitrary constant, and refer to this situation as *Scheme K.* The scheme 0 case, when the coin probability of success at node 1 is constant has been analyzed in [1].

It is easy to see that Scheme K_1 is a particular case of Scheme K_2 when $K_1 < K_2$, since for $0 \leq k \leq K_1$ the parameters q_k in Scheme K_2 can be chosen to be identical to the parameters q_k of Scheme K_1 and for $K_1 \le k \le K_2$ we can choose $q_k = q_{K_1}$ for Scheme K_2. Therefore when K increases, we expect to obtain better performance of the network (certainly by proper choosing of the parameters q_k).

3. Steady-State Distribution

Suppose node 1 uses channel access scheme $K (K \ge 2)$. We say that the network is in state 0 if there are no packets at node 1 or the first packet in queue 1 has just arrived to the head of the queue. The network is in state k, $1 \le k < K$ if the packet has been waiting at the head of the queue for exactly k slots, and in state K if it has been waiting for K slots or more. Let $P_k(m,n)$ $0 \le k \le K$, $m \ge 0$, $n \ge 0$ be the equilibrium joint probability that the network is in state k and the queue lengths at nodes 1,2 are m,n respectively. The equilibrium equations are (henceforth $\bar{q}_k = 1 - q_k$):

$$P_0(0,n) = a(0,n)P_0(0,0) + \sum_{l=0}^{n} a(0,l)P_0(0,n-l+1) \tag{3a}$$

$$+ a(0,n)\sum_{k=0}^{K} q_k P_k(1,0) \quad n \ge 1$$

$$P_0(m,0) = a(m,0)[P_0(0,0)+P_0(0,1)] + \sum_{k=0}^{K}\sum_{l=0}^{m} q_k a(l,0)P_k(m-l+1,0) \quad m \ge 0 \tag{3b}$$

$$P_0(m,n) = a(m,n)P_0(0,0) + \sum_{l=0}^{n} a(m,l)P_0(0,n-l+1) + \tag{3c}$$

$$+ \sum_{k=0}^{K}\sum_{l=0}^{m} q_k a(l,n)P_k(m-l+1,0) \quad m \ge 1, n \ge 1$$

For $0 < k < K$

$$P_k(0,n) = 0 \qquad n \ge 0 \tag{4a}$$

$$P_k(m,0) = \sum_{l=0}^{m-1} a(l,0)[\bar{q}_{k-1}P_{k-1}(m-l,0) + \bar{q}_{k-1}P_{k-1}(m-l,1)] \quad m \ge 1 \tag{4b}$$

$$P_k(m,n) = \sum_{l=0}^{m-1}\left\{\sum_{j=0}^{n-1} a(l,j)q_{k-1}P_{k-1}(m-l,n-j) + \right. \tag{4c}$$

$$\left. + \sum_{j=0}^{n} a(l,j)\bar{q}_{k-1}P_{k-1}(m-l,n-j+1)\right\} + \sum_{l=0}^{m-1} a(l,n)\bar{q}_{k-1}P_{k-1}(m-l,0) \qquad m \ge 1, n \ge 1$$

and

$$P_K(0,n) = 0 \quad n \geq 0 \tag{5a}$$

$$P_K(m,0) = \sum_{k=K-1}^{K} \sum_{l=0}^{m-1} a(l,0)[\bar{q}_{k-1}P_{k-1}(m-l,0) \tag{5b}$$

$$+\bar{q}_{k-1}P_{k-1}(m-l,1)] \quad m \geq 1$$

$$P_K(m,n) = \sum_{k=K-1}^{K} \left\{ \sum_{l=0}^{m-1} \left[\sum_{j=0}^{n-1} a(l,j)q_{k-1}P_{k-1}(m-l,n-j)+ \right.\right. \tag{5c}$$

$$\left. + \sum_{j=0}^{n} a(l,j)\bar{q}_{k-1}P_{k-1}(m-l,n-j+1) \right] +$$

$$\left. + \sum_{l=0}^{m-1} a(l,n)\bar{q}_{k-1}P_{k-1}(m-l,0) \right\} \quad m \geq 1, n \geq 1$$

Let

$$G_k(x,y) = \sum_{m=0}^{\infty} \sum_{n=0}^{\infty} P_k(m,n)x^m y^n \quad 0 \leq k \leq K \tag{6}$$

be the steady-state queue length joint generating function when the system is stable k. Then in Appendix A we show that:

$$G_0(x,y)=F(x,y)\left\{ G_0(0,0)(1-q_0 x^{-1})+[G_0(0,y)-G_0(0,0)]y^{-1} + \right. \tag{7a}$$

$$\left. + \sum_{k=0}^{K} q_k G_k(x,0)x^{-1} \right\}$$

$$G_1(x,y) = F(x,y)\left\{ [G_0(x,y)-G_0(x,0)-G_0(0,y)+ \right. \tag{7b}$$

$$\left. + G_0(0,0)](q_0+\bar{q}_0 y^{-1}) + [G_0(x,0)-G_0(0,0)]\bar{q}_0 \right\}$$

for $2 \leq k \leq K-1$

$$G_k(x,y) = F(x,y)\left\{ [G_{k-1}(x,y)-G_{k-1}(x,0)](q_{k-1}+\bar{q}_{k-1}y^{-1}) \right. \tag{7c}$$

$$\left. +G_{k-1}(x,0)\bar{q}_{k-1} \right\}$$

and

$$G_K(x,y) = F(x,y)\left\{ \sum_{k=K-1}^{K} \left[[G_k(x,y)-G_k(x,0)[(q_k+\bar{q}_k y^{-1})+ G_k(x,0)\bar{q}_k \right] \right\} \tag{7d}$$

In order to uniquely determine $G_k(x,y)$ $0 \le k \le K$, we still have to find the boundary functions $G_0(0,y)$, $G_k(x,0)$ for $0 \le k \le K$ and the constant $G_0(0,0)$.

Determination of $G_0(0,y)$:

Let $x \to 0$ in (7a). Then

$$G_0(0,y) = F(0,y)\left\{G_0(0,0)+[G_0(0,y)-G_0(0,0)]y^{-1} + \sum_{k=0}^{K} q_k P_k(1,0)\right\} \tag{8}$$

Therefore,

$$G_0(0,y) = F(0,y)\frac{G_0(0,0)(1-y^{-1})+\sum_{k=0}^{K} q_k P_k(1,0)}{1-F(0,y)y^{-1}} \tag{9}$$

Let t designate the solution of the equation $F(0,t)t^{-1}=1$ in the unit circle $|t|<1$. In Appendix B we show that such a solution exists and is unique. Since $G_0(0,y)$ is analytic for $|y|<1$, we have

$$\sum_{k=0}^{K} q_k P_k(1,0) = G_0(0,0)(t^{-1}-1) \tag{10}$$

Therefore

$$G_0(0,y) = F(0,y)\frac{G_0(0,0)(t^{-1}-y^{-1})}{1-F(0,y)y^{-1}} \tag{11}$$

This determines $G_0(0,y)$ up to the constant $G_0(0,0)$.

Determination of $G_k(x,0)$ $0 \le k \le K$

The determination of the $K+1$ boundary functions $G_k(x,0)$ $0 \le k \le K$ is much more complex. For convenience let us use the following definitions for $0 \le k \le K$

$$b_k(y) \overset{\Delta}{=} q_k + \bar{q}_k y^{-1} \tag{12}$$

$$d_k(y) \overset{\Delta}{=} \bar{q}_k - b_k(y)$$

Using these definitions we see from (7d) that:

$$G_K(x,y) = F(x,y)\frac{N(x,y)}{1-F(x,y)b_K(y)} \tag{13a}$$

where

$$N(x,y) = G_{K-1}(x,y)b_{K-1}(y)+G_K(x,0)d_K(y)+G_{K-1}(x,0)d_{K-1}(y) \tag{13b}$$

Now recursive substitution of (7c) in (13b) yields:

$$N(x,y) = G_K(x,0)d_K(y)+\sum_{k=1}^{K-1} d_k(y)B_{k+1}(y)F^{K-k-1}(x,y)G_k(x,0) \tag{14a}$$

$$+ B_1(y)F^{K-2}(x,y)G_1(x,y)$$

where

$$B_k(y) \triangleq \prod_{i=k}^{K-1} b_i(y) \quad 0 \le k \le K-1 \tag{14b}$$

$$B_K(y) \triangleq 1 \tag{14c}$$

Using (7b) in (14a) we have

$$N(x,y) = G_K(x,0)d_K(y)+\sum_{k=0}^{K-1} d_k(y)B_{k+1}(y)F^{K-k-1}(x,y)G_k(x,0)+ \tag{15}$$

$$+ F^{K-1}(x,y)\Big\{[G_0(x,y)-G_0(0,y)]B_0(y)-d_0(y)B_1(y)G_0(0,0)\Big\}$$

Now using (7a) in (15) we finally obtain:

$$N(x,y) = \sum_{k=0}^{K} D_k(x,y)G_k(x,0)+M(x,y) \tag{16a}$$

where

$$D_k(x,y) = d_k(y)B_{k+1}(y)F^{K-k-1}(x,y)+ \tag{16b}$$

$$+ B_0(y)F^K(x,y)x^{-1}q_K \quad 0 \le k \le K-1$$

$$D_K(x,y) = d_K(y)+B_0(y)F^K(x,y)x^{-1}q_K \tag{16c}$$

$$M(x,y) = F^{K-1}(x,y)\Big\{B_0(y)[y^{-1}F(x,y)-1]G_0(0,y)+ \tag{16d}$$

$$+ [(1-q_0x^{-1}-y^{-1})B_0(y)F(x,y)-d_0(y)B_1(y)]G_0(0,0)\Big\}$$

The functions $D_k(x,y)$ $0 \le k \le K$ are all known and $M(x,y)$ is known up to the constant $G_0(0,0)$.

In Appendix C we show that each boundary function $G_k(x,0)$ $1 \le k \le K$ can be expressed as a linear combination of $G_0(x,0)$ as follows

$$G_k(x,0) = C_k(x)G_0(x,0)+H_k(x) \quad 1 \le k \le K \tag{17}$$

where the functions $C_k(x)$ and $H_k(x)$ for $1 \le k \le K$ are defined in Appendix C. For simplicity we define $C_0(x) \triangleq 1$ and $H_0(x) \triangleq 0$. Substitution of (17) in (16a) yields:

$$N(x,y) = G_0(x,0)\sum_{k=0}^{K} D_k(x,y)C_k(x) + \tilde{M}(x,y) \tag{18a}$$

where

$$\tilde{M}(x,y) = M(x,y) + \sum_{k=0}^{K} D_k(x,y)H_k(x) \tag{18b}$$

Now for $|x|<1$ let $f=f(x)$ be the unique solution of $F(x,f)b_K(f)=1$ with the property $|f|<1$. The existence and uniqueness of such a solution is proved in Appendix B. Since $G_K(x,y)$ is analytic in the polydisk $|x|<1$, $|y|<1$, it is clear from (13a) that

$$N(x,f) = 0 \tag{19}$$

Therefore, from (18a)

$$G_0(x,0) = -\frac{\tilde{M}(x,f)}{\sum_{k=0}^{K} D_k(x,f)C_k(x)} \tag{20}$$

Thus $G_0(x,0)$ is determined up to the constant $G_0(0,0)$. Consequently from (17) it is clear that $G_k(x,0)$ $1\le k\le K$ are all determined up to the constant $G_0(0,0)$. Finally $G_0(0,0)$ is determined by using the normalization condition

$$\sum_{k=0}^{K} G_k(1,1) = 1. \tag{21}$$

Now that we have determined the boundary functions $G_0(0,y)$ $G_0(x,0)$ $0\le k\le K$ and the constant $G_0(0,0)$ we see from (7) that $G_k(x,y)$ $0\le k\le K$ are uniquely determined. The steady-state generating function of the queue lengths at the nodes is $G(x,y) = \sum_{k=0}^{K} G_k(x,y)$. From $G(x,y)$ any moment of the queue lengths at the nodes can, in principle, be derived and using Little's law [3] the average time delays can be obtained as well.

4. Independent Bernoulli Arrival Processes

Although the results of the previous section hold for general input processes, we now consider a particular example where the arrivals into nodes 1 and 2 are independent Bernoulli processes with rates r_1 and r_2 respectively, i.e.,

$$F(x,y) = F_1(x)F_2(y) \tag{22a}$$

where

$$F_1(x) = xr_1+\bar{r}_1 \tag{22b}$$

$$F_2(y) = yr_2+\bar{r}_2 \tag{22c}$$

and $\bar{r}$ denotes $(1-r)$. For this case we immediately find from (11) that

$$G_0(0,y) = G_0(0,0)[1+\frac{r_2}{\bar{r}_2}y] \tag{23}$$

Also in Appendix D we show that

$$G_k(x,0) = F_1^k(x)[G_0(x,0)-G_0(0,0)]\alpha_k \quad 1\le k\le K-1 \tag{24a}$$

and that

$$G_K(x,0) = \tag{24b}$$

$$\left\{[G_0(x,0)-G_0(0,0)][1-F_1(x)\bar{r}_2x^{-1}(q_0+\sum_{k=1}^{K-1}q_kF_1^k(x)\alpha_k)]+\right.$$

$$\left.+\ G_0(0,0)[1-F_1(x)]\right\}/\ F_1(x)\bar{r}_2\bar{q}_Kx^{-1}$$

where the coefficients α_k are determined in Appendix D.

Using (20) $G_0(x,0)$ is determined up to the constant $G_0(0,0)$. The constant $G_0(0,0)$ is derived in Appendix D and we obtain

$$G_0(0,0) = \bar{r}_2\left\{1-r_1\frac{\beta_K+\bar{r}_2\sum_{k=0}^{K-1}\alpha_k(\beta_kq_K-\beta_Kq_k)}{q_K(\bar{q}_K-r_2)}\right\} \tag{25}$$

where the constants β_k $0\le k\le K$ are given in Appendix D.

The expressions for the average queue lengths at the nodes are too complicated to be given here. However we shall present several numerical results in the following section.

Examples

Here we still assume independent Bernoulli arrival processes. We shall present several numerical results for Schemes 0,1,2. Although the analysis in Section 3 was restricted to $K\ge 2$, it is easy to see that Schemes 0,1 were implicitly included in the analysis, simply by choosing $q_0=q_1=q_2$ and $q_1=q_2$ respectively. For convenience let $q_K=p$. Then we obtain:

Scheme K=0

$$G_0(0,0) = \bar{r}_2\left[1-\frac{r_1\bar{p}}{p(\bar{p}-r_2)}\right] \tag{26}$$

Scheme K=1

$$G_0(0,0) = \bar{r}_2\left[1-r_1\frac{\bar{p}+(p-q_0)(\bar{p}-r_2)}{p(\bar{p}-r_2)}\right] \tag{27}$$

Scheme K=2

$$G_0(0,0) = \bar{r}_2\left[1-r_1\frac{2p\bar{p}+\bar{p}-p(\bar{q}_0+\bar{q}_1)+\bar{r}_2(p-q_0+(p-q_1)(1-q_0\bar{p}))}{p(\bar{p}-r_2)}\right] \tag{28}$$

According to Scheme 0, node 1 tries to transmit a packet, if it has any, with constant probability p. This scheme has been analyzed in [1] and all figures there corresponding to System 1 describe the performance of Scheme 0. According to Scheme 1, node 1 tries to transmit a packet that has just arrived to the head of the queue with probability q_0 and if the packet is not transmitted successfully, it is transmitted with probability p. According to Scheme 2, node 1 tries to transmit a packet that has just arrived to the head of the queue with probability q_0, if the packet does not leave the node it tries to transmit it with probability q_1 and afterwards with constant probability p.

From (27) we see that for Scheme 1, $G_0(0,0)$ is a monotonic increasing function of q_0 for all r_1, r_2 and p. Therefore $G_0(0,0)$ is maximized in Scheme 1 for $q_0=1$. In Scheme 2 we have found from (28) (by numerical search) that $q_0=1$, $q_1=0$ maximize $G_0(0,0)$ for all values of r_1, r_2 and p.

We now turn to investigate the time delays in the network. An interesting result for Scheme 1 is that the average time delay at node 2 is independent of q_0. The average time delay at node 1 and hence the total average delay in the network are decreasing functions of q_0. Fig. 2 demonstrates this behavior. Consequently, to minimize the total average delay for Scheme 1 we have to choose $q_0=1$. In Fig. 3 the total average delay T is plotted versus p for different arrival rates and $q_0=1$. As is seen from this figure, for low arrival rates, T is almost independent of p (except for very large or very small values of p). The reason is that in this case most of the packets at node 1 are successfully transmitted in the first attempt.

For Scheme 2 we found that the total average delay is minimized when q_0=1, q_1=0. From Fig. 4 we see that T is a monotonic increasing function of q_1 when q_0=1. In Fig. 5, T is plotted versus p for different arrival rates and q_0=1, q_1=0. The fact that the optimal q_1 is q_1=0 is not surprising, since if node 1 tries to transmit a packet with probability 1 (q_0=1) then unsuccessful transmission indicates that node 2 is nonempty. Therefore in order to insure successful transmission from node 2, node 1 must be silent (q_1=0) for one slot.

Comparisons between the three schemes appear in Figs. 6-10. In all these figures we choose q_0=1 in Scheme 1 and q_0=1, q_1=0 in Scheme 2. In Fig. 6 the probability for empty network $G_0(0,0)$ is plotted versus p for different arrival rates. In Figs. 7 and 8 the average time delay at nodes 1 and 2 respectively are plotted versus p for $r_1=r_2=0.1$. In Fig. 9 the total average time delay T is plotted versus p. From these figures we see that as K increases the performance of the network is improved.

In each of the three schemes the parameter p plays an important role. We can find the optimal p that minimizes the total average delay for any arrival rates. In Fig. 10 the *minimal* total average delay is plotted versus the total arrival rate - γ when $r_1=r_2=r$ (clearly $\gamma=2r$) for the three schemes. As expected we see that the performance of the network improves when K increases.

We have also treated the problem of determining the maximal throughput of the system in the case $r_1=r_2=r$. Since the condition for steady-state is that $G_0(0,0)>0$ for the three schemes, we can find the maximal throughput from (26) - (28). The results are that for Schemes 0,1,2 the total throughput $\gamma(\gamma=2r)$ must be less than 0.5, 0.591 and 0.618 respectively.

References

[1] M. Sidi and A. Segall, "Two Interfering Queues in Packet-Radio Networks", IEEE Transactions on Communications, Vol. COM-31, Jan. 1983.

[2] E.T. Copson, "Theory of Functions of a Complex Variable", London: Oxford University Press, 1948.

[3] J.D.C. Little, "A Proof for the Queueing Formula $L=\lambda W$", Operations Research, Vol. 9, 1961, pp. 383-387.

Appendix A

From (3) using (6) we obtain for $k=0$:

$$G_0(x,y) = \sum_{m=0}^{\infty}\sum_{n=0}^{\infty} x^m y^n P_0(m,n) = \tag{A.1}$$

$$= \sum_{m=1}^{\infty}\sum_{n=1}^{\infty} x^m y^n \{a(m,n)P_0(0,0)+\sum_{l=0}^{n} a(m,l)P_0(0,n-l+1) +$$

$$+ \sum_{k=0}^{K}\sum_{l=0}^{m} a(l,n)P_k(m-l+1,0)\} +$$

$$+ \sum_{m=0}^{\infty} x^m \{a(m,0)[P_0(0,0)+P_0(0,1)]+\sum_{k=0}^{K}\sum_{l=0}^{m} q_k a(l,0)P_k(m-l+1,0)\} +$$

$$+ \sum_{n=1}^{\infty} y^n \{a(0,n)P_0(0,0)+\sum_{l=0}^{n} a(0,l)P_0(0,n-l+1)+a(0,n)\sum_{k=0}^{K} q_k P_k(1,0)\}$$

Arranging (A.1) we obtain:

$$G_0(x,y)=P_0(0,0)F(x,y)+y^{-1}\sum_{m=0}^{\infty}\sum_{n=0}^{\infty}\sum_{l=0}^{n} x^m y^l a(m,l)P_0(0,n-l+1)y^{n-l+1} \tag{A.2}$$

$$+ x^{-1}\sum_{k=0}^{K}\sum_{m=0}^{\infty}\sum_{n=0}^{\infty}\sum_{l=0}^{m} x^l y^n q_k a(l,n)P_k(m-l+1,0)x^{m-l+1}$$

Therefore from (A.2) using (6) we obtain:

$$G_0(x,y)=P_0(0,0)F(x,y)+y^{-1}F(x,y)[G_0(0,y)-G_0(0,0)] + \tag{A.3}$$

$$+ x^{-1}\sum_{k=0}^{K} f(x,y)q_k[G_k(x,0)-G_k(0,0)]$$

and since $G_k(0,0)=0$ for $1 \le k \le K$, we immediately obtain (7a).

From (4) using (6) we obtain for $0<k<K$:

$$G_k(x,y) = \sum_{m=0}^{\infty}\sum_{n=0}^{\infty} x^m y^n P_k(m,n) = \tag{A.4}$$

$$= \sum_{m=1}^{\infty}\sum_{n=1}^{\infty} x^m y^n \Bigg[\sum_{l=0}^{m-1}\sum_{j=0}^{n-1} a(l,j)q_{k-1}P_{k-1}(m-l,n-j) +$$

$$+ \sum_{l=0}^{m-1}\sum_{j=0}^{n} a(l,j)\bar{q}_k-1P_{k-1}(m-l,n-j+1)+\sum_{l=0}^{m-1} a(l,n)\bar{q}_{k-1}P_{k-1}(m-l,0)\Bigg]$$

$$+ \sum_{m=1}^{\infty} x^m \Bigg\{\sum_{l=0}^{m-1} a(l,0)\Big[\bar{q}_{k-1}P_{k-1}(m-l,0)+\bar{q}_{k-1}P_{k-1}(m-l,1)\Big]\Bigg\}.$$

Arranging (A.4) yields:

$$G_k(x,y)=\sum_{m=1}^{\infty}\sum_{n=1}^{\infty}\sum_{l=0}^{m-1}\sum_{j=0}^{n-1} x^l y^j a(l,j) q_{k-1} P_{k-1}(m-l,n-j) x^{m-l} y^{n-j} + \tag{A.5}$$

$$+\sum_{m=1}^{\infty}\sum_{n=0}^{\infty}\sum_{l=0}^{m-1} x^l y^n a(l,n) \bar{q}_{k-1} P_{k-1}(m-l,0) x^{m-l} +$$

$$+y^{-1}\sum_{m=1}^{\infty}\sum_{n=0}^{\infty}\sum_{l=0}^{m-1} x^l y^j a(l,j) \bar{q}_{k-1} P_{k-1}(m-l,n-j+1) x^{m-l} y^{n-j+1} +$$

Using (6) we obtain:

$$G_k(x,y)=F(x,y)[G_{k-1}(x,y)-G_{k-1}(x,0)-G_{k-1}(0,y)+G_{k-1}(0,0)]q_{k-1} + \tag{A.6}$$

$$+F(x,y)[G_{k-1}(x,0)-G_{k-1}(0,0)]\bar{q}_{k-1} +$$

$$+y^{-1}F(x,y)[G_{k-1}(x,y)-G_{k-1}(x,0)-G_{k-1}(0,y)+G_{k-1}(0,0)]q_{k-1}$$

For $k=1$, (A.6) implies (7b). For $2\le k\le K$ using the fact that $G_k(x,0)=0$ for all x, (A.6) implies (7c). Finally (7d) is proved from (5) in the same way as (7c) was proved from (4).

Appendix B

Theorem:

For $|x|<1$ the equation

$$F(x,y)(yq_K+\bar{q}_K) = y, \tag{B.1}$$

has a unique solution in the unit circle $|y|<1$.

Proof:

Since $F(x,y)$ depends on x, then there exists $a(i,j)>0$ for some j and some $i>0$. Therefore for $|x|<1$ and $|y|=1$ we have:

$$|F(x,y)\ (yq_K+\bar{q}_K)| \le |F(x,y)| = \tag{B.2}$$

$$= |\sum_{i=0}^{\infty}\sum_{j=0}^{\infty} a(i,j) x^i y^i| \le \sum_{i=0}^{\infty}\sum_{j=0}^{\infty} a(i,j)|x|^i < \sum_{i=0}^{\infty}\sum_{j=0}^{\infty} a(i,j) = 1 = |y|.$$

Hence, applying Rouche's theorem [2], (B.1) has exactly one zero within $|y|=1$ for $|x|<1$.

Q.E.D.

Now the existence and uniqueness of t in (10) and (11) is clear if we take $x=0$ and $q_K=0$ in (B.1). The existence and uniqueness of f in (19) and (20) is directly proved by the above Theorem.

Appendix C

In this Appendix we want to show that each of the boundary functions $G_k(x,0)$, $1 \le k \le K$ can be expressed as a linear combination of $G_0(x,0)$ and a known function. In this Appendix we use the notation $G_k^{(n)}(x,y)$ and $F^{(n)}(x,y)$ to denote the n-th derivative of $G_k(x,y)$ and $F(x,y)$, respectively, with respect to y. From (7a) we see that:

$$yG_0(x,y) = F(x,y)[yg(x)+G_0(0,y)-G_0(0,0)] \tag{C.1}$$

where

$$g(x) = G_0(0,0)(1-q_0x^{-1}) + x^{-1}\sum_{k=0}^{K} q_k G_k(x,0) \tag{C.2}$$

Therefore:

$$yG_0^{(n)}(x,y)+nG_0^{(n-1)}(x,y) = F^{(n)}(x,y)[yg(x)+G_0(0,y)-G_0(0,0) + \tag{C.3}$$
$$+ nF^{(n-1)}(x,y)[g(x)+G_0^{(1)}(0,y)]+\sum_{i=2}^{n}\binom{n}{i}F^{(n-i)}(x,y)G_0^{(i)}(0,y)$$

Let $y \to 0$ in (C.3). Then

$$G_0(x,0) = F(x,0)[g(x) + P_0(0,1)] \tag{C.4}$$

and for $1 \le n \le K-1$:

$$G_0^{(n)}(x,0) = G_0(x,0)c_{n,0}(x)+h_{n,0}(x). \tag{C.5}$$

where

$$c_{n,0}(x) = F^{(n)}(x,0)/F(x,0) \tag{C.6}$$

$$h_{n,0}(x) = \frac{1}{n+1}\sum_{i=2}^{n+1}\binom{n+1}{i}i!F^{(n+1-i)}(x,0)P_0(0,i) \tag{C.7}$$

Now from (7b) we see that

$$yG_1(x,y)=F(x,y)\{[G_0(x,y)-G_0(x,0)-G_0(0,y)+G_0(0,0)](q_0y+\bar{q}_0) + \tag{C.8}$$
$$+ y\bar{q}_0[G_0(x,0)-G_0(0,0)]\}$$

Therefore:

$$yG_1^{(n)}(x,y)+nG_1^{(n-1)}(x,y)=(q_0y+\bar{q}_0)\sum_{i=0}^{n}\binom{n}{i}F^{(n-i)}(x,y)[G_0^{(i)}(x,y)- \tag{C.9}$$

$$-G_0^{(i)}(0,y)] + nq_0\sum_{i=1}^{n-1}\binom{n-1}{i}F^{n-1-i}(x,y)[G_0^{(i)}(x,y)-G_0^{(i)}(0,y)] +$$

$$+\bar{q}_0[G_0(x,0)-G_0(0,0)][yF^{(n)}(x,y)+nF^{(n-1)}(x,y)]$$

Let $y \to 0$ in (C.9). Then

$$G_1^{(n-1)}(x,0)=\frac{1}{n}\bar{q}_0\sum_{i=1}^{n}\binom{n}{i}F^{(n-i)}(x,0)[G_0^{(i)}(x,0)-i!P_0(0,i)] + \tag{C.10}$$

$$+ q_0\sum_{i=1}^{n-1}\binom{n-1}{i}F^{(n-1-i)}(x,0)[G_0^{(i)}(x,0)-i!P_0(0,i)] +$$

$$+\bar{q}_0[G_0(x,0)-G_0(0,0)]F^{(n-1)}(x,0)$$

From (C.10) we finally obtain for $0 \le n \le K-2$ that

$$G_1^{(n)}(x,0) = c_{n,1}(x)G_0(x,0)+h_{n,1}(x) \tag{C.11}$$

where

$$c_{n,1}(x) = \bar{q}_0F^{(n)}(x,0)+\frac{1}{n+1}\bar{q}_0\sum_{i=1}^{n+1}\binom{n+1}{i}F^{(n+1-i)}(x,0)c_{i,0}(x) + \tag{C.12}$$

$$+ q_0\sum_{i=1}^{n}\binom{n}{i}F^{(n-i)}(x,0)c_{i,0}(x)$$

$$h_{n,1}(x) = \frac{1}{n+1}\bar{q}_0\sum_{i=1}^{n+1}\binom{n+1}{i}F^{(n+1-i)}(x,0)[h_{i,0}(x)-i!P_0(0,i)] + \tag{C.13}$$

$$+ q_0\sum_{i=1}^{n}\binom{n}{i}F^{(n-i)}(x,0)[h_{i,0}(x)-i!P_0(0,i)] -$$

$$- \bar{q}_0G_0(0,0)F^{(n)}(x,0)$$

Now from (7c) in the same way we obtained (C.10) we have for $2 \le k \le K-1$:

$$G_k^{(n)}(x,0) = \frac{1}{n}\bar{q}_{k-1}\sum_{i=1}^{n}\binom{n}{i}F^{(n-i)}(x,0)G_{k-1}^{(i)}(x,0) + \tag{C.14}$$

$$+ q_{k-1}\sum_{i=1}^{n-1}\binom{n-1}{i}F^{(n-1-i)}(x,0)G_{k-1}^{(i)}(x,0)+\bar{q}_{k-1}G_{k-1}(x,0)F^{(n-1)}(x,0)$$

Assume that

$$G_{k-1}^{(n)}(x,0) = c_{n,k-1}(x)G_0(x,0)+h_{n,k-1}(x) \tag{C.15}$$

for $0 \le n \le K-k$.

We have proved (C.15) for $k=1,2$ in (C.5) and (C.11). Substituting (C.15) in (C.14) yields:

$$G_k^{(n)}(x,0) = c_{n,k}(x)G_0(x,0)+h_{n,k}(x) \quad 0\le n\le K-k-1 \tag{C.16}$$

where

$$c_{n,k}(x) = \frac{1}{n+1}\bar{q}_{k-1}\sum_{i=1}^{n+1}\binom{n+1}{i}F^{(n+1-i)}(x,0)c_{i,k-1}(x) + \tag{C.17}$$

$$+ q_{k-1}\sum_{i=1}^{n}\binom{n}{i}F^{(n-i)}(x,0)c_{i,k-1}(x)+\bar{q}_{k-1}c_{0,k-1}(x)F^{(n)}(x,0)$$

$$h_{n,k}(x) = \frac{1}{n+1}\bar{q}_{k+1}\sum_{i=1}^{n+1}\binom{n+1}{i}F^{(n+1-i)}(x,0)h_{i,k-1}(x) + \tag{C.18}$$

$$+ q_{k+1}\sum_{i=1}^{n}\binom{n}{i}F^{(n-i)}(x,0)h_{i,k-1}(x)+\bar{q}_{k-1}h_{0,k-1}(x)F^{(n)}(x,0)$$

Then by induction (C.15) is proved for $0\le k\le K-1$.

So we have shown that for $0\le k\le K-1$ we have:

$$G_k(x,0) = C_k(x)G_0(x,0)+H_k(x) \tag{C.19}$$

where

$$C_0(x) = 1 \ ; \ H_0(x) = 0 \tag{C.20}$$

$$C_k(x) = c_{0,k}(x) \ ; \ H_k(x) = h_{0,k}(x) \quad 1\le k\le K-1$$

Finally, from (7a) using (C.19) we obtain:

$$G_K(x,0) = C_K(x)G_0(x,0)+H_K(x) \tag{C.21}$$

where

$$C_K = [x-F(x,0)\sum_{k=0}^{K-1}q_k C_k(x)]/ q_k F(x,0) \tag{C.22}$$

$$H_k(x) = -F(x,0)[xP_0(0,1)+G_0(0,0)(x-q_0)+\sum_{k=0}^{K-1}q_k H_k(x)]/ q_k F(x,0) \tag{C.23}$$

Thus proving our claim in (17).

Appendix D

In this Appendix we consider the case of independent Bernoulli arrival processes into the nodes, i.e. $F(x,y)$ is given by (22).

From (C.6), (C.7) we obtain in this case:

$$\alpha_{n,0} \overset{\Delta}{=} c_{n,0}(x) = \begin{cases} r_2/\bar{r}_2 & n = 1 \\ 0 & 2 \le n \le K-1 \end{cases} \tag{D.1}$$

$$h_{n,0}(x) = 0 \qquad 1 \le n \le K-1 \tag{D.2}$$

Therefore from (C.12), (C.13) using the fact that $P_0(0,1) = \frac{r_2}{\bar{r}_2} G_0(0,0)$ and $P_0(0,i) = 0$ for $i \ge 2$, we obtain:

$$h_{n,1}(x) = -c_{n,1}(x) G_0(0,0) \tag{D.3}$$

$$c_{n,1}(x) = F_1(x)\alpha_{n,1} \tag{D.4}$$

where

$$\alpha_{n,1} = \bar{q}_0 F_2^{(n)}(0) + \frac{\bar{q}_0}{n+1}\sum_{i=1}^{n+1}\binom{n+1}{i} F_2^{(n+1-i)}(0)\alpha_{i,0} + q_0 \sum_{i=1}^{n}\binom{n}{i} F_2^{(n-i)}(0)\alpha_{i,0} \tag{D.5}$$

and for $2 \le k \le K-1$ we obtain by induction from (C.17), (C.18):

$$h_{n,k}(x) = -c_{n,k}(x) G_0(0,0) \tag{D.6}$$

$$c_{n,k}(x) = F_1^k(x)\alpha_{n,k} \tag{D.7}$$

where

$$\alpha_{n,k} = \frac{1}{n+1}\bar{q}_{k-1}\sum_{i=1}^{n+1}\binom{n+1}{i} F_2^{(n+1-i)}(0)\alpha_{i,k-1} + q_{k-1}\sum_{i=1}^{n}\binom{n}{i} F_2^{(n-1)}(0)\alpha_{i,k-1} + \bar{q}_{k-1} F_2^{(n)}(0)\alpha_{0,k-1} \tag{D.8}$$

If we define $\alpha_k = \alpha_{0,k}$ for $1 \le k \le K-1$, then (24) is proved. (24b) is obtained directly from (C.21).

Determination of $G_0(0,0)$

From (7a) - (7c) we obtain:

$$\sum_{k=0}^{K-1} G_k(1,1) = G_0(0,1) - q_0 G_0(0,0) + \sum_{k=0}^{K} (k+1) q_k G_k(1,0) \qquad \text{(D.9)}$$

Let $y=1$ and $x \to 1$ in (7d). Then

$$G_k(1,1) = -G_0(0,1) - q_0 \frac{\bar{r}_1}{r_1} G_0(0,0) + \qquad \text{(D.10)}$$

$$+ \sum_{k=1}^{K-1} \frac{q_k}{r_1} [1-(k+1)r_1] + \frac{q_k}{r_1}(1-Kr_1)$$

Summing (D.9) and (D.10) using the normalization condition $\sum_{k=0}^{K} G_k(1,1)=1$ we obtain:

$$r_1 = \sum_{k=0}^{K} q_k G_k(1,0) - q_0 G_0(0,0) \qquad \text{(D.11)}$$

From (7a) and (23) we obtain:

$$G_0(1,0) = G_0(0,0) + \bar{r}_2 [G_0(1,0) - G_0(0,0)] q_0 + \bar{r}_2 \sum_{k=1}^{K} q_k G_k(1,0) \qquad \text{(D.12)}$$

and from (D.11) and (D.12) we have:

$$G_0(1,0) - G_0(0,0) = r_1 \bar{r}_2 \qquad \text{(D.13)}$$

Let $x=1$ and $y \to 1$ in (7d). Then

$$G_k(1,1) = \frac{1}{\bar{q}_k - r_2} \Bigg\{ \sum_{k=0}^{K-1} \Big[\bar{q}_k - q_k \sum_{j=0}^{k} \bar{q}_j + (k+1) q_k r_2 \Big] G_k(1,0) + \qquad \text{(D.14)}$$

$$+ \Big[\bar{q}_K + q_K \big(K r_2 - \sum_{j=0}^{K-1} \bar{q}_j \big) \Big] G_k(1,0) - \bar{r}_2 G_0(0,1) +$$

$$+ q_0 G_0(0,0)(\bar{r}_2 + \bar{q}_0) \Bigg\}$$

Summing (D.9) and (D.14) using the normalization condition yields:

$$\bar{q}_K - r_2 = \sum_{k=0}^{K-1} \beta_k G_k(1,0) + G_0(0,0)[q_0(\bar{q}_0 + q_K) - q_K / \bar{r}_2] \qquad \text{(D.15)}$$

where

$$\beta_K = Kq_K\bar{q}_K + \bar{q}_K - q_K\sum_{j=0}^{K-1}\bar{q}_j \tag{D.16}$$

$$\beta_k = (k+1)q_k\bar{q}_K + \bar{q}_k - q_k\sum_{j=0}^{k}\bar{q}_j \qquad 0\le k\le K-1 \tag{D.17}$$

From (D.15) and (24a) we have

$$\bar{q}_K - r_2 = G_0(1,0)\cdot\beta_0 + r_1\bar{r}_2\sum_{k=1}^{K-1}\alpha_k\beta_k + \beta_K(r_1 - q_0 r_1\bar{r}_2 - r_1\bar{r}_2\sum_{k=1}^{K-1}\alpha_k q_k)/q_K + \tag{D.18}$$

$$+ G_0(0,0)[q_0(\bar{q}_0 + q_K) - q_K/\bar{r}_2]$$

Let $\alpha_0 = 1$. Then from (D.13) and (D.18) we obtain:

$$\bar{q}_K - r_2 = G_0(0,0)[1 - q_K/\bar{r}_2] + r_1\beta_K/q_K + \tag{D.19}$$

$$+ r_1\bar{r}_2\sum_{k=0}^{K-1}\alpha_k(\beta_k - \beta_K q_k/q_K)$$

and finally,

$$G_0(0,0) = \bar{r}_2\left[1 - r_1\frac{\beta_K + \bar{r}_2\sum_{k=0}^{K-1}\alpha_k(\beta_k q_K - \beta_K q_k)}{q_K(\bar{q}_K - r_2)}\right] \tag{D.20}$$

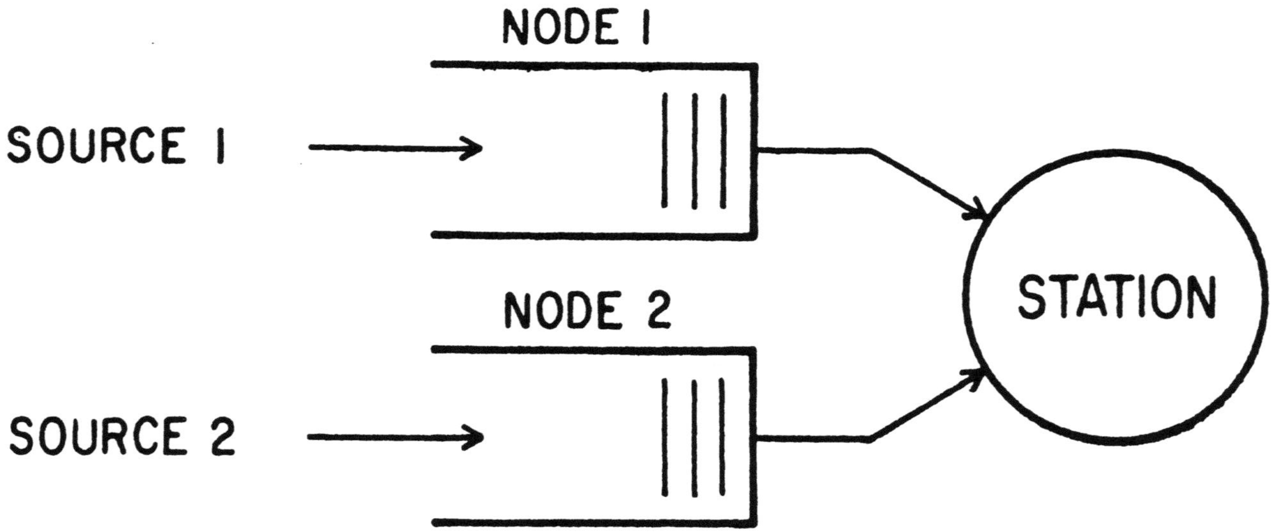

<u>Fig. 1</u> - A two-node packet-radio network.

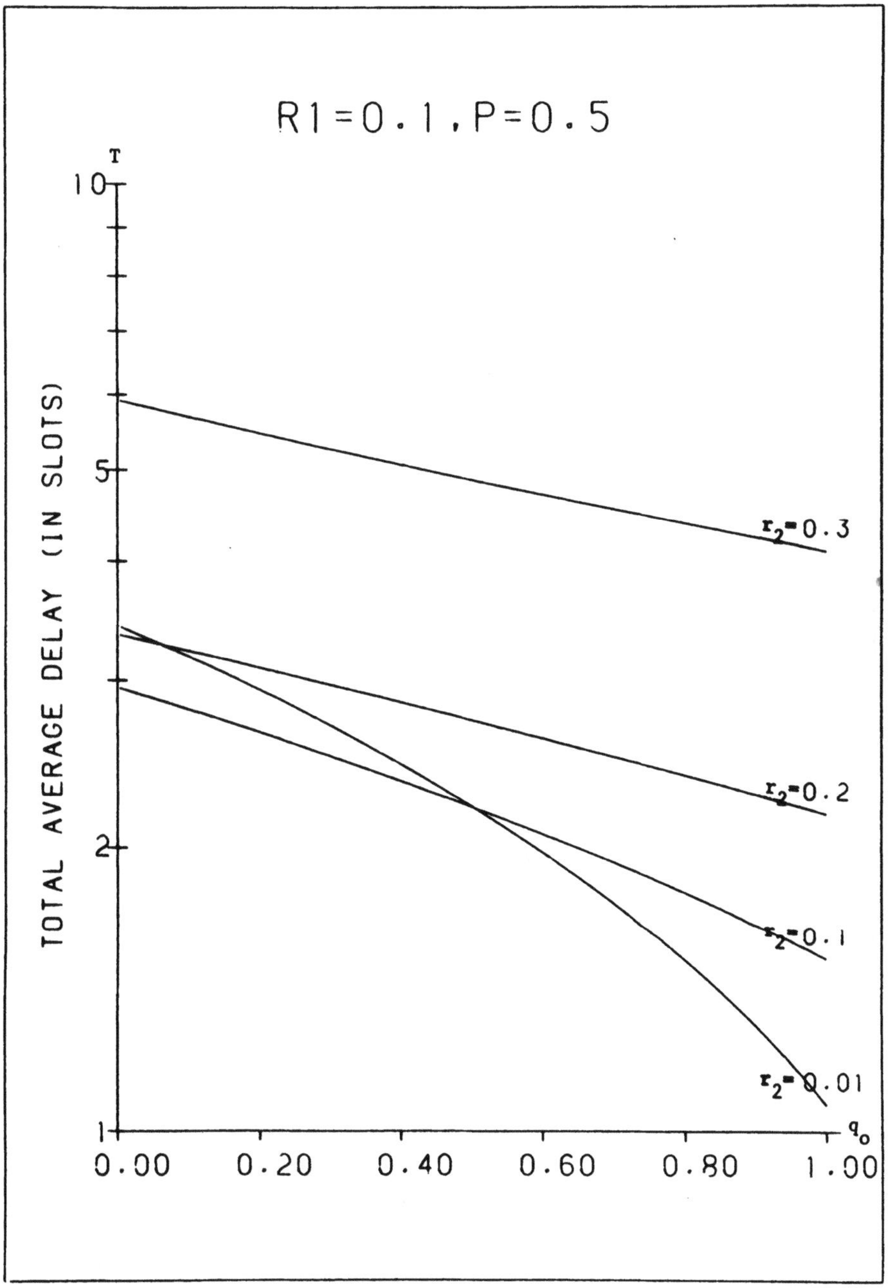

Fig. 2 - Scheme K = 1: T vs q_o

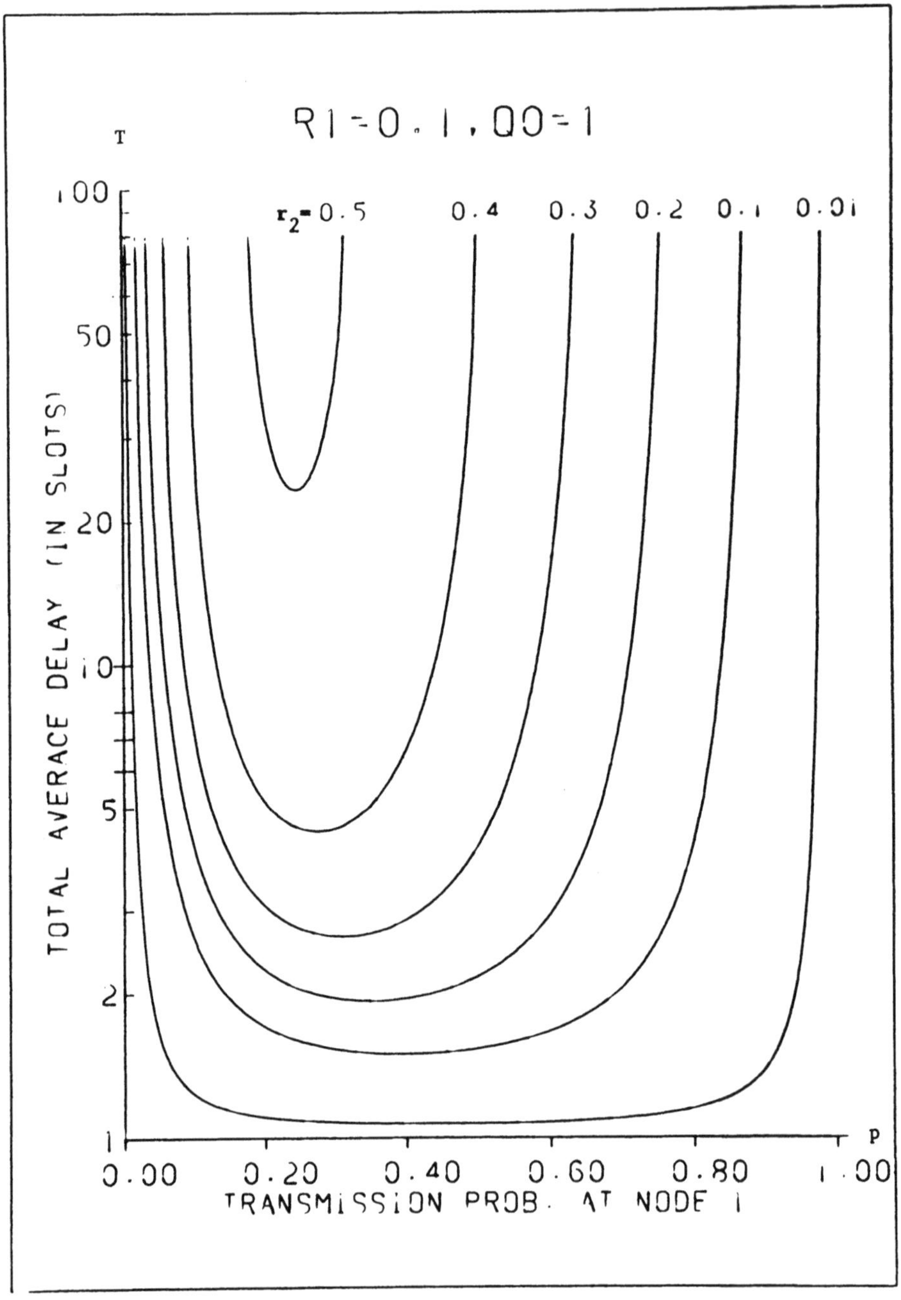

Fig. 3 - Scheme K = 1: T vs p.

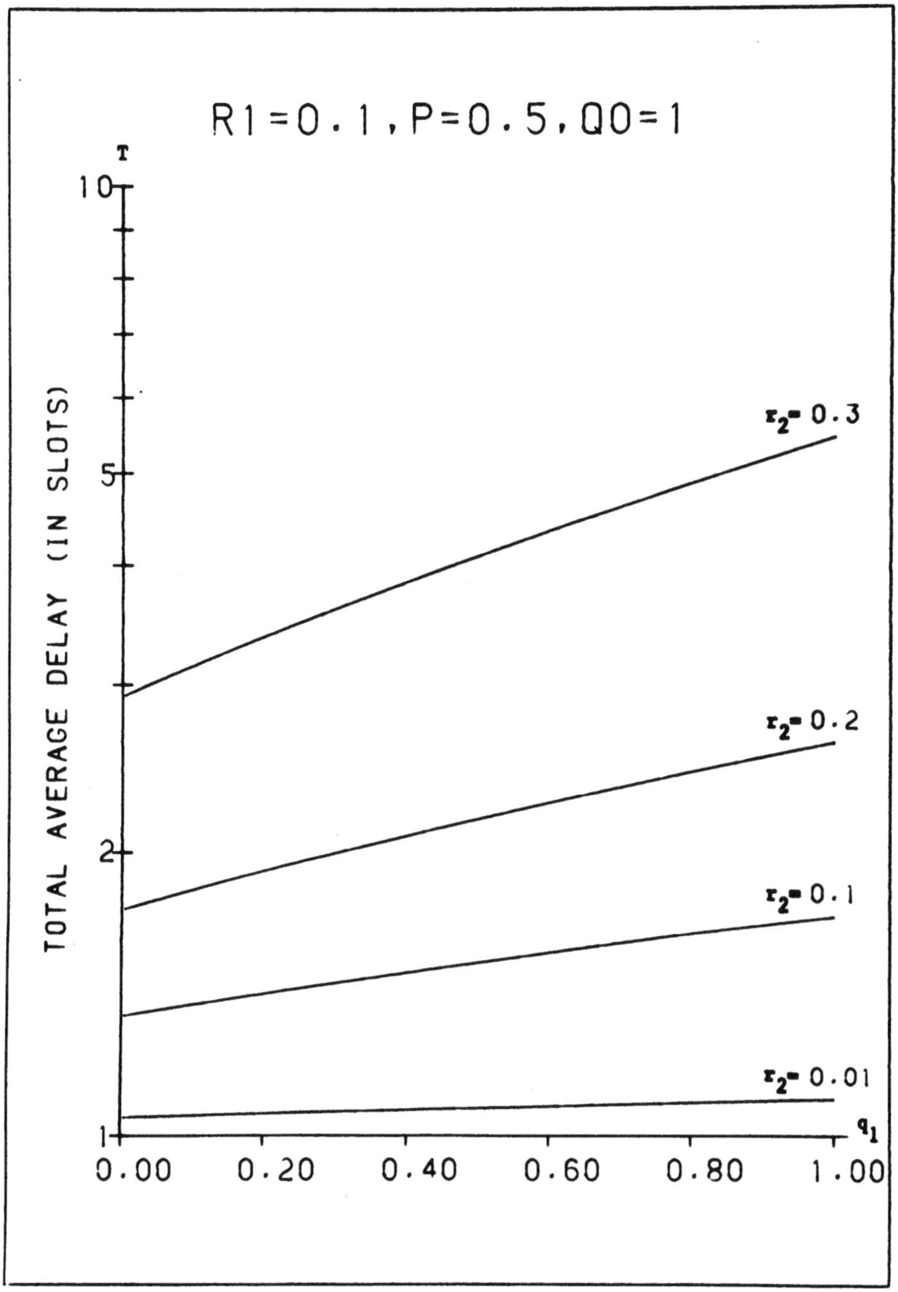

Fig. 4 - Scheme K = 2: T vs q_1.

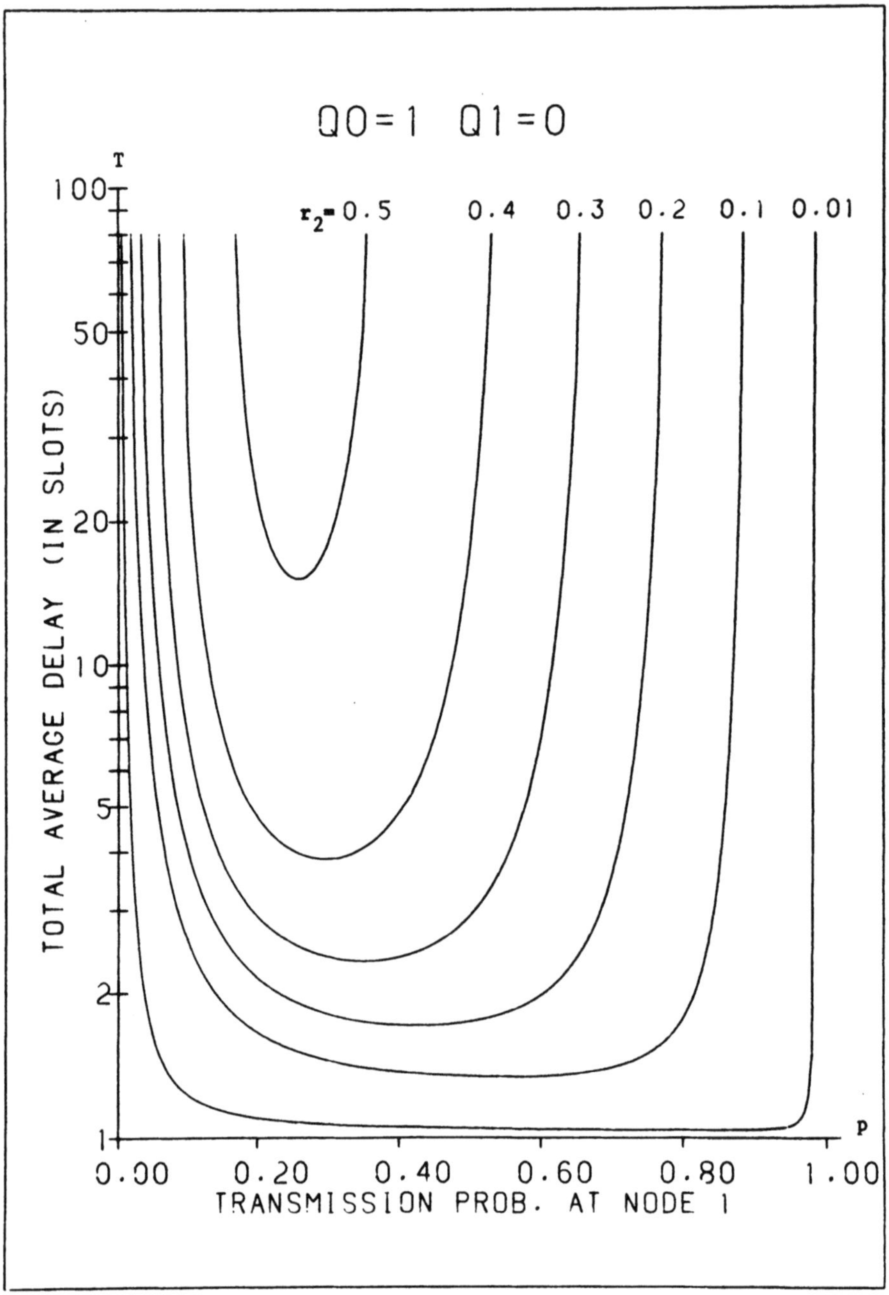

Fig. 5 - Scheme K = 2: T vs p.

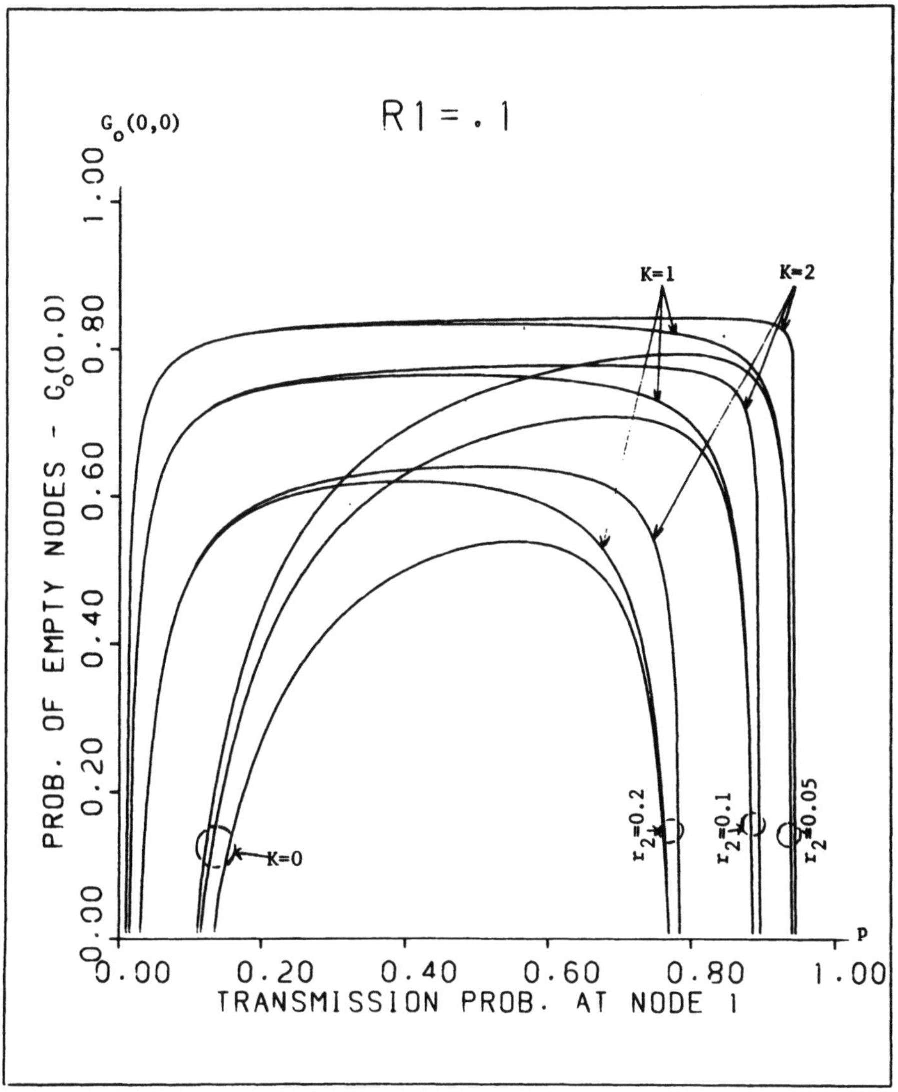

Fig. 6 - Schemes K = 0,1,2:
$G_o(0,0)$ vs $p(q_o = 1,\ q_1 = 0)$.

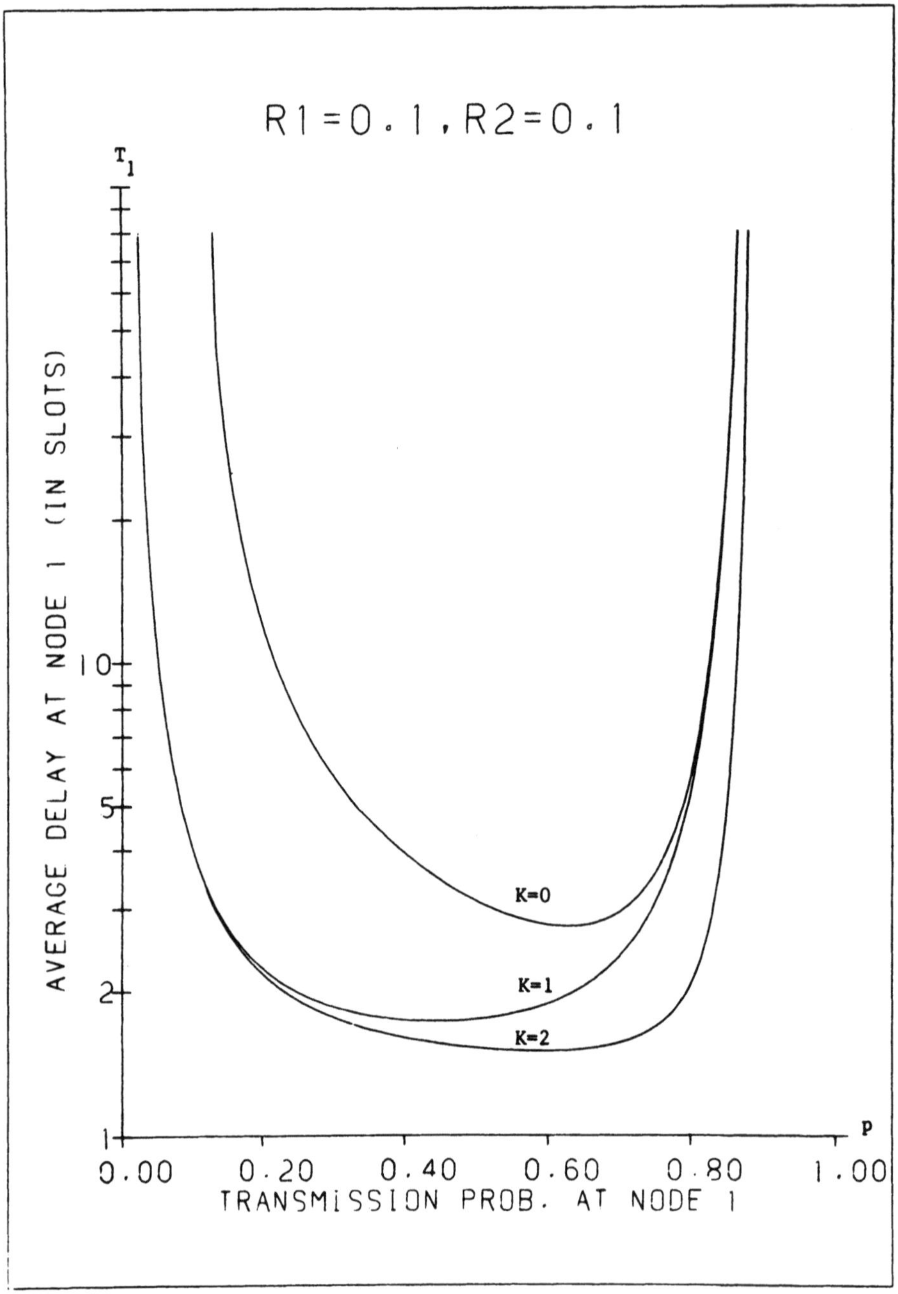

Fig. 7 - Schemes K = 0,1,2:
T_1 vs $p(q_o = 1, q_1 = 0)$.

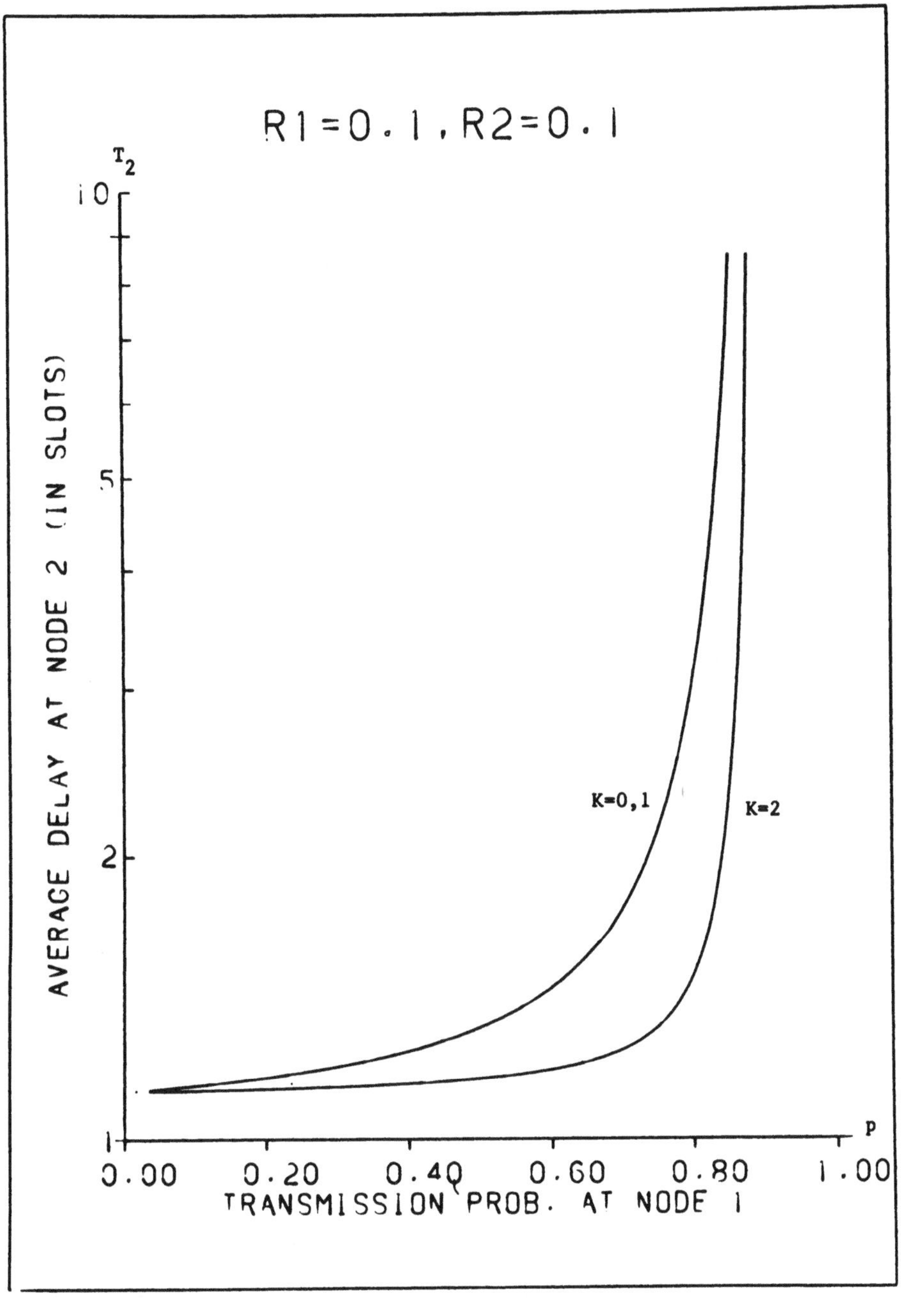

Fig. 8 - Schemes K = 0,1,2:
T_2 vs $p(q_o = 1, q_1 = 0)$.

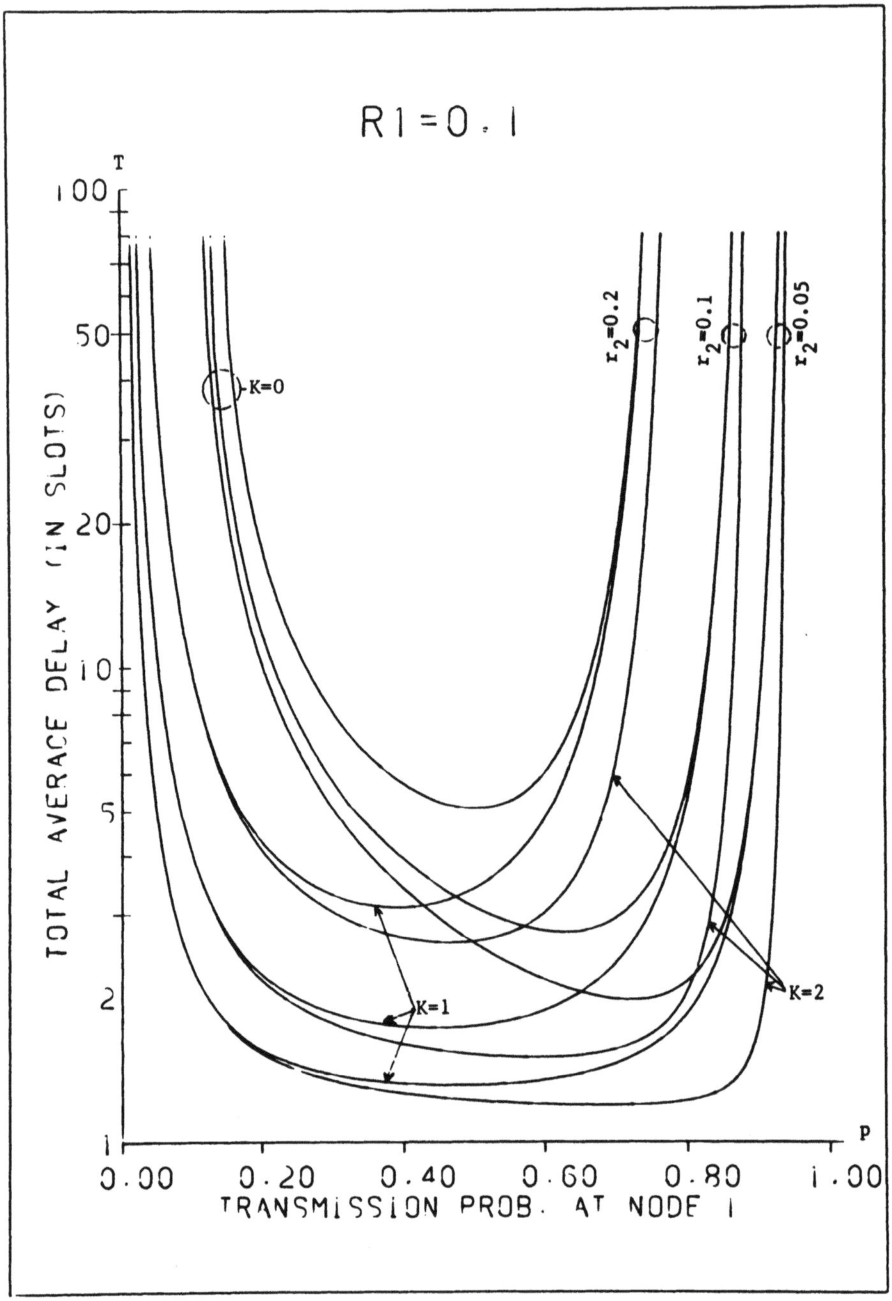

Fig. 9 - Schemes K = 0,1,2:
T vs $p(q_o = 1, q_1 = 0)$.

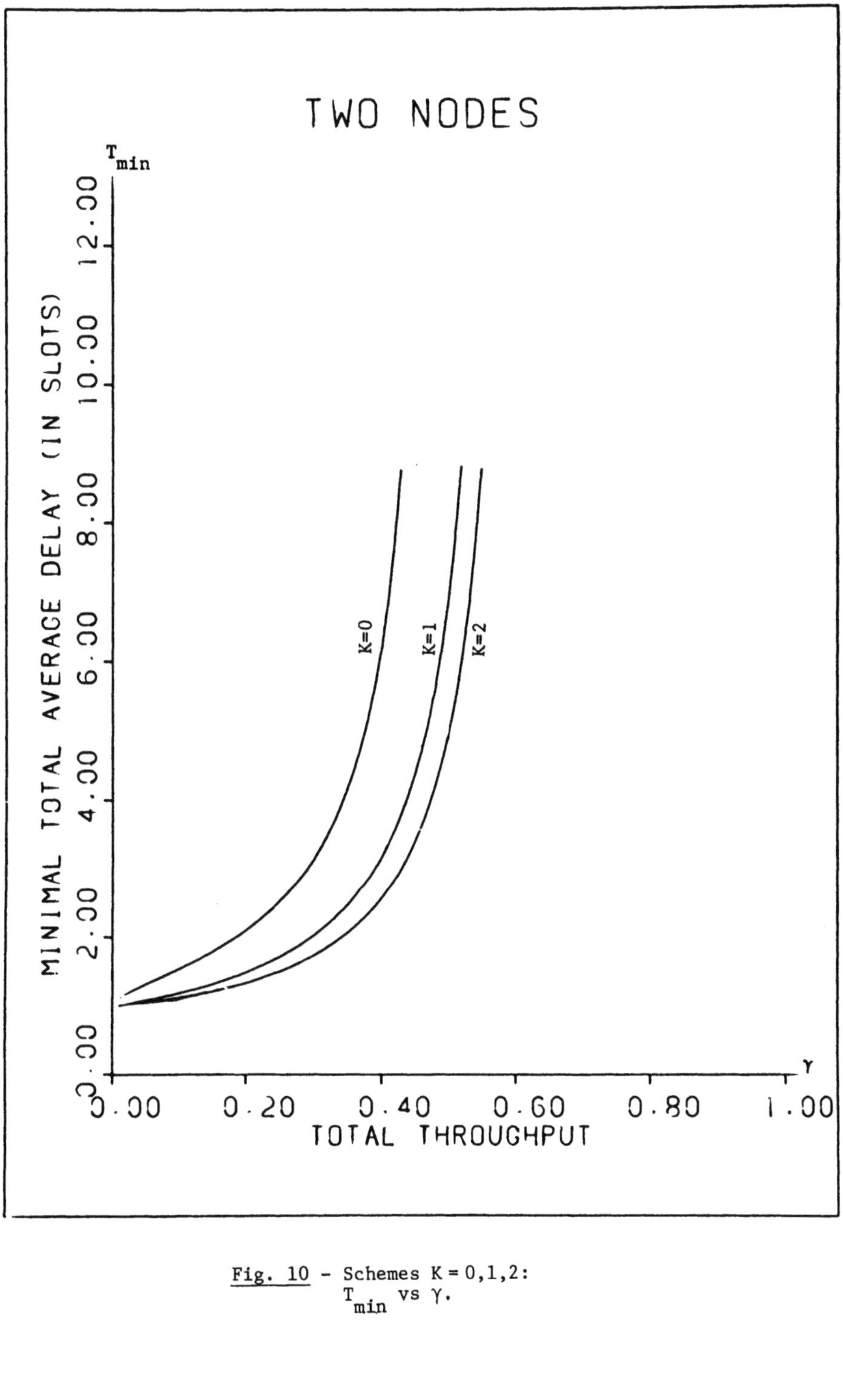

Fig. 10 - Schemes K = 0,1,2: T_{min} vs γ.

IV

QUEUES AND NETWORKS 2

FILES D'ATTENTE ET RESEAUX 2

DECOMPOSABLE STOCHASTIC NETWORKS: SOME OBSERVATIONS

R. Schassberger
Technical University of Berlin

Abstract: Global balance equations of the type $pQ = 0$, where Q is the transition intensity matrix of an irreducible ergodic Markov chain and p is the corresponding invariant probability vector, sometimes exhibit a property of so-called partial balance. In all such examples known to the author, such partial balance means that Q can be written in a nontrivial way as $Q = \sum Q_i$, where the Q_i are themselves transition intensity matrices and satisfy $pQ_i = 0$ for the mentioned p. The paper elaborates on this phenomenon.

1. Introduction

Complex stochastic networks such as those introduced in [3] and further developed in [1] and [4] have been analyzed with a certain degree of success on account of a so-called partial balance property they exhibit. Such analysis starts from a time-continuous homogeneous Markov chain describing the network state process. It is normally the first objective to obtain stationary measures for this chain, i.e. measures satisfying the so-called global balance condition

$$pQ = 0$$

or

$$p(z) \mid q(z,z) \mid = \sum_{z'\in Z} p(z')q(z',z), z\in Z, \tag{1.1}$$

where Z denotes the state space, $Q = (q(z,z'); z,z'\in Z)$ the transition or Q-matrix of the chain, and $p = (p(z); z\in Z)$ represents the stationary measure. Partial balance is said to prevail, if the stationary measure represented by p in addition to (1.1) satisfies certain collections of equations of the type

$$p(z) \mid q_i(z,z) \mid = \sum_{z'\in Z} p(z')q_i(z',z), \; z\in Z, \tag{1.2}$$

or

$$pQ_i = 0, \; i = 1,2,\ldots$$

where $Q_i = (q_i(z,z'); z,z' \in Z)$ is such that $|q_i(z,z)| \le |q(z,z)|, z \in Z$, and $0 \le q_i(z,z') \le q(z,z')$ for $z,z' \in Z$, $z' \neq z$.

The objective of the present paper is to elaborate on an observation made by the author in [7] (and presumably made by others), i.e. that, in the case of the networks mentioned above as well as for all networks that have been found to enjoy a property of partial balance, this property can be written down in such a way that the matrices Q_i in (1.2) are Q-matrices themselves, i.e. have row sums equal to zero, and that, moreover,

$$Q = \sum Q_i. \tag{1.3}$$

Thus, in all instances of partial balance known to the author, one has a decomposition of Q into "sub-Q-matrices" Q_i all satisfying $pQ_i = 0$.

This concept of decomposition is formally defined in section 2 after having been introduced by way of the Jackson network example. Two further examples are given in section 3. Section 4 studies modifications of decomposable stochastic networks obtained my means of modifying the component matrices Q_i.

Section 5 demonstrates a successful search for a decomposition in the case of a Jackson network variant involving genuine blocking. The final section, 6, contains a remark concerning a "product form" appearance of stationary measures p associated with decompositions.

2. A definition of decomposability

Consider an open Jackson network with J nodes, a Poisson input stream of rate λ, service rates $\mu_j(n_j)$ for node j in the presence there of n_j customers, and an irreducible routing matrix $R = (r_{ij})_o^J$, where r_{oj} and r_{jo} are the probabilities of an arrival joining node j and a customer departing after service at node j, respectively, and $r_{oo} = 0$. The states $\bar{n} \in Z \doteq \{\bar{n} = (n_1,\ldots,n_J) : n_1,n_2,\ldots,n_J \ge 0\}$ fluctuate according to the law of a time continuous, homogeneous Markov process whose Q-matrix, i.e. whose matrix of transition rates, be denoted by Q. The linear system of equations

$$uQ = 0 \tag{2.1}$$

is solved by $p = \{p(\bar{n}) : \bar{n} \in Z\}$,

$$p(\bar{n}) = C \prod_{j=1}^{J} (\lambda e_j)^{n_j} \left(\prod_{i=1}^{n_j} \mu_j(i) \right)^{-1} , \tag{2.2}$$

$\bar{n} \in Z$, where $e = (e_o, \ldots, e_J)$ satisfies $e = eR$, $e_o = 1$, and C is an arbitrary constant. Let $\bar{J} = \{1, \ldots, J\}$,

$T_{ij}\ \bar{n} = (n_1, \ldots, n_i - 1, \ldots, n_j + 1, \ldots, n_J)$ for $i, j \in \bar{J}$, $j \neq i$,

$T_{io}\ \bar{n} = (n_1, \ldots, n_i - 1, \ldots, n_J)$ for $i \in \bar{J}$, and $T_{ii}\ \bar{n} = \bar{n}$ for $i \in \bar{J}$.

Fix some $\bar{n} \neq (0, \ldots, 0)$ and some i with $n_i > 0$. The solution (2.2) of (2.1) also satisfies the equation

$$\begin{aligned} p(\bar{n})\mu_i(n_i) &= p(T_{io}\ \bar{n})\lambda r_{oi} + p(\bar{n})\mu_i(n_i) r_{ii} + \\ &+ \sum_{j \in \bar{J} - \{i\}} p(T_{ij}\ \bar{n})\mu_j(n_j + 1)\ r_{ji} , \end{aligned} \tag{2.3}$$

which is well known as expressing a form of partial balance within the global balance represented by the equation $pQ = 0$. Clearly, (2.3) expresses a property of the matrix Q, hence ought to be rewritten in matrix terms. Towards achieving this associate with (2.3) the equations

$$\begin{aligned} p(T_{ij}\ \bar{n})\mu_j(n_j + 1) &= p(T_{io}\ \bar{n})\lambda r_{oj} \\ &+ p(\bar{n})\mu_i(n_i) r_{ij} \\ &+ \sum_{k \in \bar{J} - \{i\}} p(T_{ik}\ \bar{n})\mu_k(n_k + 1) r_{kj} , \\ &\qquad j \in \bar{J} - \{i\}, \end{aligned} \tag{2.3'}$$

and

$$\begin{aligned} p(T_{io}\ \bar{n})\lambda &= p(\bar{n})\mu_i(n_i) r_{io} \\ &+ \sum_{j \in \bar{J} - \{i\}} p(T_{ij}\ \bar{n})\mu_j(n_j + 1) r_{jo} , \end{aligned} \tag{2.3"}$$

which are also satisfied. Letting

$$Z_{\bar{n},i} \doteq \{\bar{n}\} \cup \{T_{io}\ \bar{n}\} \cup \{T_{ij}\ \bar{n} : j \in \bar{J} - \{i\}\}$$

and $p_{\bar{n},i}$ the projection of p onto $Z_{\bar{n},i}$, the system (2.3) can be rewritten as

$$p_{\bar{n},i}\ \bar{Q}_{\bar{n},i} = 0 , \tag{2.4}$$

where $\bar{Q}_{\bar{n},i}$ is an irreducible Q-matrix on $Z_{\bar{n},i} \times Z_{\bar{n},i}$.

Equivalently, (2.4) can be written as

$$pQ_{\bar{n},i} = 0 \ , \qquad (2.5)$$

where $Q_{\bar{n},i}$ is the extension of $Q_{\bar{n},i}$ onto Z x Z by way of filling in zeros. Note that the elements of $Q_{\bar{n},i}$ do not exceed the corresponding elements of Q in absolute value, which shall be expressed henceforth by calling $Q_{\bar{n},i}$ a sub-Q-matrix of Q, and which implies that $Q-Q_{\bar{n},i}$ is also a sub-Q-matrix of Q. Clearly, $Q-Q_{\bar{n},i}$ can be broken up into a sum of infinitely many further sub-Q-matrices of Q, each corresponding to a certain choice of $\bar{n}$ and i as above, and each satisfying $pQ_{\bar{n},i} = 0$ for p given by (2.2). Call $Z_{\bar{n},i}$ the support of $Q_{\bar{n},i}$, $\bar{Q}_{\bar{n},i}$ the restriction of $Q_{\bar{n},i}$ to its support. Then, in matrix terms, the property of partial balance enjoyed by Q can be stated by saying that

$$Q = \sum Q_{\bar{n},i}$$

for certain sub-Q-matrices $Q_{\bar{n},i}$ of Q whose restrictions to their supports are irreducible, and which all allow the same stationary measure p (given by (2.2)). The sum extends over all the different $Z_{\bar{n},i}$ (note that $Z_{\bar{n},i} = Z_{T_{ij}\bar{n},j}$ for $j \neq i$). Note that the measure p is not necessarily finite here. If one visualizes the nodes as single servers operating under the LIFS preemptive resume discipline, then (2.5) can be said to represent the fact that freezing all but the motion of one customer yields a partially balanced motion.

To the best of our knowledge, wherever in the literature a stochastic network has been analyzed by means of a global balance equation analogous to (2.1) and has been found to enjoy a form of partial balance, the latter can be described in matrix terms analogous to the above. The general setup would be as follows: The network state space, Z, is countable, and the states fluctuate in time according to the law of a homogeneous irreducible Markov chain with Q-matrix Q. There is a countable collection $\{Q_i \ ; i \in I\}$ of sub-Q-matrices of Q, I denoting the corresponding set of indices. Each Q_i is irreducible on its support $Z_i \subset Z$, and

$$Q = \sum Q_i .$$

Finally, there is a (not necessarily finite) measure on Z represented by $p = (p(z);\ z \in Z)$ such that $p(z) > 0$ for all $z \in Z$ and that

$$pQ_i = 0$$

for all $i \in I$. The stochastic network, or - simply - the matrix Q, shall then be called <u>decomposable</u> into the collection $\{Q_i ; i \in I\}$ with associated measure p.

3. Two mini-network examples

The purpose of this section is to present two examples of decomposable networks which are not of the Jackson type and do not feature such properties as quasireversibility or insensitivity (see Kelly, [5], for these notions).

3.1 Example

A network consisting of nodes 1 and 2 receives a Poisson arrival stream of rate λ to node 1. This node can only hold one customer and, while doing so, provides service at the exponential rate μ_1. Arrivals finding node 1 busy are lost with probability $(1-\alpha)$ and go to node 2 with probability α. Node 2 is an infinite server working at the exponential rate μ_2 per customer. Upon having received a service at node 2 a customer enters node 1 if the latter is empty, otherwise leaves the network with probability β while running through another service at node 2 with probability $1-\beta$. This model can be used for instance to investigate the influence of repeated calls in telephone traffic theory (see Cohen, [2]).

The state space Z is given here by $Z = \{(n_1,n_2): n_1 = 0,1;\ n_2 \geq 0\}$, where n_i represents the number of customers present at node i, $i = 1,2$. The linear system of equations $uQ = 0$ is given explicitly by

$$(\lambda+n_2\mu_2)\, u(0,n_2) = \mu_1 u(1,n_2)$$

and

$$\begin{aligned}(\lambda\alpha+n_2\mu_2\beta+\mu_1)\, u(1,n_2) &= \lambda\alpha u(1,n_2-1)\\ &+ (n_2+1)\mu_2 u(0,n_2+1)\\ &+ \lambda u(0,n_2)\\ &+ (n_2+1)\mu_2 u(1,n_2+1)\end{aligned}$$

for $n_2 \geq 0$, where $u(1,-1) = 0$.

The system is solved by

$$p(1,n_2) = C \prod_{i=1}^{n_2}\left(\frac{\lambda\alpha}{\mu_2}\left[i\left(\beta + \frac{\mu_1}{\lambda+i\mu_2}\right)\right]^{-1}\right)$$

and
$$p(0,n_2) = \frac{\mu_1}{\lambda+n_2\mu_2} p(1,n_2),$$

$n_2 \geq 0$. This solution satisfies, for each fixed $n_2 \geq 0$, the system of equations

$$(\lambda+n_2\mu_2)u(0,n_2) = \mu_1 u(1,n_2),$$
$$\lambda\alpha u(1,n_2-1) = n_2\mu_2 u(0,n_2)$$
$$+ n_2\mu_2\beta u(1,n_2),$$
$$(\mu_1+n_2\mu_2\beta)u(1,n_2) = \lambda\alpha u(1,n_2-1) + \lambda u(0,n_2).$$

The system defines a sub-Q-matrix Q_{n_2} of Q. Thus $Q = \sum_{n_2=0}^{\infty} Q_{n_2}$, and $pQ_{n_2} = 0$ for all $n_2 \geq 0$.

3.2 Example

In [5], chapter 6.3, Kelly discusses some so-called flow models which are all decomposable in the sense of section 2. A simple three-node model of this kind is depicted in Fig. 1. Each node can hold one customer, only, the customers arriving at rate ν at node 1. Nodes 1 and 2 work at rate λ when holding a customer, node 3 at rate μ. Arrivals finding node 1 busy are lost, services in node 1 and 2 ending at busy times of nodes 2 and 3, resp., are repeated. The state space is $\{0,1\}^3$. The Q-matrix is given by

	000	001	010	100	011	101	110	111
000	$-\nu$	0	0	ν	0	0	0	0
001	μ	$-(\mu+\nu)$	0	0	0	ν	0	0
010	0	λ	$-(\lambda+\nu)$	0	0	0	ν	0
100	0	0	λ	$-\lambda$	0	0	0	0
011	0	0	μ	0	$-(\mu+\nu)$	0	0	ν
101	0	0	0	μ	λ	$-(\mu+\lambda)$	0	0
110	0	0	0	0	0	λ	$-\lambda$	0
111	0	0	0	0	0	0	μ	$-\mu$

$Q =$

For $\lambda = \mu+\nu$ the system $uQ = 0$ is solved by

$$p(n_1,n_2,n_3) = C\left(\frac{\nu}{\mu}\right)^{n_1+n_2+n_3}, \quad (n_1,n_2,n_3) \in \{0,1\}^3,$$

where C is an arbitrary constant. A decomposition in the form $Q = Q_1+Q_2+Q_3+Q_4$, with $pQ_i = 0$ is given by

$$\bar{Q}_1 = \begin{array}{c} \\ 000 \\ 001 \\ 010 \\ 100 \end{array} \begin{array}{c} \begin{array}{cccc} 000 & 001 & 010 & 100 \end{array} \\ \left[\begin{array}{cccc} -\nu & 0 & 0 & \nu \\ \mu & -\mu & 0 & 0 \\ 0 & \mu & -\mu & 0 \\ 0 & 0 & \mu & -\mu \end{array}\right] \end{array},$$

$$\bar{Q}_2 = \begin{array}{c} \\ 001 \\ 010 \\ 011 \\ 101 \end{array} \begin{array}{c} \begin{array}{cccc} 001 & 010 & 011 & 101 \end{array} \\ \left[\begin{array}{cccc} -\nu & 0 & 0 & \nu \\ \nu & -\nu & 0 & 0 \\ 0 & \mu & -\mu & 0 \\ 0 & 0 & \mu & -\mu \end{array}\right] \end{array},$$

$$\bar{Q}_3 = \begin{array}{c} \\ 010 \\ 100 \\ 110 \\ 101 \end{array} \begin{array}{c} \begin{array}{cccc} 010 & 100 & 110 & 101 \end{array} \\ \left[\begin{array}{cccc} -\nu & 0 & \nu & 0 \\ \nu & -\nu & 0 & 0 \\ 0 & 0 & -\mu & \mu \\ 0 & \mu & 0 & -\mu \end{array}\right] \end{array},$$

$$\bar{Q}_4 = \begin{array}{c} \\ 011 \\ 101 \\ 110 \\ 111 \end{array} \begin{array}{c} \begin{array}{cccc} 011 & 101 & 110 & 111 \end{array} \\ \left[\begin{array}{cccc} -\nu & 0 & 0 & \nu \\ \nu & -\nu & 0 & 0 \\ 0 & \nu & -\nu & 0 \\ 0 & 0 & \mu & -\mu \end{array}\right] \end{array},$$

where $\bar{Q}_i$ is the restriction of Q_i to its support, $i = 1,\ldots,4$.

4. Modifications of decomposable networks

Throughout this section let Q, Z, Q_i, Z_i, and p be as in the definition of decomposability of section 2. Thus $Q = \sum Q_i$, and $pQ_i = 0$ for all i. The purpose of this section is to present some examples of modifications of Q by way of substituting for some or all of the Q_i new matrices Q_i, again satisfying $pQ_i' = 0$.

4.1 Skipping

Put $q(z) \doteq |q(z,z)|, z \in Z$, and assume that $\sum_{z \in Z} p(z)q(z) < \infty$. Let

$$s(z,z') = \begin{cases} \frac{1}{q(z)} \; q(z,z') & , \; z' \neq z, z, z' \in Z \\ 0 & , \; z' = z, z \in Z, \end{cases}$$

and call the matrix $S = (s(z,z'))$ the jump matrix of Q. Note that $u(z) = \sum_{z' \in Z} u(z')s(z',z)$, $z \in Z$, where $u(z) = p(z)q(z)$, $z \in Z$.

Let Z_o be a nonempty subset of Z and denote by $s_o(z,z')$, $z \in Z$, $z' \in Z_o$, the probability that, given start in z, the next visit of the discrete-time Markov chain with transition law S to Z_o is at state z'. Thus, for $z \in Z, z' \in Z_o$,

$$s_o(z,z') = s(z,z') + \sum_{z'' \in Z - Z_o} s(z,z'')s_o(z'',z').$$

Multiplying this equation with $p(z)q(z)$ $(=u(z))$ and then summing over $z, z \in Z$, yields

$$u(z) = \sum_{z' \in Z_o} u(z')s_o(z',z), \quad z \in Z_o .$$

This implies that

$$pQ_o = 0,$$

where the Q-matrix Q_o is defined on Z x Z by

$$q_o(z,z') = \begin{cases} q(z)s_o(z,z') & , \; z' \neq z, z, z' \in Z_o \\ q(z)(s_o(z,z)-1) & , \; z' = z, z \in Z_o \\ 0 & , \; \text{otherwise.} \end{cases}$$

This matrix shall be said to result from Q by skipping $Z - Z_o$. Thus skipping preserves the stationary measure p. Instead of modiying Q itself in this way one can apply the skipping idea to some or all of its sub-Q-matrices Q_i, skipping some nonempty subset $Z_i - Z_{io}$ of Z_i and again preserving p.

For an example consider the Jackson network of section 2 and disallow the states of some nonempty genuine subset $Z - Z_o$ of Z. Replace the matrices $Q_{\bar{n},i}$ by the matrices $Q_{\bar{n},i,o}$ obtained from the former by skipping $Z_{\bar{n},i,o} \doteq (Z - Z_o) \cap Z_{\bar{n},i}$, letting $Q_{\bar{n},i,o} = Q_{\bar{n},i}$ if $Z_{\bar{n},i,o} = \emptyset$. Since the jump matrix of $Q_{\bar{n},i}$ is, after a suitable state permutation, identical with the routing matrix R, the skipping rule here is that, whenever out of some allowable state $z \in Z_o$ a customer selects his next node in such a way that a forbidden state would result, he only performs an imaginary jump to that node, spending no time there but jumping on immediately according to R and performing further imaginary jumps until a state from Z_o is obtained again.

That such skipping preserves p was shown by Pittel ([6]) in similar fashion for a closed Jackson network and by others (see e.g. Kelly, [5]) by allowing infinitely large service rates in their networks. More complicated skipping rules preserving p as given by (2.2) can be set up, for instance, by forbidding only certain transitions into a state rather than disallowing this state altogether, choosing the $Z_{\bar{n},i,o}$ correspondingly.

4.2 Changing speeds, stalling

Replacing the Q_i by c_iQ_i, $c_i \geq 0$, clearly preserves p. In case of positive c_i this represents changes of speeds, whereas $c_i = 0$ implies that the transitions represented by Q_i are forbidden. Again consider the Jackson network example of section 2 and disallow some genuine nonempty subset $Z - Z_o$ of states. Replace all $Q_{\bar{n},i}$ by $c_{\bar{n},i}Q_{\bar{n},i}$, where $c_{\bar{n},i} = 1$ if $Z_{\bar{n},i} \cap (Z-Z_o) = \emptyset$, $c_{\bar{n},i} = 0$ otherwise. This yields a Q-matrix $Q_o = \sum c_{\bar{n},i}Q_{\bar{n},i}$ with support Z_o, and $pQ_o = 0$. Towards interpreting the corresponding motion consider some $Z_{\bar{n},i}$ such that $Z_{\bar{n},i} \cap (Z-Z_o) \neq \emptyset$ as well as $Z_{\bar{n},i} \cap Z_o \neq \emptyset$. Let $z \in Z_{\bar{n},i} \cap Z_o$, implying that z is an allowable state. Putting $c_{\bar{n},i} = 0$ now means interrupting in state z all those activities (services, interarrival-procedure) whose transition intensities are represented by $Q_{\bar{n},i}$, i.e. which might lead to a forbidden state via the closed flow represented by $Q_{\bar{n},i}$. These activities are resumed at the instant at which some allowable activity in state z ceases and produces a new state outside $Z_{\bar{n},i}$. Thus one might call this a rule of <u>stalling</u>. Both, the skipping and the stalling rules are measures one can take in order to prevent blocking.

4.3 Reversing, n-stepping

The reversed Markov law obtained from Q_i is represented by the Q-matrix $Q_{ir} = (q_{ir}(z,z'); z,z' \in Z_i)$, where

$$q_{ir}(z,z') = \frac{p(z')q_i(z',z)}{p(z)}, \quad z,z' \in Z_i .$$

Replacing Q_i by Q_{ir} preserves p. Another modification preserving p is obtained by replacing the jump matrices S_i of Q_i by iterates $S_i^{n_i}$, $n_i \geq 1$.

5. A blocking example

The purpose of this section is to show in which way the concept of decomposition under study might be useful for obtaining stationary distributions. Suppose that Z and Q are as in section 2 and that a decomposition of Q as $Q = Q_1 + Q_2$ with associated supports Z_1, Z_2 and associated stationary measure p is given. By assumption, Q_1 and Q_2 are irreducible on Z_1 and Z_2, resp., and hence $Z_1 \cap Z_2 \neq \emptyset$, or else Q would not be irreducible. In general, $|Z_1 \cap Z_2| > 1$. Suppose that $z_1, z_2 \in Z_1 \cap Z_2$. Then the quotient $p(z_1)/p(z_2)$ is already determined by the systems of equations $uQ_1 = 0$, whence the fact that the system $uQ_2 = 0$ yields the same quotient must be considered a lucky coincident. An attempt to find a decomposition as above might start from certain "natural" Z_1 and Z_2 and then proceed by trying for suitable Q_1 and Q_2 (for which there are still many candidates). This would normally be a difficult enterprise if Q were entirely fixed. If, however, one is allowed to suitably choose certain parameters on which Q depends, then success comes more easily. This shall now be demonstrated by imposing a blocking condition on a closed Jackson network.

The network to be considered is defined, to start with, by means of the rates $\mu_j(n_j)$ and the routing matrix R as in section 2, but node O, there the outside world, now becomes an ordinary server working at rates $\mu_o(n_o)$ with $\mu_o(n_o) > 0$ if the number n_o of customers present at node O is positive, and $\mu_o(0) = 0$. The network has a population of a fixed number, N, of customers. Nodes $j, j \in \bar{J} \doteq \{1,\dots,J\}$, can hold the entire population, whereas node O has a capacity of only M customers, $1 \leq M < N$. Whenever node O holds M customers and another one wishes to enter it from node $j, j \in \bar{J}$, the latter node becomes blokked and stays that way at least until that moment at which node O finishes its next service. At any instant at which node O, holding M customers, finishes a service, it sends that customer to one of the other nodes and takes in a customer from one of the blocked nodes, if any. The exchange depends on the state $(\bar{n};B) \doteq (n_1,\dots,n_J;B)$ of the system just prior to that instant, B denoting the collection of blokked nodes. The exchange has with probability $r_{oi,jo}(\bar{n},B)$ the customer from node O being transferred to node $i, i \in \bar{J}$, the one in node j, $j \in B$, being transferred to node O. In the presence of blocking the routing among nodes of $\bar{J}$ is still according to R. For the sake of simpler notation it is assumed that $r_{ii} = 0$ for all i.

According to this description the state process is a time-continuous, homogeneous Markov chain with a known Q matrix. The exchange probabilities shall be considered the parameters of Q.

Let $Z_i = \{(\bar{n};B): i-1 \leq |B| \leq i\}$, $i = 0,\dots,J$, and let the Q-matrices Q_i with supports Z_i be defined by the system (i) of equations as follows:

System (0) is given by

$$u(\bar{n};\emptyset) \sum_{j=o}^{J} \mu_j(n_j)$$

$$= \sum_{i=o}^{J} \sum_{j=o}^{J} u(T_{ij}\bar{n};\emptyset)\mu_j(n_j+1)r_{ji}$$

for $\bar{n} = (n_1,n_2,\dots,n_J)$ with $n_1+\dots+n_J > N-M$ and by

$$u(\bar{n};\emptyset)\lambda = \sum_{j\in\bar{J}} u(T_{oj}\bar{n};\emptyset)\mu_j(n_j+1)r_{jo}$$

for $\bar{n} = (n_1,\dots,n_J)$ with $n_1+\dots+n_J = N-M$,

where $\lambda = \mu_o(M)$, $n_o = N-(n_1+\dots+n_J)$, and $T_{ij}\,\bar{n}$ is defined as in section 2.

System (i), $1 \leq i \leq J$, is given by

$$u(\bar{n};B) \sum_{j\in\bar{J}-B} \mu_j(n_j)$$

$$= \sum_{i\in\bar{J}} \sum_{j\in\bar{J}-B} u(T_{ij}\,\bar{n};B)\mu_j(n_j+1)r_{ji} \tag{i.1}$$

$$+ \sum_{i\in\bar{J}} \sum_{j\in\bar{J}-B} u(T_{ij}\,\bar{n};\ B\cup\{j\})\lambda r_{oi,jo}(T_{ij}\bar{n},B\cup\{j\})$$

and

$$u(\bar{n};B')\lambda = \sum_{j\in B'} u(\bar{n};B'-\{j\})\mu_j(n_j)r_{jo} \tag{i.2}$$

for $\bar{n} = (n_1,\dots,n_J)$ with $n_1+\dots+n_J = N-M$, $B\subset\bar{J}$ with $|B| = i-1$, $B'\subset\bar{J}$ with $|B'| = i$.

Clearly, $Q = \sum_{j=o}^{J} Q_j$, and the Q_j are irreducible on their supports.

System (0) is solved by

$$p(\bar{n};\emptyset) = C_o \prod_{j=o}^{J} e_j^{n_j} \left(\prod_{n=1}^{n_j} \mu_j(n)\right)^{-1}$$

for $n = (n_1,\dots,n_M)$ with $n_1+\dots+n_M \geq N-M$, where C_o is an arbitrary constant and $e = (e_o,\dots,e_J)$ is the probability vector satisfying $e = eR$. This is the only solution of system (0), and systems (i),

$1 \leq i \leq J$, have unique solutions (up to factors C_i) too. Thus the exchange probabilities will have to be chosen in such a way as to make these solutions compatible. Now, by (i.2), these solutions are given by

$$p(\bar{n};B) = p(\bar{n};\emptyset)\lambda^{-|B|}|B|! \prod_{j\in B} (\mu_j(n_j)r_{jo})$$

for all $(\bar{n},B)$ with $B \neq \emptyset$. Inserting this into (i.1) and observing that

$$p(T_{ij}\bar{n};\emptyset) = p(\bar{n};\emptyset) \frac{e_j}{e_i} \frac{\mu_i(n_i)}{\mu_j(n_j+1)},$$

one obtains that the equations

$$\begin{aligned} \sum_{j\in\bar{J}-B} \mu_j(n_j) = & \sum_{i\in\bar{J}-B} \frac{\mu_i(n_i)}{e_i} \sum_{j\in\bar{J}-B} e_j r_{ji} \\ & + \sum_{i\in B} \frac{\mu_i(n_i-1)}{e_i} \sum_{j\in\bar{J}-B} e_j r_{ji} \\ (*) \qquad & + (1+|B|) \sum_{i\in\bar{J}-B} \frac{\mu_i(n_i)}{e_i} \sum_{j\in\bar{J}-B} e_j r_{jo} r_{oi,jo}(T_{ij}\bar{n},B\cup\{j\}) \\ & + (1+|B|) \sum_{i\in B} \frac{\mu_i(n_i-1)}{e_i} \sum_{j\in\bar{J}-B} e_j r_{jo} \times \\ & \times r_{oi,jo}(T_{ij}\bar{n},B\cup\{j\}) \end{aligned}$$

have to be satisfied for all $(\bar{n},B)$, where $\mu_i(-1) = 0$ and, for $n_i = 0$, $r_{oi,jo}(T_{ij}\bar{n},B\cup\{j\}) = 0$.

Consider the special case of the central server depicted in Fig. 2. One exchange rule is obtained here by setting

$r_{oi,jo}(\bar{n},B) = \frac{1}{|B|}$ for $j\in B$ and $i = j$, and $= 0$ otherwise, for all $(\bar{n};B)$ with $B \neq \emptyset$. This rule satisfies (*). Another exchange rule for this case is defined by putting, for all $(\bar{n};B)$ with $B \neq \emptyset$,

$r_{oi,jo}(\bar{n},B) = \frac{1}{|B|} r_{oi}(\bar{n},B)$ for $j\in B, i\in\bar{J}-B$ and $= 0$ otherwise, where then $\sum_{i\in\bar{J}-B} r_{oi}(\bar{n},B) = 1$. This turns (*) into the conditions

$$0 = \sum_{j\in\bar{J}-B} \mu_j(n_j) \left(1 - \frac{1}{r_j} \sum_{i\in\bar{J}-B} r_i r_{oj}(T_{ji}\bar{n}, B\cup\{i\})\right)$$

for all $(\bar{n};B)$ with $B \neq \emptyset$. If, for instance,

$r_1 = .. = r_J = \frac{1}{J}$ and $r_{oi}(\bar{n},B) = \frac{1}{J-|B|}$ for all $i\in\bar{J}-B$ and all $(\bar{n};B)$ with $B \neq \emptyset$, then these conditions are satisfied.

Clearly, other special cases will allow nicely interpretable exchange rules for which the $p(\bar{n},B)$ given above represent the steady-state law of this blocking problem.

6. Product form distributions

All the specific measures p figuring in section 2-5 can be said to have a "product form" appearance. This can be seen as a trivial consequence of the decomposition these measures are associated with.

Within the general setup at the end of section 2 let $z_o \in Z$ be some fixed state (usually a "distinguished" state such as the state $(0,\ldots,0)$ in the Jackson model of section 2). For every $z \in Z$ there will then exist finite "paths" $(z_o,i_o,z_1,i_1,\ldots,z_n,i_n,z)$ such that $z_o \in Z_{i_o}$, $z_1 \in Z_{i_o} \cap Z_{i_1},\ldots,z_n \in Z_{i_{n-1}} \cap Z_{i_n}$, and $z \in Z_{i_n}$. Thus the quotients $C_{i_k} \doteq p(z_{k+1})/p(z_k)$, $0 \leq k \leq n-1$, and $C_{i_n} \doteq p(z)/p(z_n)$ are determined by the equations $u = uQ_{i_k}$, $0 \leq k \leq n$, respectively.

In terms of these quotients one has

$$p(z) = p(z_o) \prod_{k=o}^{n} C_{i_k} ,$$

i.e. a product form appearance of $p(z)$ with factors solely determined by the closed partial flows represented by the matrices Q_{i_k}.

For instance, regarding (2.2), let $z_o = (0,\ldots,0)$ and $z = (n_1,0,\ldots,0)$. One path of the kind described above is given by $(z_o,i_o,z_1,\ldots,z_{n_1-1},i_{n_1-1},z)$, where $z_k = (k,0,\ldots,0)$ and $i_k = ((k+1,0,\ldots,0),1)$. By (2.2),

$$p(n_1,0,\ldots,0) = C \prod_{k=o}^{n_1-1} \left(\frac{\lambda e_1}{\mu_1(k+1)}\right) = C \prod_{k=o}^{n_1-1} C_{i_k} ,$$

where

$$C_{i_k} = \frac{p(k+1,0,\ldots,0)}{p(k,0,\ldots,0)} = \frac{\lambda e_1}{\mu_1(k+1)} .$$

These observations were already made in [7] and are repeated here for completeness' sake.

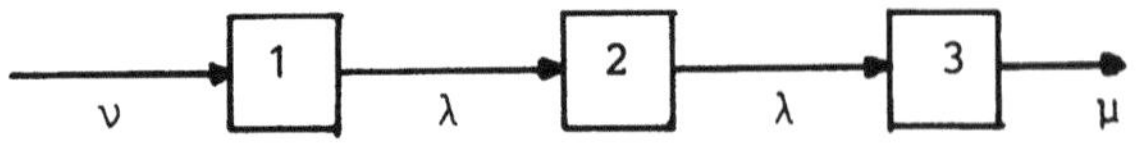

Fig. 1

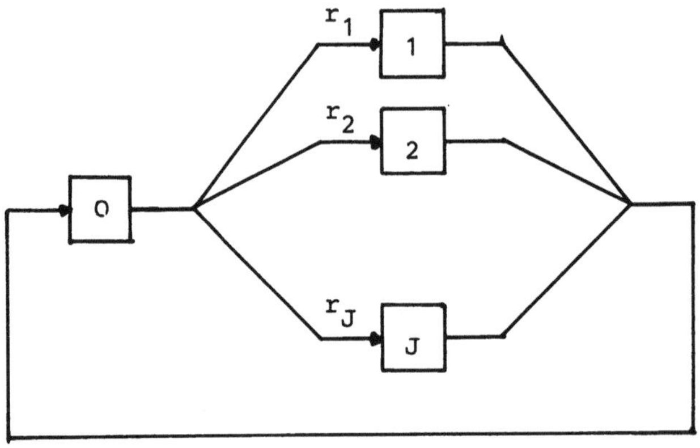

Fig. 2

References

[1] Baskett,F.,Chandy,K.M.,Muntz,R.R.,Palacios,F.G.; Open, closed and mixed networks of queues with different classes of customers. J.A.C.M. 22, 248-260, 1975

[2] Cohen,J.W.; Basic problems of telephone traffic theory and the influence of repeated calls. Philips Telecomm. Rev. 18, 49-100, 1957

[3] Jackson,J.R.; Jobshop-like queuing systems. Management Science 10, 131-142, 1963

[4] Kelly,F.P.; Networks of queues. Adv. Appl. Prob. 8, 416-432, 1975

[5] Kelly,F.P.; Reversibility and stochastic networks. Wiley, N.Y., 1979

[6] Pittel,B.; Closed exponential networks of queues with saturation. The Jackson-type stationary distribution and its asymptotic analysis. Math. of Oper. Res. 4, 357-378, 1979

[7] Schassberger,R.; A definition of discrete product form distributions. Zeitschr. für Oper. Res. 23, 189-195, 1979.

Networks of queues

Part I: Job-local-balance and the adjoint process

Part II: General routing and service characteristics

Arie Hordijk and Nico van Dijk

Institute of Applied Mathematics and Computer Science,
University of Leiden,
Wassenaarseweg 80,
Postbus 9512,
2300 RA Leiden
The Netherlands
tel. 071-148333

Abstract

In these papers we study networks of queues with general routing- and service-characteristics.
It is extensively analysed which models do satisfy a partial balance property, called the job-local-balance property. The key to our analysis is the adjoint process. Roughly speaking, the original process satisfies the job-local-balance property if and only if the adjoint process is reversible.

Part I: Job-local-balance and the adjoint process

1. INTRODUCTION

Stationary distributions for networks of queues are important for the performance analysis in computer science. Generally speaking one can say that only for networks of queues which satisfy the so-called partial balance property the stationary distributions have been obtained. It is known that networks with state independent routing matrix and symmetric queues satisfy the partial balance property (cf. [8]). However, in many practical models the routing probabilities do depend on the workload at the various nodes e.g. the models in which for heavily loaded routes there is a rerouting of jobs and/or models of networks with blocking phenomena.
The service characteristics of symmetric queues are only dependent on the total numbers of jobs. Again in many practical models there are different classes of jobs and the service-characteristics do depend on the numbers of jobs in the various classes e.g. in models where there is priority of the jobs of one class above the jobs of another class. In this paper we analyse models with a very general routing and rather general service-characteristics.

Our purpose is to specify those models which do satisfy partial balance and to point out the models where one may not expect partial balance. Since all kinds of partial balance have appeared we firstly give a definition (2.1) of the partial balance we study. It is natural to call this Job-local-balance (JLB). The key to our analysis is the adjoint process which we introduce in section 2. Roughly speaking the original satisfies JLB if and only if the adjoint process is reversible. To analyse whether a process is reversible we have available the Kolmogorov criterion (K-criterion). Applying the K-criterion we obtain necessary and sufficient conditions for JLB. In section 3 we show the necessity of symmetric queues for JLB in networks as described in [1], [2] and [8] (proposition 3.1). Reversibility of the network is shown to imply a reversible routing. Also in section 3 another application of the adjoint process is given in the analysis of a cyclic-three-centre model with blocking. In the main theorem of section 2 (theorem 2.5) among other things we show that if the adjoint process is reversible then the stationary distributions for the original and the adjoint process are equal. For reversible processes there is an algorithm to compute the stationary distribution. This algorithm leads in a straightforward way to the famous product-form solution in the standard models. We trust that the algorithm can be helpful in other models too.

In section 4 we introduce the general network. The routing is as general as possible. The service-characteristics at node j depend in a very general way on the jobmarks of the jobs present at node j. The jobmarks are 3-tuples (j,r,s) where j resp. r resp. s denote the node- resp. job- resp. service-number. Almost all stochastic networks do fit in this model. In section 4 we mainly do the prepatory work for applying theorem 2.5, more precisely we deduce the adjoint process. In the sections 5 and 6 we then apply the K-criterion. In section 5 the "routing" is analysed, in section 6 the service-characteristics are studied.

Our notation suggests to closed models. However, in sections 2 and 4 we show how the analogous results for open models can be obtained in the structure of this paper.

Proposition 4.1 already shows that a strict waiting on strict priority model cannot satisfy JLB. Only pseudo-priority is possible to model if JLB must hold, as stated in Corollary 4.1. Some examples with pseudo-priority are given later.

The JLB-property for Jacksontype networks and networks with number dependent blocking but a reversible routing (cf. [7], [9],[10]) directly results from the reversibility of their local throughputs (proposition 5.1). In view of JLB the restriction of a reversible routing on one hand seems necessary (example 5.1), on the other hand allows for more general blockings satisfying a blocking-invariant-condition at each node (proposition 5.2). The condition is satisfied by class-dependent blocking, but usually destroyed if blocking for jobs of one class depends on jobs of other classes (Corollary 5.3 and example 5.2).

In section 6 an invariance-property for service characteristics is shown to be sufficient for JLB (proposition 6.1).
This invariance-property is almost necessary for JLB in closed networks (propostion 6.2) and necessary for JLB in open networks. A counter-example is given.
In section 7 we overview several examples with blocking, pseudo-priority and state-dependent routing.
The first example is known in teletraffic literature (see [ti]).

Our analysis is particularly useful for detecting which models do or do not satisfy JLB, unfortunately most models do not, since JLB is equivalent to the famous insensitivity-property (cf. [1], [3], [4], [11]). The property states that the stationary distribution depends on the service-time distributions only by their means. Hence, our method explores the insensitivity phenomenon.

In the rest of this section we introduce the notions of JLB and the adjoint process with the help of two simple models.

In order to introduce job-local-balance let us recall the celebrated Jackson model of a network of queues (cf. [7]). Say there are N nodes in the network each with one server and an infinite waiting room. Let $\pi(\overline{n})$, with $\overline{n} = (n_1, n_2, \ldots, n_N)$ denote the stationary probability on the event that n_i customers are at node i, i = 1,...,N. Then Jackson found that

$$(1.1) \qquad \pi(\overline{n}) = c \prod_{i=1}^{N} \left(\frac{\lambda_i}{\mu_i}\right)^{n_i},$$

where c is a normalizing constant and λ_i resp. μ_i^{-1} is the throughput resp. the mean service time at node i, i = 1,...,N. This particular form of the stationary distribution is often called the product form. As is well-known the $\pi(\overline{n})$'s are not only solutions of the set of linear equations called the global balance equations. They are also solutions of subsystems in fewer unknown variables called the partial balance equations. In a verbal setting they read; "the probability flux out of a state due to a transition of a certain job equals the probability flux into that state due to a transition of the same job". Therefore we call this job-local-balance (JLB). In formula we have

$$(1.2) \qquad \pi(\overline{n})\, \mu_j = \sum_{i=1}^{N} \pi(\overline{n} + e_i - e_j)\mu_i p_{ij}, \qquad j = 1,\ldots,N$$

where $\overline{n} + e_i - e_j$ denotes the state $\overline{n}^*$ with $n_\ell^* = n_\ell$, $\ell \neq i,j$, $n_i^* = n_i+1$ and $n_j^* = n_j-1$.

It is well-known that the stationary probabilities for this model depend on the distributions of the service-times only through their means μ_i^{-1}, i = 1,...,N. This phenomenon is called insensitivity of the stationary distribution.

In this paper we analyse which models do satisfy the JLB-property. Our key for this study is a new notion which we call the adjoint process. Before we introduce this adjoint process in all generality we will introduce it for a special model, the cyclic-three-centre model (cf. [5]).

Consider a network with three nodes 1,2,3 and a cyclic routing i.e. after a service completion at node i a job requests service at node [i+1]mod 3, i = 1,2,3. Let $1 - b_i(n_1,n_2,n_3)$ denote the blocking probability at node i when there are n_1 resp. n_2 resp. n_3 jobs present at node 1 resp. node 2 resp. node 3. The blocking-protocol is as follows, if the service at node i is completed and the job requests service at node [i+1]mod 3 and if it is blocked there, then the job will get a new-service-requirement at node i. We assume that the service-distributions are exponential with parameter μ_i at node i, i = 1,2,3. There are various blocking-protocols studied in the literature. Note that since the service-times are exponentially distributed there is a close connection with the protocol in which service is interrupted at node i when blocking occurs at node [i+1]mod 3.
The network is assumed to be a closed network, hence no jobs leave or enter the network. Say the total number of jobs in the network is M.

This network of queues is a continuous-time Markov chain with transition rates,

$$(1.3)\quad \begin{cases} q((n_1,n_2,n_3),(n_1-1,n_2+1,n_3)) = \mu_1 b_2(n_1,n_2,n_3) \\ q((n_1,n_2,n_3),(n_1,n_2-1,n_3+1)) = \mu_2 b_3(n_1,n_2,n_3) \\ q((n_1,n_2,n_3),(n_1+1,n_2,n_3-1)) = \mu_3 b_1(n_1,n_2,n_3) \\ q((n_1,n_2,n_3),(n_1,n_2,n_3)) = \mu_1(1-b_2(n_1,n_2,n_3)) \\ \quad + \mu_2(1-b_3(n_1,n_2,n_3)) + \mu_3(1-b_1(n_1,n_2,n_3)) \\ \text{and all other transition rates are equal to zero.} \end{cases}$$

For this model we say that the stationary distribution π satisfies JLB' if for each $\bar{n} = (n_1,n_2,n_3)$, and with

$\bar{n}_{13} = (n_1-1,n_2,n_3+1), \bar{n}_{21} = (n_1+1,n_2-1,n_3), \bar{n}_{32} = (n_1,n_2+1,n_3-1)$

$$(1.4)\quad \begin{cases} \pi(\bar{n})q((n_1,n_2,n_3),(n_1-1,n_2+1,n_3)) = \pi(\bar{n}_{13})q((n_1-1,n_2,n_3+1),(n_1,n_2,n_3)) \\ \pi(\bar{n})q((n_1,n_2,n_3),(n_1,n_2-1,n_3+1)) = \pi(\bar{n}_{21})q((n_1+1,n_2-1,n_3),(n_1,n_2,n_3)) \\ \pi(\bar{n})q((n_1,n_2,n_3),(n_1+1,n_2,n_3-1)) = \pi(\bar{n}_{32})q((n_1,n_2+1,n_3-1),(n_1,n_2,n_3)) \end{cases}$$

In order to analyse whether these equations have a solution we introduce another set of transition rates, denoted by $\bar{q}$. The new transition rates define a Markov chain, it is this Markov chain which we call the adjoint process.

We define $\bar{q}$ to be equal to q for the transitions from (n_1,n_2,n_3) to $(n_1-1,n_2+1,n_3),(n_1,n_2-1,n_3+1)$ and (n_1+1,n_2,n_3-1). For the other transitions we define,

$$(1.5)\quad \begin{cases} \bar{q}((n_1,n_2,n_3),(n_1-1,n_2,n_3+1)) = q((n_1,n_2,n_3),(n_1-1,n_2+1,n_3)) \\ \bar{q}((n_1,n_2,n_3),(n_1+1,n_2-1,n_3)) = q((n_1,n_2,n_3),(n_1,n_2-1,n_3+1)) \\ \bar{q}((n_1,n_2,n_3),(n_1,n_2+1,n_3-1)) = q((n_1,n_2,n_3),(n_1+1,n_2,n_3-1)) \\ \text{and all other transition rates } \bar{q} \text{ are equal to zero.} \end{cases}$$

If we combine the relations (1.4) and (1.5) we find that π satisfies JLB if and only if π satisfies,

$$(1.6)\quad \pi(\bar{n})\ \bar{q}(\bar{n},\bar{m}) = \pi(\bar{m})\ \bar{q}(\bar{m},\bar{n}) \qquad \text{for all states } \bar{n},\bar{m}\ .$$

However, this is precisely the condition quaranteeing that the adjoint process is reversible (cf. [8]). Roughly speaking we found that JLB is equivalent to reversibility of the adjoint process. Moreover, if the adjoint process is reversible then the stationary distribution for both processes are equal. The advantage of using the adjoint process is twofold;

(i) with Kolmogorov's criterion it can be checked whether the adjoint process is reversible,

(ii) If the adjoint process is reversible then there is an algorithm to compute the stationary distribution.

Remark.

In a closed network there are of course always the standard methods to solve a finite set of linear equations. However, in practical closed models the number of equations is prohibitive for these methods. In open models the number of equations is often infinite. Our algorithm leads straightforward to the product form in the models for which the product form solution holds. We trust that the algorithm can be helpful in other models too. However, until now we have not analysed its complexity.

2. JOB-LOCAL-BALANCE AND THE ADJOINT PROCESS

In this section we give a formal derivation of the adjoint process and we derive its relation with the JLB-property.

Let L be a countable set of so-called jobmarks.

We consider a continuous-time Markov process with state space S.

A state $E \in S$ is a finite set of jobmarks, say $E = \{\ell_1,\dots,\ell_K\}$ with $\ell_i \in L$, $i = 1,\dots,K$ and K some natural number .

The state $\{\ell_1,\dots\ell_K\}$ denotes that precisely K jobs are present in the system having jobmarks $\ell_1,\dots,\ell_K$.

If $E = \{\ell_1,\dots,\ell_K\}$ and $\ell \in L$ then we let $E+\ell$ denote $\{\ell_1,\dots,\ell_K,\ell\}$. Further $q(E(1),E(2))$ denotes the transition rate for a transition from state $E(1)$ to state $E(2)$.

Our notation can be used for closed as well as open models.
For a closed model the number of jobs in the system will always be fixed, say M, so that we only need to consider states with precisely M jobmarks. For an open model we make the convention that the system always contains precisely one "dummy"-job with jobmark 0, and that a state is always presented as $E = \{\ell_1,\ldots,\ell_K,0\}$ with $\ell_i \neq 0$, $\ell_i \in L$, $i = 1,\ldots,K$. The "dummy" jobmark 0 is added to let $q(E+0,E+\ell')$ denote the transition rate for an arrival into the system of a job with jobmark ℓ' and $q(E+\ell',E+0)$ for a departure out of the system of such a job.

All results in the sequel are valid for closed as well as open models, unless explicitly stated else. For the reader however, it is convenient only to think of a closed model with a fixed number of jobs, say M.

Throughout this paper we assume that,

I : If the states $E(1),E(2)$ differ in more than one jobmark then $q(E(1),E(2)) = 0$,

II: The Markov chain is irreducible on a set $V \subset S$ and there exists a stationary distribution π on V.

<u>Remarks</u>.

1. A Markov chain is called irreducible on a set $V \subset S$ if for any $E,E' \in V$, $E \neq E'$, there exist $E(1),\ldots,E(n)$ such that $q(E,E(1))q(E(1),E(2))\ldots q(E(n-1),E(n))q(E(n),E')) > 0$, and for any $E \in V$, $E' \notin V$, $q(E,E') = 0$.

2. In this paper we exclude irregular Markov chains, so with probability one there will only be a finite number of transitions in a finite time-interval. Further, note that it follows from the irreducibility of the Markov chain that the stationary distribution is unique.

The stationary distribution π satisfies on V the global-balance-equations

$$(2.1) \qquad \pi(E) \sum_{E'} q(E,E') = \sum_{E'} \pi(E')q(E',E)$$

<u>Definition:</u> A distribution p on an irreducible set V has the job-local-balance-property (JLB) on V, if for each E and $E + \ell \in V$,

$$(2.2) \qquad p(E+\ell) \sum_{\ell'} q(E+\ell,E+\ell') = \sum_{\ell'} p(E+\ell')q(E+\ell',E+\ell)$$

A Markov chain has the JLB-property if the stationary distribution satisfies (2.2). Note that for open models ℓ and ℓ' can be taken 0 .

If p is a solution of the local-balance-equations (2.2) then it satisfies also the global-balance-equations. From assumption II it then follows,

$$p = \pi \text{ on } V.$$

For fixed E the number of unknown variables in (2.2) is equal to $|L|$, the number of different jobmarks. The number of unknown variables in (2.1) is equal to $|V|$. So in general the equations (2.2) are easier to solve and quite often an explicit solution can be given. However, many queueing models do not satisfy the JLB-property and hence one solution of (2.2) for all $E + \ell$ does not exist. But if it exists then (2.2) is preferable above (2.1).

Now we analyse the feasibility of (2.2) for all E. For fixed E let $V_i(E)$, $i = 1,\dots,m(E)$ denote the irreducible sets in V with respect to the Markov chain with transition rates $q(E+\ell,E+\ell')$, $\ell,\ell' \in L$. Let V(E) be the state space of this Markov chain restricted to V, then $V(E) = V \cap \{E + \ell \mid \ell \in L\}$.

Proposition 2.2: If π satisfies the JLB-property, then for any E:

$$V(E) = \bigcup_{i=1}^{m(E)} V_i(E).$$

Proof. The only thing to prove is that the Markov chain on V(E) for fixed E has no transient states. The stationary distribution for this Markov chain is $c\,\pi\,(E+\ell)$, $\ell \in L$ and $c = [\sum_{\ell \in L} \pi(E+\ell)]^{-1}$. Since $\pi(E+\ell) > 0$ for all $\ell \in L$, all states are positive recurrent. □

Let $V(\ell;E)$ denote the irreducible subset of V(E) which contains $E + \ell$. If $E + \ell$ is transient then we notate $V(\ell;E) = \emptyset$.
The following example shows that although the Markov chain on V is irreducible the Markov chain on V(E) for some E can be reducible.

Example. Consider a closed system with 3 nodes, 1,2,3, and 2 jobs; one job 1 of class 1 and one job 2 of class 2.
At all nodes servers of a same type, say s, are available, and any job present at a node only uses one server at a time.
Job 2 is always being served and routes from node 1 to 2 and backwards.
Job 1 is stopped being served at node 1 as long as job 2 is present at node 2, and at node 3 as long as job 2 is present at node 1. Otherwise, it is always served.
Job 1 routes from node 1 to 2 and backwards as long as job 2 is at node 1, but routes from node 3 to 2 and backwards as long as job 2 is at node 2.

Let jobmark (i,1,s) represent that job 1 is present at node i, i = 1,2,3 and has service-number s, (j,2,s) is defined analogously for job 2.
If we shortly notate state {(i,1,s),(j,2,s)} by (i,j) then the positive transition rates are depicted in the figure with an arc.

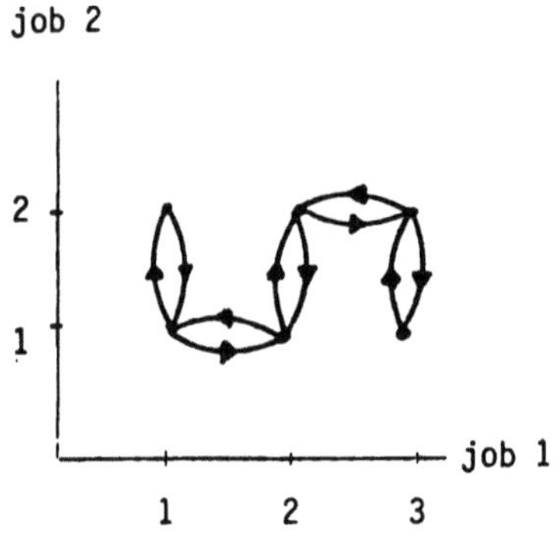

We have $L = \{(\ell,R,s) \mid R = 1 \text{ and } \ell \in 1,2,3 \text{ or } R = 2 \text{ and } \ell \in 1,2\}$,

$V = \{\{(i,1,s),(j,2,s)\} \mid i = 1,2,3,\ j = 1,2\}$.

for $i = 1,2,3$ and

$E=\{(i,1,s)\}:m(E)=1$ and $V((j,2,s);E)=\{\{(i,1,s),(1,2,s)\},\{(i,1,s),(2,2,s)\}\} j=1,2$

for

$E=\{(1,2,s)\}:m(E)=2$ and $V((i,1,s);E)=\{\{(1,1,s),(1,2,s)\},\{(2,1,s),(1,2,s)\}\}, i=1,2$

$V((3,1,s);E)= \{(3,1,s),(1,2,s)\}$

and for

$E=\{(2,2,s)\}:m(E)=2$ and $V((1,1,s);E)=\{(1,1,s),(2,2,s)\}$

$V((i,1,s);E)=\{\{(2,1,s),(2,2,s),\{(3,1,s),(2,2,s)\}\}, i=2,3.$

Hence for $E = \{(1,2,s)\}$ or $\{(2,2,s)\}$ the Markov chain on $E + L$ is reducible. Note, however, that this Markov chain has no transient states. □

Proposition 2.2 implies that for any E and ℓ, $V(\ell,E) \neq \emptyset$ if the JLB-property holds. Therefore we have that for any E and $i = 1,\dots,m(E)$ the following set of equations has at most one solution.

$$(2.3) \quad \begin{cases} x(\ell) \sum_{\ell'} q(E+\ell,E+\ell') = \sum_{\ell'} x(\ell')q(E+\ell',E+\ell) \\ \sum_{\ell'} x(\ell') = 1, \\ \text{where } \ell,\ell' \text{ are such that } E+\ell,E+\ell' \in V_i(E) \end{cases}$$

Assumption III: For any E and ℓ, $V(\ell;E) \neq \emptyset$, and
for any E and $i = 1,\ldots,m(E)$ the system (2.3) has a
unique solution, denoted by $\pi(E+\ell;V_i(E))$, $\ell \in L$.

The existence of a solution of (2.3) is always satisfied if $|L| < \infty$.
The next proposition together with proposition 2.2 shows that assumption
III is valid if the JLB-property holds.
We use the standard notation $\pi(E|V')$ for the conditional distribution
of π on V' i.e.:

$$\pi(E|V') = \pi(E)\cdot\Big[\sum_{E'\in V'} \pi(E')\Big]^{-1} \tag{2.4}$$

Proposition 2.3: The stationary distribution π has the JLB-property
if and only if

$$\pi(\,\cdot\,|\,V_i(E)) = \pi(\,\cdot\,;\,V_i(E)) \quad \text{for all } E \text{ and } i = 1,\ldots,m(E). \tag{2.5}$$

Proof. If π satisfies (2.2) then $\pi(\,\cdot\,|\,V_i(E))$ is the solution of (2.3), hence (2.5) holds. If (2.5) is valid then $\pi(\,\cdot\,|\,V_i(E))$ is a solution of (2.3) and consequently $\pi(\cdot)$ satisfies (2.2). □

As already said in many practical models the equations (2.3) can be
solved explicitly. We therefore assume that the $\pi(\,\cdot\,;\,V_i(E))$, $E \in V$,
$i = 1,\ldots,m(E)$ are known. Now we define transition rates of an adjoint
process expressed in these $\pi(\,\cdot\,;\,V_i(E))$'s.

Definition 2.4: A Markov chain on V with transition rates $\overline{q}$ which
satisfy the following relations will be called an adjoint process.

$$
(2.6)\quad \begin{cases} \pi(E+\ell;V_i(E))\bar{q}(E+\ell,E+\ell') = \pi(E+\ell';V_i(E))\bar{q}(E+\ell',E+\ell) \\ \qquad \text{for } E+\ell,E+\ell' \in V_i(E),\ E \in V,\ i = 1,\dots,m(E) \\ \qquad \bar{q}(E+\ell,E+\ell') > 0 \quad \text{if} \quad q(E+\ell,E+\ell') > 0, \text{ and} \\ \qquad \bar{q}(E(1),E(2)) = 0 \qquad \text{else} \end{cases}
$$

It is clear that there are in general many adjoint processes, since the $\bar{q}$ are not uniquely defined. However, the quotients $\bar{q}(E(1),E(2))/\bar{q}(E(2),E(1))$ are uniquely defined for all $E(1), E(2) \in V_i(E)$. We will use these quotients in the sequel. From (2.6) we conclude that the original and adjoint process have the same irreducible sets.

By constructing an adjoint process it is convenient that one of the two $\bar{q}(E+\ell,E+\ell')$, $\bar{q}(E+\ell',E+\ell)$ can be given an arbitrary positive value. For example in the cyclic-three-centre model of section 7 we chose the $\bar{q}$'s equal to the q's for the transitions from (n_1,n_2,n_3) to (n_1-1,n_2+1,n_3), (n_1,n_2-1,n_3+1) and (n_1+1,n_2,n_3-1).

Let us recall that a Markov chain with transition rates q is reversible if and only if the stationary distribution satisfies (cf. [8])

$$
(2.7)\qquad \pi(E(1))\, q(E(1),E(2)) = \pi(E(2))\, q(E(2),E(1)) \quad \forall (E(1),E(2)).
$$

In the sequel we also use the notations $\pi(\ell;E)$ resp. $\pi(\ell|E)$ for $\pi(E+\ell;V_i(E))$ resp. $\pi(E+\ell|V_i(E)|$ with $V_i(E) = V(\ell;E)$. The constant c in explicit expressions for stationary distributions denotes always the normalizing constant.

We are now ready to state the relations between the original process, the adjoint processes, the stationary distributions of these processes, and the JLB-property.

Theorem 2.5:

(i) If a Markov chain is reversible then there exists an adjoint process equal to it.

(ii) If some adjoint process is reversible then the original process has the JLB-property.

(iii) If the original process has the JLB-property then any adjoint process is reversible. Moreover, the stationary distributions of the original process and all adjoint processes are equal.

(iv) The original process has the JLB-property if and only if the stationary distribution satisfies for any adjoint transition rates $\bar{q}$ and any reference state $E(0) \in V$,

$$(2.8) \qquad \pi(E) = c \cdot \frac{\bar{q}(E(0),E(1))}{\bar{q}(E(1),E(0))} \frac{\bar{q}(E(1),E(2))}{\bar{q}(E(2),E(1))} \cdot \ldots \cdot \frac{\bar{q}(E(m),E)}{\bar{q}(E,E(m))}$$

for any path $E(1),\ldots,E(m)$, $E \in V$ such that the denominators of the quotients are positive.

Proof.

(i) If the stationary distribution π satisfies (2.7) then certainly (2.2) holds. Hence by proposition 2.3,

$$(2.9) \qquad \pi(\ell;E) = \pi(E+\ell)\Big[\sum_{E+\ell' \in V(\ell;E)} \pi(E+\ell')\Big]^{-1}$$

The relations (2.7) and (2.9) give

$$\pi(\ell;E)q(E+\ell,E+\ell') = \pi(\ell';E)q(E+\ell',E+\ell)$$

Hence the transition rates q are solutions of the equations (2.6). Consequently, the original process is also an adjoint process.

(ii) If the $\bar{q}$-process is reversible then its stationary distribution, say $\bar{\pi}$, satisfies

$$(2.10)\qquad \bar{\pi}(E+\ell)\bar{q}(E+\ell,E+\ell') = \bar{\pi}(E+\ell')\bar{q}(E+\ell',E+\ell).$$

From the definition of $\pi(\ell;E)$, $\ell \in L$, the relations (2.3) and (2.6), it then follows that for all E, and $E+\ell \in V$,

$$(2.11)\qquad \pi(\ell;E) = \bar{\pi}(E+\ell)\Big[\sum_{E+\ell' \in V(\ell;E)} \bar{\pi}(E+\ell')\Big]^{-1}$$

The relations (2.3) and (2.11) give that $\bar{\pi}$ is a solution of (2.2). Therefore $\bar{\pi}$ satisfies the global balance equations for the original process and consequently, $\bar{\pi}$ is the stationary distribution of the original process and this process has the JLB-property.

(iii) If the original process has the JLB-property then by proposition 2.3 relation (2.5) is valid. The relations (2.4),(2.5) and (2.6) give for any adjoint transition rates $\bar{q}$,

$$(2.12)\qquad d\,\pi(E+\ell)\bar{q}(E+\ell,E+\ell') = d\,\pi(E+\ell')\bar{q}(E+\ell',E+\ell)$$

with $d = \Big[\sum_{E+\ell' \in V(\ell;E)} \pi(E+\ell')\Big]^{-1}$, for all $E+\ell$ and $E+\ell' \in V(\ell;E)$

Recollecting assumption I and the definition (2.6), (2.12) yields

$$(2.13)\qquad \pi(E)\bar{q}(E,E') = \pi(E')\bar{q}(E',E) \qquad \text{for all } E,E'.$$

By (2.7) relation (2.13) says that the adjoint process is reversible and π is its stationary distribution.

(iv) To show the if-part, suppose π satisfies (2.8) for some $\bar{q}$-transition rates and reference state $E(0) \in V$. From Kolmogorov's criterion (see [8] p.21) it then follows that the adjoint process is reversible and its stationary distribution is equal to π. From assertion (ii) it then follows that the original process has the JLB-property.
In order to proof the only-if part suppose that the original process has the JLB-property. Then from assertion (iii) it follows that any of the adjoint processes is reversible and its stationary distribution is equal to that of the original process. Applying again Kolmogorov's criterion yields relation (2.8). □

Remarks.

1. Note that since V is irreducible there always exist at least one path $E(1),\ldots,E(m)$ in V such that all denominators in (2.8) are positive. Further, the only paths for which (2.8) will be positive are obtained if $E(0),E(1) \in V_0$; $E(1),E(2) \in V_1;\ldots;E(m)$, $E \in V_m$, with V_i, $i = 0,\ldots,m$ irreducible subsets.
2. Also without using an adjoint process we can obtain a criterion for JLB analogous to relation (2.8), but in terms of $\pi(\ ;E(i))$. This can be seen by using proposition 2.3.
However, especially to simplify the algorithmic procedure as well as for further applications (see section 3 for instance), it appears to be useful to have a characterization in terms of transition rates $\bar{q}$. For instance, by representing positive transition rates $\bar{q}$ with directed arcs we obtain a directed graph. From that we may learn the basic paths or cycles for verifying (2.8) for all paths, (see [5]), or simple paths or cycles which lead to necessary conditions for JLB (see section 3).
3. In [12] it is shown that a suitable decomposition of the generator implies a product form expression closely related to our formula (2.8).

3. CYCLIC-THREE-CENTRE-MODEL; SYMMETRIC DISCIPLINES.

In this section we give two applications of theorem 2.5.
Firstly we return to the cyclic-three-centre model introduced in section 1 and we derive the relation, which the blocking probabilities must satisfy in order to have the JLB-property.
Secondly, we focus on a network of queues with service-characteristics as defined in [1], [2] and [8] and show that they have to be symmetric if the JLB-property is valid.

3.1. Cylic-three-centre-model

If we take the adjoint process as defined in section 1 we know from theorem 2.5 that this process is reversible if and only if the original process has the JLB-property. The adjoint process is reversible if and only if Kolmogorov's criterion is satisfied. Now applying this criterion we found (cf. [5] theorem 7.3) under a weak regularity condition that the adjoint process is reversible if and only if

$$\begin{aligned} & b_1(n_1,n_2,n_3)b_2(n_1,n_2+1,n_3-1)b_3(n_1-1,n_2+1,n_3) \\ & = b_1(n_1-1,n_2+1,n_3)b_2(n_1,n_2,n_3)b_3(n_1,n_2+1,n_3-1) \ . \end{aligned} \tag{3.1}$$

Examples of models satisfying (3.1) together with the derivation and a discussion of (3.1) can be found in [5].

3.2. Symmetric disciplines

We consider a network of queues with nodes $j = 1,\dots,N$.
The characteristic of the service at node j is parametrized through the 3-tuple (f_j,φ_j,δ_j) where,

(i) $f_j(x)$ is the service capacity when x jobs are present

(ii) $\varphi_j(x|i)$, $i = 1,\dots,x$ is the fraction of the total service capacity given to the job at the i-th place

(iii) $\delta_j(x|i)$, $i = 1,\dots,x+1$ is the probability that an arriving job at node j will be assigned to place i.

We call f resp. φ resp. δ the capacity- resp. sharing- resp. placing-function. It is well-known that many of the practical queueing networks can be parametrized in this way (cf. [1], [2], [8]).

We have to specify what happens when a job enters or leaves a node. Assume the shift protocol is followed. The protocol prescribes that if x jobs are present at a certain node and the job which occupies the i-th place leaves the node then the jobs at the places $i+1,\dots,x$ shift to the places $i,\dots,x-1$; and if an arriving job is assigned to place ℓ then the jobs at the places $\ell,\dots,x$ shift to the places $\ell+1,\dots,x$.

Let p_{ij} denote the probability that a job which leaves node i requires its next service at node j, $i,j = 1,\dots,N$.

Further the service-requirement for any job at node j is exponentially distributed with parameter μ_j, $j = 1,\dots,N$.

The derivation we give below is for a closed model with say M jobs. However, a completely similar derivation can be given for an open model.

For the derivation we need the following assumptions:

(a) The service-capacity is positive i.e. $f_j(x) > 0$ if $x > 0$ $j = 1,\dots,N$

(b) There is no blocking i.e.
$$\sum_{i=1}^{x+1} \delta_j(x|i) = 1 \text{ and } \sum_{i=1}^{x} \varphi_j(x|i) = 1,\ j = 1,\dots,N$$

(c) The routing matrix p_{ij}, $i,j = 1,\dots,N$ is irreducible.

Let λ_j, $j = 1,\dots,N$ be the unique solution, up to a constant, of

$$(3.2) \qquad \lambda_j = \sum_{i=1}^{N} \lambda_i p_{ij} \ , \ j = 1,\dots,N.$$

The set of jobmarks L are 3-tuples (j,r,s) where j resp. r resp. s denotes the node- resp. job- resp. place-number of the job. Let E denote a state with x_i jobs at node i, $i = 1,\dots,N$. Say $(j,r,s) \in E$ then we write $E - (j,r,s) + (j',r,s')$ for the state only differing from E by that job r is at the s'-th place at node j'. Remark that in the notation $E - (j,r,s) + (j',r,s')$ the jobs at node j resp. j' with place-numbers $s+1,\dots,x_j$ resp. $s',\dots,x_{j'}$ are reshifted to $s,\dots,x_j-1$ resp. $s'+1,\dots,x_{j'}+1$. Now note that the results of section 2 are still valid if assumption I of section 2 is changed in: If the states E(1),E(2) differ in more than one jobmark up to reshifting then $q(E(1),E(2)) = 0$.

Then, using assumption (b) and (c) we conclude from proposition 2.2 that JLB implies for $E \in V$, with x_j the number of jobs at node j, $j = 1,\dots,N$,

$$\varphi_j(x_j|s) > 0 \Rightarrow \delta_j(x_j-1|s) > 0 \qquad s = 1,\dots,x_j.$$

and

$$\delta_j(x_j-1|s) > 0 \Rightarrow \varphi_j(x_j|s) > 0 \qquad s = 1,\dots,x_j.$$

Hence, in view of JLB the following assumption is justified,

$$\text{(d)} \qquad \delta_j(x-1|s) = 0 \Leftrightarrow \varphi_j(x|s) = 0 \quad \text{for all } j,x \text{ and } s = 1,\dots,x.$$

Now let $E + (j,r,s) \in V$ with E a set of jobmarks with x_i jobs at node i, $i = 1,\dots,N$. Then the set $V(\ell;E)$ as introduced in section 2 here becomes with $\ell = (j,r,s)$ and by using (a) - (d),

$$V((j,r,s);E) = \begin{cases} E + (j,r,s) & \text{if } \varphi_j(x_j+1|s) = 0 \\ \{E + (j',r,s') | \delta_{j'}(x_{j'}|s') > 0\} & \text{if } \varphi_j(x_j+1|s) > 0 \end{cases}$$

A small problem now arises by that a state $E + (j,r_1,s) + (j,r_2,s+1)$ representing that a job r_1 resp. r_2 occupies place s resp. s+1 at some node j, can have two different positive transitions into the state $E + (j,r_1,s+1) + (j,r_2,s)$ since reshifting is taken into account. To cope with such difficulties, i.e. ; changes of different jobmarks may cause a same change of the state, we slightly extend the results of section 2 by:

Write $q(\ell,\ell';E)$ for the transition rate from $E+\ell$ into $E+\ell'$ caused by the change of a jobmark ℓ into ℓ'. Define the JLB-property as given by (2.2) but with $q(E+\ell,E+\ell')$ replaced by $q(\ell,\ell';E)$. Note that such a definition of JLB still corresponds to the verbal representation given on p. 6. Proceeding as in section 2 but with $\bar{q}(E + \ell;E + \ell')$ replaced by $\bar{q}(\ell,\ell';E)$ in (2.6) we conclude that JLB holds if and only if

for any $E+\ell \in V$: $\pi(E+\ell)\bar{q}(\ell,\ell';E) = \pi(E+\ell')\bar{q}(\ell',\ell;E)$.

As in section 2,

this (strong) reversibility is satisfied if and only if the $\bar{q}$-rates satisfy the Kolmogorov criterion.

To derive the $\bar{q}$-rates we first write the system (2.3), with omission of the normalizing condition for our network of queues as,

$$(3.3) \qquad x((j,r,s)) \sum_{(j',r,s')} f_j(x_j+1)\varphi_j(x_j+1|s)\, \mu_j p_{jj'} \delta_{j'}(x_{j'}|s')$$

$$= \sum_{(j',r,s')} x((j',r,s')) f_{j'}(x_{j'}+1)\varphi_{j'}(x_{j'}+1|s')\mu_{j'} p_{j'j} \delta_j(x_j|s)$$

By using (b),(d) and (3.2), it is easily checked that (3.3) has the following explicit solution, with c a normalizing constant on $V((j,r,s);E)$

$$(3.4) \qquad x((j,r,s)) = \begin{cases} 1 & \text{if } \varphi_j(x_j+1\ s) = 0 \\ c\,\dfrac{\lambda_j\ \delta_j(x_j|s)}{\mu_j\ f_j(x_j+1)\varphi_j(x_j+1|s)} & \text{if } \varphi_j(x_j+1|s) > 0 \end{cases}$$

It follows directly from (2.6) and (3.4) that for any adjoint process the transition rates $\bar{q}$ satisfy,

$$(3.5) \qquad \frac{\bar{q}((j_1,r,s_1),(j_2,r,s_2);E)}{\bar{q}((j_2,r,s_2),(j_1,r,s_1);E)} =$$

$$\frac{\lambda_{j_2}\ \delta_{j_2}(x_{j_2}|s_2)\mu_{j_1}\ f_{j_1}(x_{j_1}+1)\varphi_{j_1}(x_{j_1}+1|s_1)}{\lambda_{j_1}\ \delta_{j_1}(x_{j_1}|s_1)\mu_{j_2}\ f_{j_2}(x_{j_2}+1)\varphi_{j_2}(x_{j_2}+1|s_2)}$$

for $E + (j_2,r,s_2) \in V((j_1,r,s_1);E)$ and $\varphi_{j_1}(x_{j_1}+1|s_1) > 0$.

Let us choose now a particular adjoint process by defining,

$$(3.6) \qquad \bar{q}((j_1,r,s_1),(j_2,r,s_2);E) =$$
$$\lambda_{j_2}\ \delta_{j_2}(x_{j_2}|s_2)\mu_{j_1}\ f_{j_1}(x_{j_1}+1)\varphi_{j_1}(x_{j_1}+1|s_1)$$
for all (j_2,r,s_2), and $\bar{q}(E(1),E(2)) = 0$. else.

Note that (2.6) is satisfied with $\pi((j,r,s);E) = x(j,r,s))$ and $\bar{q}(E+\ell,E+\ell')$ replaced by $\bar{q}(\ell,\ell';E)$. Further, according to (d) the transition rate $\bar{q} = 0$ if $E + (j_2,r,s_2) \notin V((j_1,r,s_1);E)$.

Next we will derive a condition on the service characteristics which is necessary for the JLB-property.

Consider a state E with x_j jobs at node j then according to (b) for at least one place-number $s \in \{1,2,\ldots,x_j\}$: $\varphi_j(x_j|s) > 0$.
If $\varphi_j(x|s') = 0 \;\; \forall s' \neq s$ then assumptions (b) and (d) directly yield:

$$(3.7) \qquad \varphi_j(x_j|s) = \delta_j(x_j-1|s) \qquad s = 1,\ldots,x_j.$$

Service-discipline (f_j,φ_j,δ_j) satisfying (a), (b) and (3.7) for all x_j is called <u>symmetric</u> (see [8]).
The following proposition shows that under assumptions (a) - (d) the JLB-property requires symmetric disciplines.

<u>Proposition 3.1</u>. If a distribution satisfies the JLB-property on V then the disciplines are symmetric on V, i.e.;

(3.7) is satisfied for any $j = 1,\ldots,N$ and $x \in \mathbb{N}$ such that there exists a state $E \in V$ with $x_j = x$.

<u>Proof</u>. Consider some node j and let $x_j = x$.
If $\varphi_j(x|s) > 0$ for only one place-number $s \in \{1,\ldots,x\}$ then as already given above, (3.7) is satisfied. Hence, suppose that for some s and $s' = s + k$ with $k \geq 1$ and with $s, s' \in \{1,\ldots,x\}$:
$\varphi_j(x|s) > 0$ and $\varphi_j(x|s') > 0$.

Let $r_0,r_1,\ldots,r_k$ be the job-numbers of the jobs at the places $s,s+1,\ldots,s+k$ at node j in state E. Then we denote $[(r_0,s_{i_0}),(r_1,s_{i_1}),\ldots,(r_k,s_{i_k})]$ for the state such that job r_ℓ is at place $s_{i_\ell} \in \{s,s+1,\ldots,s+k\}$ at node j , $\ell=0,1,\ldots,k$ and all other jobs have the same jobmarks as in state E.

According to section 2 the JLB-property here requires that the K-criterion is satisfied for the adjoint rates $\bar{q}$ as defined by (3.6).
We will apply the K-criterion on the cycles:

$[(r_0,s),(r_1,s+1),\ldots,(r_{k-1},s+k-1),(r_k,s')] \rightarrow$

$[(r_0,s'),(r_1,s),(r_2,s+1),\ldots,(r_k,s+k-1)] \rightarrow$

$[(r_0,s+k-1),(r_1,s'),(r_2,s),(r_3,s+1),\ldots,(r_k,s+k-2)] \rightarrow$

$\ldots\ldots \rightarrow$

$[r_0,s+1),(r_1,s+2),\ldots,(r_{k-2},s+k-1),(r_{k-1},s'),(r_k,s)] \rightarrow$

$[(r_0,s),(r_1,s+1),\ldots,(r_{k-1},s+k-1),(r_k,s')]$

and the same cycle in opposite direction.

By comparing the corresponding $\bar{q}$-rates given by (3.6) we conclude from the K-criterion:

$$(3.8)\quad [\delta_j(x-1|s')\varphi_j(x|s)]^{k+1} = [\delta_j(x-1|s)\varphi_j(x|s')]^{k+1}$$

Since s and s' can be taken arbitrarily such that $s' = s + k$, $k \geq 1$ and $\varphi(x|s) > 0$ and $\varphi(x|s') > 0$ we conclude from (3.8):

$$(3.9)\quad \frac{\delta_j(x-1|s_1)}{\varphi_j(x\ \ |s_1)} = \frac{\delta_j(x-1|s_2)}{\varphi_j(x\ \ |s_2)} \qquad \text{for all } s_1,s_2 \in \{1,2,\ldots,x\} \text{ such that } \varphi(x|s_1) > 0 \text{ and } \varphi(x|s_2) > 0.$$

Relation (3.9) together with assumptions (b) and (d) directly gives relation (3.7) for $x_j = x$.

This holds for any $j = 1,\ldots,N$ and x such that there exists $E \in V$ with $x_j = x$. □

For open networks proposition 3.1 and its proof remain valid. In [2] is shown the necessity of (3.7) for station balance of an isolated queue. In our setting station balance must be interpreted as JLB and an isolated queue can be considered as an open network with one queue. Hence, the result of [2] is included.

On the other hand, symmetric disciplines can be shown to be sufficient for JLB. This can be proved by verifying Kolmogorov's criterion for the $\bar{q}$-rates However, it is easier to assume the well-known product-form solution (cf. [1],[4],[8]) for the stationary distribution and to show that this solution satisfies the JLB-equations (2.2).

As a last result we show that a network with characteristics satisfying (a) - (d) and which is reversible necessarily has a reversible routing. From theorem 2.5 part (i) we conclude that for a reversible process,

$$(3.10) \qquad \frac{q((j_1,r,s_1),(j_2,r,s_2);\,E)}{q((j_2,r,s_2),(j_1,r,s_1);E)} = \frac{\bar{q}((j_1,r,s_2),(j_2,r,s_2);\,E)}{\bar{q}((j_2,r,s_2),(j_1,r,s_1);\,E)}$$

for any adjoint $\bar{q}$-rates, and such that the denominators are positive. From assumptions (a), (b) and (c) we learn that for each $\tilde{E} \in V$, any j_1 and j_2 always exist s_1 and s_2 such that

$$(3.11) \qquad \varphi_{j_1}(x_{j_1}|s_1) > 0 \qquad \text{and} \qquad \delta_{j_2}(x_{j_2}|s_2) > 0.$$

With $\bar{q}$ as defined by (3.6) we have,

$$q((j_1,r,s_1),(j_2,r,s_2);\,E) = p_{j_1j_2}\, f_{j_1}(x_{j_1}+1)\varphi_{j_1}(x_{j_1}+1 \mid s_1)\, \mu_{j_1}\, \delta_{j_2}(x_{j_2}|s_2)$$

$$\bar{q}((j_1,r,s_1),(j_2,r,s_2);\,E) = \lambda_{j_2}\, f_{j_1}(x_{j_1}+1)\varphi_{j_1}(x_{j_1}+1|s_1)\, \mu_{j_1}\, \delta_{j_2}(x_{j_2}|s_2).$$

Hence, with $E = \tilde{E} - (j_1,r,s_1)$, (3.11) and assumption (d), we derive from (3.10)

$$(3.12) \qquad \frac{\lambda_{j_2}}{\lambda_{j_1}} = \frac{P_{j_1j_2}}{P_{j_2j_1}} \quad \text{if positive or } P_{j_1j_2} = P_{j_2j_1} = 0 \text{ , } \forall\, j_1,j_2 \text{ .}$$

<u>Conclusion:</u> A reversible queueing network always has a reversible routing matrix.

Part II: General routing and service characteristics

INTRODUCTION

This part II is the continuation of part I. For an introduction to its results we refer to the introduction of part I. For convenience the sections are numbered consecutively with part I.

In section 3 we dealt with queueing networks with a fixed routing and service disciplines only depending on place numbers. However, in many practical models a more general description appears to be wanted.
The routing of a job from one node to another may depend on the loadings at the different nodes. For instance, a saturated node may give rise, to overflow, modelled by rerouting, or to a blocking. Further, one can have different classes of jobs such that jobs of one class have priority over jobs of another class in being served. Also different types of servers may be available at a node.

In this second part we shall consider networks of queues with

a state dependent routing and blocking, and

at each node a general service characteristic which only depends on the jobs present at that node.

Having such a general parametrization one can hardly expect to obtain the stationary distribution.
By means of section 2, however, we can for such a network of queues

analyse whether the JLB-property can be valid or not, and

compute the stationary distribution if it is valid.

Further, since JLB is necessary for insensitivity of the stationary distribution with respect to the service distributions, we hereby also analyse whether this phenomenon may occur. For instance, we show that a strict priority model can only be insensitive under very special conditions.

In section 4 we give sufficient conditions only on the system characteristics such that the results of section 2 can be applied. Particularly, we

derive appropriate adjoint processes and give the consequences of checkin the Kolmogorov criterion for the adjoint transition rates. Further, for checking this criterion we give a decomposition of the adjoint rates in routing and service characteristics. These components will be separately analysed in sections 5 and 6.
We first consider a closed network. At the end of section 4 is shown how to derive analogous results for open models.

4. NETWORKS OF QUEUES WITH GENERALIZED ROUTING AND SERVICE CHARACTERISTIC

In this section we will determine the irreducible subsets and the adjoint transition rates as defined in section 2 for a closed network with a general parametrization. For doing so we do not make any of the assumptions I,II and III of section 2 a priori. However, we will derive conditions on the parametrization such that these assumptions are satisfied indeed.

Consider a closed network with N nodes and M jobs.

A jobmark of a jobs is a 3-tuple (j,r,s) where j resp. r resp. s denotes the node- resp. job- resp. service-number of the job.
The job-number of a job is never changed. If a job is accepted at a node then it will be assigned a service-number.

Two different protocols can be considered for changing the service-numbers of the jobs present at a node when some job leaves or enters that node.

In the constant-protocol a job does not change its service-number up to its next service completion, and several jobs may have a same service-number, for instance representing a type of server.

In the shift-protocol the service-numbers are place-numbers 1,...,x if x jobs are present, and these numbers shift after an arrival or departure as described in section 3 or [8].

The service characteristic at node j is determined by (f_j,φ_j,δ_j) as:

If n jobs are present at node j with pairs of job-and service-number $(r_1,s_1),\ldots,(r_n,s_n)$, then

(i) $f_j((r_1,s_1),\ldots,(r_n,s_n))$ is the total capacity at node j

(ii) $\varphi_j((r_1,s_1),\ldots,(r_n,s_n)|(r_i,s_i))$ is the fraction of this capacity given to the job with number-pair (r_i,s_i), $i = 1,\ldots,n$.

(iii) $\delta_j((r_1,s_1),\ldots,(r_n,s_n)|(r,s))$ is the probability that a job given it is accepted at node j and given it has a job-number r, is assigned a service-number s.

In order to specify what happens after a job has completed a service we introduce routing and blocking characteristics (p,b) as follows:

If a job with job-number r has completed a service at node i and if E denotes the set of jobmarks of the other jobs, then

$$\text{(iv)}\quad \begin{cases} p_{ij}(r;E) \text{ is the probability that it requests a new service at node j,} \\ b_{ij}(r;E) \text{ is the probability that this request at node j is accepted.} \end{cases}$$

Let $\bar{p}_{ij}(r;E)$ be the <u>general routing</u> which denotes that the job indeed routes from node i to j, then we have:

$$\text{(4.1)}\quad \begin{cases} \bar{p}_{ij}(r;E) = p_{ij}(r;E)b_{ij}(r;E), \text{ and} \\ \sum_{j=1}^{N} p_{ij}(r;E)[1-b_{ij}(r;E)] \text{ is the probability that it is "blocked".} \end{cases}$$

Since a strict waiting model can never satisfy JLB, see proposition 4.1, we assume the following <u>blocking-protocol</u>:

A job, say with jobmark (j,r,s), which has completed its service but which is "blocked", retains its jobmark i.e., remains at node j with service-number s, and gets a new service.

Note that such a blocking can not be modelled by rerouting if not $\delta_i(E_i|(r,s)) = \varphi_i(E_i+(r,s)|(r,s)) = 1$, where E_i denotes the jobs at node i.

First let us shortly illustrate some possibilities of the parametrization.

1. Strict priority model.

   Let $c_1, c_2, \dots c_t$ be a partition of the job-numbers, and let class c_{i_1} have priority of service over class c_{i_2} at node j if $i_1 < i_2$. Take as service numbers 1,2,...,t and define

   $\delta_j((r_1,s_1),\dots,(r_n,s_n)|(r_{n+1},s)) = 1 \quad$ if $r_{n+1} \in c_s$.

   $\varphi_j((r_1,s_1),\dots,(r_n,s_n)|(r_i,s_i)) = 0 \quad$ if $s_i > \min(s_1,\dots,s_n)$,

   i.e. all capacity is given to the jobs with highest priority.

2. State-dependent routing; Overflow

   The example of section 2 has a state dependent routing and can be given a parametrization as given in this section.

   A model with rerouting for overflow is given by, jobs route from node 1 to 3 via node 2; however if node 2 is saturated, say with n_2 jobs, then jobs arriving at node 2 are rerouted via node 2'. i.e.;

   $p_{12}(r;E) = 1$ resp. $p_{12'}(r;E) = 1$ if the number of jobs at node 2 is smaller resp. not smaller than n_2.

3. Blocking

   One example has already been given by the cyclic-three-centre model

   Another more natural blocking model occurs if

   $b_{ij}(r;E) = b_j((r_1,s_1),\dots,(r_n,s_n)|r)$ i.e.;

   the blocking probability only depends on the jobs present at the node to which the job requests.

The following assumptions are made (cf. section 3).

For each $j,(r_1,s_1),\dots,(r_n,s_n)$ and r holds,

(a) $f_j((r_1,s_1),\dots,(r_n,s_n)) > 0 \quad$ for $n \geq 1$.

(b) $\sum_s \delta_j((r_1,s_1),\dots,(r_n,s_n)| \quad (r,s)) = 1$

$$\sum_{i=1}^{n} \varphi_j((r_1,s_1),\dots,(r_n,s_n)|(r_i,s_i)) = 1$$

The service distributions are exponential with parameter $\mu_j(r)$ for a job with job-number r at node j, j = 1,...,N. Further, for any E we notate E_i for the set of number-pairs $(r_1,s_1),\ldots,(r_n,s_n)$, denoting the job- and service-numbers of the jobs present at node i.

First let us remark that since the service distributions are exponential assumption I of section 2 is satisfied for the constant protoçol. For the shift protocol it is also satisfied if interpreted up to reshifting.

As noted in section 3 the results of section 2 remain valid if we replace $q(E+\ell,E+\ell')$ resp. $\bar{q}(E+\ell,E+\ell')$ by $q(\ell',\ell';E)$ resp. $\bar{q}(\ell,\ell';E)$ to represent a transition rate from $E+\ell$ into $E+\ell'$ only due to a change of the jobmark ℓ in ℓ'.

In view of this we will use the latter notations $q(\ell,\ell';E)$ and $\bar{q}(\ell,\ell';E)$ Further, we will only notate for the constant protocol. The reader, however, can directly give analogous results and derivations for shift protocols.

As first step in analysing JLB for the network here introduced, we must determine the irreducible subsets $V(\ell;E)$, as introduced in section 2, with $\ell = (j,r,s)$. We therefore first analyse some conditions on the service-characteristics and the general routing which are necessary for JLB.

The transition rate from any $E+(j_1,r,s_1)$ into $E+(j_2,r,s_2)$ due to a transition of job r, is given by

(4.2) $$q((j_1,r,s_1),(j_2,r,s_2);E) =$$
$$\mu_{j_1}(r)\, f_{j_1}(E_{j_1}+(r,s_1))\,\varphi_{j_1}(E_{j_1}+(r,s_1)\,|\,(r,s_1))\,\bar{p}_{j_1j_2}(r;E)\,\delta_{j_2}(E_{j_2}\,|\,(r,s_2))$$

<u>Definition</u>: Job (j,r,s) is isolated on E if for all $E+(j',r,s')$

$$q((j,r,s),(j',r,s');E) = q((j',r,s'),(j,r,s);E) = 0.$$

Now we are able to give a first necessary condition for JLB.

Proposition 4.1 Let the JLB-property hold on some irreducible set V and let $E+(j,r,s) \in V$, then

$$(4.3) \quad \begin{cases} \varphi_j(E_j+(r,s)|(r,s)) > 0 \text{ and } \delta_j(E_j|(r,s)) > 0, \\ \text{or} \\ (j,r,s) \text{ is isolated on } E. \end{cases}$$

Proof. Suppose (j,r,s) is not isolated on E, then we have;

(i) $\exists j',s'$ such that $E+(j',r,s') \in V$ and

$$\varphi_{j'}(E_{j'}+(r,s')|(r,s'))\ \bar{p}_{j'j}(r;E)\ \delta_j(E_j|(r,s)) > 0,$$

and/or (ii) $\exists j',s'$ such that

$$\varphi_j(E_j+(r,s)|(r,s))\ \bar{p}_{jj'}(r;E)\ \delta_{j'}(E_{j'}|(r,s')) > 0.$$

However, according to relation (2.1) JLB implies : (i) $\Leftrightarrow$ (ii) .

Hence we must have, $\delta_j(E_j|(r,s)) > 0$ and $\varphi_j(E_j+(r,s)|(r,s)) > 0$. □

Since in a strict priority model as long as a higher priority job is present a job of lower priority receives no capacity i.e. $\varphi = 0$, prpposition 4.1 implies

Corollary 4.1 A strict priority model can only satisfy JLB if

(i) a lower priority job is not accepted at a node if a higher priority job is already present there, and

(ii) a lower priority job is only being served as long as a higher priority is not present at the same node.

Now consider the routing matrix $(\bar{p}_{ij}(r;E))$.

In general this matrix may be defective and may have

a set of transient nodes $T(r;E)$, and

irreducible subsets of regular nodes, say $R_i(r;E)$, $i = 1,\dots,m(r;E)$.

Hereby we make the convention also to call node j transient if $\bar{p}_{jj'}(r;E) = 0$ for all $j' = 1,\dots,N$. Further write

$R(j;(r;E)) = R_i(r;E)$ if $j \in R_i(r;E)$ for some i

$R(j;(r;E)) = \emptyset$ if $j \in T(r;E)$ and

$R(r;E) = \{j | j \in R_i(r;E)$ for some $i \in \{1,\dots,m(r,E)\}\}$

The next proposition shows the consequences of JLB for transient and regular nodes.

Therefore first remark that (j,r,s) is isolated on E if

$\bar{p}_{jj'}(r;E)\varphi_j(E_j+(r,s)|(r,s)) = \delta_j(E_j|(r,s))\ \bar{p}_{j'j}(r;E) = 0 \qquad \forall\ j'$

Proposition 4.2. Let the JLB-property hold on some irreducible set V, and let E + (j,r,s) $\in$ V, then

$$(4.4)\qquad \begin{cases} \varphi_j(E_j+(r,s)|(r,s)) > 0 \Leftrightarrow \delta_j(E_j|(r,s)) > 0 & \text{if } j \in R(r;E) \\ (j,r,s) \text{ is isolated on E} & \text{if } j \in T(r;E) \end{cases}$$

Proof. Let j $\in$ R(r;E) then for some $j_1,j_2 \in \{1,\dots,N\}$ $\bar{p}_{j_1 j}(r;E) > 0$ and $\bar{p}_{jj_2}(r;E) > 0$. Now the statement directly results by using (4.3) and the characterization of an isolated job given above. For j $\in$ T(r;E) the statement results from proposition 2.2. □

In view of JLB the propositions 4.1 and 4.2 justify the following assumption

(c) Condition (4.4) is satisfied for each E + (j,r,s).

Let $\bar{V}((j,r,s);E)$ denote the subset on $\{E+(j',r,s')|(j',s')\}$ which is irreducible w.r.t. $q(\cdot,\cdot;E)$ and which contains E+(j,r,s). Then under assumptions (a) - (c) we have

Proposition 4.3 For each E + (j,r,s)

$$\bar{V}((j,r,s);E) = \begin{cases} E + (j,r,s) \text{ if } j \in T(r;E) \quad \text{or} \quad \varphi_j(E_j + (r,s)|(r,s)) = 0 \\ \{E + (j',r,s')|\ j' \in R(j;(r;E)) \text{ and } \delta_{j'}(E_{j'}|(r,s')) > 0\} \text{ else} \end{cases}$$

Proof. By using assumption (c) first conclude that

for (j,r,s) not isolated on E we must have

$\varphi_j(E_j + (r,s)|(r,s)) > 0,\ \delta_j(E_j|(r,s)) > 0$ and $R(j;(r;E)) \neq \emptyset$.

Now let j' $\in$ R(j;(r,E)) and $\delta_{j'}(E_{j'}|(r,s')) > 0$, then assumptions (a), (b) and (c) yield:

$\exists(j_1,s_1),(j_2,s_2),\dots,(j_n,s_n)$ such that

$\bar{p}_{j\,j_1}(r;E)\ \bar{p}_{j_1 j_2}(r;E) \dots \bar{p}_{j_{n-1}j_n}(r;E)\ \bar{p}_{j_n j'}(r;E) > 0$, and

$\delta_{j_i}(E_{j_i}|(r,s_i)) > 0$ and $\varphi_{j_i}(E_{j_i}|(r,s_i)|(r,s_i)) > 0,\ i = 1,\dots,n.$

From this together with relation (4.2) we conclude:
$E + (j',r,s') \in \bar{V}((j,r,s);E)$. On the other hand, for any $E + (j',r,s') \in \bar{V}((j,r,s);E)$ we must necessarily have,

$$\delta_{j'}(E_{j'}|(r,s')) > 0 \quad \text{and} \quad j' \in R(j;(r;E)). \qquad \square$$

The next corollary shows that indeed an irreducible set V exists and that one can even choose any initial state.

<u>Corollary 4.3</u>. For any state $E(0)$ exists an irreducible set V with $E(0) \in V$.

<u>Proof</u>. A state can only change due to some jobmark at a time. Let $E(0)$ be transient, then for at least one jobmark $\ell \in E(0)$ it is also transient on $\{E(0)-\ell+\ell' | \ell' \in L\}$ w.r.t. $q(\ell,\ell';E(0)-\ell)$, so that $\bar{V}(\ell;E(0)-\ell) = \emptyset$. However, according proposition 4.3, $\bar{V}(\ell;E(0) - \ell) \neq \emptyset$. $\square$

From now on let V be some irreducible set. Then for $V((j,r,s);E)$ as defined in section 2 w.r.t. V we obtain
$V((j,r,s);E) = \bar{V}((j,r,s);E)$ if $E + (j,r,s) \in V$.

The system of linear equations (2.3), with omission of the normalizing condition, for the general network of queues as given in this section becomes,

$$(4.5): \quad \sum_{(j',r,s')} x(j,r,s) \quad f_j(E_j+(r,s))\, \varphi_j(E_j+(r,s)|(r,s))\, \mu_j(r)\, \bar{p}_{jj'}(r;E)\, \delta_{j'}(E_{j'}|(r,s'))$$

$$= \sum_{(j',r,s')} x(j',r,s')\, f_{j'}(E_{j'}+(r,s'))\, \varphi_{j'}(E_{j'}+(r,s')|(r,s'))\mu_{j}'(r)\, \bar{p}_{j'j}(r;E)\, \delta_j(E_j|(r,s)).$$

Note that the transition rates out of (j,r,s) as well as into (j,r,s) which correspond to a blocking are omitted in (4.5) since they give an equal contribution to the left and right hand side.

In order to derive appropriate adjoint transition rates $\bar{q}$, let us first suppose that the adjoint rates $\bar{q}$ are chosen such that,

$$(4.6)\quad \begin{cases} \bar{q}((j,r,s),(j',r,s');E) = \bar{q}((j',r,s'),(j,r,s);E) = 0 \quad \text{if} \\ q((j,r,s),(j',r,s');E) = q((j',r,s'),(j,r,s);E) = 0 \end{cases}$$

Then in verifying (2.8) or Kolmogorov's criterion (see [8] p. 21) for the adjoint transition rates $\bar{q}$ we can conclude;

we do not need to consider any cycle with a transition rate $\bar{q}((j',r,s'),(j,r,s);E)$ or $\bar{q}((j,r,s),(j',r,s');E)$ if (j,r,s) is isolated on E.

Proposition (4.3) on the other hand shows that for solving (4.5) on $V((j,r,s);E)$ we do not have to concern transient nodes of $(\bar{p}_{ij}(r;E))$.

From the Perron Frobenius theorem we know that for each i, i = 1,...,m (r;E) exists a unique normalized solution $\lambda_j(r;E)$ on $R_i(r;E)$ of,

$$(4.7)\quad x(j) \sum_{j'\in R_i(r;E)} \bar{p}_{jj'}(r;E) = \sum_{j'\in R_i(r;E)} x(j')\, \bar{p}_{j'j}(r;E)$$

If the routing matrix $(\bar{p}_{ij}(r;E))$ does not depend on E, then $\lambda_j = \lambda_j(r;E)$ is usually called the throughput at node j. We therefore call $\lambda_j(r;E)$ the <u>local throughput</u> at node j on E.

Now under assumptions (a) - (c), proposition 4.3 and relation (4.7), it is easily checked that the solution of (4.5) on $V((j,r,s);E)$ is given by

(4.8):

$$x(j,r,s) = \begin{cases} 1 \quad \text{if} \quad (j,r,s) \text{ is isolated} \\ c\,\dfrac{\lambda_j(r;E)\delta_j(E_j|(r,s))}{\mu_j(r)\ f_j(E_j+(r,s))\varphi_j(E_j+(r,s)|(r,s))} \quad \text{else} \end{cases}$$

where $\lambda_j(r;E)$ is the normalized solution of (4.7)

for $R_i(r;E) = R(j;(r\ ;E))$ and

where c is a normalizing constant on $V((j,r,s);E)$.

To exclude the case that c = 0 if an infinite number of service-numbers can be chosen, the following assumption is made,

(d) $\sum_s \delta_j(E_j|(r,s))/[f_j(E_j+(r,s))\ \varphi_j(E_j+(r,s)|(r,s))] < \infty$

for any r, j and E_j.

Now note that under assumptions (a) - (d), we have shown by proposition (4.3) together with (4.8) that assumption III of section 2 is satisfied.

Relations (4.8) and (2.6) imply that adjoint rates-$\bar{q}$ can be defined by

(4.9) $\bar{q}((j_1,r,s_1),(j_2,r,s_2);E) =$

$\lambda_{j_2}(r;E)\ \delta_{j_2}(E_{j_2}|(r,s_2))$

$\mu_{j_1}(r)\ f_{j_1}(E_{j_1}+(r,s_1))\ \varphi_{j_1}(E_{j_1}+(r,s_1)|(r,s_1))$ for all

$E+(j_2,r,s_2) \in V((j_1,r,s_1);E)$ and $\bar{q}(E(1),E(2)) = 0$ else.

First note that condition (4.6) indeed is satisfied.
Further recollect that an irreducible set V exists and that assumptions I and III of section 2 are satisfied.

It can also be shown that any stationary distribution on V has to satisfy (2.1) by making the following natural assumption:

(e) $\sup_s [f_j(E+(r,s))] < \infty$ for any E, j and r.

Let us consider the $\bar{q}$-rates defined by (4.9).
By using the results of section 2 and assumption (e) we can now make the following conclusions for our network of queues:

i) if a stationary distribution exists then it satisfies JLB if and only if the $\bar{q}$-rates satisfy the K-criterion.

ii) if the $\bar{q}$-rates satisfy the K-criterion on V then π as defined in (2.8) is the unique stationary distribution on V, if

(4.11) the normalizing constant c on V in (2.8) is positive.

Consequently, by checking the K-criterion for the $\bar{q}$-rates we not only analyse whether JLB holds or not, but also in case it holds we can check the existence of a unique stationary distribution and give its computation.

By combining (4.6) and proposition (4.3) we can conclude:

For checking the K-criterion for the $\bar{q}$-rates defined by (4.9) it suffices only to consider cycles with jobmarks changing in non-isolated irreducible subsets of the form $V((j,r,s);E)$. The corresponding subsets of the general routing $\bar{p}$ are irreducible subsets of the form $R(j;(r;E))$.

This conclusion together with (4.5) implies that the K-criterion holds if the K-criterion holds for $\bar{q}_1$-rates on V_R and for $\bar{q}_2$-rates on V, where

$$(4.12)\quad \begin{cases} \bar{q}_1((j_1,r,s_1),(j_2,r,s_2);E) = \lambda_{j_2}(r;E), \\ \bar{q}_2((j_1,r,s_1),(j_2,r,s_2);E) = f_{j_1}(E_{j_1}+(r,s_1))\,\varphi_{j_1}(E_{j_1}+(r,s_1)\,|\,(r,s_1)) \\ \qquad\qquad \mu_{j_1}(r)\;\delta_{j_2}(E_{j_2}\,|\,(r,s_2)) \end{cases}$$

and $V_R = \{E \mid E \in V$ and for some $(j,r,s) \in E$ and $i : j \in R_i(r;E-(j,r,s))\}$.

Conversely, the $\bar{q}$-rates do not satisfy the K-criterion if only one of the $\bar{q}_1$- or $\bar{q}_2$-rates satisfies the K-criterion.

One can devise pathological examples such that the $\bar{q}$-rates satisfy the K-criterion whereas neither the $\bar{q}_1$- nor $\bar{q}_2$-rate does. However, in realistic models this seldom occurs since mostly the $\bar{q}_1$-rates only depend on the number of jobs of various classes present in the nodes, say the "workloads", and the $\bar{q}_2$-reates have only to do with the priorities and sharing within the nodes.

Further, in checking the K-criterion for the $\bar{q}_1$-rates we always implicitly assume only to focus on V_R .

In separately analysing the $\bar{q}_1$- and $\bar{q}_2$-rates one can interpret

> $\bar{q}_1$-rates represent adjoint rates of a network with a general routing as given, but with characteristics representing infinite server queues i.e.,
>
> for any j,r,x and $E_j = \{(r_1,1),(r_2,2),\dots,(r_x,x)\} : f_j(E_j) = x$
>
> $\varphi_j(E_j|(r_i,i)) = 1/x \quad i = 1,\dots,x \qquad \delta_j(E_j|(r,i)) = 1/x+1 \quad i = 1,\dots,x+1.$
>
> $\bar{q}_2$-rates represent adjoint rates of a network with service characteristics as given, but with a symmetric general routing i.e.; $\bar{p}_{ij}(r;E) = 1/N \quad \forall i,j,r,E.$

For open networks analogous results can be obtained. One way of showing this is by using a limiting argument for closed networks with M tending to infinity, see [4] and [5].

However, it can also be shown directly in the setting of this section by using the following interpretation and notation :

> Suppose that jobs of a class r arrive at the network according a Poisson process with parameter $\gamma_0(r)$, $r = 1,\dots,M$.
>
> A job of class r always has job-number r.
>
> If the jobmarks of jobs present are denoted by E and a job of class r arrives, then $p_{0j}(r;E)$ resp. $b_{0j}(r;E)$ is the probability that it requests resp. is accepted at node j. A blocked job does not enter the network. Analogously define $p_{j0}(r;E)$ resp. $b_{j0}(r;E)$.

Now introduce a pseudo node 0, a fixed state E_0, jobmarks (0,r,0), and

$$\delta_0(E_0|(r,0)) = 1$$

$$f_0(E_0+(r,0))\ \varphi_0(E_0+(r,0)|(r,0)) = 1 \quad \text{and} \quad \mu_0(r) = \gamma_0(r).$$

As stated in section 2, the results of section 2 remain valid for open models Hence, with the notation given above we can rewrite all results of this section with always taking into account jobmark (0,r,0) for the q- and $\bar{q}$-rates and pseudo node 0 for the general routing $\bar{p}$.

5. LOCAL THROUGHPUTS; K_1-CRITERION

In this section only the $\bar{q}_1$-rates will be focussed on.

Therefore let us first reconsider condition (4.4).

Suppose that any job accepted at a node will always be served, i.e.; $\varphi_j(E_j+(r,s)|(r,s)) > 0 \;\forall\, j,r,s,E$ such that $E+(j,r,s) \in V$, and assume

(5.1.1) $\delta_j(E_j|(r,s)) > 0 \Leftrightarrow \varphi_j(E_j+(r,s)|(r,s)) > 0 \quad \forall\, j,r,s,E.$

Then condition (4.4) requires;

$$\begin{aligned} &\forall E+(j,r,s) \in V \\ (5.1.2)\quad &\exists E+(j',r,s') \in V \quad \text{with} \quad \bar{p}_{j'j}(r;E) > 0 \Leftrightarrow \\ &\exists E+(j',r,s') \in V \quad \text{with} \quad \bar{p}_{jj'}(r;E) > 0 \end{aligned}$$

i.e. $\bar{p}$ does not have any transient state on V.

On the other hand one directly checks that (5.1.1) and (5.1.2) imply (4.4). Further the (weak) irreducibility condition (5.1.2) is a realistic necessary condition for (4.4) if the routing and blocking do not explicitly depend on service-numbers.

In view of these arguments and section 4 we define the

<u>K_1-criterion</u>: The $\bar{q}_1$-rates satisfy the Kolmogorov-criterion and (5.1.2) holds.

Let us first give two positive results for models with state-independent routing. In view of section 4 these results more or less explain the JLB-property for

Jackson-type networks (cf. [7], [9]), and

networks with blocking but reversible routing (cf. [9], [10]).

Therefore let $n_j(E)$ denote the number of jobs of E at node j and introduce the following conditions:

(I): The routing is state-independent and irreducible, i.e.;

$p_{ij}(r;E) = p_{ij}(r) = \forall i,j,r,E$ and $\forall r$

exists a unique normalized solution $\lambda_j(r)$, $j = 1,\ldots,N$ of

(5.2.1) $\lambda_j(r) = \sum_{i=1}^{N} \lambda_i(r)\, p_{ij}(r)$ $\qquad j = 1,\ldots,N.$

(R): As (I) and $(p_{ij}(r))_{i,j=1,\ldots,N}$ is reversible $\forall$ r i.e.;

(5.2.2) $\lambda_j(r)\, p_{ij}(r) = \lambda_i(r)\, p_{ij}(r)$ $\qquad \forall$ i,j,r.

Proposition 5.1. Let condition (I) be satisfied, then

(i) The K_1-criterion holds and $\lambda_j(r;E) = \lambda_j(r)$ $\forall$ j,r,E if

there is no blocking i.e.; $b_{ij}(r;E) = 1$ $\forall$ i,j,r,E

(ii) The K_1-criterion holds and $\lambda_j(r,E) = \lambda_j(r)\, b_j(n_j)$ if

the reversibility condition (R) holds and

$b_{ij}(r;E) = b_j(n_j)$ if $n_j(E) = n_j$.

Proof.

(i) System (4.5) for this model becomes (5.2.1). Relation (5.1.2) directly results from (I) and $b_{ij}(r;E) = 1$ $\forall$i,j,r,E.

The Kolmogorov-criterion is satisfied since $\lambda_j(r;E) = \lambda_j(r)$ $\forall$ j,r,E,; that means state-independent.

(ii) Again by system (4.5) together with the uniqueness on irreducible subsets for $\bar{p}_{ij}(r;E)$ we conclude that on V_R, defined in section 4, $\lambda_j(r;E) = c\, \lambda_j(r)\, b_j(n_j)$, with c a normalizing constant on the irreducible subset. Hereby one can easily verify the Kolmogorov criterion on V_R. Condition (5.1.2) results by that if

$\bar{p}_{ij}(r;E) = p_{ij}(r)\, b_j(n_j(E)) > 0$ for $E+(i,r,s) \in V$ and $E+(j,r,s') \in V$

then: $p_{ji}(r) > 0$ according to (R), and $b_i(n_i(E)) > 0$

since either $i = j$ and $b_j(n_j(E)) > 0$ or $i \neq j$ and on V

E + (i,r,s) can be reached out of E+(j,r,s').

Hence, $\bar{p}_{ji}(r;E) = p_{ji}(r)\, b_i(n_i(E)) > 0$. □

Next we shall analyse whether the conditions of (ii) are necessary for the K_1-criterion.
The following example illustrates that with the blocking as given in (ii), hence only depending on the number of jobs at a node,
the reversibility of the routing is almost necessary for the K_1-criterion.

Example 5.1. Consider the cyclic-three-centre model with

$$b_{ij}(r;E) = b_j(n_j(E)) \text{ and } b_j(n_j) = 0 \text{ if } n_j \geq N_j \text{ and } 1 \text{ else.}$$

Further, we assume service-characteristics satisfying (a),(b),(d),(e) and (5.1.1.), for instance corresponding to infinite server queues.
Let $N_j < M$, $j = 1,2,3$ and $N_1 + N_2 + N_3 > M$
Using assumption (b) and corollary 4.3 one can prove the existence of an irreducible set V containing states
$E+(1,r,s_1)$, $E+(2,r,s_2)$ and $E+(3,r,s_3)$ for some r,s_1,s_2,s_3
and fixed E with $n_1(E) \leq N_1-1$ and $n_2(E) = N_2$.

One easily checks that $\bar{p}_{1j}(r;E) = 0$ for $j = 2,3$ but $\bar{p}_{31}(r;E) = 1$.
Hence, node 3 is transient
for $\bar{p}(r;E)$ and the K_1-criterion fails in condition (5.1.2).

Heuristically, this results by that if node 2 is saturated then at node 1 jobs can only enter but not leave. □

More generally, in [5] we have shown the necessity of (R) for an upper-limit-blocking together with a product-form and JLB.

Conversely, even under (R) the K_1-criterion is not satisfied if the blocking at a node for jobs of one class strictly depends on jobs of another class.
This is shown by

Example 5.2. Consider 2 nodes with cyclic routing. $p_{12}(r;E) = p_{21}(r;E) = 1$.
$\forall r,E$. The jobs are partitioned into two classes C_1 and C_2.
Let $n_j^i(E)$ denote the number of jobs of E of class C_i at node j, then

$b_{12}(r;E) = 0$ if $r \in C_1$ and $n_2^2(E) \geq N_2$,

$b_{ij}(r;E) = 1$ else.

Further, let service-characteristics be given as in example 5.1 and $|C_2| \geq N_2$.

Again, by using assumption (b) and corollary 4.3 one can prove the existence of an irreducible set V containing states $E+(1,r,s_1)$ and $E+(2,r,s_2)$ for some r,s_1,s_2 and fixed E such that $r \in C_1$ and $n_2^2(E) = N_2$.

Then one easily checks $\bar{p}_{12}(r;E) = 0$, but $\bar{p}_{21}(r;E) = 1$, if $r \in C_1$.

Hence, node 2 is transient for $\bar{p}(r;E)$ if $r \in C_1$, and the K_1-criterion is not satisfied.

Heuristically, if node 2 contains too many jobs of class C_2 then jobs of class 1 can only leave but not enter node 2. □

So even under a reversible routing JLB gives restrictions on blocking. Nevertheless, some extension of (ii) can be shown.

Therefore consider the blocking of the form:

$$(5.3)\quad \begin{cases} b_{ij}(r;E) = b_i^{out}(r|E_i)\, b_j^{in}(r|E_j) & \text{such that} \\ b_i^{out}(r|E_i) = 0 \Leftrightarrow b_i^{in}(r|E_i) = 0 & \text{if } \exists E + (i,r,s) \in V, \\ b_i^{out}(r|E_i - (r,s)) > 0 \text{ for some } (r,s) \in E_i & \forall i \text{ and } E_i. \end{cases}$$

Notate $b_i(r|E_i) = b_i^{in}(r|E_i)/b_i^{out}(r|E_i)$ if $b_i^{out}(r|E_i) > 0$

For this blocking we have,

<u>Proposition 5.2</u> Let the reversibility condition (R) hold and let for any j, n and $E_j = \{(r_1,s_1),\ldots,(r_n,s_n)\}$

$$(5.4)\quad \prod_{k=0}^{n-1} b_j(r_{j_{k+1}}|(r_{j_1},s_{j_1}),\ldots,(r_{j_k},s_{j_k}))\, \lambda_j(r_{j_{k+1}}) \text{ be invariant}$$

for all permutations $(j_1,\ldots,j_n) \in (1,\ldots,n)$ for which it is defined, then the K_1-criterion holds and the process with $\bar{q}_1$-rates is reversible with stationary distribution on V;

(5.5) $$\pi_1(E) = c \prod_{j=1}^{N} \bar{\pi}_j(E_j), \text{ with } \bar{\pi}_j(E_j) \text{ defined by (5.4).}$$

Proof. First let us show condition 5.1.2. From (R) and (5.3) we conclude for any $E + (i,r,s) \in V$ and $E + (j,r,s') \in V$:

$$\bar{p}_{ij}(r;E) = p_{ij}(r)\ b_i^{out}(r|E_i)\ b_j^{in}(r|E_j) > 0, \quad \Leftrightarrow$$

$$\bar{p}_{ji}(r;E) = p_{ji}(r)\ b_i^{in}(r|E_i)\ b_j^{out}(r|E_j) > 0.$$ This shows (5.1.2).

From system (4.5) and (R) we now derive $\forall i,r,E$

$\lambda_i(r;E) = \lambda_i(r)\ b_i(r|E_i)\, c$ if $b_i^{out}(r|E_i) > 0$ and 1 else, c a constant

For the process with $\bar{q}_1$-rates the only positive transition rates are given by

$\bar{q}_1((j_1,r,s_1),(j_2,r,s_2);E) > 0$ if $E + (j_2,r,s_2) \in V((j_1,r,s_1);E)$.

Now first note that according to (5.3) for at least one permutation the product in (5.4) is positive. With $\lambda_i(r;E)$ as given above and by choosing appropriate permutations one can show:

$$\pi_1(E + (j_1,r,s_1))\ \lambda_{j_2}(r;E) \quad = \quad \pi_1(E + (j_2,r,s_2))\ \lambda_{j_1}(r;E)$$

for any $E + (j_2,r,s_2) \in V((j_1,r,s_1);E)$.
By expression (4.12) for $\bar{q}_1$ and the remark on positiveness given above, this shows the reversibility with stationary distribution π_1.
Hence, the $\bar{q}_1$-rates satisfy the Kolmogorov-criterion which together with (5.1.2) proves the K_1-criterion. □

Further, the blocking-invariance property (5.4) has interpretation:

"It does not matter in which order jobs arrive at and can leave a node j if we consider $b_j^{in}(\cdot|\cdot)$ as arrival and $b_j^{out}(\cdot|\cdot)$ as departure rates."

The next corollary and example shows that (5.4) allows for blocking only depending on the jobs of its own class, or blocking which may depend on jobs of another class by special coupling.

<u>Corollary</u>. Let jobs be partitioned to job-numbers in classes $C_1,\dots,C_t$ and for E let $n_j^i(E)$ denote the number of jobs of class C_i at node j,

then the invariance-condition (5.4) holds if $\forall$ E+(j,r,s) $\in$ V

$$b_j^{in}(r|E_j) = b_j^k(n_j^k(E_j)) \quad \text{for } r \in C_k \quad \text{and } b_j^{out}(r|E_j) = 1.$$

<u>Proof</u>. First note that (5.3) is satisfied even if

$b_j^k(N_j^k) = 0$ since then $\exists$ no state E $\in$ V with $n_j^k(E_j) > N_j^k$.

The proof now results by that a product (5.4) always factorizes to the different classes as

$$(5.6) \qquad \bar{\pi}_j(E_j) = \prod_{k=1}^{t} \prod_{i=1}^{n_j^k(E_j)} b_j^k(i-1) \qquad \square$$

Below we illustrate that although a strict dependence of classes as in example 5.2 will not yield JLB a special dependence will do.

<u>Example 5.3</u>. Reconsider example 5.2 by changing only the blocking as;

$b_{ij}(r;E)$ satisfies (5.3) with

$b_2^{in}(r|E_2) = 0$ if $r \in C_1$ and $n_2^2(E_2) \geq N_2$ and

$b_2^{out}(r|E_2) = 0$ if $r \in C_1$ and $n_2^2(E_2) \geq N_2$.

$b_{ij}(r;E) = 1$ else.

Verification of condition (5.4) for node 1 and node 2 is direct. Since also (R) holds the conditions of proposition 5.2 are satisfied. Note that as in corollary 4.3 we have <u>pseudo-priority</u> by that

Jobs of class 1 are only accepted

but also only served

if the number of jobs of class 2 is not too large. $\square$

If the routing is not reversible and blocking is involved then, in general, we can not characterize when the K_1-criterion holds.
To conclude this section we give positive results for two specific models, the cyclic-three-centre model and an overflow model.

Example 5.4. The cyclic-three-centre model described in section 1 or [5] can be modelled in the notation of section 4 with

$$\bar{p}_{ij}(r;E) = \begin{cases} b_j(\bar{n} + e_i) & \text{for } \bar{n} = (n_1,n_2,n_3) \text{ with } n_j = n_j(E), \\ & j = [i+1] \text{ mod3}. \\ 0 \text{ else} & \end{cases}$$

The system (4.7) here becomes, with i+1 and i-1 taken mod3; $\forall i,r,E$

$$\lambda_i(r;E)\ b_{i+1}(\bar{n}+e_i) = \lambda_{i-1}(r;E)\ b_i(\bar{n}+e_{i-1}) \quad \text{with } \bar{n} = \bar{n}(E).$$

As in [5] one can prove under a "convexity" assumption for V:

The K_1-criterion holds $\Leftrightarrow$ Relation(3.1)is satisfied.

Example 5.5. Consider a network with nodes 1,2 and 2'.
The routing and blocking probabilities are given by ;

if $n_2(E) < N_2$ then $\bar{p}_{1,2}(r;E) = \bar{p}_{2,1}(r;E) = 1$ but $\bar{p}_{2',j}(r;E) = 0 \quad j = 1,2$

if $n_2(E) = N_2$ then $\bar{p}_{1,2'}(r;E) = 1$ and $\bar{p}_{2',1}(r;E) = 1$

Remark that a job at node 2' is isolated on E if $n_2(E) < N_2$

Further, system (4.7) here gives:

$\lambda_i(r;E) = 1$ for $i = 1,2$ if $n_2(E) < N_2$ and

$\lambda_i(r;E) = 1$ for $i = 1,2'$ if $n_2(E) = N_2$.

Now one can conclude that the K_1-criterion is satisfied either by verifying the Kolmogorov criterion or by directly showing reversibility of the $\bar{q}_1$-process with stationary distribution $\pi_1(E) = c$ on V. □

6. SERVICE CHARACTERISTICS; K_2-CRITERION

In this section only the $\bar{q}_2$-rates, given by (4.12), will be analysed. As in section 5, by reconsidering proposition 4.2, we here conclude the necessity of (5.1.1) for JLB if the routing satisfies (5.1.2). Also recall the sufficiency of (5.1.1) with (5.1.2) for condition (4.4). In view of these facts and section 4 we define the

K_2-criterion: The $\bar{q}_2$-rates satisfy the Kolmogorov criterion and

(5.1.1): $\qquad \varphi_j(E_j + (r,s)\,|(r,s)) > 0 \Leftrightarrow \delta_j(E_j\,|(r,s)) > 0 \quad \forall j,r,s \text{ and } E_j.$

First consider the constant protocol. Then resembling to proposition 5.2 the following proposition shows an invariance-condition on the service-characteristics which is sufficient for the K_2-criterion.

Proposition 6.1. Suppose that for any i,n and $E_i = \{(r_1,s_1),\dots,(r_n,s_n)\}$

$$(6.1)\quad \prod_{k=1}^{n} \frac{\delta_i((r_{j_1},s_{j_1}),\dots,(r_{j_{k-1}},s_{j_{k-1}})\,|(r_{j_k},s_{j_k}))}{f_i((r_{j_1},s_{j_1}),\dots,(r_{j_k},s_{j_k}))\,\varphi_i((r_{j_1},s_{j_1}),\dots,(r_{j_k},s_{j_k})\,|(r_{j_k},s_{j_k}))\,\mu_i(r_{j_k})}$$

is invariant for all permutations $(j_1,\dots,j_n) \in (1,\dots,n)$ for which it is positive, then

the K_2-criterion holds and the process with $\bar{q}_2$-rates is reversible with stationary distribution on V:

$$(6.2)\quad \pi_2(E) = c \prod_{i=1}^{N} \bar{\bar{\pi}}_i(E_i), \text{ with } \bar{\bar{\pi}}_i(E_i) \text{ defined by (6.1).}$$

The condition of this proposition at node i is called in accordance with [4] :
The service-invariance-property (SIP) at node i .
It can be interpreted by :
It does not matter in which order jobs
are accepted at and can leave from node i, if we interpret
$\delta_i(\cdot|\cdot)$ as arrival - and $f_i(\cdot)\,\varphi_i(\cdot|\cdot)\,\mu_i(\cdot)$ as departure rates.

Proof of proposition 6.1. First remark that according to (a), (b) and (5.1.1) for at least one permutation all denominators and numerators in (6.1) are positive. Hence, by using the invariance properties at the nodes, one can indeed consistently define positive probabilities $\pi_2(E)$ as given by (6.2)

Consider states $E + (j_1,r,s_1) \in V$ and $E + (j_2,r,s_2) \in V$. Then,

either by using (5.1.1)

or by taking appropriate permutations can be proven

$$(6.3)\quad \begin{cases} \pi_2(E + (j_1,r,s_1))f_{j_1}(E_{j_1}+(r,s_1))p_{j_1}(E_{j_1}+(r,s_1)|(r,s_1))\mu_{j_1}(r)\delta_{j_2}(E_{j_2}|(r,s_2)) \\ = \\ \pi_2(E + (j_2,r,s_2))f_{j_2}(E_{j_2}+(r,s_2))p_{j_2}(E_{j_2}+(r,s_2)|(r,s_2))\mu_{j_2}(r)\delta_{j_1}(E_{j_1}|(r,s_1)). \end{cases}$$

Further, for the process with $\bar{q}_2$-rates the only positive transition rates are given by $\bar{q}_2((j,r,s),(j',r,s');E) > 0$ if $E+(j',r,s') \in V((j,r,s);E)$. With this remark, expression (4.12) for $\bar{q}_2$-rates and relation (6.3) we can conclude the reversibility of the $\bar{q}_2$-process with stationary distribution π_2. This yields the Kolmogorov criterion. □

For shift protocols the proposition remains valid if in (6.1) we replace $(s_{j_1},\ldots,s_{j_n})$ by $(\tilde{s}_{j_1},\ldots,\tilde{s}_{j_n})$ where,

$\tilde{s}_{j_k}$ is deduced from $\{(r_1,s_1),\ldots,(r_n,s_n)\}$ and $(j_1,\ldots,j_n)$ by:
If the job with job-number r_{j_k} is a k-th arriving job at the node and is assigned service-number $\tilde{s}_{j_k}$ at his arrival, $k = 1,\ldots,n$, then after the arrivals of $r_{j_1},\ldots,r_{j_n}$ the shifted service-number of the job with job-number r_k is given by s_k, $k = 1,\ldots,n$.

In the rest of this section we only give the results for constant protocols. However, these results can be given analogously for shift protocols, if shifting is taken into account.

Let us analyse the necessity of the invariance-property for JLB. In the following counter-example the invariance-property does not hold at some node, whereas the K_2-criterion is still satisfied, or equivalently the $\bar{q}_2$-process is still reversible.

Counter-example 6.1. Consider a network with 2 nodes and 2 jobs having job-numbers r_1, r_2. The service-numbers are s_1, s_2 and $\bar{s}_2$. The service-characteristics are given by

$$\delta_i(\ |(r_j,s_j)) = \varphi_i((r_j,s_j)|(r_j,s_j)) = 1 \qquad i = 1,2 \ \text{ and } \ j = 1,2$$

$$f_i((r_j,s_j)) = 1 \ \text{ if } \ (i,j) \neq (1,2) \ \text{ and } \ f_1((r_2,s_2)) = 2.$$

$$\delta_1((r_2,s_2)|(r_1,s_1)) = 1 \qquad \varphi_1((r_1,s_1),(r_2,s_2)|(r_j,s_j)) = \tfrac{1}{2} \quad j = 1,2$$
$$\delta_1((r_1,s_1)|(r_2,s_2)) = 1 \qquad f_1((r_1,s_1),(r_2,s_2)) = 3$$

$$\delta_2((r_1,s_1)|(r_2,\bar{s}_2)) = 1 \qquad \varphi_2((r_1,s_1),(r_2,\bar{s}_2)|(r_2,\bar{s}_2)) = 1$$
$$\delta_2((r_2,s_2)|(r_1,s_1)) = 1 \qquad \varphi_2((r_1,s_1),(r_2,s_2)|(r_1,s_1)) = 1$$

$$f_2((r_1,s_1),(r_2,\bar{s}_2)) = f_2((r_1,s_1),(r_2,s_2)) = 1.$$

The invariance-property fails at node 1 since for $(r_1,s_1),(r_2,s_2)$ the two possible products in (6.1) are equal to $\frac{2}{3}$ resp. $\frac{1}{3}$.
The invariance-property holds at node 2 since $(r_1,s_1),(r_2,s_2)$ resp. $(r_1,s_1),(r_2,\bar{s}_2)$ can only be reached in one way such that (6.1) is positive.

Out of $E(0) = \{(1,r_1,s_1),(1,r_2,s_2)\}$ one can only reach and return from either

$$E_1(1) = \{(2,r_1,s_1),(1,r_2,s_2)\} \quad \text{and} \quad E_1(2) = \{(2,r_1,s_1),(2,r_2,\bar{s}_2)\} \quad \text{or}$$
$$E_2(1) = \{(1,r_1,s_1),(2,r_2,s_2)\} \quad \text{and} \quad E_2(2) = \{(2,r_1,s_1),(2,r_2,s_2)\}.$$

One directly checks that the $\bar{q}_2$-process on $V = \{E(0);\ E_1(i),\ i = 1,2; E_2(i), i=1,2\}$ is reversible with stationary probabilities

$$\pi_1[E_1(1)] = 3/2\ \pi[E(0)] \qquad \pi_1[E_1(2)] = 3\ \pi_1[E(0)]$$
$$\pi_1[E_2(i)] = 3/2\ \pi[E(0)] \quad i = 1,2 \ \text{ and } \ \pi_1[E(0)] = 2/17$$

□

In the proposition below we only give simple conditions under which the invariance property is necessary for the K_2-criterion or equivalently JLB for the service-process. Obviously several extensions of these conditions can be given.

Therefore write

$$\bar{\delta}_i(E_i|(r,s)) = \frac{\delta_i(E_i|(r,s))}{f_i(E+(r,s))\ \varphi_i(E_i+(r,s)|(r,s))\mu_i(r)} \quad \text{if positive}$$

Proposition 6.2. Let the K_2-criterion be satisfied. Then we have,

(i) The SIP holds for each node for ℓ jobs with $\ell \le K$ if $N = K+1$, with N the number of nodes

i.e. For $E \in V$ with $E_i = \{(r_1,s_1),\ldots,(r_n,s_n)\}$ any $\ell \le K$ and fixed $\bar{j}_1,\ldots,\bar{j}_{n-\ell} \in \{1,\ldots,n\}$ with $\bar{j}_s \ne \bar{j}_{s'}$ if $s \ne s'$:

the products in (6.1) if positive are invariant

for all permutations $(\bar{j}_1,\ldots,\bar{j}_{n-\ell}, j_{n-\ell+1},\ldots,j_n) \in (1,\ldots,n)$

(ii) The SIP holds for each node if for some node i' and any

$E \in V$ and $E_{i'} = \{(r_1,s_1),\ldots,(r_n,s_n)\}$.

the products in (6.1) for $i = i'$ are positive and invariant

for all permutations $(j_1,\ldots,j_n) \in (1,\ldots,n)$.

(iii) The SIP holds for each node if for some node i'

the service-characteristic at node i' does not depend on job-numbers,

i.e. $\delta_{i'}(r_1,s_1),\ldots,(r_{k-1},s_{k-1})|(r_k,s_k)) = g_{i'}(s_1,\ldots,s_{k-1}|s_k)$

$\forall\ (r_1,s_1),\ldots,(r_k,s_k)$ and some function $g_{i'}$

and

the stationary distribution π_2 of the service process does not depend on the job-numbers of the jobs at node i',

i.e. $\pi(E) = \pi(\tilde{E})$ for $E,\tilde{E} \in V$ if $\tilde{E}_j = E_j$ for $j \ne i'$ and

$E_{i'} = \{(r_1,s_1),\ldots,(r_n,s_n)\}$

$\tilde{E}_{i'} = \{(r_{j_1},s_1),\ldots,(r_{j_n},s_n)\}$ for some $(j_1,\ldots,j_n) \in (1,\ldots,n)$.

Proof. First remark that the K_2-criterion requires for fixed $E(0)$, $E \in V$

$$(6.4)\qquad \frac{\bar{q}_2(E(0),E(1))}{\bar{q}_2(E(1),E(0))}\,\frac{\bar{q}_2(E(1),E(2))}{\bar{q}_2(E(2),E(1))}\cdot\ \ldots\ \cdot\frac{\bar{q}_2(E(m),E)}{\bar{q}_2(E,E(m))} = c(E(0),E)$$

for any E(1),...,E(m) for which all denominators are positive, and where $\bar{q}_2(E,\tilde{E})$ is defined by (4.12). Further by (4.12):

$$\text{(6.5)}\qquad \frac{\bar{q}_2((j_1,r,s_1),(j_2,r,s_2);\tilde{E})}{\bar{q}_2((j_2,r,s_2),(j_1,r,s_1);\tilde{E})} = \frac{\bar{\delta}_{j_2}(\tilde{E}_{j_2}|(r,s_2))}{\bar{\delta}_{j_1}(\tilde{E}_{j_1}|(r,s_1))} \qquad \text{if positive.}$$

(i) Let $E \in V$ and consider $E_i = \{(r_1,s_1),\ldots,(r_n,s_n)\}$ and $\ell \le K$. Say $i = K+1$. Take some permutation $(\bar{j}_1,\ldots,\bar{j}_{n-\ell},j_{n-\ell+1},\ldots,j_n) = (j_1,\ldots,j_n)$ such that the product in (6.1) is positive.
Say $(r_{j_{n-\ell+1}},s_{j_{n-\ell+1}}),\ldots,(r_{j_n},s_{j_n}) = (\tilde{r}_1,\tilde{s}_1),\ldots,(\tilde{r}_\ell,\tilde{s}_\ell)$.

With assumption (b) one can prove
$\exists(s'_1,\ldots,s'_\ell)$ and E(0) such that

$E_k(0) = \{(r_{\bar{j}_1},s_{\bar{j}_1}),\ldots,(r_{\bar{j}_{n-\ell}},s_{\bar{j}_{n-\ell}})\}$ $\qquad k = K+1$
$E_k(0) = E_k + (\tilde{r}_k,s'_k)$ and $\delta_k(E_k(0)|(\tilde{r}_k,s'_k)) > 0$, $k = 1,\ldots,\ell$.

Hence $E(0) \in V$. Now starting from E(0) we let successively for $k = 1,..,\ell$ change jobmark $(k,\tilde{r}_k,s'_k)$ in $(K+1,\tilde{r}_k,\tilde{s}_k)$.
By taking the corresponding quotients in (6.4) as given by (6.5) and repeating this procedure for any permutation as given yields the result.

(ii) Let $E \in V$ and $E_i = \{(r_1,s_1),\ldots,(r_n,s_n)\}$ with $i \neq i'$.
Let $(j_1,\ldots,j_n) \in (1,\ldots,n)$ such that (6.1) is positive, and say
$(r_{j_1},s_{j_1}),\ldots,(r_{j_n},s_{j_n}) = (\tilde{r}_1,\tilde{s}_1,),\ldots,(\tilde{r}_n,\tilde{s}_n)$.
Again with assumption (b) one can prove:
$\exists$ $E(0) \in V$ and $s'_1,\ldots s'_n$ with $E_i(0) = \emptyset$, $E_j(0) = E_j$ for $j \neq i,i'$.
$E_{i'}(0) = E_{i'} + (\tilde{r}_n,s'_1)+ \ldots +(\tilde{r}_1,s'_n)$.

Starting from E we let successively for $k = 1,\ldots,n$ change jobmark $(i,\tilde{r}_k,\tilde{s}_k)$ into $(i',\tilde{r}_k,s'_{n-k})$. By interchanging E(0) and E in (6.4), taking the corresponding quotients in (6.4) given by (6.5), repeating the procedure for any permutation and using the invariance-condition at node i' gives the result.

(iii) Let E and E(0) be given as in (ii) for some $(j_1,\dots,j_n) \in (1,\dots,n)$, i.e.

$$E_{i'}(0) = E_{i'} + (r_{j_n},s'_1)+\dots+(r_{j_1},s'_n)$$

According to (6.4) and (6.5) and the condition of (iii) we have

$$\frac{\pi(E)}{\pi(E(0))} = \prod_{k=1}^{n} \bar{\delta}_i((r_{j_1},s_{j_1}),\dots,(r_{j_{k-1}},s_{j_{k-1}})|(r_{j_k},s_{j_k}))/c(E_{i'}(0))$$

where $c(E_{i'}(0))$ is some constant not depending on $(r_{j_n},\dots,r_{j_1})$.

Since $(j_1,\dots,j_n) \in (1,\dots,n)$ can be taken arbitrarily and $\pi(E(0))$ does not depend on the job-numbers at node i', this gives the proof. □

Condition (ii) is satisfied for instance if jobs are partitioned to job-numbers in classes $C_1,\dots,C_t$ and at jode i' any job of a class C_k is always assigned a fixed service-number s_k, $k = 1,\dots,t$.
By recollecting from section 4 how to model open networks from this we can conclude:

In an open network with a service process satisfying JLB all nodes have the service-invariance-property.

To conclude this section we give an example showing that the SIP allows for some pseudo-priority in service.
For some further examples we refer to [4].

Example 6.2.

Suppose 1 job has job-number 2 and all other jobs have job (class)-number 1.
Let the characteristic (f,φ,δ) of some node be determined as follows.

The service-numbers are place-numbers $1,\dots,x$ if x jobs are present, and change according the shift protocol and by

$$\delta((1,1),\dots,(1,x)|(2,x+1)) = 1$$

$$\varphi((1,1),\dots,(1,x),(2,x+1)|(2,x+1)) = \tau/[\tau+x]$$

$$\varphi((1,1),\dots,(1,x),(2,x+1)|(1,j)) = 1/[\tau+x] \qquad j = 1,\dots,x$$

$$f((1,1),\dots,(1,x),(2,x+1)) = \tau+x$$

$\delta((1,1),\dots,(1,x)|(1,j)) = 1/[x+1] \quad j = 1,\dots,x+1 \qquad f((1,1),\dots,(1,x)) = x$

$\varphi((1,1),\dots,(1,x)|(1,j)) = 1/x \qquad j = 1,\dots,x$

$\delta((1,1),\dots,(1,x),(2,x+1)|(1,j)) = 1/[x+1] \quad j = 1,\dots,x,\ x+2$

and for

$E = (1,1),\dots,(1,x),(2,x+1),(1,x+2),\dots,(1,x+1+y)$ with $y \geq 1$:

$\delta(E|(1,j)) = 1/[x+y+1] \quad j = 1,\dots,x+2+y \quad j \neq x+1$

$\varphi(E|(1,j)) = 1/[x+y] \quad j = 1,\dots,x+1+y \quad j \neq x+1$

$\varphi(E|(2,x+1)) = 0$

$f(E) = x + y$

First note that (f,φ,δ) satisfies (a),(b),(d),(e) and (5.1.1). Further for $E = \{(r_1,s_1),\dots,(r_n,s_n)\}$ the products of (6.1) are equal to

$$\bar{\bar{\pi}}(E) = \begin{cases} 1 & \text{if } r_i = 1, \quad i = 1,\dots,n \\ 1/\tau & \text{if } r_{i_0} = 2 \quad \text{for some } i_0 \end{cases}$$

for all permutations such that (6.1) is defined.

Hence, the service-invariance-property is satisfied for this node.

The jobs of class 1 are always served with unit capacity whereas job 2 is only served at rate τ as long as no other job of class 1 is present behind job 2.

So jobs of class 1, have strict-priority over job 2 if placed behind job 2.

Particularly by letting τ tend to 0 this characteristic approximates a strict waiting of and priority over job 2. Obviously extension to more jobs of class 2 is possible.

7. EXAMPLES WITH JLB

First recollect that any example satisfying the K_1- and K_2-criterion also satisfies JLB with stationary probabilities decomposing to the routing - and service-characteristics and with c a normalizing constant as

$$(7.1) \qquad \pi(E) = c\, \pi_1(E)\, \pi_2(E)$$

Particularly for examples satisfying the conditons of propositions 5.2 and 6.1 and with $\bar{\pi}_i$ and $\bar{\bar{\pi}}_i$ given by (5.4) and (6.1) it factorizes as

$$(7.2) \qquad \pi(E) = c \prod_{i=1}^{N} \bar{\pi}_i(E_i)\, \bar{\bar{\pi}}_i(E_i)$$

To conclude with we review several examples which have the JLB- or equivalently the Insensitivity property.

BLOCKING; REVERSIBLE ROUTING.

1.
This example is adapted from the teletraffic literature [6] and shows a generalization of the generalized Engset Formula.

It refers to a system with a limited number of sources M, say numbered 1,...,M, each of which has an individual holding and idle time distribution. After a holding time a source becomes idle if it finds a free line out of a total n lines by <u>hunting</u> over k lines. Further the busy lines are assumed to be random distributed and $M \geq n \geq k$.
Hence the <u>call congestion</u> for a source r if i lines are busy is given by
$\binom{i}{k} / \binom{n}{k}$

In our setting we model this by a closed network with nodes 1 and 2.
Node 1 has M available places and generates jobs for node 2.
Node 2 has only n available servers. The number of jobs is M.

The job-numbers are 1,...,M.

The service-numbers are 1,...,M at node 1 and 1,...,n at node 2.

Both nodes have the infinite-server-characteristics $(f_\infty, \varphi_\infty, \delta_\infty)$:

$f_\infty(x) = x$, $\varphi_\infty(x|j) = \delta_\infty(x-1|j) = 1/x \quad j = 1,\dots,x$.

The general routing is determined as

$$b_{12}(r;E) = 1 - \binom{i}{k}\Big/\binom{n}{k} \qquad \text{if } n_2(E) = i$$

$$b_{21}(r;E) = p_{12}(r;E) = p_{21}(r;E) = 1.$$

Further let job r have a mean service-time $\rho_j(r)$ at node j, j = 1,2. Since the routing is reversible and $b_2(i)$ only depends on i we obtain from proposition 5.1, 5.2 and 6.1 and relation (7.2):

The network has the JLB-property and for

$E = \{(j_1,1,s_1),(j_2,2,s_2),\dots,(j_M,M,s_M)\}$ with $n_2(E) = n_2 \le n$

the stationary probability is given by

$$\pi(E) = c\left[\prod_{i=1}^{M} \rho_{j_i}(i)\right]\left[\prod_{i=0}^{n_2-1} \left[1-\binom{i}{k}\Big/\binom{n}{k}\right]\right]$$

According to the equivalence between JLB and Insensitivity we conclude: The formula holds for arbitrary holding and idle time distributions with mean holding- resp. mean idle time $\rho_1(r)$ resp. $\rho_2(r)$ for source r.

The next example shows that blocking of jobs of one class may strictly depend on jobs of another class (also see example 5.2).

2.
Jobs of class 1 and 2 arrive at an open queue according independent Poisson processes with parameters γ_1 and γ_2.

The total number of available lines is M_3, however, jobs of class 1 only have accession via M_1 lines and jobs of class 2 only have accession via M_2 lines. Suppose $M_1 + M_2 > M_3$. Then the acceptation probabilities are

$$b_{01}(r;E) = \begin{cases} 1 \text{ if } n_1^r(E) < M_r \text{ and } n_1^1(E) + n_1^2(E) < M_3 \\ 0 \text{ else} \end{cases}$$

Further with $p_{01}(r;E) = p_{10}(r;E) = b_{10}(r;E) = 1$ and characteristic $(f_\infty, \varphi_\infty, \delta_\infty)$ as given in 1, again propositions 5.2 and 6.1 imply that JLB is satisfied here.

PSEUDO-PRIORITY

Example 5.3 together with infinite-server-characteristics
shows JLB with priority on being accepted
Example 6.2 together with infinite-server-characteristics
shows JLB with priority on being served

OVERFLOW

Example 5.5 together with infinite server characteristics
shows JLB with overflow by "by-passing" a saturated node

Another form of overflow can be "jumping-over" a saturated node as if an infinite capacity is given instantaneously.

Example:

Consider an open network with 2 nodes in tandem. A job is blocked at node 1 if $n_1(E) = N_1$ and then jumps over node 1 to node 2. Hence we have

$\bar{p}_{01}(r;E) = 1$ if $n_1(E) < N_1$, $\bar{p}_{12}(r;E) = \bar{p}_{20}(r;E) = 1$ and
$\bar{p}_{02}(r;E) = 1$ if $n_1(E) = N_1$.

Again with any characteristic satisfying the SIP the network also has the JLB-property and exhibits the insensitivity-phenomenon.

References

[1] Baskett, F., Chandy, M., Muntz, R. and Palacios, J., (1975). Open, closed and mixed networks of queues with different classes of customers. J.A.C.M., 22, 248-260.

[2] Chandy, K.M., Howard, J.H. and Towsley, D.F., (1977). Product form and local balance in queueing networks. J.A.C.M., 24, 250-263.

[3] Cohen, J.W., (1979). The multiple phase service network with generalized processor sharing. Acta Informatica 2, 245-284.

[4] Hordijk, A. and Van Dijk, N.,(1981). Stationary probabilities for networks of queues. Proceedings of the Boca-Raton-meeting.

[5] Hordijk, A. and Van Dijk, N., (1981). Networks of queues with blocking. Performance 81 (ed. K.J. Kylstra). North-Holland, 51-65.

[6] Iversen, V.B. (1981). The A-formula, Teleteknik, 64-79.

[7] Jackson, J.R., (1963). Jobshop-like queueing systems. Mgnt. Sci. 10, 131-142.

[8] Kelly, F.P., (1979). Reversibility and stochastic networks. Wiley.

[9] Kingman, J.F.C., (1969). Markov population processes. J. Appl. Prob. 6, 1-18.

[10] Pittel, B., (1979). Closed Exponential Networks of Queues with Saturation. The Jackson-type Stationary Distribution and its Asymptotic Analysis. Math. of Operations Res. Vol. 4, 357-378.

[11] Schassberger, R., (1978). The insensitivity of stationary probabilities in networks of queues. Adv. Appl. Prob., 10, 906-912.

[12] Schassberger, R., (1979). A definition of discrete product form distributions. Zeitschrift für Op. Res. 23, 189-195.

V

DIFFUSION APPROXIMATIONS

APPROXIMATION PAR LES DIFFUSIONS

Some Diffusion Approximations With State Space Collapse

Martin I. Reiman

Bell Laboratories
Murray Hill, New Jersey 07974

ABSTRACT

The known analytical solutions for queueing systems arising as models for computer system performance evaluation require either specific distributional assumptions or special service disciplines. Using diffusion approximations, one can obtain both insights into, and approximations for, systems with more general characteristics. The study of diffusions, which are essentially continuous path Markov processes, reduces most questions of interest to the study of certain differential equations.

In a manner similar to the way the central limit theorem allows a normal approximation for a sum of random variables, the stochastic processes occurring in queueing systems (e.g. queue length and workload) can be approximated by diffusion processes. In some cases the dynamics of the queueing system leads to a collapse in dimensionality of the state space of the associated diffusion approximation. A rigorous approach to diffusion approximations, via heavy traffic limit theorems, provides the approximating diffusion, along with an indication of conditions under which it will provide a reasonable approximation.

In this paper we focus on several examples where the diffusion approximation, after collapse, is one dimensional reflected Brownian motion, which is a much studied process. In particular, we consider: (i) priority queues, (ii) a system where customers join the shortest queue, and (iii) networks with one 'bottleneck' station. We show, for these systems, that the associated queue length processes collapse to one dimension in heavy traffic, under the standard normalization for such limits. It is then straightforward to show that the limit process is one dimensional reflected Brownian motion.

1. Introduction and Summary

Many queueing models arising in the performance analysis of computer systems have aspects which have shown themselves to be intractable to exact analysis. If seeking analytical results, one is then faced with the choice of making simplifying assumptions in order to make the model more tractable (which is essentially an approximation in the space of models), or keeping the complexities of the model and using an approximate analytical technique. One of the popular approximation techniques involves the use of diffusion processes, principally reflected Brownian motion. The use of diffusion approximations reduces most questions related to system performance to the solution of a partial or ordinary differential equation.

In this paper, we focus on several queueing systems where the diffusion approximations have a state space of lower dimension than the original process. These diffusion approximations are justified by heavy traffic limit theorems which are proven in this paper. In all cases we treat here the limit diffusion process is one dimensional reflected Brownian motion (RBM). RBM is a much studied process, and as a result, expressions are available for transient behavior, first passage times and stationary behavior. (See Cox and Miller [3] for these results.)

In the next section we introduce a probability space containing the underlying random variables used to construct all of the stochastic processes in the paper, and prove some functional central limit theorems which will be used in the sequel. In section 3 we prove heavy traffic limit theorems for the GI/G/1 queue. We also discuss some generalizations of the GI/G/1 queue, and describe how to use heavy traffic limit theorems to obtain diffusion approximations for stable queueing systems.

Sections 4 through 6 present three examples of queueing systems which 'collapse' in heavy traffic (after normalization) to diffusion limits which 'live' on a lower dimensional space than the original system. In section 4 we consider priority queues. The next example, covered in section 5, is a queueing system where entering customers join the shortest queue. In section 6 we consider our final example, a two station queueing network with one 'bottleneck' station, which we interpret to mean that only one station enters heavy traffic.

The appendix provides a definition of weak convergence, along with results from that area which are used in this paper.

Equations and theorems are numbered by section. Thus equation (3.5) refers to equation (5) of section 3. For equations (and theorems) in the same section no prefix is used, so that if equation (5) is referred to in section 3, it is again equation (5) of section 3 which is meant.

We use the abbreviations $BM(a,b)$ and $RBM(a,b)$ respectively to denote Brownian motion and reflected Brownian motion with drift a and variance b.

2. Technical Preliminaries

In this section we define the probability spaces supporting all of the stochastic processes studied in this paper. In section 2.1 we present sequences of interarrival times, service times, and routing indicators which serve as the building blocks for all other stochastic processes in this paper. In section 2.2 we consider a sequence of probability spaces of the type defined in section 2.1 and prove functional central limit theorems for normalized versions of the basic stochastic processes. These results, which are generalizations of central limit theorems, form the basis for proving the heavy traffic limit theorems of later sections.

2.1 The Basic Probability Space

We assume that we have a probability space (Ω,F,P) containing all of the random variables necessary for our queueing processes. In particular, we have independent sequences of IID random variables $\{u_i^j, i \geq 1\}$, $\{v_i^k, i \geq 1\}$, and $\{\phi_i^k, i \geq 1\}$ for $1 \leq j \leq J$ and $1 \leq k \leq K$, where J and K are positive integers. The elements of the sequences $\{u_i^j, i \geq 1\}$ and $\{v_i^k, i \geq 1\}$ are nonnegative, and we define

$$\lambda_j = (E[u_1^j])^{-1} \quad , \quad a_j = var(u_1^j)$$

$$\mu_k = (E[v_1^k])^{-1} \quad , \quad s_k = var(v_1^k) \; .$$

The u_i^j's function as interarrival times, while the v_i^k's will be service times. The ϕ_i^k's function as network routing indicators and have a probability distribution given by

$$P\{\phi_1^k = l\} = p_{kl} \quad , \quad 1 \leq l \leq K \; ,$$

and

$$P\{\phi_1^k = 0\} = 1 - \sum_{l=1}^{K} p_{kl} \; .$$

We will need several stochastic processes which are defined using the above sequences. Let $T_0^j = 0$ and

$$T_l^j = \sum_{i=1}^{l} u_i^j \quad , \quad 1 \leq j \leq J \quad , \quad l \geq 1 \; .$$

For $t \geq 0$, and $1 \leq j \leq J$, let

$$A_j(t) = \max \{l \geq 0 : T_l^j \leq t\}$$

and

$$S_j(t) = \begin{cases} 0, & v_1^j > t \\ \max\{l \geq 1 : \sum_{i=1}^{l} v_i^j \leq t\}, & v_1^j \leq t \; . \end{cases}$$

Finally, let

$$S_{jm}(t) = \sum_{i=1}^{S_j(t)} 1_{\{\phi_i^j = m\}}$$

for $1 \leq j,m \leq K$.

2.2 Functional Central Limit Theorems

We present some functional central limit theorems for the processes defined in the last section. These results are generalizations of standard central limit theorems and form building blocks for the proofs which come later.

We assume that we have a sequence of probability spaces (Ω_n, F_n, P_n), $n \geq 1$, similar to (Ω, F, P) of the last section in that each probability space has defined on it all of the random variables introduced in the last section. The following conditions will be needed for proving the limit theorems:

$$\lambda_j(n) \to \lambda_j \quad \text{and} \quad a_j(n) \to a_j \quad , \quad 1 \leq j \leq J \, , \tag{1}$$

$$\mu_k(n) \to \mu_k \quad \text{and} \quad s_k(n) \to s_k \quad , \quad 1 \leq k \leq K \, , \tag{2}$$

as $n \to \infty$, where the limits are all assumed to be finite. In addition, we assume that

$$\sup_{n \geq 1} E\{(u_1(n))^{2+\epsilon}\} < \infty \quad \text{for some } \epsilon > 0 \tag{3}$$

and

$$\sup_{n \geq 1} E\{(v_1(n))^{2+\epsilon}\} < \infty \quad \text{for some } \epsilon > 0 \; . \tag{4}$$

We assume that $P = \{p_{ij}, 1 \leq i,j \leq K\}$ does not depend on n.

We introduce the normalizations

$$\hat{A}_j^{(n)}(t) = n^{-1/2}[A_j^{(n)}(nt) - nt\lambda_j(n)] \, , \tag{5}$$

$$\hat{S}_k^{(n)}(t) = n^{-1/2}[S_k^{(n)}(nt) - nt\mu_k(n)] \, , \tag{6}$$

$$\hat{S}_{km}^{(n)}(t) = n^{-1/2}[S_{km}^{(n)}(nt) - nt\mu_k(n)p_{km}] \, , \tag{7}$$

$$\Phi_{km}^{(n)}(t) = n^{-1/2} \sum_{i=1}^{\lfloor nt \rfloor} (1_{\{\phi_i^k = m\}} - p_{km}) \, , \tag{8}$$

$$\xi_k^{(n)} = n^{-1} S_k^{(n)}(nt) \, , \tag{9}$$

$$X_k^{(n)}(t) = n^{-1/2} \sum_{i=1}^{\lfloor nt \rfloor} (v_i^k(n) - \mu_k^{-1}(n)) \, , \tag{10}$$

$$\alpha_j^{(n)}(t) = n^{-1} A_j^{(n)}(nt) \quad , \quad \text{and} \tag{11}$$

$$\theta_j^{(n)}(t) = n^{-1} T^j_{\lfloor \lambda_j(n) nt \rfloor} \tag{12}$$

for $1 \leq j \leq J$, $1 \leq k \leq K$, $0 \leq t \leq 1$, and $n \geq 1$.

The following two theorems characterize the asymptotic (as $n \to \infty$) behavior of the normalized stochastic processes defined in (5)-(12). The mode of convergence is weak convergence, denoted by =>. Weak convergence is relative to a metric space containing the sample paths of the stochastic processes. The space D consists of all right continuous functions with left limits. A further discussion of weak convergence is contained in the appendix.

Theorem 1.

a) If (1) and (3) hold, then

$$\hat{A}_j^{(n)} => \hat{A}_j = \mathrm{BM}(0, \lambda_j^3 a_j) ,$$

$$\alpha_j^{(n)} => \lambda_j e \quad , \quad \text{and}$$

$$\theta_j^{(n)} => e$$

in D for $1 \leqslant j \leqslant J$, where e is the identity function, i.e. $e(t) = t, 0 \leqslant t \leqslant 1$.

b) If (2) and (4) hold, then

$$\hat{S}_k^{(n)} => \hat{S}_k = \mathrm{BM}(0, \mu_k^3 s_k) ,$$

$$X_k^{(n)} => X_k = \mathrm{BM}(0, s_k) \quad , \quad \text{and}$$

$$\xi_k^{(n)} => \mu_k e$$

in D for $1 \leqslant k \leqslant K$.

Proof. The first assertion of a) follows immediately from the Renewal Process FCLT. The other assertions are immediate consequences of the first. The first assertion of b) again follows from the Renewal Process FCLT, the second assertion from Prohorov's theorem, and the third is an immediate consequence of the first.

□

Theorem 2. If (2) and (4) hold, then

$$\Phi_{kl}^{(n)} => \Phi_{kl} = \mathrm{BM}(0, p_{kl}(1-p_{kl})) \quad , \quad \text{and}$$

$$\hat{S}_{kl}^{(n)} => \hat{S}_{kl} = \mathrm{BM}(0, \mu_k p_{kl}(1-p_{kl}) + p_{kl}^2 u_k^3 s_k)$$

in D for $1 \leqslant k, l \leqslant K$.

Proof. The first assertion actually does not need the hypotheses, and is a consequence of Prohorov's theorem. To obtain the second result, note that

$$\hat{S}_{kl}^{(n)}(t) = \Phi_{kl}^{(n)}(\xi_k^{(n)}(t)) + p_{kl} \hat{S}_k^{(n)}(t) . \tag{13}$$

Let $\Phi_{kl}^{(n)} \circ \xi_k^{(n)}(t) = \Phi_{kl}^{(n)}(\xi_k^{(n)}(t))$. Then $\Phi_{kl}^{(n)} \circ \xi_k^{(n)} \Rightarrow BM(0,\mu_k p_{kl}(1-p_{kl}))$ by the first assertion of this theorem and the random time change theorem. The second term on the right hand side of (13) converges by Theorem 1, and the independence of the two terms yields the result.

□

3. The GI/G/1 Queue

In this section we state and prove heavy traffic limit theorems for the GI/G/1 queue. The GI/G/1 queue has a single server with an unlimited waiting space. Customers arrive in a renewal process and have IID service times. We will focus on the virtual waiting time (unfinished work) process, waiting times, and the queue length process. The heavy traffic limit theorems state that properly normalized sequences of the above processes converge weakly to RBM. The proofs of the limit theorems also bring out important relationships between the processes which hold in heavy traffic. The methods and relationships used in these proofs appear again in the later results of this paper.

We also present some extensions of the above results (without proofs) to systems with superimposed renewal process input and multiple servers. These results, due to Iglehart and Whitt [7], contain the prior results of this section as special cases. The proofs in this section do not require consideration of modified queueing systems and as a result are simpler than those of [7]. Whitt [10] contains results, similar to ours, which do not require modified queueing systems.

Finally, we close this section with a brief explanation of how one uses heavy traffic limit theorems to obtain a diffusion approximation for a stable queueing system.

3.1 Processes of Interest

All of the processes of interest to us can be constructed on the probability space described in section 2.1. For simplicity, we take $J = K = 1$, so we will drop the associated indices. In addition, we have no need for routing indicators.

The virtual waiting time (or unfinished work) process, $U = \{U(t), t \geq 0\}$, can be defined as follows, cf. Benes [1]. For $t \geq 0$, let

$$L(t) = \sum_{i=1}^{A(t)} v_i \ , \qquad (1a)$$

$$V(t) = L(t) - t \ , \quad \text{and} \qquad (1b)$$

$$U(t) = V(t) - [\inf_{0 \leq s \leq t} \{V(s)\} \wedge 0] \, . \qquad (1c)$$

If the arrival process is turned off at t, the system would next be empty at $t + U(t)$.

Assuming a FIFO discipline, the kth customer to enter the system, who arrives at T_k, has a waiting time (queue delay plus service time) of

$$W(k) = U(T_k) . \tag{2}$$

Let $Q(t)$ denote the number of customers in the system at time $t \geqslant 0$. Finally, let $\rho = \lambda/\mu$ denote the traffic intensity.

3.2 Heavy Traffic Limit Theorems

Using the sequence of probability spaces and the corresponding functional central limit theorems introduced in section 2.2, we prove heavy traffic limit theorems for the processes introduced in the last section.

Let

$$\hat{V}^{(n)}(t) = n^{-1/2} V^{(n)}(nt) , \tag{3}$$

$$\hat{Q}^{(n)}(t) = n^{-1/2} Q^{(n)}(nt) , \tag{4}$$

$$\hat{W}^{(n)}(t) = n^{-1/2} W^{(n)}(\lfloor \lambda nt \rfloor) , \tag{5}$$

and

$$\hat{U}^{(n)}(t) = n^{-1/2} U^{(n)}(nt) , \tag{6}$$

for $0 \leqslant t \leqslant 1$ and $n \geqslant 1$. In addition, let

$$c(n) = \sqrt{n}\ (\lambda(n) - \mu(n)) . \tag{7a}$$

We will need the assumption

$$c(n) \to c \quad \text{as} \quad n \to \infty \quad , \quad -\infty \leqslant c < \infty . \tag{7b}$$

Limit theorems for the case $c = +\infty$, which we do not deal with here, were considered in [7].

We first prove a heavy traffic limit theorem for the virtual waiting time process, whose representation (1) makes it the easiest of the three to deal with.

Theorem 1. If (2.1)-(2.4) and (7b) hold, with $c > -\infty$, then

$$\hat{U}^{(n)} \Rightarrow \hat{U} = \text{RBM}(\frac{c}{\mu}, \lambda(a+s)) \ \text{ in } D .$$

Proof. Combining (2.5), (2.10), (2.11), (1), and (3),

$$\hat{V}^{(n)}(t) = X^{(n)} \circ \alpha^{(n)}(t) + \mu^{-1}(n)[\hat{A}^{(n)}(t)+c(n)t] \tag{8}$$

for $0 \leqslant t \leqslant 1$ and $n \geqslant 1$. It follows that

$$\hat{V}^{(n)} => \hat{V} = \mathrm{BM}\left(\frac{c}{\mu}, \lambda[a+s]\right) \quad \text{in } D ,$$

as in the proof of Theorem 2.2. Let $f:D \to D$ be defined for $x \in D$ as $f(x) = y$, where $y(t) = x(t) - [\inf_{0 \leqslant s \leqslant t} \{x(s) \wedge 0\}]$, $0 \leqslant t \leqslant 1$. It has been shown (cf. [12]) that f is continuous, so that on applying the continuous mapping theorem we obtain

$$\hat{U}^{(n)} = f(\hat{V}^{(n)}) => f(\hat{V}) = \hat{U} = \mathrm{RBM}\left(\frac{c}{\mu}, \lambda[a+s]\right) . \qquad \square$$

We turn now to be the case $c = -\infty$, which corresponds to the situation where the traffic intensity approaches unity from below 'too slowly' (or, in fact, may approach some limit strictly less than unity). Because of the normalization, the limits in this case are all zero. This result is a simple example of state space collapse.

Theorem 2. If (2.1)-(2.4) and (7b) hold, with $c = -\infty$, then $\hat{U}^{(n)} => 0$ in D.

Proof. We wish to show that, for any $\epsilon > 0$,

$$P\{\sup_{0 \leqslant t \leqslant 1} \hat{U}^{(n)}(t) > \epsilon\} \to 0 \quad \text{as } n \to \infty .$$

Fix $\epsilon > 0$ and let

$$\tau_n = \inf\{t \geqslant 0;\ \hat{U}^{(n)}(t) > \epsilon\}$$

$$\tau_n^{\bullet} = \sup\{t \leqslant \tau_n :\ \hat{U}^{(n)}(t) \leqslant \epsilon/2\} .$$

During $[\tau_n^{\bullet}, \tau_n]$ the server is continuously serving, so that for $\tau_n^{\bullet} \leqslant t \leqslant \tau_n$

$$\hat{U}^{(n)}(t) = \hat{U}^{(n)}(\tau_n^{\bullet}-) + \hat{V}^{(n)}(t) - \hat{V}^{(n)}(\tau_n^{\bullet}-) .$$

Therefore,

$$P\{\sup_{0 \leqslant t \leqslant 1} \hat{U}^{(n)}(t) > \epsilon\} \leqslant P\{\sup_{0 \leqslant s \leqslant t \leqslant 1} \hat{V}^{(n)}(t) - \hat{V}^{(n)}(s-) > \epsilon/2\} . \tag{9}$$

Let

$$\bar{V}^{(n)}(t) = \hat{V}^{(n)}(t) - \mu^{-1}(n)c(n)t$$

As in the proof of Theorem 1,

$$\bar{V}^{(n)} => \bar{V} = \mathrm{BM}(0,\lambda(a+s)) .$$

We can rewrite (9) as

$$P\{\sup_{0\leqslant t\leqslant 1} \hat{U}^{(n)}(t) > \epsilon\} \leqslant P\{\sup_{0\leqslant s\leqslant t\leqslant 1} \bar{V}^{(n)}(t)-\bar{V}^{(n)}(s)+\mu^{-1}(u)c(n)[t-s] > \epsilon/2\} . \qquad (10)$$

Let $Z^{(n)}(t) = \sup_{\substack{0\leqslant s\leqslant 1\\ 0\leqslant u\leqslant t}} [\bar{V}(s+n)-\bar{V}(s)]$. From the transient distribution of Brownian motion, it is clear that $Z(t) => 0$ as $t \to 0$. This implies that, given $\eta > 0$, there exists a δ such that if $t < \delta$, then $P\{Z(t) > \epsilon/2\} < \eta/4$. The fact that $Z(1)$ is a proper random variable implies that there exists a $K < \infty$ such that $P\{Z(1) > K\} < \eta/4$. Let $v = \delta/2$ and take N such that for $n > N$, $\mu^{-1}(n)c(n)v < -K$,

$$|P\{Z^{(n)}(v) > \epsilon/2\} - P\{Z(v) > \epsilon/2\}| < \eta/4 ,$$

$$\text{and } |P\{Z^{(n)}(1) > \epsilon/2 - \mu^{-1}(n)c(n)v\} - P\{Z(1) > \epsilon/2-\mu^{-1}(n)c(n)v\}| < \eta/4 .$$

We thus have

$$P\{\sup_{0\leqslant s\leqslant t\leqslant 1} \bar{V}(t)-\bar{V}^{(n)}(s) + \mu^{-1}(n)c(n)[t-s] > \epsilon/2\}$$

$$\leqslant P\{Z^{(n)}(v) > \epsilon/2\} + P\{Z^{(n)}(1) > \epsilon/2 - c(n)\mu^{-1}(n)\ v\} < \eta ,$$

completing the proof.

□

We now consider the waiting time process.

Theorem 3. If (2.1)-(2.4) and (7b) hold, then

$$\hat{W}^{(n)} => \hat{U} \text{ in } D .$$

Proof. From (2) and (5), we have

$$\hat{W}^{(n)}(t) = n^{-1/2}W^{(n)}(\lfloor \lambda nt \rfloor)$$

$$= n^{-1/2}[U^{(n)}(T_{\lfloor \lambda nt \rfloor})] . \qquad (11)$$

Note that

$$n^{-1/2}U^{(n)}(T_{\lfloor \lambda nt \rfloor}) = \hat{U}^{(n)} \circ \theta^{(n)}(t) ,$$

or

$$\hat{W}^{(n)}(t) = \hat{U}^{(n)} \circ \theta^{(n)}(t) .$$

By the random time change theorem, $\hat{U}^{(n)} \circ \theta^{(n)} => \hat{U}$ in D.

□

The standard manner for proving the heavy traffic limit theorem for the queue length process ([7],[9]) involves the use of a representation of the queue length in terms of the interarrival and service time sequences. We prove the result here using Theorem 1, showing the relationship between the queue length and virtual waiting time processes in heavy traffic.

Theorem 4. If (2.1)-(2.4) and (7b) hold, then

$$\hat{Q}^{(n)} => \lambda \hat{U} \quad \text{in } D .$$

Proof. Let $\nu^{(n)}(t)$ be the arrival time of the customer in service at time t in the nth system. If the server is idle, let $\nu^{(n)}(t) = t$. Then

$$A^{(n)}(t) - A^{(n)}(\nu^{(n)}(t)) \leqslant Q^{(n)}(t) \leqslant A^{(n)}(t) - A^{(n)}(\nu^{(n)}(t)-) ,$$

or

$$|\hat{Q}^{(n)}(t) - (\hat{A}^{(n)}(t) - \hat{A}^{(n)}(\zeta^{(n)}(t)) + \sqrt{n}\lambda(u)(t-\zeta^{(n)}(t))| \leqslant$$

$$\hat{A}^{(n)}(\zeta^{(n)}(t)) - \hat{A}^{(n)}(\zeta^{(n)}(t)-) , \tag{12}$$

where $\zeta^{(n)}(t) = n^{-1}\nu^{(n)}(nt)$. For $t \geqslant 0$, $n \geqslant 1$, we have

$$U^{(n)}(\nu^{(n)}(t)) \leqslant t - \nu^{(n)}(t) \leqslant U^{(n)}(\nu^{(n)}(t)) + v^{(n)}_{A^{(n)}(\nu^{(n)}(t))} . \tag{13}$$

From [7], Lemma 3.3, we know that

$$\sup_{1 \leqslant k \leqslant \lambda n} n^{-1/2} v_k \xrightarrow{P} 0 \text{ as } n \to \infty . \tag{14}$$

Combining (13) and (14), it follows that

$$\sup_{0 \leqslant t \leqslant 1} |\hat{U}^{(n)} \circ \zeta^{(n)}(t) - \sqrt{n}\ (t-\zeta^{(n)}(t))| \xrightarrow{P} 0 . \tag{15}$$

Since $\hat{U}^{(n)} \Rightarrow \hat{U}$ and $\zeta^{(n)}(t) \leqslant t$, $n^{-1/2}\hat{U}^{(n)} \circ \zeta^{(n)} \Rightarrow 0$ and hence

$$\zeta^{(n)} \Rightarrow e \quad \text{in } D . \tag{16}$$

We can rewrite (12) as

$$\begin{aligned} |\hat{Q}^{(n)}(t)-\lambda\hat{U}^{(n)}(t)| \leqslant {} & [\hat{A}^{(n)}(t)-\hat{A}^{(n)} \circ \zeta^{(n)}(t)] \\ & - \lambda[\hat{U}^{(n)} \circ \zeta^{(n)}(t) - \sqrt{n}\ (t-\zeta^{(n)}(t))] \\ & + \lambda[\hat{U}^{(n)} \circ \zeta^{(n)}(t)-\hat{U}^{(n)}(t))] \\ & + [\hat{A}^{(n)}(\zeta^{(n)}(t))-\hat{A}^{(n)}(\zeta^{(n)}(t)-)] . \end{aligned} \tag{17}$$

Using (16) and the random time change theorem, the first and third bracketed terms on the right hand side of (17) converge weakly to the zero functional. The second bracketed term is taken care of by (15), and the fourth by Theorem 2.1, so we can conclude that $\hat{Q}^{(n)} - \lambda\hat{U}^{(n)} \Rightarrow 0$, yielding the result by the converging together theorem.

□

3.3 Extensions

For use in the rest of the paper we will need an extension of the results of the last section to a system with several servers, and an input which is the superposition of several renewal processes. This system was studied by Iglehart and Whitt [7], who obtained the results we need. The extension of our proofs to a superposition input process is straightforward, but the multiple server system would require some additional work, since the representation (1) no longer holds.

Consider a queue with J renewal processes superimposed for input and K servers, where we use the probability space described in section 2.1. We carry over the notation introduced earlier in this section for queue lengths, waiting time, and unfinished work. Iglehart and Whitt [7] showed the following.

Theorem 5. If (2.1)-(2.4) and (7b) hold, then

a) If $c > -\infty$, then $\hat{Q}^{(n)} \Rightarrow \hat{Q} = \text{RBM}\left(c, \sum_{j=1}^{J} \lambda_j^3 a_j + \sum_{k=1}^{K} \mu_k^3 s_k\right)$ in D. If, in addition, $K = 1$ or all service sequences are identically distributed, then

$$\hat{U}^{(n)} \Rightarrow (K\mu)^{-1}\hat{Q}$$

$$\text{and} \quad \hat{W}^{(n)} \Rightarrow (K\mu)^{-1}\hat{Q} \quad \text{in } D .$$

b) If $c = -\infty$, then $\hat{Q}^{(n)} \Rightarrow 0$ in D. If, in addition, $K = 1$ or all service sequences are identically distributed, then

$$\hat{U}^{(n)} \Rightarrow 0$$

$$\text{and} \quad \hat{W}^{(n)} \Rightarrow 0 \quad \text{in } D .$$

3.4 Approximating a Stable Queue

Given a GI/G/1 queue (or an extension, as described in the last subsection), the appropriate choice among Theorems 1, 3, 4, and 5 can be used to obtain a diffusion approximation. The heavy traffic limit theorem can be used to justify the claim that for a queue with ρ close to 1, the distribution of the normalized queue length process is close to that of the appropriate RBM. The limit theorem does not yield information on how close ρ must be to 1 to obtain a certain degree of accuracy in the approximation. This information would come from results on the rate of convergence.

The use of a heavy traffic limit theorem requires an identification of the limit diffusion process and a determination of the appropriate index, n, to be associated with the given queue. The index n is not directly related to physical properties of the queue. It is, however, tied to the drift through (7). One way to determine n is as follows. If we have a family of queues indexed by $0 \leqslant \rho < 1$, we can restate all of our limit theorems in terms of $\rho \to 1$. The results are the same as setting $n = (1-\rho)^{-2}$. From (7) we then obtain $c = -\mu$, which is thus the natural choice of drift for our approximating diffusion. The choice of variance is not uniquely determined by the limit theorem, since multiplying or dividing the variance term by ρ yields the same limit as $\rho \to 1$. The most reasonable approach is simply to use the variance term that appears in the limit theorem. The astute reader may have noted that the limit variance given in theorems 1, 3, and 4 may not match that in theorem 5 when $\rho < 1$, so even this technique leads to ambiguity. In any case, given that one has settled on a choice of drift and variance for the limit diffusion, one then 'undoes' the normalization to obtain the approximation. As an example, we consider the virtual waiting time process. From (6), using $n = (1-\rho)^{-2}$, we obtain

$$U(t) \overset{d}{\approx} (1-\rho)^{-1}\hat{U}(t(1-\rho)^2) .$$

4. Priority Queues

In this section we consider a priority queue with two classes of customers. High priority customers have preemptive (resume) priority over low priority customers so that, from the point of view of high priority customers, the low priority customers do not exist. As a result, when the system enters heavy traffic, the high priority customers see a system which is not in heavy traffic, so the associated (normalized) processes vanish in the limit. This 'collapse' leaves only low priority customers, so that the limit process is one dimensional. Although the high priority customers vanish from the queue in the heavy traffic limit, they exert an effect on the resulting limits for low priority customers. We examine this effect by comparing two simple priority queues with feedback.

Most of the results in this section have been obtained previously. Whitt [11] obtained heavy traffic limit theorems for the queue length and unfinished work processes. Harrison [6] proved a heavy traffic limit theorem for the virtual waiting time process for low priority customers, which we do not consider here (we consider instead actual waiting times).

4.1 Processes of Interest

We will again be using the probability space of section 2.1 to construct the priority queueing system. We take $J = K = 2$. The index 1 refers to high priority customers, who have preemptive (resume) priority over low priority, represented by index 2.

The unfinished work process can be defined as in section 3.1. Let

$$L_j(t) = \sum_{i=1}^{A_j(t)} v_i^j ,$$

$$L(t) = L_1(t) + L_2(t) ,$$

$$V_1(t) = L_1(t) - t ,$$

$$V(t) = L(t) - t ,$$

$$U_1(t) = V_1(t) - \left[\inf_{0 \leqslant s \leqslant t} \{V_1(s)\} \wedge 0 \right] , \text{ and}$$

$$U(t) = V(t) - \left[\inf_{0 \leqslant s \leqslant t} \{V(s)\} \wedge 0 \right] ,$$

for $j = 1,2$ and $t \geqslant 0$.

The discipline is work conserving, so $U(t)$ is the unfinished work in the system at time t. High priority customers are never affected by low priority customers, so $U_1(t)$ is the unfinished high priority work in the system. We can then let

$$U_2(t) = U(t) - U_1(t) , \quad t \geqslant 0 ,$$

represented the unfinished low priority work.

The kth high priority customer has a waiting time of

$$W_1(k) = U_1(T_k^1) .$$

Let

$$B(t,Z) = \inf\{s \geqslant t: V_1(s) - V_1(t) + Z \leqslant 0\} + t . \tag{1}$$

The kth low priority customer has a waiting time of (see [6])

$$W_2(k) = B(T_k^2, U(T_k^2)) .$$

Let $Q_j(t)$ denote the number of type j ($j = 1,2$) customers in the system at time t.

4.2 Heavy Traffic Limit Theorems

We again use the sequence of probability spaces introduced in section 2.2. Let

$$\hat{Q}_j^{(n)}(t) = n^{-1/2} Q_j^{(n)}(nt) , \tag{2}$$

$$\hat{W}_j^{(n)}(t) = n^{-1/2} W_j^{(n)}(\lfloor \lambda_j nt \rfloor) \tag{3}$$

$$\hat{U}_j^{(n)}(t) = n^{-1/2} U_j^{(n)}(nt), \quad \text{and} \tag{4}$$

$$\hat{U}^{(n)}(t) = n^{-1/2} U^{(n)}(nt) , \tag{5}$$

for $1 \leqslant j \leqslant 2$, $0 \leqslant t \leqslant 1$, and $n \geqslant 1$. In addition, let

$$d_1(n) = \sqrt{n}\,(\rho_1(n)-1), \quad \text{and}$$

$$d_2(n) = \sqrt{n}\,(\rho(n)-1)$$

where $\rho_j(n) = \lambda_j(n)/\mu_j(n)$, $1 \leqslant j \leqslant 2$, and $\rho(n) = \rho_1(n)+\rho_2(n)$.

We will need the assumptions

$$d_j(n) \to d_j \quad as \quad n \to \infty, \quad -\infty \leqslant d_j < \infty , \tag{6}$$

$$\text{and} \quad \rho_1(n) \to \rho_1 \quad as \quad n \to \infty, \quad 0 \leqslant \rho_1 \leqslant 1 . \tag{7}$$

The following theorem is an immediate consequence of the results of section 3.2.

Theorem 1. If (2.1)-(2.4) and (6) hold, with $d_1 = -\infty$, then

$$\hat{U}^{(n)} \Rightarrow \hat{U} = \mathrm{RBM}(d_2, \sum_{j=1}^{2} \lambda_j[s_j + \rho_j^2 a_j])$$

$$\hat{U}_1^{(n)} \Rightarrow 0 ,$$

$$\hat{U}_2^{(n)} \Rightarrow \hat{U} ,$$

$$\hat{Q}_1^{(n)} \Rightarrow 0 , \quad \text{and}$$

$$\hat{W}_1^{(n)} \Rightarrow 0 \quad \text{in } D .$$

We now consider the waiting times of low priority customers. Let

$$\hat{B}^{(n)}(t) = n^{-1/2} B(nt, U^{(n)}(nt)) , \quad o \leqslant t \leqslant 1 , \quad n \geqslant 1 . \tag{8}$$

From (1), (3), and (8) we have

$$\hat{W}_2^{(n)}(t) = \hat{B}^{(n)}(\theta_2^{(n)}(t)) .$$

Defining

$$\overline{V}_1^{(n)}(t) = \hat{V}_1^{(n)}(t) + \sqrt{n}t\,[1 - \rho_1(n)] , \tag{9}$$

we have $\overline{V}_1^{(n)} \Rightarrow \overline{V}_1 = \mathrm{BM}(0,\lambda_1[s_1 + \rho_1^2 a_1])$ in D as $n \to \infty$, as in the proof of Theorem 2.2. Combining (1), (8), and (9), we obtain

$$\hat{B}^{(n)}(t) = \inf\{s \geqslant 0: \overline{V}_1^{(n)}(t + n^{-1/2}s) - \overline{V}_1^{(n)}(t) + \hat{U}^{(n)}(t) \leqslant [1 - \rho_1(n)]s\} .$$

Theorem 2. If (2.1)-(2.4) and (7) hold, with $\rho_1 < 1$, then

$$\hat{W}_2^{(n)} \Rightarrow (1-\rho_1)^{-1}\hat{U} \quad \text{in } D .$$

Proof. Note that $\rho_1(n) \to \rho_1 < 1$ implies that $c_1(n) \to -\infty$ as $n \to \infty$, so that Theorem 1 holds. Let $\hat{B}_b^{(n)}(t) = \hat{B}^{(n)}(t) \wedge n^{1/4}$. If

$$\sup_{0 \leqslant t \leqslant 1} \sup_{0 \leqslant s \leqslant n^{1/4}} |\overline{V}_1^{(n)}(t + n^{-1/2}s) - \overline{V}_1^{(n)}(t)| \leqslant [1-\rho_1(n)]\epsilon ,$$

then

$$\sup_{0 \leqslant t \leqslant 1} |\hat{B}_b^{(n)}(t)[1-\rho_1(n)] - \hat{U}^{(n)}(t)| \leqslant \epsilon .$$

The fact that $\overline{V}_1^{(n)} \Rightarrow \overline{V}_1$ implies that

$$\sup_{0 \leq t \leq 1} |\hat{B}_b^{(n)}(t)[1-\rho_1(n)] - \hat{U}^{(n)}(t)| \xrightarrow{P} 0 \quad as \quad n \to \infty ,$$

which implies that

$$\sup_{0 \leq t \leq 1} |\hat{B}^{(n)}(t)[1-\rho_1(n)] - \hat{U}^{(n)}(t)| \xrightarrow{P} 0 \quad n \to \infty .$$

The result follows from Theorem 2.1 and an application of the random time change theorem.

□

We are now ready to treat the low priority queue length process.

Theorem 3. If (2.1)-(2.4) and (7) hold, with $\rho_1 < 1$, then

$$\hat{Q}_2^{(n)} => \lambda_2(1-\rho_1)^{-1}\hat{U} \text{ in } D .$$

Proof. Let $\nu_2^{(n)}(t)$ be the arrival time of the 'oldest' low priority customer in the system. If there are none in the system, set $\nu_2^{(n)}(t) = t$. We then have

$$A_2^{(n)}(t)-A_2^{(n)}(\nu_2^{(n)}(t)) \leq Q_2^{(n)}(t) \leq A_2^{(n)}(t)-A_2^{(n)}(\nu_2^{(n)}(t)-) .$$

In addition,

$$B(\nu_2^{(n)}(t),U^{(n)}(\nu_2^{(n)}(t)-)) \leq t-\nu_2^{(n)}(t) \leq B(\nu_2^{(n)}(t),U^{(n)}(\nu_2^{(n)}(t)))$$

for $t \geq 0$ and $n \geq 1$. The proof now proceeds like that of Theorem 3.3.

□

4.3 A Priority Queue with Feedback

Although the high priority queue length vanishes (after normalization) in heavy traffic, the high priority customers still exert an effect which shows up in heavy traffic. A striking example of this effect can be seen by comparing the sojourn time in two simple priority queues with feedback.

In system A, customers enter with low priority. After completing service, they feed back with high priority. After its second service completion, the customer leaves the system. In system B, customers enter with high priority and then switch to low priority upon feeding back. Both processes are constructed on the same probability space, as described in section 2.1. In particular, the sequence $\{v_i^1, i \geq 1\}$ is used as high priority service times and $\{v_i^2, i \geq 1\}$ is used for low priority service times. In addition, both systems have the same arrival process.

Let $W_A(k)$ and $W_B(k)$ denote, respectively, the sojourn times of the kth customer to enter system A and system B. Adopting our standard heavy traffic setting, we let

$$\hat{W}_A^{(n)}(t) = n^{-1/2} W_A^{(n)}(\lfloor \lambda n t \rfloor) \ , \quad \text{and}$$

$$\hat{W}_B^{(n)}(t) = n^{-1/2} W_B^{(n)}(\lfloor \lambda n t \rfloor) \ , \quad 0 \leqslant t \leqslant 1 \ , \ n \geqslant 1 \ .$$

System A is simply a GI/G/1 queue where the kth customer has service time $v_k^1 + v_k^2$. This becomes clear upon noting that no high priority customer can enter the system during a low priority service, so upon its first service completion the customer immediately reenters the server for its second service. From Theorem 3.2 we thus have

$$\hat{W}_A^{(n)} => \hat{W}_A = \text{RBM}(c/\mu, \ \mu[a+s_1+s_2]) \ \text{ in } D \ ,$$

where $\mu = (\mu_1^{-1}+\mu_2^{-1})^{-1}$. Let $\bar{w}_A$ denote the mean of the stationary distribution of $\hat{W}_A$. (It exists if and only if $c < 0$, which we assume here.) Then

$$\bar{w}_A = \frac{1}{2} \, |c|^{-1} \mu^2 (a+s_1+s_2)$$

The analysis of system B requires the results of this section. We have $W_B(k) = W_{B,1}(k) + W_{B,2}(k)$, where $W_{B,1}(k)$ denotes the time from the kth customer's entrance until its feedback, and $W_{B,2}(k)$ denotes the time from its feedback until it leaves the system. It can be verified that Theorems 1 and 2 apply here, allowing us to conclude that

$$\hat{W}_{B,1}^{(n)} => 0$$

and
$$\hat{W}_{B,2}^{(n)} => (1-\rho_1)^{-1} \hat{W}_A \quad \text{in } D \ .$$

We therefore have

$$\hat{W}_B^{(n)} => \hat{W}_B = (1-\rho_1)^{-1} \hat{W}_A \quad \text{in } D \ .$$

Letting $\bar{w}_B$ denote the mean of the stationary distribution of $\hat{W}_B$, we have

$$\bar{w}_B = \frac{1}{2} \, (1-\rho_1)^{-1} |c|^{-1} \mu^2 (a+s_1+s_2) \ .$$

and
$$\frac{\bar{w}_B}{\bar{w}_A} = \frac{1}{1-\rho_1} \ .$$

We can use the relationship $\hat{W}_B = (1-\rho_1)^{-1} \hat{W}_A$ to compare other aspects of the processes as well (higher moments, first passage times, etc.).

5. Shortest Queue System

In this section we consider a queueing system with two servers, each with its own queue, where entering customers join the shorter queue at the time of their arrival. We make no assumption about which queue is joined if both are equal, as it is irrelevant in the heavy traffic limit. We also allow for 'dedicated' side traffic to either or both queues, where these customers do not have any choice over which queue to join.

In the heavy traffic limit the two dimensional queue length process collapses to the line where both queue lengths are equal. In addition, the sum of the two queue lengths behaves as a two server system with one queue.

Using these results, and the results of section 3, we compare the dynamic control of the shortest queue system with two similar systems having static control, where the customers who can choose which queue to join do so without regard to the actual queue lengths. In one system customers flip a coin to determine which queue to join, while in the other customers alternate between the two queues. (This assumes a certain symmetry; a more complex cyclic choice scheme is needed when the symmetry does not exist.)

The heavy traffic limit of the shortest queue system was first considered by Foschini and Salz [5], who restricted their attention to a system with Poisson arrivals and exponential service times. They showed that the properly normalized stationary distribution of the shortest queue system has the same heavy traffic limit as that of the two server queue. In addition, they compared the mean delay in the shortest queue system with that of a two-server two-queue system where customers allowed to make a choice flip a state independent coin to determine which queue to join. Foschini [4] extended these results to systems with more than two servers where separate arrival streams have a different set of queues to choose from.

5.1 Processes of Interest

In this system there are three arrival streams ($J=3$) and two servers ($K=2$). Arrival stream A_1 exclusively feeds server 1, and arrival stream A_2 exclusively feeds server 2. Customers in arrival stream A_3 join the shorter of the two queues at the moment of their arrival. As far as the heavy traffic limit is concerned, it is irrelevant which queue is joined when they are equal, so we do not make any assumption about this. The service times at server k come from the sequence $\{v_i^k, i \geqslant 1\}$, $1 \leqslant k \leqslant 2$. For simplicity we assume that $\mu_1 = \mu_2$. Indeed, if $\mu_1 \neq \mu_2$ a better strategy from the customers' point of view would be to join the queue with the shortest expected wait, which in this case would not be the shortest queue. The results obtained here can be extended to the case $\mu_1 \neq \mu_2$.

Let $Q_j(t)$ denote the number of customers in server j's queue (including the customer in service, if any) at time $t \geqslant 0$, $1 \leqslant j \leqslant 2$. Let $W_j(k)$ be the waiting time (including service) of the kth type j to enter the system, $k \geqslant 1$, $1 \leqslant j \leqslant 3$. We will need to consider the number of customers served by server j during $[0,t]$, $D_j(t)$. If we let $B_j(t)$ denote the accumulated busy time of server j during $[0,t]$, we have $D_j(t) = S_j(B_j(t))$, $t \geq 0$, $1 \leq j \leq 2$. Finally, let $U_j(t)$ denote server j's unfinished work at time t.

For the proof, it will be helpful to have a two server, single queue system defined on the same probability space as the shortest queue system. Let $Q(t)$ denote the number of customers in this system at time t. Note that, since in the single queue system neither server is idle as long as there is work in the queue,

$$Q(t) \leq Q_1(t) + Q_2(t) \quad , \quad t \geq 0 . \tag{1}$$

If we let $\tilde{B}_j(t)$ denote the accumulated busy time of server j in the single queue system during $[0,t]$, we can write

$$Q_1(t) + Q_2(t) - Q(t) = [S_1(\tilde{B}_1(t)) - S_1(B_1(t))]$$

$$+ [S_2(\tilde{B}_2(t)) - S_2(B_2(t))] \quad , \quad t \geq 0 . \tag{2}$$

The only time a server in the shortest queue system will be idle while the corresponding server in the single queue system is working is when all of the customers are in the other queue. We can therefore write

$$0 \leq \tilde{B}_j(t) - B_j(t) \leq \sup_{0 \leq s \leq t} |U_1(s) - U_2(s)| \tag{3}$$

for $1 \leq j \leq 2$, $t \geq 0$.

5.2 Heavy Traffic Limit Theorems

Working on the sequence of probability spaces defined in section 2.2, let

$$\hat{Q}_k^{(n)}(t) = n^{-1/2} Q_k^{(n)}(nt) , \tag{4}$$

$$\hat{W}_j^{(n)}(t) = n^{-1/2} W_j^{(n)}(\lfloor \lambda_j nt \rfloor) , \tag{5}$$

$$\text{and } \hat{Q}^{(n)}(t) = n^{-1/2} Q^{(n)}(nt) \tag{6}$$

for $1 \leq j \leq 3$, $1 \leq k \leq 2$, $0 \leq t \leq 1$, and $n \geq 1$. In addition, let $c(n) = \sqrt{n}\ (\lambda_1(n) + \lambda_2(n) + \lambda_3(n) - 2\mu(n))$. We will need the assumption

$$c(n) \to c \quad \text{as} \quad n \to \infty \quad , \quad -\infty \leq c < \infty . \tag{7}$$

We also require that $|\lambda_1 - \lambda_2| < \lambda_3$, so that there are enough customers with a choice to equalize the queues.

The first result shows that the two dimensional queue length process collapses to one dimension in heavy traffic.

Theorem 1. If (2.1)-(2.4) and (7) hold, then

$$\sup_{0\leq t\leq 1} |\hat{Q}_1^{(n)}(t) - \hat{Q}_2^{(n)}(t)| \xrightarrow{P} 0 \quad \text{as} \quad n \to \infty .$$

Proof. We need to show that for any $\epsilon > 0$,

$$P\{\sup_{0\leq t\leq 1} |\hat{Q}_1^{(n)}(t) - \hat{Q}_2^{(n)}(t)| > \epsilon\} \to 0 \quad \text{as} \quad n \to \infty .$$

Fix $\epsilon > 0$ and let

$$\tau_n = \inf\{t \geq 0: |\hat{Q}_1^{(n)}(t) - \hat{Q}_2^{(n)}(t)| > \epsilon\}$$

and

$$\tau_n^{\cdot} = \sup\{t \leq \tau_n: |\hat{Q}_1^{(n)}(t) - \hat{Q}_2^{(n)}(t)| \leq \epsilon/2\} .$$

Assume that $\hat{Q}_1^{(n)}(\tau_n) > \hat{Q}_2^{(n)}(\tau_n)$. The other case can be handled in an identical manner. For $\tau_n^{\cdot} \leq t \leq \tau_n$, all type 3 arrivals join server 2's queue, so that

$$\hat{Q}_1^{(n)}(t) - \hat{Q}_2^{(n)}(t) = \hat{Q}_1^{(n)}(\tau_n^{\cdot}-) - \hat{Q}_2^{(n)}(\tau_n^{\cdot}-)$$

$$-n^{-1/2}[D_1^{(n)}(nt) - D_1^{(n)}(n\tau_n^{\cdot}-)] + n^{-1/2}[D_2^{(n)}(nt) - D_2^{(n)}(n\tau_n^{\cdot}-)]$$

$$+ [\hat{A}_1^{(n)}(t) - \hat{A}_1^{(n)}(\tau_n^{\cdot}-)] - [\hat{A}_2^{(n)}(t) - A_2^{(n)}(\tau_n^{\cdot}-)]$$

$$- [\hat{A}_3^{(n)}(t) - \hat{A}_3^{(n)}(\tau_n^{\cdot}-)] + (t-\tau_n^{\cdot})\sqrt{n}\,[\lambda_1(n)-\lambda_2(n)-\lambda_3(n)] .$$

Server 1 is never idle during $[\tau_n^{\cdot},\tau_n]$, so that

$$D_1^{(n)}(nt) - D_1^{(n)}(n\tau_n^{\cdot}-) = S_1^{(n)}(B_1^{(n)}(n\tau_n^{\cdot})+nt-n\tau_n^{\cdot}) - S_1^{(n)}(B_1^{(n)}(n\tau_n^{\cdot}-)) . \tag{9}$$

Server 2 may be idle during the interval, yielding

$$D_2^{(n)}(nt) - D_2^{(n)}(n\tau_n^{\cdot}-) \leq S_2^{(n)}(B_2^{(n)}(n\tau_n^{\cdot})+nt-n\tau_n^{\cdot}) - S_2^{(n)}(B_2^{(n)}(n\tau_n^{\cdot}-)) . \tag{10}$$

Combining (8), (9), and (10) we can write

$$P\{\sup_{0\leq t\leq 1} \hat{Q}_1^{(n)}(t) - \hat{Q}_2^{(n)}(t) > \epsilon\} \leq P\{\sup_{0\leq s\leq t\leq 1}\ \sup_{0\leq u,v\leq s} - [\hat{S}_1^{(n)}(u+t-s) - \hat{S}_1^{(n)}(u)]$$

$$+ [\hat{S}_2^{(n)}(v+t-s) - \hat{S}_2^{(n)}(v)] + [\hat{A}_1^{(n)}(t) - \hat{A}_1^{(n)}(s)] - [\hat{A}_2^{(n)}(t) - \hat{A}_2^{(n)}(s)]$$

$$- [\hat{A}_3^{(n)}(t) - \hat{A}_3^{(n)}(s)] + (t-s)\sqrt{n}\,[\lambda_1(n)-\lambda_2(n)-\lambda_3(n)] > \epsilon/2\} . \tag{11}$$

Using an argument similar to the proof of Theorem 3.2, it can be shown that the right hand side of (11) converges to zero as $n \to \infty$.

□

Now it simply remains to identify the one dimensional limit process.

Theorem 2. It (2.1)-(2.4) and (7) hold, then

$$\hat{Q}_1^{(n)} + \hat{Q}_2^{(n)} \Rightarrow \hat{Q} = \text{RBM}\left(c, \sum_{j=1}^{3} \lambda_j^3 a_j + \sum_{k=1}^{2} \mu^3 s_k\right)$$

in D.

Proof. From Theorem 3.5,

$$\hat{Q}^{(n)} \Rightarrow \hat{Q} \quad \text{in } D .$$

From (1) and (2) we have

$$|\hat{Q}_1^{(n)}(t) + \hat{Q}_2^{(n)}(t) - \hat{Q}^{(n)}(t)| = [\hat{S}_1^{(n)}(n^{-1}\tilde{B}_1^{(n)}(nt)) - \hat{S}_1^{(n)}(n^{-1}B_1^{(n)}(nt))]$$

$$+ [\hat{S}_2^{(n)}(n^{-1}\tilde{B}_2^{(n)}(nt)) - \hat{S}_2^{(n)}(n^{-1}B_2^{(n)}(nt))]$$

$$+ \mu(n)n^{-1/2}[\tilde{B}_1^{(n)}(nt) - B_1^{(n)}(nt)] + \mu(n)n^{-1/2}[\tilde{B}_2^{(n)}(nt) - B_2^{(n)}(nt)] . \tag{12}$$

$0 \leqslant t \leqslant 1$, $n \geqslant 1$. Note that

$$\sup_{0 \leqslant t \leqslant 1} |\hat{U}_1^{(n)}(t) - \hat{U}_2^{(n)}(t)| \leqslant \max_{1 \leqslant j \leqslant 2} \sup_{1 \leqslant i \leqslant A^{(n)}(n)} \sup_{0 \leqslant t \leqslant n} n^{-1/2} \sum_{l=i}^{i+|Q_1^{(n)}(t)-Q_2^{(n)}(t)|} v_l^j(n) , \tag{13}$$

where $A^{(n)}(n) = A_1^{(n)}(n) + A_2^{(n)}(n) + A_3^{(n)}(n)$.

Substituting (2.10) into (13),

$$\sup_{0 \leqslant t \leqslant 1} |\hat{U}_1^{(n)}(t) - \hat{U}_2^{(n)}(t)| \leqslant \max_{1 \leqslant j \leqslant 2} \sup_{0 \leqslant s \leqslant n^{-1}A^{(n)}(n)} \sup_{0 \leqslant t \leqslant 1} [X_j^{(n)}(s+n^{-1/2}|\hat{Q}_1^{(n)}(t) - \hat{Q}_2^{(n)}(t)|)$$

$$- X_j^{(n)}(s) + \mu^{-1}(n)|\hat{Q}_1^{(n)}(t) - \hat{Q}_2^{(n)}(t)|] . \tag{14}$$

Using Theorem 1 and the random time change theorem, the right hand side of (14) converges to zero in probability as $n \to \infty$. Combining this with (3),

$$\sup_{0 \leqslant t \leqslant 1} n^{-1/2}[\tilde{B}_j^{(n)}(nt) - B_j^{(n)}(nt)] \xrightarrow{P} 0 \text{ as } n \to \infty , \ 1 \leqslant j \leqslant 2 . \tag{15}$$

From (12) and (15) and the random time change theorem,

$$\sup_{0 \leqslant t \leqslant 1} |\hat{Q}_1^{(n)}(t) + \hat{Q}_2^{(n)}(t) - \hat{Q}^{(n)}(t)| \xrightarrow{P} 0 \quad \text{as} \quad n \to \infty .$$

By the converging together theorem, $\hat{Q}_1^{(n)} + \hat{Q}_2^{(n)} => \hat{Q}$ in D.

□

The following result treats waiting times.

Theorem 3. If (2.1)-(2.4) and (7) hold, then

$$\hat{W}_j^{(n)} => (\lambda_1+\lambda_2+\lambda_3)^{-1}\hat{Q} \quad \text{in } D \text{ for } 1 \leqslant j \leqslant 3 .$$

Proof. We can write

$$\min_{1 \leqslant k \leqslant 2} \left[\sum_{i=D_k^{(n)}(T_l^k(n))+2}^{D_k^{(n)}(T_l^k(n))+Q_k^{(n)}(T_l^k(n))} v_i^k(n) \right] \leqslant W_j^{(n)}(l) \leqslant \max_{1 \leqslant k \leqslant 2} \left[\sum_{i=D_k^{(n)}(T_l^k(n))+1}^{D_k^n(T_l^k(n))+Q_k^{(n)}(T_l^k(n))} v_i^k(n) \right] \tag{16}$$

for $1 \leqslant j \leqslant 3$, $l \geqslant 1$. Using the definition (2.10),

$$n^{-1/2} \sum_{i=D_k^{(n)}(nt)+m}^{D_k^{(n)}(nt)+Q_k^{(n)}(nt)} v_i^k(n) = X_k^{(n)} \left(\frac{D_k^{(n)}(nt) + Q_k^{(n)}(nt)}{n} \right) - X_k^n \left(\frac{D_k^{(n)}(nt)+m}{n} \right)$$

$$+ \mu^{-1}(n) \left[\hat{Q}_k^{(n)}(t) - \frac{m}{\sqrt{n}} \right] . \tag{17}$$

By the random time change theorem, the right hand side of (17) converges weakly to $\frac{1}{2\mu} \hat{Q} = (\lambda_1+\lambda_2+\lambda_3)^{-1}\hat{Q}$ in D as $n \to \infty$. The result follows from Theorem 2.1 and the random time change theorem.

□

5.3 A Comparison of Static and Dynamic Queue Selection

Using the results of the last section, it is possible to compare the performance of the shortest queue system to a system where the type 3 customers choose a server without observing the queue length. We will assume that $\lambda_1 = \lambda_2$, which makes the static selection systems simple. In one system, the type 3 customers flip a fair coin to determine which queue to join. In the other system, type 3 customers alternate between the queues (i.e. the choice sequence is 1,2,1,2,1,2,...).

We focus on the waiting time of a type 3 customer in each of these systems. Let $W_S(k)$ be the waiting time of the kth type 3 customer in the shortest queue system. Again adopting the heavy traffic setting, let

$$\hat{W}_S^{(n)}(t) = n^{-1/2} W_S^{(n)}(\lfloor \lambda_3 nt \rfloor) \quad , \quad 0 \leqslant t \leqslant 1 \ , \ n \geqslant 1 \ .$$

We know from Theorem 3 that

$$\hat{W}_S^{(n)} \Rightarrow \hat{W}_S = \mathrm{RBM}\left[\frac{c}{\lambda}, \lambda^{-2}\left[\sum_{j=1}^{3} \lambda_j^3 a_j + \sum_{k=1}^{k} \mu^3 s_k\right]\right] \quad \text{in } D \ ,$$

where $\lambda = \lambda_1 + \lambda_2 + \lambda_3$.
Assuming that $c < 0$ so that $\hat{W}_S$ has a stationary distribution, we let $\bar{w}_S$ be its mean. We then have

$$\bar{w}_S = \frac{1}{2} (|c|\lambda)^{-1} \left[\sum_{j=1}^{3} \lambda_j^3 a_j + \mu^3 \sum_{k=1}^{2} s_k\right] .$$

To analyze the other systems, we use theorem 3.5. If we wanted more than just the mean of the waiting time, we would need the joint distribution of the two dimensional queue length process. A heavy traffic limit theorem for these two dimensional processes is a special case of the results in Reiman [9]. We avoid the additional complexity of these results by considering only means, which can be obtained from results for the marginal distributions of the two queues. Let $W_{C,j}(k)$ and $W_{A,j}(k)$ be the waiting time of the kth type 3 customer to join queue j in the coin toss and alternating systems, respectively. In addition, let

$$\hat{W}_{C,j}^{(n)}(t) = n^{-1/2} W_{C,j}^{(n)}(\lfloor \tfrac{1}{2} \lambda_3 nt \rfloor)$$

$$\text{and} \quad \hat{W}_{A,j}^{(n)}(t) = n^{-1/2} W_{A,j}^{(n)}(\lfloor \tfrac{1}{2} \lambda_3 nt \rfloor)$$

for $1 \leqslant j \leqslant 2$, $0 \leqslant t \leqslant 1$, and $n \geqslant 1$.

To use theorem 3.5, we need to know the mean and variance of the interarrival times of type 3 customers to a particular queue (its the same for both queues) in each system. The arrival rate in both cases is clearly $\frac{1}{2} \lambda_3$. In the alternating system, an interarrival time is simply the sum of 2 'old' interarrival times, so the variance is $2a_3$. In the coin toss system, an interarrival time is the sum of a geometrically distributed (with parameter 1/2) number of 'old' interarrival times (with the geometric distribution shifted to have a minimum value of 1), so the variance is $2[\lambda_3^{-2}+a_3]$. Using these constants in theorem 3.5, we find that

$$\hat{W}_{C,j}^{(n)} \Rightarrow \hat{W}_{C,j} = \mathrm{RBM}\left(\frac{c}{\lambda}, \lambda^{-2}[4\lambda_j^3 a_j + \lambda_3 + \lambda_3^3 a_3 + 4\mu^3 s_j]\right)$$

$$\text{and } \hat{W}_{A,j}^{(n)} \Rightarrow \hat{W}_{A,j} = \mathrm{RBM}\left(\frac{c}{\lambda}, \lambda^{-2}[4\lambda_j^3 a_j + \lambda_3^3 a_3 + 4\mu^3 s_j]\right) \quad in\ D .$$

Let $\bar{w}_{A,j}$ and $\bar{w}_{C,j}$ be the means of the stationary distributions of $\hat{W}_{A,j}$ and $\hat{W}_{C,j}$ respectively. Then

$$\bar{w}_{A,j} = \frac{1}{2}(|c|\lambda)^{-1}[4\lambda_j^3 a_j + \lambda_3^3 a_3 + 4\mu^3 s_j]$$

$$\text{and } \bar{w}_{C,j} = \frac{1}{2}(|c|\lambda)^{-1}[4\lambda_j^3 a_j + \lambda_3 + \lambda_3^3 a_3 + 4\mu^3 s_j]$$

If we let $\bar{w}_A$ and $\bar{w}_C$ denote the mean waiting time of a type 3 customer arriving to a stationary system (in the heavy traffic limit) for the alternating and coin toss systems respectively, then

$$\bar{w}_A = \frac{1}{2}(\bar{w}_{A,1} + \bar{w}_{A,2}) = \frac{1}{2}(|c|\lambda)^{-1}[2\lambda_1^3 a_1 + 2\lambda_2^3 a_2 + \lambda_3^3 a_3 + 2\mu^3(s_1+s_2)]$$

$$\text{and } \bar{w}_C = \frac{1}{2}(\bar{w}_{C,1} + \bar{w}_{C,2}) = \frac{1}{2}(|c|\lambda)^{-1}[2\lambda_1^3 a_1 + 2\lambda_2^3 a_2 + \lambda_3 + \lambda_3^3 a_3 + 2\mu^3(s_1+s_2)] .$$

It is clear that $\bar{w}_S \leqslant \bar{w}_A < \bar{w}_C$. For the special case where all distributions are exponential, $\frac{\bar{w}_C}{\bar{w}_s} = 2$, as was shown in [5], and $\frac{\bar{w}_A}{\bar{w}_S} = 2 - \frac{1}{2}\lambda^{-1}\lambda_3$. Other cases can be compared using the expressions for $\bar{w}_A$, $\bar{w}_C$, and $\bar{w}_S$.

6. A Network With One Bottleneck Station

We now consider a queueing network with two stations where one of the stations is a 'bottleneck'. We interpret this to mean that only one of the stations (the bottleneck) goes into heavy traffic. As a result the two dimensional queue length process collapses, in the limit, to a one dimensional diffusion where the non-bottleneck station has disappeared.

The queueing network model assumes that each station has a renewal input process (one of them may be identically zero), and a sequence of IID service times. After completing service at a station, the customer flips a coin to determine whether to go to the other station or to leave the system. This model is a special case of that treated in Reiman [9], where it was assumed that all stations enter heavy traffic. It was shown there that properly normalized queue length and sojourn times converge weakly to reflected Brownian motion on the nonnegative orthant (quadrant for 2 stations).

6.1 Processes of Interest

We use the probability space of section 2.1 to construct the queueing network. There are two stations in the network, so $J = K = 2$. The process $A_j = \{A_j(t),\, t \geqslant 0\}$ is the arrival process to station j, $1 \leqslant j \leqslant 2$. We allow for the possibility that one of A_1 or A_2 is identically zero, in which case we set the corresponding parameters to zero. The sequence $\{u_i^k,\, i \geqslant 1\}$ is used for service times and the sequence $\{\phi_i^k,\, i \geqslant 1\}$ for routing indicators at station k, $1 \leqslant k \leqslant 2$. After the ith service completion at station k, the associated customer is routed to station ϕ_i^k if $\phi_i^k = 1$ or 2, and out of the network if $\phi_i^k = 0$. For simplicity, when we deal with queue lengths we will assume that customers cannot immediately feed back to the station at which they are being served. This is without loss of generality. The queue length process is invariant to where in the queue the customer feeds back, so if we feed the customer to the head of the queue, self feedback is equivalent to having a new service time distribution which is a geometric mixture of the original service time distribution. Therefore, if a system has self feedback, we can change the service time distribution accordingly and treat it as if there were no self feedback. Although the above invariance does not hold for waiting times, where order of service is important, by continually permuting customer labels in the proper manner (to obtain the 'correct' service order), results for waiting times in a network with self feedback can be obtained from the above system without self feedback.

Let $Q_j(t)$ be the number of customers in station j, $1 \leqslant j \leqslant 2$, at time $t \geqslant 0$. Let $W_j(l)$ be the sojourn time at station j of the lth customer to enter station j, during that visit.

We need to solve the traffic equation in order to define traffic intensities for the stations in the network. To this end, let $\boldsymbol{\gamma} = (\gamma_1, \gamma_2)$ be given by $\gamma = \lambda(I-P)^{-1}$, where $\boldsymbol{\lambda} = (\lambda_1, \lambda_2)$, and $(I-P)^{-1}$ exists when $p_{12}p_{21} < 1$, which we assume. We can then define $\rho_j = \gamma_j/\mu_j$, $1 \leqslant j \leqslant 2$.

For the proof of Theorem 3 it will be helpful to have a different, but distributionally equivalent, construction of the network. In this construction, routing indicators for use after service completion at station 2 come from two separate sequences. Specifically, the ith customer to enter station 2 from outside the network uses routing indicator ϕ_i^2, while the ith customer completing service at station 1 receives routing indicator $\bar{\phi}_i^2$. Note that customers who leave the network after completing service at station 1 receive a routing indicator for station 2 even though they never use it. This apparent 'waste' simplifies the notation. The sequence $\{\bar{\phi}_i^2,\, i \geqslant 1\}$ is an independent (distributionally identical) copy of $\{\phi_i^2,\, i \geqslant 1\}$.

Let $\nu_2(t)$ denote the time at which the customer in service at station 2 at time t arrived to station 2, on this visit. (Set $\nu_2(t)=t$ if station 2 is empty.) We can write

$$Q_1(t) = A_1(t) + \sum_{i=1}^{A_2(\nu_2(t)-)} 1_{\{\phi_i^2=1\}} + \sum_{i=1}^{S_1(B_1(\nu_2(t))-)} 1_{\{\phi_i^1=2,\bar{\phi}_i^2=1\}} - S_1(B_1(t)) \,, \tag{1}$$

where $B_1(t)$ is the accumulated busy time of the server at station 1 during $[0,t]$, for $t \geqslant 0$.

We will also need a single station feedback queue, constructed alongside the above network (using the same random variables). The arrival process is the superposition of A_1 and those customers in A_2 whose routing indicators are 1. The service time of the ith customer is v_1^i, and upon completing service, customer i feeds back if $\phi_i^1 = 2$ and $\tilde{\phi}_i^2 = 1$. Letting $Q(t)$ denote the number of customers in the system at time t, we have

$$Q(t) = A_1(t) + \sum_{i=1}^{A_2(t)} 1_{\{\phi_i^2=1\}} + \sum_{i=1}^{S_1(B(t))} 1_{\{\phi_i^1=2,\tilde{\phi}_i^2=1\}} - S_1(B(t)) \tag{2}$$

for $t \geqslant 0$, where $B(t)$ is the accumulated busy time of the server during $[0,t]$.

We have

$$0 \leqslant B(t) - B_1(t) \leqslant \sup_{1 \leqslant l \leqslant A(t)} \sup_{0 \leqslant s \leqslant t} \sum_{i=l}^{l+Q_2(s)} \tilde{v}_i^1 \tag{3}$$

for $t \geqslant 0$, where

$$A(t) = A_1(t) + \sum_{i=1}^{A_2(t)} 1_{\{\phi_i^2=1\}}$$

and $\tilde{v}_i^1$ represents the 'total' service time of the ith customer at station 1, which includes all of its feedbacks. This total service time is simply a sum of a geometrically distributed (with parameter $p_{12}p_{21}$) number of elements of the sequence $\{v_i^1, i \geqslant 1\}$. Exactly which elements are used for each customer is notationally cumbersome to indicate, and since it is irrelevant for our purposes we do not attempt it.

6.2 Heavy Traffic Limit Theorems

We now consider a sequence of systems of the type described in the last section. Let

$$\hat{Q}_j^{(n)}(t) = n^{-1/2} Q_j^{(n)}(nt) \quad , \text{ and}$$

$$\hat{W}_j^{(n)}(t) = n^{-1/2} W_j^{(n)}(\lfloor \gamma_j nt \rfloor)$$

for $1 \leqslant j \leqslant 2$, $0 \leqslant t \leqslant 1$, and $n \geqslant 1$. In addition, let

$$d_j(n) = \sqrt{n}\ (\rho_j(n)-1) \ .$$

We will assume that station 1 enters heavy traffic and that station 2 remains in light traffic. This corresponds to

$$d_1(n) \to d_1 \quad , \quad -\infty < d_1 < \infty \, , \tag{4}$$

$$\text{and} \quad d_2(n) \to -\infty \quad , \quad \text{as} \quad n \to \infty \, . \tag{5}$$

The first result shows that the queue at station 2 vanishes in the limit.

Theorem 1. If (2.1)-(2.4), (4) and (5) hold, then $\hat{Q}_2^{(n)} \Rightarrow 0$ in D.

Proof. Let $c_2(n) = \sqrt{n}\ (\lambda_2(n)+\mu_1(n)p_{12}-\mu_2(n))$, $n \geqslant 1$. A simple manipulation yields $c_2(n) = d_2(n)[\mu_2(n)(1-p_{21})] - \mu_1(n)d_1(n)$. From (4) and (5) it can then be seen that $c_2(n) \to -\infty$ as $n \to -\infty$.

We need to show that for $\epsilon > 0$, $P\{\sup_{0 \leqslant t \leqslant 1} \hat{Q}_2^{(n)}(t) > \epsilon\} \to 0$ as $n \to \infty$. Fix $\epsilon > 0$ and let

$$\tau_n = \inf\{t \geqslant 0\colon \hat{Q}_2^{(n)}(t) > \epsilon\}$$

$$\text{and} \quad \tau_n^{\cdot} = \sup\{t \leqslant \tau_n\colon \hat{Q}_2^{(n)}(t) \leqslant \epsilon/2\} \quad , \quad n \geqslant 1 \, .$$

We then have (for $\tau_n^{\cdot} \leqslant t \leqslant \tau_n$)

$$\begin{aligned} \hat{Q}_2^{(n)}(t) - \hat{Q}_2^{(n)}(\tau_n^{\cdot}-) &\leqslant [\hat{A}_2^{(n)}(t) - \hat{A}_2^{(n)}(\tau_n^{\cdot}-)] \\ &+ [\hat{S}_{12}^{(n)}(n^{-1}B_1^{(n)}(n\tau_n^{\cdot})+t-\tau_n^{\cdot}) - \hat{S}_{12}^{(n)}(n^{-1}B_1^{(n)}(n\tau_n^{\cdot}-))] \\ &- [\hat{S}_2^{(n)}(n^{-1}B_2^{(n)}(n\tau_n^{\cdot})+t-\tau_n^{\cdot}) - \hat{S}_2^{(n)}(n^{-1}B_2^{(n)}(n\tau_n^{\cdot}-))] \\ &+ (t-\tau_n^{\cdot})c_2(n) \, . \end{aligned}$$

The proof can now be completed in the same way as that of Theorem 5.1.

□

As before, the waiting time process vanishes when the queue length process does.

Theorem 2. If (2.1)-(2.4), (4), and (5) hold, then

$$\hat{W}_2^{(n)} \Rightarrow 0 \quad \text{in } D \, .$$

Proof. This result follows from Theorem 1, using an argument similar to that used in the proof of Theorem 5.3.

□

We can now determine the limit of the queue length process at station 1.

Theorem 3. If (2.1)-(2.4), (4) and (5) hold, then

$$\hat{Q}_1^{(n)} => \hat{Q} = \text{RBM}(d_1\mu_1[1-p_{12}p_{21}], \lambda_1^3 a_1 + \lambda_2 p_{12}(1-p_{12}) + p_{12}^2\lambda_2^3 a_2$$

$$+ \mu_1[1-p_{12}p_{21}]p_{12}p_{21} + \mu_1^3 s_1[1-p_{12}p_{21}]^2)$$

in D.

Proof. Using (3), Theorem 1, and an argument similar to that used in Theorem 5.2 it can be shown that

$$\sup_{0 \leqslant t \leqslant 1} n^{-1/2}[B^{(n)}(nt) - B_1^{(n)}(nt)] \xrightarrow{P} 0 \quad \text{as} \quad n \to \infty . \tag{6}$$

We have

$$n^{-1/2}[A_2^{(n)}(nt) - A_2^{(n)}(\nu_2^{(n)}(nt))] = \hat{A}_2^{(n)}(t) - \hat{A}_2^{(n)}(n^{-1}\nu_2^{(n)}(nt))$$

$$+ \lambda_2(n) \left(\frac{nt - \nu_2^{(n)}(nt)}{\sqrt{n}} \right) \tag{7}$$

for $0 \leqslant t \leqslant 1$, $n \geqslant 1$. From Theorem 2 it is clear that

$$\sup_{0 \leqslant t \leqslant n} n^{-1/2}(t - \nu_2^{(n)}(t)) \xrightarrow{P} 0 \quad \text{as} \quad n \to \infty . \tag{8}$$

Using (8), the right hand side of (7) converges to zero in probability. The argument for $[S_1^{(n)}(B(t)) - S_1(B_1(\nu_2^{(n)}(t)))]$ is similar (using (6) in addition to (8)), so we can conclude that

$$\sup_{0 \leqslant t \leqslant 1} |\hat{Q}_1^{(n)}(t) - \hat{Q}^{(n)}(t)| \xrightarrow{P} 0 \quad \text{as} \quad n \to \infty .$$

To complete the proof, we need $\hat{Q}^{(n)} => \hat{Q}$ in D which follows from Theorem 3.5 upon noting that, as described in the last section, we can replace the feedback queue by one without feedback. Let $\tilde{\mu}$ and $\tilde{s}$ denote the service rate and service time variance of the nonfeedback replacement queue, respectively. We then have

$$\tilde{\mu} = (1-p_{12}p_{21})\mu$$

$$\text{and } \tilde{s} = p_{12}p_{21}(1-p_{12}p_{21})^{-2}\mu^{-1} + s_1(1-p_{12}p_{21})^{-1} . \qquad \square$$

The following result follows from Theorem 3, using an argument similar to the proof of Theorem 5.3.

Theorem 4. If (2.1)-(2.4), (4) and (5) hold, then

$$\hat{W}_1^{(n)} \Rightarrow \gamma_1^{-1}\hat{Q} \quad \text{in} \quad D\,.$$

Although we will not prove it here, it can be shown that the heavy traffic limit of the sojourn time of a customer who visits station 1 l times is simply the customers first waiting time at station 1 multiplied by l. The key to this result is the fact that $n^{-1/2}\hat{W}_1^{(n)} \Rightarrow 0$, which means that on the normalized time scale, in the limit, the customer spends no time in the network. See [9] for this result in a network where all stations enter heavy traffic.

APPENDIX

Weak Convergence

The object of this section is to provide an easily accessible reference for the fundamental results from weak convergence theory used in this paper. A more complete reference to this material is Billingsley [2].

We begin with a definition of weak convergence. As a setting, we have a metric space S with metric m and Borel sets S. In addition, we have a sequence of probability measures $\{P_n,\, n \geqslant 1\}$ and a probability measure P on $(S,\mathbf{S})$. The sequence $\{P_n,\, n \geqslant 1\}$ is said to converge weakly to the probability measure P (denoted by $P_n \Rightarrow P$) if

$$\lim_{n\to\infty} \int_S f dP_n = \int_S f dP$$

for all bounded continuous real valued functions f on S.

We will be dealing with random elements rather than directly with probability measures. A random element, X, is a measurable map from a probability space (Ω,F,P) into S. The random element induces a probability measure P on $(S,\mathbf{S})$ defined as

$$P(A) = P(X \in A) \quad \text{for} \quad A \in \mathbf{S}\,.$$

A sequence of random elements $\{X_n,\, n \geqslant 1\}$ is said to converge weakly to a random element X (denoted by $X_n \Rightarrow X$) if $P_n \Rightarrow P$.

We can now state two important results from weak convergence. Assume we have two sequences of random elements defined on a common domain. In addition, assume that (S,m) is separable. The following result is Theorem 4.1 in [2].

Converging Together Theorem. If $X_n \Rightarrow X$ and $m(X_n, Y_n) \Rightarrow 0$, then $Y_n \Rightarrow X$.

Let S' be another metric space, with Borel sets $\mathbf{S}'$. If h is a measurable map from S into S', each probability measure P on $(S, \mathbf{S})$ induces a unique probability measure P' on $(S', \mathbf{S}')$ given by

$$P'(A) = P(h^{-1}A) \quad \text{for} \quad A \in \mathbf{S}' .$$

Let D_h denote the set of discontinuities of h. The following result is Theorem 5.1 in [2].

Continuous Mapping Theorem. If $X_n \Rightarrow X$ and $P(D_h) = 0$, then $h(X_n) \Rightarrow h(X)$.

There are two metric spaces in which we are most interested, as they are the function spaces containing the sample paths of our stochastic processes. The space $C[0,1]$, which we abbreviate to C, is the space of continuous real valued functions on $[0,1]$. The space $D[0,1]$, abbreviated to D, is the space of right continuous real valued functions on $[0,1]$ having left limits (the only discontinuities are jumps). The space D contains the sample paths of the queue length and delay processes, while C contains the sample paths of the limit diffusion processes. Clearly $C \subset D$. To complete the definition of C we give its metric

$$\rho(x,y) = \sup_{0 \leq t \leq 1} |x(t) - y(t)| \quad \text{for} \quad x,y \in C .$$

To complete the triple $(C, m, \mathbf{C})$, we let $\mathbf{C}$ be the Borel sets of C. With the metric ρ, C is a complete, separable metric space. The metric on D is more complicated to define, as there are several candidates. See Billingsley [2] and Whitt [12] for more information. Our main concern with the metric on D, $d(x,y)$, is in satisfying the hypotheses of the converging together theorem. It is therefore sufficient for our purposes to know that $d(x,y) \leq \rho(x,y)$ for $x,y \in D$.

We now turn to the two main functional central limit theorems that we will be using. The first is the functional generalization of the Lindberg-Feller central limit theorem (CLT). (In fact, setting $t=1$ below yields the Lindeberg-Feller CLT.) Assume that, for each $n \geq 1$, we have a sequence of IID random variables $\{x_i^n, i \geq 1\}$ with $E[x_1^n] = 0$ and $var(x_1^n) = \sigma_n^2$. Assume

$$\sigma_n^2 \to \sigma^2 \quad \text{as} \quad n \to \infty \ , \ 0 < \sigma^2 < \infty \ , \tag{1}$$

and

$$\sup_{n \geq 1} E\{|x_1^n|^{2+\epsilon}\} < \infty \quad \text{for some } \epsilon > 0 \ . \tag{2}$$

Condition (2) implies the standard Lindeberg condition. In practice, (2) usually holds with $\epsilon = 1$. Let

$$Y_n(t) = n^{-1/2} \sum_{i=1}^{\lfloor nt \rfloor} x_i^n \quad , \quad 0 \leqslant t \leqslant 1 \quad , \quad n \geqslant 1 \; .$$

The following result is Theorem 3.1 of Prohorov [8].

Prohorov's Theorem. If (1) and (2) hold, then $Y_n \Rightarrow Y$ in D, where $Y = \{Y(t), 0 \leqslant t \leqslant 1\}$ is Brownian motion with zero drift and variance σ^2.

For the next result, which is a functional central limit theorem (FCLT) for renewal processes, we again assume that we have sequences of IID random variables $\{x_i^n, i \geqslant 1\}$ for $n \geqslant 1$. Assume that $x_i^n \geqslant 0$ for $i,n \geqslant 1$ and let $\mu_n^{-1} = E[x_1^n] > 0$, and $\sigma_n^2 = var(x_1^n)$. Let

$$S_n(t) = \begin{cases} 0, & x_1^n > t \\ \max\left\{l : \sum_{i=1}^{l} x_i^n \leqslant t\right\}, & x_1^n \leqslant t \end{cases}$$

and $Z_n(t) = n^{-1/2}[S_n(nt) - nt\mu_n]$, $0 \leqslant t \leqslant 1$, $n \geqslant 1$. The following result is the generalization of Theorem 17.3 of [2] to the setting of Prohorov's Theorem.

Renewal Process FCLT. If (1) and (2) hold, then

$$Z_n \Rightarrow Z \quad \text{in } D \; ,$$

where $Z = \{Z(t), t \geqslant 0\}$ is Brownian motion with zero drift and variance $\mu^3\sigma^2$.

We conclude this section with a result on random time changes. Let D_0 be the set of elements ϕ of D which are nondecreasing and satisfy $0 \leqslant \phi(t) \leqslant 1$ for $0 \leqslant t \leqslant 1$. For $x \in D$, let $x \circ \phi(t) = x(\phi(t))$ for $0 \leqslant t \leqslant 1$. Assume we have a probability space (Ω, F, P) on which $X \in D$ and $\Phi \in D_0$ are defined, and in addition we have a sequence of probability spaces $\{(\Omega_n, F_n, P_n), n \geqslant 1\}$ each element of which has an $X_n \in D$ and $\Phi_n \in D_0$ defined on it. The following result is proved in section 17 of [2].

Random Time Change Theorem. If $(X_n, \Phi_n) \Rightarrow (X, \Phi)$ in D, and $P\{X \in C\} = P\{\Phi \in C\} = 1$, then $X_n \circ \Phi_n \Rightarrow X \circ \Phi$ in D.

REFERENCES

[1] V. Benes (1963). *General Stochastic Processes in the Theory of Queues.* Addison-Wesley, Reading, Mass.

[2] P. Billingsley (1968). *Convergence of Probability Measures.* John Wiley and Sons, New York.

[3] D. R. Cox and H. D. Miller (1965). *The Theory of Stochastic Processes.* Wiley, New York.

[4] G. J. Foschini (1977). On heavy traffic diffusion analysis and dynamic routing in packet switched networks. *Computer Performance.* North Holland, Amsterdam, 499-514.

[5] G. J. Foschini and J. Salz (1978). A basic dynamic routing problem and diffusion. *IEEE Trans. on Comm.* **26**, 320-327.

[6] J. M. Harrison (1973). A limit theorem for priority queues in heavy traffic. *J. Appl. Probl.* **10**, 907-912.

[7] D. L. Iglehart and W. Whitt (1970). Multiple channel queues in heavy traffic, I and II. *Adv. Appl. Prob.* **2**, 150-177 and 355-364.

[8] Yu. V. Prohorov (1956). Convergence of random processes and limit theorems in probability theory. *Theor. Probability Appl.* **1**, 157-214.

[9] M. I. Reiman (1983). Open queueing networks in heavy traffic. *Math of O.R.* to appear.

[10] W. Whitt (1968). Weak convergence theorems for queues in heavy traffic. Ph.D thesis, Cornell University.

[11] W. Whitt (1971). Weak convergence theorems for priority queues: preemptive-resume discipline. *J. Appl. Prob.* **8**, 74-94.

[12] W. Whitt (1980). Some useful functions for functional limit theorems. *Math of O.R.* **5**, 67-85.

BOUNDARY CONDITIONS FOR DIFFUSION APPROXIMATIONS TO QUEUEING PROBLEMS

René Boel

Lab. for System Dynamics
Rijksuniversiteit Gent
Grote Steenweg Noord 2
B 9710 Zwijnaarde (Gent)
Belgium

and: Department of Systems Engineering
A.N.U., Canberra, A.C.T., Australia

Abstract

One of the main difficulties of the diffusion approximation to queues is that it underestimates the probability of an empty queue. This is due to the use of instantaneous reflection at the boundary. In this paper the use of diffusion processes with delayed reflections is suggested. A survey of the available mathematical literature on this topic is given. A method is suggested to give a rational choice of the stickyness of the boundary, based on how long the queue is empty in an interval up to the first buffer overflow. Some applications (ergodic theorems and recursive estimation) and some problems for further research are outlined.

1. Introduction

The mathematical analysis of queues (especially of the transient phenomena considered in control applications) is very difficult because at empty queue and at full buffer, complicated boundary conditions are added to otherwise fairly straightforward difference equations. Several authors have tried to simplify the analysis by using a diffusion approximation [1,2,3]. In the one dimensional case this leads to a diffusion process with reflection at the "empty queue" boundary and, in case of finite waiting room, at the "full buffer" boundary. The arrival rate $\lambda(Q_t)$ and the departure rate $\mu(Q_t)$ determine the drift $f(x_t)$ and the variance $\sigma^2(x_t)$ in the stochastic differential equation

$$dx_t = f(x_t)\,dt + \sigma^2(x_t)\,dw_t + d\ell_t^1 + d\ell_t^2 \qquad (1)$$

Here ℓ_t^i is the local time of x_t, a measure of the amount of time the process x_t spends on the boundary: $\{x=0\}$ for $i = 1$, $\{x=B\}$ for $i = 2$:

$$\ell_t^1 = \lim_{\varepsilon\downarrow 0} \frac{\sigma^2(0)}{2\varepsilon} \text{ Lebesgues measure } \{0 \leq s \leq t : 0 \leq x_s < \varepsilon\}$$

$$= \lim_{\varepsilon\downarrow 0} \sigma^2(0)\varepsilon . \# \{\text{downcrossing of } x_s \text{ from } \varepsilon \text{ to } 0,\ 0 \leq s \leq t\}. \qquad (2)$$

(see McKean [4], Ito - McKean [5], Williams [6]). If $\sigma(x)$ is continuous at the boundary then the above definitions specify a reflected diffusion process. This corresponds to the differential generator $(A,\mathcal{D}(A))$ for $Ag(x) = f(x)\, g'(x) + \frac{\sigma^2(x)}{2} g''(x)$ for $g \in \mathcal{D}(A) = \{g \in C_0^2,\ g'(0) = 0\}$. Ito and McKean [5,§2] show that

x_t spends no time (in the sense of Lebesgue measure) at the origin, with probability one. However in §4 we will see that if $\sigma(0) \neq \lim_{x \downarrow 0} \sigma(x)$, then x_t can spend a non-negligible amount of time at the boundary. This is very important for our models since then the atom $P(x_t = 0 \mid x_o) > 0$ has to be taken into account.

Consider a queueing network with the vector of queue lengths Q_t^n, corresponding to the mean interarrival time $1\ \lambda_i^n$ and mean service time $1\mu_i^n$ at the i th queue. Reiman [7] has shown that the diffusion approximation $Z_t = \lim_{n\to\infty} Q_{nt}^n/\sqrt{n}$ is a reflected Ornstein-Uhlenbeck process in the multi-dimensional case, assuming renewal processes as arrivals and general independent service times (with $\lambda_i^n/\mu_i^n \to 1$ as $n \to \infty$). Reiman assumes λ and μ independent of the queue lengths Q_t (a vector), but this can be generalized, as long as $f(x)$ and $\sigma^2(x)$ turn out to be sufficiently smooth.

While theoretical investigations always lead to an instantaneous reflection boundary condition, $f'(0)=0$ (i.e. $P(x_t=0|x_o) = 0$), several applications studies have shown that the stationary probabilities obtained via diffusion approximation give more accurate results if one modifies the model as follows (see Gelenbe-Mitrani [9, chpt. 4.2 and 4.3]). Once x_t hits the boundary $x_t=0$, it remains there for an exponentially distributed time (mean $1/\lambda$), and then jumps to an interior point (corresponding to queue length 1). Instead of a reflecting boundary condition this introduces a complicated (non-local) boundary condition in the differential generator of the Markov process x_t. Simulation has shown that especially for small N, the stationary probability $P(Q_t=N)$ is a lot more accurate in this case. This is heuristically obvious because this model gives a more accurate description of an idle period. Theoretically it cannot be justified since in the time scale ($nt \to \infty$) of the diffusion approximation, the idle period should be negligible.

In a paper on diffusion approximations to the content of a reservoir behind a dam, Harrison and Lemoine [10] consider a different arrival rate for empty reservoirs and non-empty reservoirs. This amounts to $f(x)$ and $\sigma^2(x)$ having a discontinuity at the origin. It turns out that the resulting diffusion approximation x_t sticks to the boundary. Its differential generator $(A,\mathcal{D}(A))$ is : $Ag(x) = f.g'(x) + \frac{\sigma^2}{2}g''(x)$ for $x \in (0,\infty)$ with domain $\mathcal{D}(A) = \{g \in C_o^2 : \frac{f'(0)}{\rho} = \lim_{x\downarrow 0} (f.g'(x) + \frac{\sigma^2}{2} g''(x))\}$. This corresponds to a diffusion process with delayed reflection, discussed in detail by Gihman and Skorohod [11, § 24]. It implies that while $\{0 \leq s \leq t : x_s = 0\}$ is still a set which contains no intervals (i.e. whenever $x_{t1} = 0$, $x_{t2} = 0$, there is an s, $t_1 \leq s \leq t_2$, s.t. $x_s > 0$), it has positive Lebesgue measure with probability one. Thus $P(x_t = 0 \mid x_o) > 0$ is an atom to be taken into account. Harrison and Lemoine do not give any details on the corresponding queueing problem. Nevertheless one expects that stationary probabilities for short queue lengths will be more accurate using this delayed reflection diffusion approximation.

Indeed, as explained earlier, the boundary conditions are what makes the analysis of queues difficult. Therefore in this paper we will compare different mathematical treatments of diffusions with delayed reflection (terminology of Gihman and Skorohod [11]) or sticky diffusions (terminology of Stroock and Varadhan [12], as applied to queueing networks in [8]). These will be compared with the sticky diffusion approximation of Harrison and Lemoine [10] and with a diffusion which jumps away from the boundary [9]. The mathematical problem is difficult because two limiting operations are carried out at the same time. The diffusion approximation takes the infinite time scale nt for $n \to \infty$, while the boundary condition considers the boundary layer $\{0 \leq Q^n_{nt}/\sqrt{n} \leq \varepsilon\}$ for $\varepsilon \downarrow 0$. Depending on the order in which the limits are taken one can obtain a reflection or an absorption or (if ε and n depend on each other) something in between, i.e. delayed reflection.

In order to make this modified boundary behaviour lead to better approximations, one has to choose in a rational way the parameter ρ, the stickyness, a measure of how much time is spent on the boundary. Intuitively the boundary behaviour depends on how long the queue remains near the boundary ($L_t = m\,(0 \leq s \leq t : Q_s = 0)$ or $m\,(0 \leq s \leq t : 0 \leq Q_s \leq \varepsilon_o)$)before it makes a large excursion, i.e. before $T = \inf\,\{t > 0 : Q_t = K\}$ where K is the buffer size in case of a finite waiting room, or some heuristically chosen number considered an undesirably long queue. In § 2 some joint distributions for these random variables will be calculated. Both Markovian techniques and martingale methods (cf. Boel [12], Kennedy [18], Rosenkrantz [19]) will turn out to be useful.

In § 4 the corresponding relations between the local time ℓ^1_t (a measure of how long x_t remains at zero) and the hitting time $\tau = \inf\,\{t > 0 : x_t = \mu\}$ when the upper boundary is first reached, will be discussed. Some results are already available via semi-group and stochastic calculus methods, and can be found in [Ito -McKean, 5] and in [Gihman and Skorohod, 11]. It is hoped that further explicit calculations will lead to a good measure of relations such as $E(\ell^1_\tau|\tau)$, depending on the stickyness ρ^1. Comparison of these with $E(L_T|T)$ should then lead to a rational choice of the parameters of the sticky diffusion, the drift f (or f(x)), the variance σ^2(or $\sigma^2(x)$) and the stickyness ρ^1 at the origin. Similar calculations are possible for the case of a full buffer.

This preliminary version of the paper has two purposes. Firstly, to extract from the mathematical literature some relevant results on boundary behaviour of diffusions; and secondly, to explain qualitatively a method which, after further calculations and verifications, will hopefully lead to a rational choice of the stickyness parameter of this diffusion. In § 5 some results from Gihman and Skorohod [11] on stationary densities and some applications to recursive non-linear filtering will be outlined, in order to illustrate that the method may be of more than just academic interest.

2. Boundary behaviour of a queueing system

Since this paper is only a first attempt at applying the ideas explained in the introduction, we consider only the simple case of an M/M/1/K queue with finite buffer. This means that the arrival rate is $\lambda . 1_{Q_{t-} < K} = \lambda_t$ and the departure rate $\mu . 1_{Q_{t-} > 0} = \mu_t$. The probability distributions of the point processes A_t (arrivals) and D_t (departures) are specified by the statements that $A_t - \int_0^t \lambda_s ds$ and $D_t - \int_0^t \mu_s . ds$ are (P, F_t) martingales. Since $Q_t = Q_o + A_t - D_t$, this is all the information necessary to analyze the M/M/1/K-queue.

Note that for an M/M/1 queue (infinite buffer, $K = \infty$) similar expressions to what follows can be derived, by imposing an artificial "long waiting line" boundary K_o. The interpretation of the results will be quite different though.

Consider for the M/M/1/K queue, intervals $[R_n, R_{n+1}]$ determined by visits of Q_t (a finite-state Markov process) to the states 0 or K : $R_{n+1} = \inf\{t > R_n + S_n : Q_t = 0 \text{ or } Q_t = K\}$ where $R_n + S_n$ is the first time Q_t is different from Q_{R_n} (S_n is the exponentially distributed time until an arrival after the queue has become empty or until a departure after the waiting room has become full). During $[R_n + S_n, R_{n+1})$, the arrivals and departures form independent Poisson process with fixed rates λ and μ (since $0 < Q_t < K$). Hence (Q_{R_n}, R_n) is a semi-Markov process.

Consider now the queue during one renewal interval $[R_n + S_n, R_{n+1})$, denoting $R_n + S_n$ as time origin, $R_{n+1} - R_n - S_n = R$, $Q_{R_n + S_n} = Q_o$. Since $A_t - \lambda t$ and $D_t - \mu t$ are independent martingales on $[0,R]$, stochastic calculus shows that $z_1^{Q_t} . z^{D_t} . e^{-[(z_1 - 1)\lambda - (\frac{z}{z_1} - 1)\mu]t}$ is a martingale in $[0,R]$. Applying optional sampling at R (justified in [13]) gives an explicit formula $f(s,z,Q_o,Q_R)$ for

$$E(z^{D_R} e^{-sR} | Q_o, Q_R) = \frac{1 - \rho^K}{\rho^{K-Q_o} - \rho^{Q_R}} \cdot \frac{Z_1(s,z)^{Q_o - K} - Z_1(s,z)^{-Q_R} . g(s,z,Q_o)}{1 - Z_1(s,z)^{-K}}$$

where $\rho = \frac{\lambda}{\mu}$, $Z_1(s,z) = \frac{s+\lambda+\mu}{2\lambda}(1 - a(s,z))$, $a(s,z) = \sqrt{1 - \frac{4\lambda\mu z}{(s+\lambda+\mu)^2}}$

$$g(s,z,Q_o) = (\frac{\mu z}{s+\lambda+\mu})^{Q_o} \cdot d_{K-Q_o-1}(s,z) + (\frac{\lambda}{s+\lambda+\mu})^{K-Q_o} \cdot d_{Q_o-1}(s,z)$$

$$d_n(s,z) = \frac{1}{2^{n+1}} [(1+\frac{1}{a(s,z)})(1+a(s,z))^n + (1-\frac{1}{a(s,z)})(1-a(s,z))^n]$$

In the next paragraph we will be interested in the conditional distribution of R given $Q_o = 1$, $Q_R = 0$ or K.) From the above formula we can calculate for example:

$$E(R \mid Q_o = 1, Q_R = K) = \frac{\rho^K - 1}{\rho^{K-1} - \rho^K} \cdot \frac{\partial}{\partial s} \left(\frac{Z_1(s,1)^{1-K} - Z_1(s,1)^{-K} \cdot g(s,1,1)}{1 - Z_1(s,1)^{-K}} \right) \Bigg|_{s=0}$$

L' Hopitals' rule has to be applied twice which makes the result rather complicated.

Consider as an application the boundary behaviour near the "empty queue", i.e. consider an interval $[R_n,R_k]$ starting with $Q_{R_n} = 0$ and such that $k = \inf\{\ell > n : Q_{R_\ell} = K\}$ is the first time a large excursion (to a full buffer) occurs. In this paragraph set for simplicity $R_n = 0$, $R_k = T$. During the larger renewal interval $[R_n,R_k)$ (denoted $[0,T)$) we are also interested in the following random variables:

$N = k - n$ = number of times the origin $Q_t = 0$ is visited during $[0,T)$

$$L_T = \int_0^T 1_{Q_{s-} = 0} \cdot ds = m(0 \leq s < T : Q_s = 0)$$

= amount of time the queue is empty during the renewal interval $[0,T)$.

An operationally interesting variable will be $\frac{L_T}{T} \times 100\%$, the percentage of time the service station is inactive. As explained in the introduction the interaction (that is the joint distribution) of L_T, T and N is a measure of the stickyness of the system.

N is obviously geometrically distributed. Indeed for each ℓ, $n \leq \ell < k$, at $T_\ell + S_\ell$ a new small interval starts with initial state $Q_{R_\ell + S_\ell} = 1$, and has probabilities $p_K(1)$, resp. $p_0(1)$ to end the large renewal interval $[0,T)$ (i.e. $\ell = k-1$) or to continue it. Hence $P(N=n) = (\frac{\rho-\rho^K}{1-\rho^K})^n (\frac{1-\rho}{1-\rho^K})$.

Now T consists of the following intervals, whose lengths are independent given N:

N+1 exponentially distributed interarrival times S_ℓ (mean $\frac{1}{\lambda}$); $L_T = \sum_{\ell=0}^{N} S_\ell$

N renewal intervals as considered in the previous paragraph, with $Q_{R_\ell + S_\ell} = 1$,

$Q_{R_{\ell+1}} = 0$, and characteristic function $f(s,1,1,0)$ $(=f(s,1,0)$ from here on, since $z = 1$ always).

1 renewal interval as above, but with $Q_{R_{k-1}} + S_{k-1} = 1$, $Q_{R_k} = K$, and characteristic function $f(s,1,K)(=f(s,1,1,K))$

Since L_T and $T-L_T$ are independent one can write:

$$E(e^{-uL_T-v(T-L_T)} | N = n) = (\frac{\lambda}{\lambda+u})^{n+1}.f(v,1,0)^n\ f(v,1,K)$$

Combining this with the geometric distribution for N and using a change of variables $s_1 = u-v$, $s_2 = v$, gives

$$h(z,s_1,s_2) = E(z^N e^{-s_1L_T-s_2T}) =$$

$$= \frac{1-\rho^{K-1}}{\rho^{K-1}-\rho^K} \cdot \frac{(1-\rho)\ .(Z_1(s_2,1)-g(s_2,1,1)}{(1-\rho^{K-1})(\lambda+s_1+s_2)(Z_1(s_2,1)^K-1)-(\rho-\rho^K)\lambda.z(Z_1(s_2,1)^K g(s_2,1,1)-Z_1(s_2,1))}$$

In principle, we can now calculate by inversion (numerically if necessary) $p_T(t)$ and $p_{L_T,T}(\ell,t)$. This gives for example the interesting operational characteristic $E(\frac{L_T}{T}) = \int \frac{E(L_T|T=t)}{t} \cdot p_T(t)\ dt$, which one hopes would be a measure of the stickyness at the boundary 0 of the system.

To illustrate the power of a combination of martingale and renewal methods, we consider an alternative method to calculate : $E(R \mid Q_\circ, Q_R)$. On $[0,R)$ both

$Q_t-(\lambda-\mu)t$ and $Q_t^2 + 2(\lambda-\mu)t\ Q_t - (\lambda+\mu)t - (\lambda-\mu)t^2$

are martingales, as can easily be verified using stochastic calculus on the martingale increments $dA_t - \lambda dt$ and $dD_t - \mu dt$ (observe in particular that $E((dQ_t)^2 \mid Q_t) = (\lambda + \mu)\ dt$). Applying optional sampling at R gives:

$$E\ Q_R - (\lambda - \mu)\ E(R|Q_\circ) = Q_\circ = K.p_K(Q_\circ) - (\lambda - \mu)\ E\ (R|\ Q_\circ)$$

$$K^2.p_K(Q_\circ) - 2\ (\lambda - \mu)\ K.E\ (R|\ Q_\circ,\ Q_R = K) - (\lambda + \mu)\ E\ (R|\ Q_\circ) - (\lambda- \mu)E(R^2|Q_\circ)=Q_\circ^2$$

Combine this with $E\ (R\ |Q_\circ) = p_K(Q_\circ)E(R|Q_\circ,Q_R=K)+p_\circ(Q_\circ).E(R|Q_\circ,Q_R = 0)$ and we have three equations in four unknowns. The fourth equation can be obtained as follows (cf. [13]). To calculate $E(R^2|Q_\circ)$, restart at the first arrival or departure. A new interval $\tilde{R}$ is generated with the same distribution but initial state $Q_\circ-1$ or $Q_\circ+1$, after an exponentially distributed time.

$$E\ (R^2 \mid Q_\circ) = \frac{2}{\lambda\mu} + \frac{K(p_K\ (Q_\circ-1) + p_K(Q_\circ+1)) - 2\ Q_\circ}{(\lambda-\mu)(\lambda+\mu)}$$

$$+ \frac{\lambda}{\lambda + \mu} \cdot E\,(R^2 \mid Q_o+1) + \frac{\mu}{\lambda + \mu}\, E\,(R^2 \mid Q_o-1)$$

This difference equation, with boundary conditions $E\,(R^2 \mid 0) = E\,(R^2 \mid K) = 0$, can be solved explicitly. Unfortunately the results are so complicated that I have not even been able yet to verify whether they agree with the results obtained for $E\,(R \mid Q_o, Q_R) = \frac{\partial}{\partial s} f\,(s,Q_o,Q_R) \mid_{s = 0}$.

In summary, in principle conditional means such as $E\,(L_T \mid T)$, $E\,(T \mid N)$, etc. can be calculated. They should correlate with the stickyness of a diffusion model. Statistics for them seem easy to obtain from monitoring a real system, which should allow estimation of the stickyness parameter. Further analysis is in progress.

3. Diffusion processes with accessible boundary

In this chapter we summarize briefly some results on the boundary behaviour of a diffusion process on the state space S :

$$dx_t = f\,(x_t)\,dt + \sigma(x_t)\,d\,w_t,\quad x_t \in \text{int}(S) \tag{1}$$

Details can be found in Gihman and Skorohod [11, chapter 5] for a stochastic differential equation approach and in Breiman [14, chpt. 16] and Wong [15, & 5.5] for a semi-group approach.

Allowing drift f(x) and variance σ(x) to be state dependent is useful to describe controlled queues (λ_t and μ_t dependent on the queue length Q_t). Since $Q_t \geq 0$, a diffusion approximation to a queue will have as state space $S = [0, \infty)$ or $[0,1]$ (after rescaling the finite buffer case). To simplify the notation consider the following further rescaling. Breiman [14, § 16.5] shows that there exists a strictly increasing, continuous function $\mu\,(x)$ such that for any $J=(a,b) \in \text{int}(S)$

$$\tau(J) = \inf\,(t : x_t \notin J),\quad p^+(x,J) = P(x_{\tau(J)} = b \mid x_o = x) = \frac{\mu(x) - \mu(a)}{\mu(b) - \mu(a)}$$

Then $\tilde{x}_t = \mu(x_t)$ is a Markov process on its natural scale. It has at each time equal probabilities to go up or down, i.e. $\tilde{f}\,(\tilde{x}) = 0$. From Wong [15, § 5.3] we see that

$$\mu(x) = \int_c^x \exp\left(-\int_c^y \frac{2f(z)}{\sigma^2(z)}\,dz\right) dy,\quad c \in S.$$

$$d\,\tilde{x}_t = \tilde{\sigma}\,(\tilde{x}_t)\,dw_t,\quad \tilde{\sigma}\,(x) = \sigma\,(x)\,.\,\exp\left(-\int_c^x \frac{2f(z)}{\sigma^2(z)}\,dz\right)$$

Further on we always work with processes in their natural scale.

Since in any reasonable model for a queue, Q_t will reach its finite boundaries (0 and K) in a finite time with probability one, we expect the same to hold for its diffusion approximation. Such boundaries are called accessible (Breiman [14, § 5.6], Wong [15, § 5.5]. It has been shown that, with J and τ(J) as longer. They then take the limit $\frac{1}{\sqrt{n}} Q^n_{nt}$ and obtain a sticky diffusion with $\rho = (\frac{\lambda}{\lambda'} - 1)^{-1}$. This rather artificial choice $\lambda' < \lambda$, has an important consequence as explained in the next section :

If : $\rho = \infty$ or $m(\{0\}) = 0$ $(\lambda = \lambda')$ then $P(x_t = 0 | x_o) = 0$

If $0 < \rho < \infty$ or $m(\{0\}) > 0$ $(\lambda' < \lambda)$ then $P(x_t = 0 | x_o) > 0$.

Intuitively one can think of another justification of the sticky boundary case. If one considers an M/M/1/K queue and takes the usual diffusion approximation [7] $\frac{1}{\sqrt{n}} Q^n_{nt}$, then K is transformed into $\frac{K}{\sqrt{n}} \to 0$, and the boundary effects become negligible. Thus the two limiting operations occurring in the boundary condition of the differential generator have to be taken into account simultaneously : $x \downarrow 0$ and $nt \uparrow 0$. To make the boundary conditions influence the results one has to take a slower time scale when x_t is small. This it will turn out in the next section is exactly what happens in the construction of a slowly reflected Brownian motion.

4. Local time for reflected Brownian motion

For uncontrolled queues one can take drift and variance to be state-independent. Hence in this section, by rescaling, we limit ourselves to Brownian motion restricted to the state space $S = [0,\infty)$(or$[0,1]$, but for ease of notation all results are developed for the boundary 0 only). We summarize results from Ito and McKean [5, chapter 2 and 4] and from Gihman and Skorohod [11, chapter 5].

First consider an instantaneously reflected Brownian motion, $x_t = |w_t|$, $d\,x_t = \operatorname{sgn} w_t \cdot dw_t + d\ell_t = d\tilde{w}_t + d\ell_t$ where $\tilde{w}_t$ is another Brownian motion and $\ell_t = \ell_t(0)$ is the local time at the origin. This is a special case of the local time at a:

$$\ell_t(a) = \lim_{\varepsilon \downarrow 0} \frac{1}{2\varepsilon} m\{(s : 0 \leq s \leq t, a \leq x_s \leq a + \varepsilon)\}$$

$$= \lim_{\varepsilon \downarrow 0} \varepsilon . \#\{\text{downcrossings of } x_s \text{ from } a + \varepsilon \text{ to } a,\ 0 \leq s \leq t\}.$$

A lot is known about the weird character of the increasing, continuous, non-differentiable process ℓ_t, which is almost everywhere constant. The joint distribution of $|w_t|$ and ℓ_t is fairly simple:

$$P(|w_t| \in dx,\ \ell_t \in dy) = 2 . \frac{}{\sqrt{2\pi}\, t^{3/2}} . e^{-\frac{(x+y)^2}{2t}} . dx\, dy$$

Unfortunately, even for simple r.v.'s such as $T_a = \inf\{t > 0 : |w_t| = a\}$, the

$$\mathcal{D}(A) = \{g \in C^2 [0,\infty) : f(0)\ g'(0) = \lim_{x \downarrow 0} \frac{g''(x)}{2}\}$$

Remark: Gihman and Skorohod [11] actually construct the diffusion process

$$d\,x_t = f(x_t)\,dt + \sigma(x_t)\,dw_t$$

with delayed reflection, by using the above method. The existence of a solution is shown if $f(x)$, $\sigma(x)$ are Lipschitz in $(0,\infty)$, $f(0) > 0$, $\sigma(0) = 0$, $\sigma(x) > 0$ in $(0,\infty)$. Note that $f(x)$ and $\sigma(x)$ have a jump at 0.

We can now compare with the results of § 3. A diffusion with delayed reflection is the same as the sticky Brownian motion of Harrison and Lemoine [10], if we take $f(0) = \rho$. Moreover, in the classification of Breiman [14, § 16.1] it corresponds to the choice $0 < m(0) = \frac{1}{f(0)} < \infty$, i.e. a slowly reflecting process.

The stochastic differential equation representation also allows us to associate it to the sticky diffusion process with boundary constructed by Stroock and Varadhan [12]. They solve:
$d\,x_t = f(x_t)\,dt + \sigma(x_t)\,dw_t + d\ell_t$ where the local time ℓ_t is specified by

$$\int_0^t 1_{x_s = 0}.\ ds = \rho.\ \ell_t.$$

Thus we can associate this x_t to the process with delayed reflection by taking $f(x)$ the same for $x > 0$, while $f(0)$ of [11] corresponds to $\frac{1}{\rho}$ of [12] and $d\ell_t = f(0).1_{x_s=0}.\ ds.$

Note that ρ in Stroock and Varadhan [12] corresponds to ρ^{-1} in Harrison and Lemoine [10].

It is very important to note that the process with delayed reflection has positive probability of being on the boundary, i.e. $P(x_t = 0 \mid x_o = 0) > 0$. Indeed (Breiman [14, problem 16.11]):

$E\,(\text{measure}\,\{t : x_t = 0,\ 0 < t < T_c\} \mid x_o = 0) = \frac{c}{\rho}$. Nevertheless ℓ_t is still a continuous, non-differentiable process which has no intervals where it increases strictly (i.e. $\{x_t = 0\}$ contains no intervals and no isolated points w.p.1). The conditional distribution of x_t thus has an atom at 0. Nevertheless we can define a density w.r.t. the speed measure $m\,(dy)$ (which has an atom $m\,(\{0\})$ at the origin). This idea is developed in Ito and McKean [5, § 4.11]:

$$P\,(x_t \in dx \mid x_o = y) = p\,(t,y,x)\ m\,(dx)$$

where $0 \leq p\,(t,y,x) = p\,(t,x,y)$ is continuous and satisfies

$$\frac{\partial}{\partial t}\,p\,(t,x,y) = \frac{1}{2}\,\frac{\partial^2}{\partial x^2}\,p\,(t,x,y)$$

above, there existsa speed measure

$$m(x,J) = E(\tau(J) \mid x_o = x) = \int_J G_J(x,y) \frac{dy}{\sigma^2(y)}$$

where $G_J(x,y) = \frac{(\min(x,y) - a)(b - \max(x,y))}{b - a}$ and where the measure $m(dy) = dy/\sigma^2(y)$ measureshow long it takes $\sigma(x_t)\, dw_t$ to move up or down by $\frac{dy}{2}$. The boundary 0 is accessible if $\mu(0) > -\infty$ and $\int_{(0,c)} y. \frac{dy}{\sigma^2(y)} < \infty$ for some c (0,1)

because one can show then that $P(\tau(0,c) > t \mid x_o = x) = \alpha < 1$ for some $x \in (0,c), t>0$. Similarly at the boundary 1 we have accessibility if $\mu(1) < \infty$ and $\int_{(c,1)} (1-y) \frac{dy}{\sigma^2(y)} < \infty$.

A second requirement for a good model is that the queue can move away from a boundary after reaching it. This means that $\sigma(y)$ can not tend to 0 too fast. We therefore have to assume that the boundary is regular; i.e. we impose the further conditions:

$$\int_{(0,c)} \frac{dy}{\sigma^2(y)} < \infty \quad , \quad \int_{(c,1)} \frac{dy}{\sigma^2(y)} < \infty$$

Note that the speed measure is defined so far only for interior points of S. For a regular boundary 0, there are now different ways of defining $m(\{0\})$. Breiman [14, § 16.7] gives the following classification:

$m(\{0\}) = \infty$: 0 is absorbing

$m(\{0\}) = 0$: 0 is instantaneously reflecting.

$0 < m(\{0\}) < \infty$: 0 is slowly reflecting (or has delayed reflection, is sticky (Stroock-Varadhan [12])

Since $\frac{1}{m(dy)} \approx \frac{\sigma^2(y)}{dy}$ measures how long it takes to move away from the value y we see that if $\sigma(0) > 0$ then $m(\{0\}) = 0$ $(dy \downarrow 0)$ and at 0 the particle has infinite speed and moves away from the boundary immediately. On the other hand if $m(\{0\}) = \infty$ then, intuitively, it will take infinitely long to move away from 0, i.e. 0 is absorbing boundary. Note that in the intermediary case of slow reflection $\sigma(0) = 0$ is still necessary. The particle moves away from 0 with finite speed, i.e. it drifts away and $f(0) > 0$ while $f(x) = 0$ had been assumed for all $x > 0$. Hence the functions $f(x)$ and $\sigma(x)$ are discontinuous, except at an instantaneously reflecting boundary.

Another interpretation of these boundary conditions is obtained via the boundary conditions for the corresponding differential generators. This is treated very thoroughly by Ito and Mc Kean [5], with the most general results in

distribution of ℓ_{T_a} does not seem to be known. Indeed for $\tilde{T}_a = \inf\{t>0 : w_t = a\}$ (Ito-McKean [5, p.72]) $E(e^{-s\ell_{\tilde{T}_a}} \mid w_0 = 0) = \frac{1}{1+as}$, i.e. $\ell_{\tilde{T}_a}$ is exponentially distributed. However because of the correlation between $\tilde{T}_a$ and $\tilde{T}_{-a}$, it does not seem possible to obtain the distribution of ℓ_{T_a}.(1) In order to tie up with the result of §2, we should calculate the joint distribution of T_1 and ℓ_{T_1} and compare it with the measured correlation of T and L_T. Hopefully, just as in §2 a combination of renewal arguments and optional sampling for martingales will lead to explicit solutions. Some results in this direction can be found in [5, § 5.2]. If $0 \leq a < x \leq y < b$, T = first passage time of $x_t = |w_t|$ to a via $b = \inf\{t > 0 : x_t = a, \sup_{0 \leq s \leq t} x_s \leq b\}$, then :

$$E(\ell_{T_a \wedge T_b}(y) \mid x_0 = x) = \frac{(x-a)(b-y)}{b-a}$$

$$E(\ell_T(y) \mid x_0 = x) = b-a$$

It is clear from the definition of the local time that measure $(s : 0 \leq s \leq t, x_s = 0) = 0$ w.p.1. This contradicts the fact that for queues the set $(s : 0 \leq s \leq t, Q_s = 0)$ has positive measure , with probability 1. We try then to increase the amount of time x_t spends at the boundary $\{x_t=0\}$, following Gihman and Skorohod [11, § 24] in the construction of a process with delayed reflection.

Let $\tilde{x}_t = |w_t|$ have the local time $\tilde{\ell}_t$. Then $k(t) = t + \frac{\tilde{\ell}_t}{f(0)}$ has an inverse τ_t such that $t = \tau_t + \frac{\tilde{\ell}_{\tau_t}}{f(0)}$. Define $x_t = \tilde{x}_{\tau_t}$. Then verify that 0, where $d x_t = \operatorname{sgn} w_t . dw_t = d\tilde{w}_t$ for $x_t > 0$, where $\tilde{w}_t$ is a new Brownian motion. At the boundary, Gihman and Skorohod obtain:

$$\int_0^t 1_{x_s=0} . ds = \frac{\tilde{\ell}_{\tau_t}}{f(0)}$$

Moreover the differential generator of this process x_t is: $Ag(x) = \frac{1}{2} g''(x)$ with the boundary condition specified by

---

(1) Note added in proof: see [20] for a conjecture that ℓ_{T_a} is nevertheless exponential.

with boundary condition

$$p^+ (t,0^+,y) = \lim_{z \downarrow 0} \frac{p(t,z,y) - p(t,0,y)}{z}$$

$$= m (\{ 0 \}) . \lim_{z \downarrow 0} \frac{1}{2} \frac{\partial^2}{\partial x^2} p(t,x,y) \Big|_{x=z}$$

In [5, § 5.4] one finds the important interpretation :

$$p (t,x,y) = \frac{\partial}{\partial t} E (\ell_t(y) \mid x_o = x) .$$

Using similar ideas, one can find for bounded, continuous functions $g(x)$ and $\bar{J} \subset S$, $\tau(J) = \inf \{t > 0 : x_t \in J\}$, J open, that

$$E(\int_0^{\tau(J)} g(x_t) \, dt \mid x_o = x) = \int_J G_J(x,y) \, g (y) \, m (dy)$$

If one could extend this formula to allow J closed, i.e. $J = [0,1)$ e.g. then it would allow calculation of properties correlating approximately ($f(x)$ continuous excludes $f(x) = 1_{x=0}$ or $1_{x<\varepsilon}$) $\ell_T(0)$ and T, our ultimate goal.

One further type of process has to be considered. From $x_o > 0$, x_t starts as a diffusion. Let T_1 be the first time x_t hits the boundary 0. Then $x_t = 0$ for $t \in [T_1, T_1 + S_1)$ where S_1 is an exponentially distributed random variable, independent of what happened up to T_1. At $T_1 + S_1$, the process jumps to a new state in the interior of the state space. Let p (dx) be the distribution of $x_{T_1 + S_1}$ independent of what happened up to $T_1 + S_1$. From $x_{T_1 + S_1}$ at $T_1 + S_1$ the process starts anew as a diffusion with the same characteristics as in the first interval. Continue this construction indefinitely. Gelenbe and Mitrani [9, chapter 4] used this as a diffusion approximation for a queue, with $p\,(dx) = 1_{1 \in dx}$ (i.e. always jump to 1) and $E\, S_1 = \frac{1}{\lambda}$ = the mean interarrival time. This process clearly gives a good description of the queue at small queue lengths, but its backward generator is a partial differential - difference equation for which few analytical results can be expected. Since the purpose of diffusion approximations is to obtain analytical results for stationary, and even for transient solutions, this may be a serious disadvantage of such jump-type models.

5. Applications

While it does not seem possible yet to calculate expressions such as $E\, (\ell_\tau \mid \tau)$ even for an instantaneously reflected Brownian motion, several explicit applications are possible even for a general diffusion with delayed reflection. To include controlled queues consider:

$d\,x_t = f(x_t)\,dt + \sigma(x_t)\,dw_t$, $x_t \geq 0$, with delayed reflection at the boundary $x_t = 0$, specified by $f(0) > 0$, $\sigma(0) = 0$. For simplicity of notation the following formulas are written down in natural scale: $\tilde{f}(x) = 0$, $x > 0$.

Gihman and Skorohod [11] calculate the ergodic distribution $F(y) = \lim_{t\to\infty} P(x_t \leq y \mid x_0 = x)$, assuming 0 is a regular boundary:

$$F(y) = \begin{cases} 0 & y \leq 0 \\ \dfrac{1 + 2\,f(0) \int_0^y \frac{dz}{\sigma^2(z)}}{1 + 2\,f(0) \int_0^\infty \frac{dz}{\sigma^2(z)}} & y > 0 \end{cases}$$

In particular:

$$\lim_{T\to\infty} \frac{1}{T} \int_0^T P(x_t = 0)\,dt = \lim_{T\to\infty} \frac{1}{T} \int_0^T 1_{x_t = 0}\cdot dt = \frac{1}{1 + 2\,f(0) \int_0^\infty \frac{dz}{\sigma^2(z)}}$$

for any measurable, integrable function $g(x)$:

$$\lim_{T\to\infty} \frac{1}{T} \int_0^T g(x_t)\,dt = \frac{g(0) + 2\,f(0) \int_0^\infty \frac{g(z)\,dz}{\sigma^2(z)}}{1 + 2\,f(0) \int_0^\infty \frac{dz}{\sigma^2(z)}}$$

It is also possible to use such a diffusion process with delayed reflection as the unobserved state in an estimation problem. The observations could be of the form

a) $dy_t = h(x_t)\,dt + d\,v_t$, signal plus Brownian motion

b) $dy_t = \lambda(x_t)\,dt + d\,m_t$, y_t a point process with rate dependent on the state.

In either case we can apply Girsanov theorem, semi-martingale representation and backward differential generators, i.e. all the tools necessary to obtain the unnormalized linear estimate (cf. Davis and Marcus [16] for a), Brémaud [17] for b) :

$$\pi_t(k) = E(k(x_t) \mid \mathcal{Y}_t) = \frac{E_0(L_t . k(x_t) \mid \mathcal{Y}_t)}{E_0(L_t \mid \mathcal{Y}_t)}$$

where L_t is a likelihood ratio (in case a)

Chapter 4. Consider a process $d\,x_t = \sigma\,(x_t)\,dw_t$. Then for $g \in C^2[0,\infty)$ the differential generator G is

$$G(g)(x) = \frac{\sigma^2(x)}{2}\,.\,g''(x) \text{ for } x > 0 \text{ subject to :}$$

$$\lim_{x\downarrow 0} \frac{g(x)-g(0)}{x} = g^+(0) = m(\{0\}).(Gg)(0) \text{ if } E_o(e^{-T_o^+}) = 1$$

$$\text{or } G(g)(0) = -\chi.g(0) \text{ if } E(e^{-T_o^+}) = 0$$

$$\text{where } E(e^{-T_o^+}) = \lim_{\varepsilon\downarrow 0} E(e^{-T_\varepsilon} | x_o = 0)$$

is 0 or 1 depending on whether x_t takes infinite or zero time to leave $x_t = 0$ (these are the only 2 possibilities). χ actually is the inverse of the probability of killing the process, and in the simple absorption case the boundary condition becomes $g(0) = 0$ ($\chi = \infty$). This happens if $T_o^+ = \infty$ w.p.1. On the other hand, if $E_o(e^{-T_o^+}) = 1$ and if $m(\{0\}) = 0$ then the boundary condition becomes $g^+(0) = 0$, i.e. the reflection condition well known for a reflected Brownian motion. For $E(e^{-T_o^+}) = 1$ and $m(\{0\}) = \infty$, $(Gg)(0) = \lim_{x\downarrow 0} \frac{\sigma^2(x)}{2}\,.\,\frac{g(x)-g(0)}{x^2} = 0$, an extra boundary condition to be imposed at an absorbing boundary, is obtained. For $E_o(e^{-T_o^+}) = 1$ and $0 < m(0) < \infty$ we obtain a slowly reflecting boundary. $\sigma(0) = 0$ is in any case necessary at an absorbing or slowly reflecting boundary, since otherwise $g^+(0)$ has to be unbounded (remember that g is continuously differentiable at 0).

Compare G now with the backward operator for what Harrison and Lemoine [10] call a sticky Brownian motion:

$$Ag(x) = \frac{\sigma^2}{2}\,.\,g''(x) \qquad , x > 0$$

$$Ag(0) = \rho.g'(0) = \lim_{x\downarrow 0} \frac{\sigma^2}{2}\,.\,g''(x)$$

The "stickyness" measure ρ can hence be identified with $\frac{1}{m(\{0\})}$. A physical interpretation is now possible by considering how Harrison and Lemoine [10] obtain their result. They take a diffusion approximation to a symmetrical random walk with exponential holding times (mean $\frac{1}{\lambda}$) for $Q_t \in \{1,2,\ldots\}$. However, if $Q_t = 0$ then jumps to 1 occur after an exponential holding time (mean $\frac{1}{\lambda'}$), which is on the average longer. That is $\lambda' < \lambda$ and the particle sticks to the boundary

$L_t = \exp(-\int_0^t h(x_s)\, dy_s + \frac{1}{2}\int_0^t h^2(x_s)\, ds))$ and π_t a measure satisfying

$d\pi_t(k) = \pi_t(Ak)\, dt + \pi_t\, (h.k)\, .\, d\nu_t$; π_t can be interpreted as the dual of an unnormalized conditional density of the state given the observations. A is the backward operator and $d\nu_t = dy_t - \pi_t\,(h)\, dt$ the innovation. Because of the delayed reflections, there is a positive conditional probability that $x_t = 0$. Hence $\pi_t(k)$ as measure will not be absolutely continuous w.r.t. Lebesgue measure dx, but it will be absolutely continuous w.r.t. the speed measure m(dx):

$\pi_t(k) = \int_{[0,\infty)} q\,(t,x)\, k(x)\, m\,(dx)$. Then q (t,x) will satisfy the equation

$d_t\, q\,(t,x) = A^*\, q(t,x)\, dt + h(x)\, q(t,x)\, dy_t$ where A^* is the forward operator, the dual of A. Its explicit calculation, including its boundary condition involving $m\,(\{\,0\,\}) = \frac{1}{f(0)}$, is tedious but possible. See [20] for further details.

6. Conclusions

This paper has tried to show that diffusion processes with delayed reflections are potentially useful tools for modelling queues. A method is suggested for choosing the appropriate stickyness parameter. A lot of further work is needed to obtain the explicit results required. It should be noted that, using results of Stroock and Varadhan, most of the above results can be extended to the multi-dimensional cases.

7. References

1. D. Gaver and G. Shedler : Multiprogramming system performance via diffusion approximations, IBM Res. Rep. RJ-938, 1971.
2. H. Kobayashi : Application of the diffusion approximation to queueing networks, parts I and II, *J.A.C.M. 21* (1974), p. 316-328; p. 459-469.
3. W. Whitt : Weak convergence for queues in heavy traffic, Ph.D. Thesis, Dept. of Op. Res., Cornell University, 1968.
4. H. McKean : *Stochastic Integrals*, Academic Press, 1969.
5. K. Ito and H. McKean : *Diffusion processes and their sample path*, Springer, 2nd printing, 1974.
6. D. Williams : *Diffusions, Markov Processes and Martingales*, vol 1, Wiley, 1979.

7. M. Reiman : Queueing networks in heavy traffic, Ph.D. dissertation, Dept. of Op. Res., Stanford University, 1977.

8. R. Boel and M. Kohlmann : A control problem in a manifold with nonsmooth boundary, to appear in Proceedings of the 2nd Bad Honnef Workshop on Stochastic Differential Systems, 1982.

9. E. Gelenbe and I. Mitrani : *Analysis and Synthesis of Computer Systems*, Academic Press, 1980.

10. M. Harrison and A.J. Lemoine : Sticky Brownian motion as the limit of storage processes, *J. Appl. Prob. 18* (1981), p. 886-905.

11. I. Gihman and A.V. Skorohod : *Stochastic Differential equations*, Springer, 1972.

12. D. Stroock and S. Varadhan : Diffusion, processes with boundary conditions, *Comm. Pure and Appl. Math, 24* (1971), p. 147-225.

13. R. Boel: Martingale methods for the semi-Markov analysis of queues with blocking, *Stochastics, 5* (1981), p. 115-133.

14. L. Breiman : *Probability*, Addison Wesley, 1968.

15. E. Wong: *Stochastic Processes in Information and Dynamical Systems*, McGraw Hill, 1971.

16. M. Davis and S. Marcus : An introduction to nonlinear filtering, in : *Stochastic Systems : the mathematics of filtering and identification and applications*, M. Hazewinkel and J. Willems, eds., p. 53-75, Reidel, 1981.

17. P. Brémaud : La méthode des semi-martingales en filtrage quand l'observation est un processus ponctuel marqué, *Séminaire de Probabilités, X*, Lecture Notes in Mathematics, vol. 511, Springer, 1976.

18. D. Kennedy : Some martingales related to cumulative sum tests and single server queues, *Stochastic Processes and their Applications, 4* (1976), p. 261-269.

19. W. Rozenkrantz : Some martingales associated with queueing and storage processes, to appear in *Zeit.für Wahrscheinlichkeitstheorie*.

20. R. Boel : Some comments on control and estimation problems for diffusions in bounded regions, to appear in : Proceedings of the ENST-CNET Symposium on Estimation, February 1983.

WEAK CONVERGENCE OF A SEQUENCE OF QUEUEING AND STORAGE PROCESSES TO A SINGULAR DIFFUSION.

WALTER A. ROSENKRANTZ
Department of Mathematics and Statistics
University of Massachusetts
Amherst, MA 01003

1. INTRODUCTION

It has been known for a long time that heavy traffic limit theorems in queueing theory are but a special case of the so-called diffusion approximation in Physics and Genetics. Take for example Kingman's (1962) heavy traffic approximation for the stationary waiting time distribution for a sequence of GI/GI/1 queues $Q(\alpha)$ depending on a parameter α. Denote the waiting time, excluding service, of the nth customer by $W(n,\alpha)$ and let $U(n,\alpha) = S(n,\alpha) - T(n,\alpha)$ where $S(n,\alpha)$ = service time of the nth customer and $T(n,\alpha)$ = inter arrival time between the nth and $(n+1)^{st}$ customer and assume $E(U(n,\alpha)) = -\alpha\sigma$, variance of $U(n,\alpha) = \sigma^2$, $\alpha > 0$. Then we have the following Theorem 1 (Kingman (1962)):

$$\lim_{n\to\infty} P((\alpha/\sigma)W(n,\alpha) \leq x) = 1-\exp(-2x),\ 0 \leq x < \infty\ ,\ \text{provided} \lim_{n\to\infty,\alpha\to 0} \alpha^2 n = \infty.$$

Somewhat later Kingman (1965) presented a more elegant but heuristic proof of this result which justifies referring to such a theorem as a diffusion approximation. It is worthwhile sketching the heuristic proof of Theorem 1 here, referring the reader to Rosenkrantz (1980) for a rigorous proof as well as an estimate of the rate of convergence. To begin with, one notes that

$$(1.1)\qquad F_{n,\alpha}(x) = P((\alpha/\sigma)W(n,\alpha) \leq x) = P(\sup_{0\leq t\leq\alpha^2 n} y_{n,\alpha}(t) \leq x)$$

where $y_{n,\alpha}(t)$ is a certain stochastic process with continuous paths. One can then show, formally at least, that

$$(1.2)\qquad \lim_{n\to\infty,\alpha\to 0} y_{n,\alpha}(t) = y(t)$$

where $y(t) = w(t) - t$. Here $w(t)$ is the standard 1-dimensional Wiener process and so $y(t)$ is the Wiener process with negative drift. It follows at once from (1.2) that

$$(1.3)\qquad \lim_{n\to\infty,\alpha\to 0} P(\sup_{0\leq t\leq\alpha^2 n} y_{n,\alpha}(t) \leq x) = P(\sup_{0\leq t<\infty} y(t) \leq x)$$

and an easy calculation, see e.g. Karlin-Taylor (1975), p.361, yields the result that $P(\sup_{0\leq t<\infty} y(t) \leq x) = 1 - \exp(-2x),\ 0 \leq x < \infty$.

Another and simpler example of a heavy traffic limit theorem is the following: let $N_n(t)$ denote the queue size of an M/M/1 queue with arrival rate λ_n, mean service time distribution μ_n^{-1} and traffic intensity $\rho_n = \lambda_n/\mu_n$. Assume $\lambda_n = \mu_n - \delta n^{-1/2}$ for some $\delta > 0$, so $0 < \rho_n < 1$ and denote by σ_n^2 the variance of the service time distribution which in this case equals μ_n^{-2}.

THEOREM 2: Assume $\lambda = \lim_{n\to\infty} \lambda_n = \lim_{n\to\infty} \mu_n = \mu$ so $\lim \rho_n \uparrow 1$, and $\lim_{n\to\infty} \sigma_n^2 = \sigma^2$; then $\lim_{n\to\infty} N_n(nt)/\sqrt{n} = y(t)$ where $y(t)$ is the Wiener process on $R^+ = [0,\infty)$ with variance $\lambda + \sigma^2\mu^3$, negative drift δ and reflected at the origin. Theorem 2 has been extended in many ways and by many authors including Iglehart and Whitt. The survey article by Whitt (1974) is a useful reference for the reader interested in these developments.

In each of the heavy traffic limit theorems cited above the limit process has turned out to be the Wiener process with a negative drift satisfying, where appropriate, a reflecting boundary condition. Recently Yamada (1982) has given a diffusion approximation for a sequence of storage processes $X_n(t)$ where the limit process $Y(t)$ is no longer a Wiener process with a negative drift but is instead a Bessel process with negative drift. This result is of more than routine interest. It shows for example that the set of possible limit processes that can occur in queueing and storage theory is a much larger class than Theorems 1, 2 and the survey article by Whitt (1974) would lead us to believe existed. In addition Yamada's theorem (a precise version of which will be stated below as Theorem 3) offers a challenge to the traditional methods by which such limit theorems are usually proved. In particular, neither the Trotter-Kato-Kurtz method cf Kurtz (1969) nor the martingale method of Papnicolaou, Stroock and Varadhan (1977) are directly applicable to this limit theorem because of some nontrivial technical problems of independent interest and the solutions of which are also of independent interest. It is the purpose of this paper to give a new and simpler proof of Yamada's theorem using some results due to Brezis, Rosenkrantz and Singer, with an appendix by P. D. Lax, (1971) which, restated in the more modern terminology of today, implies that the martingale problem for the operator corresponding to the Bessel process with drift has a unique solution - see Stroock-Varadhan (1979) and Ikeda-Watanabe (1981) for a general discussion of these ideas. It turns out however that the estimates we needed to make the martingale methods work already imply the strong convergence of the semigroups in the sense of Trotter-Kato - see Theorem 4 below. These as well as other results from Functional Analysis are collected in an appendix. We shall also use the standard notations: $C_o(R^+) = \{f$: f bounded and continuous on $R^+ = [0,\infty)$ and $\lim_{x\to\infty} f(x) = 0\}$, $f^{(\ell)}(x) = \ell$th derivative of f, $C_o^k(R^+) = \{f \in C_o(R^+): f^{(\ell)} \in C_o(R^+),\ 1 \le \ell \le k\}$. We make $C_o(R^+)$ into a Banach space in the usual way by giving it the norm $\|f\| = \sup_{0\le x<\infty} |f(x)|$. The symbol ∎ denotes the end of a proof.

2. STATEMENT AND PROOF OF YAMADA'S DIFFUSION APPROXIMATION.

Let $X(t)$ denote the content of a dam at time t (also called a storage process) with release rate $r(x)$ and random cumulative imput $A(t)$ which is assumed to be a compound Poisson process. The jump rate λ is assumed to be finite and the cumulative distribution of the size of the jump is denoted by $F(y)$. Cinlar-Pinsky (1972) have shown that $X(t)$ may be realized as the unique solution of the stochastic integral equation

(2.1) $$X(t) = X(0) - \int_0^t r(x(x))ds + A(t), \text{ where}$$

$A(t) = \sum_{i=1}^{N_\lambda(t)} S_i$ where the S_i are i.i.d. with common distribution F and $N_\lambda(t)$ is a Poisson process with intensity λ. The release rate $r(x)$ is assumed to be a non-negative, non-decreasing function with domain $R^+ = [0,\infty)$, $r(0) = 0$. From now on we also assume that $\bar{r} = \lim_{x\to\infty} r(x)$ is finite. We set $\mu_i = \int_0^\infty y^i dF(y)$, $\rho = \lambda\mu_1$ and $k = \sqrt{\lambda\mu_2}$.

Following Yamada (1982) we make the following hypotheses:

(2.2) $$X_n(t) = X_n(0) - \int_0^t r_n(X_n(s))ds + A_n(t), \quad n = 1,2,\ldots$$

is a sequence of storage processes with release rates $r_n(x)$, $A_n(t) = \sum_{i=1}^{N_{\lambda_n}(t)} S_i^n$, $P(S_i^n \leq y) = F_n(y)$ satisfying the normalization conditions:

(2.3) $$\bar{r}_n \geq \rho_n, \quad \rho_n = \lambda_n\mu_1^n, \quad \mu_i^n = \int_0^\infty y^i dF_n(y)$$

(2.4) $$x(\bar{r}_n - r_n(x)) \to c < \infty, \quad \text{as } x \to \infty, \quad n \to \infty$$

(2.5) $$\lim_{n\to\infty} n^{1/2}(\bar{r}_n - \rho_n)/k_n = d,$$

(2.6) $$\lim_{n\to\infty} k_n = k > 0, \quad k_n^2 = \lambda_n\mu_2^n$$

(2.7) $$\sup_{n,x\geq 0} x(\bar{r}_n - r_n(x)) = M < \infty$$

(2.8) $$X_n(0) = x_n, \quad \lim_{n\to\infty} x_n/k_n\sqrt{n} = x \quad .$$

(2.9) $$\lim_{c\to\infty} \int_{\{y>c\}} y^2 dF_n(y) = 0 \quad \text{uniformly in } n.$$

From these conditions it is easy to see that each of the following sequences is bounded: $\{\mu_2^n\}$, $\{\mu_1^n\}$, $\{\lambda_n\}$, $\{\rho_n\}$ and $\{\bar{r}_n\}$. For example (2.9) implies that $\{\mu_2^n\}$ is a bounded sequence and *a fortiori* so is $\{\mu_1^n\}$. This together with (2.6) implies $\{\lambda_n\}$ is bounded and the other statements are proved in a similar fashion.

THEOREM 3 (Yamada): Set $Y_n(t) = X_n(nt)/k_n\sqrt{n}$ and assume conditions (2.3) through (2.9) hold and that $\lim_{n\to\infty} Y_n(0) = x$. Then $Y_n(t)$ converges weakly to a Bessel process with negative drift $Y(t)$, starting at x. $Y(t)$ is a (Markov) diffusion process on $R^+ = [0,\infty)$ whose infinitesimal generator is given by

$$(2.10) \qquad Gf(x) = (1/2)f''(x) + (c/k^2)(f'(x)/x) - df'(x).$$

Remarks: This is not the form in which Yamada states his theorem. Specifically, he shows that $Y(t) = \sqrt{Z(t)}$ where $Z(t)$ is the unique solution to the stochastic integral equation:

$$(2.11) \qquad Z(t) = Z(0) + \int_0^t (K - 2d\sqrt{Z(s)})ds + 2\int_0^t \sqrt{Z(s)}dw(s)$$

where $K = 1 + 2c/k^2$ and w is the standard Wiener process. Thus $Z(t)$ satisfies the stochastic differential equation

$$(2.12) \qquad \begin{cases} dZ(t) = (K - 2d\sqrt{Z(t)})dt + 2\sqrt{Z(t)}dw(t) \\ \qquad\quad = b(Z(t))dt + a(Z(t))dw(t) \quad \text{with} \\ b(x) = (K - 2d\sqrt{x}), \quad x \geq 0 \quad \text{and} \quad a(x) = 2\sqrt{x} \end{cases} .$$

Notice that neither $a(x)$ nor $b(x)$ (when $d \neq 0$) are Lipschitz continuous and so the existence of a unique solution to the stochastic differential equation (2.12) is not a trivial matter. The existence of a unique solution is however a consequence of a more general result due to Okabe and Shimizu (1975). Before proceeding to our own proof let us sketch the idea behind Yamada's proof. He first shows that the processes $Y_n(t)$ are tight in $D[0,T]$ and that if $Y(t)$ is any limit then $Z(t) = Y(t)^2$ solves the martingale problem:

$$(2.13) \qquad f(Z(t)) - f(Z(0)) - \int (K - 2d)\sqrt{Z(s)})f'(Z(s))ds$$
$$- 2\int_0^t \sqrt{Z(s)}f''(Z(s))ds \qquad \text{is a zero mean}$$

martingale for every $f \in C_K^2(R)$. $C_K^2(R)$ is the set of twice continuously differentiable functions, with compact support. This shows that every weak limit solves the martingale problem (2.13) which, thanks to the results of Okabe-Shimuzu, op. cit, is known to have a unique solution. The proof that $Z(t)$ is a solution to the martingale problem (2.13) is almost 5 pages long and the proof that the processes $\{Y_n(t)\}$ form a tight sequence is nearly 6 pages long. It is the purpose of this paper to give an alternative proof of this result which we believe to be easier to follow and is also somewhat shorter. First we shall give a heuristic proof and put in the (tedious) details elsewhere.

We begin by observing that $Y_n(t)$ is for each n a Markov process on the half line $R^+ = [0,\infty)$ with infinitesimal generator G_n given by

$$
(2.14)\quad \begin{cases} G_n f(x) = -(\sqrt{n}/k_n) r_n(k_n\sqrt{n}\cdot x) f'(x) + n\lambda_n \int_0^\infty [f(x+y) - f(x)]dH_n(y) \\ \qquad\qquad \text{for } x > 0 \qquad \text{and} \\ G_n f(0) = n\lambda_n \int_0^\infty [f(y) - f(0)]dH_n(y). \\ \text{Here } H_n(y) = F_n(k_n\sqrt{n}\cdot y). \end{cases}
$$

See for example Cinlar-Pinsky (1972), Harrison-Resnick (1976) or Rosenkrantz (1981) where the operators G_n and their domains (both strong and weak) are discussed in some detail.

DEFINITION: $D(G) = \{f \in C_0^2(R^+) : f'(0) = 0\}$, where the operator G is defined at (2.10).

Later on, in Appendix A, we will show that $D(G)$ is the domain of the strong infinitesimal generator of the semi group $T(t)f(x) = E_x(f(Y(t)))$. Of course, characterizing $D(G)$ is not, in general, an easy matter but in the special case $d = 0$ this was already done by Brezis et al. (1971). The extension of their results to the case $d \neq 0$ is carried out in this paper by showing that the operator $Cf(x) = -df'(x)$ is relatively bounded with respect to the Bessel operator

$$
(2.15)\qquad Bf(x) = (1/2)f''(x) + (\gamma/x)f'(x), \quad \gamma > -1/2,
$$

in the sense of Kato (1976) cf. Appendix A. With these preliminaries out of the way we can give a quick heuristic proof of Yamada's theorem by deriving the

LEMMA 1: For every $f \in D(G)$ and $x > 0$ we have $\lim_{n\to\infty} G_n f(x) = Gf(x)$; the convergence is uniform on every interval of the form $[\delta,\infty)$, $\delta > 0$ and $\sup_n \|G_n f(x)\| < \infty$.

PROOF: Using the Taylor expansion $f(x+y) - f(x) = f'(x)y + (1/2)f''(x)y^2 + R(x,y)$ where $R(x,y) = (1/2)(f''(\xi(y)) - f''(x))$ and $x \leq \xi(y) \leq x + y$, we see that

$$
n\lambda_n \int_0^\infty [f(x + y) - f(x)]dH_n(y) =
$$

$$
n\lambda_n f'(x) \int_0^\infty y dH_n(y) + (1/2)n\lambda_n f''(x) \int^\infty y^2 dH_n(y) + R(n)
$$

where $|2R(n)| \leq n\lambda_n \int_0^\infty |[f''(\xi(y)) - f''(x)]|y^2 dH_n(y)$. In a moment we will show that $\lim_{n\to\infty} R(n) = 0$. On the other hand $\int_0^\infty y dH_n(y) = \mu_1^n/k_n\sqrt{n}$ and $\int_0^\infty y^2 dH_n(y) = \mu_2^n/k_n^2 n$ so

$$
(2.16)\qquad G_n(f(x) = [-(\sqrt{n}/k_n)r_n(k_n\sqrt{n}\cdot x) + (\sqrt{n}\rho_n/k_n)]f'(x) + (1/2)f''(x) + R(n),
$$

since $n\lambda_n\mu_1^n/k_n\sqrt{n} = \sqrt{n}\rho_n/k_n$ and $n\lambda_n\mu_2^n/k_n^2 n = 1$ - see (2.3) and (2.6). Adding and subtracting the term $(\sqrt{n}/k_n)\bar{r}_n f'(x)$ to the right hand side of (2.16) we obtain

$$G_n f(x) = (\sqrt{n}/k_n)(\bar{r}_n - r_n(k_n\sqrt{n}\cdot x))f'(x) + (\sqrt{n}/k_n)(\rho_n - \bar{r}_n)f'(x)$$

$$+ (1/2)f''(x) + R(n).$$

For $x > 0$ we have $(\sqrt{n}/k_n)(\bar{r}_n - r_n(k_n\sqrt{n}\cdot x))f'(x) = (k_n\sqrt{n}\cdot x/k_n^2)(\bar{r}_n - r_n(k_n\sqrt{n}\cdot x))f'(x)/x$ consequently (2.4)(2.6),(2.7) imply that for $x > 0$ $\lim_{n\to\infty} (\sqrt{n}/k_n)(\bar{r}_n - r_n(k_n\sqrt{n}\cdot x))f'(x) = (c/k^2)f'(x)/x$ and the convergence is uniform on the interval $[\delta,\infty)$. Hypothesis (2.7) implies that the term is uniformly bounded in n and x. Similarly condition (2.5) implies $\lim_{n\to\infty} (\sqrt{n}/k_n)(r_n - \bar{\rho}_n)f'(x) = -df'(x)$. Thus the lemma will be proved if we can show that $\lim_{n\to\infty} R(n) = 0$, where

$$|2R(n)| \leq n\lambda_n \int_0^\varepsilon |f''(\xi(y)) - f''(x)|y^2 dH_n(y) + n\lambda_n \int_\varepsilon^\infty |f''(\xi(y)) - f''(x)|y^2 dH_n(y).$$

Now for ε small enough $|f''(\xi(y)) - f''(x)| < \delta$ and this together with the fact that $n\lambda_n \int_0^\varepsilon y^2 dH_n(y) \leq n\lambda_n \int_0^\infty y^2 dH_n(y) = 1$ implies that the first summand in the expression above can be made arbitrarily small. As for the second summand a change of variable yields the formula $n\lambda_n \int_\varepsilon^\infty y^2 dH_n(y) = (\lambda_n/k_n^2) \int_{k_n\sqrt{n}\cdot\varepsilon}^\infty z^2 dF_n(z)$ which goes to zero by hypothesis (2.9) and the fact that both λ_n and k_n^2 are bounded. □

It is easy to see that $\lim_{n\to\infty} G_n f(0) \neq Gf(0)$. Because $Gf(0) = (1/2)f''(0) + (c/k^2)f''(0) - df'(0) = (1/2 + c/k^2)f''(0)$ since $f'(0) = 0$ and $f \in C_0^2(R^+)$ implies $f''(0) = \lim_{x\to 0} \frac{f'(x)}{x}$. On the other hand (by (2.14)) $G_n f(0) = n\lambda_n \int_0^\infty (f(y) - f(0))dH_n(y)$ and using a two term Taylor expansion as before we get that

$\lim_{n\to\infty} G_n f(0) = (1/2)f''(0)$. Thus the only time $G_n f(x)$ converges $Gf(x)$ for all $x \in R^+$ is in the special case $c = 0$. i.e. when the limiting process $Y(t)$ is the Wiener process with a negative drift reflected at the origin. This phenomneon of convergence of the generators except at certain exceptional points is quite common and occurs even in the example of Theorem 2 - cf. Burman (1979) p.17. Nevertheless, it has been observed by several authors including Papanicolaou, Stroock, Varadhan (1975), Burman (1979) that weak convergence of $Y_n(t)$ to $Y(t)$ can be proved, provided one can show that the occupation time of the exceptional set by the process $Y_n(t)$ can be made arbitrarily small as $n \to \infty$. In the present context we must estimate $\int_0^T I_{[0,\delta]}(Y_n(s))ds$ which is the occupation time of the set $[0,\delta]$ by the process $Y_n(t)$.

LEMMA 2: Under the hypotheses of Theorem 3 there exists for any $\varepsilon > 0$ a $\delta > 0$ such that

$$\limsup_{n\to\infty} E_x\left(\int_0^T I_{[0,\delta]}(Y_n(s))ds \right) \leq \varepsilon. \tag{2.17}$$

Setting aside the proof of (2.17) for the moment let us show that this implies strong convergence of the semi groups.

THEOREM 4: Under the hypotheses of Theorem 3

$$\lim_{n\to\infty} \|E_x(f(Y_n(t)) - E_x f(Y(t))\| = \lim_{n\to\infty} \|T_n(t)f(x) - T(t)f(x)\| = 0, \tag{2.18}$$

where the convergence is uniform for $t \in$ compact subsets of R^+. Before proceeding to the proof of Theorem 4 we need a result due to the author, Rosenkrantz (1981), characterizing the domains $D(G_n)$ of the integro-differential operators G_n defined at (2.14).

LEMMA 3: Let G_n denote the strong infinitesimal generator of the normalized storage processes. Then

Case 1: If $r_n(x) = \bar{r}_n$, $x > 0$, $r_n(0) = 0$ we have

$$D(G_n) = \{f \in C_0^1(R^+) : f'(0) = 0\} \tag{2.19}$$

$$\text{Case 2:}\quad D(G_n) = \{f \in C_0(R^+) : r_n(x)f'(x) \in C_0(R^+),\ \lim_{x\to 0} r_n(x)f'(x) = 0\}\ . \tag{2.20}$$

PROOF: This theorem is proved in exactly the same way as Theorem 4.6 on p. 219 of Rosenkrantz (1981). □

Clearly $D(G) \subset D(G_n)$ and hence for every $f \in D(G)$ we have the representation

$$T_n(t)f(x) - T(t)f(x) = \int_0^t T_n(t - s)(G_n - G)T(s)f(x)ds, \tag{2.21}$$

cf. Burman (1979) p. 14, formula 2.2.

We pause to introduce some notation: If $g(x)$ is a function set $g_\delta(x) = g(x)$ if $x \in [\delta,\infty)$ and 0 otherwise and put $\tilde{g}_\delta(x) = g(x) - g_\delta(x)$; so $g_\delta(x) + \tilde{g}_\delta(x) = g(x)$. Thus $(G_n - G)T(s)f(x) = [(G_n - G)T(s)f]_\delta(x) + \widetilde{[(G_n - G)T(s)f]}_\delta(x)$ and therefore

$$\|T_n(t)f(x) - T(t)f(x)\| \leq \int_0^t \|[(G_n - G)T(s)f]_\delta(x)ds + \left\|\int_0^t T_n(t-s)\widetilde{[(G_n - G)T(s)f]}_\delta(x)ds\right\|$$

since $T_n(t)$ is a contraction semi group.

For $f \in D(G)$ the *apriori* estimate (A.8) and Lemma 1 together imply $\lim_{n\to\infty} (G_n - G)T(s)f(x) = 0$ uniformly on $[\delta,\infty)$ and uniformly in s, $0 \le s \le t$. Consequently $\lim_{n\to\infty} \int_0^t \|[(G_n - G)T(s)f]_\delta(x)\| ds = 0$. Similarly, $[(G_n - G)T(s)f]_\delta(x) \neq 0$ only on the set $[0,\delta]$ and since by Lemma 1 and (A.8) $\|G_nT(s)f\|$ and $\|GT(s)f\|$ are both uniformly bounded we conclude $\left|\int_0^t T_n(t-s)[(G_n - G)T(s)f]_\delta(x)ds\right| = \left|E_x\int_0^t [(G_n - G)T(s)f]_\delta(Y_n(t-s))ds\right| \le c'E_x\left(\int_0^t I_{[0,\delta]}(Y_n(s))ds\right)$ where $c' = \sup_{n,0\le s\le t} \{\|G_nT(s)f(x)\| + \|GT(s)f(x)\|\}$. We now apply Lemma 2 and choose δ so small that $\limsup_{n\to\infty} E_x\left(\int_0^t I_{[0,\delta]}(Y_n(s))ds\right) \le \varepsilon \cdot c^{-1}$ from which it follows at once that $\limsup_{n\to\infty} \|T_n(t)f(x) - T(t)f(x)\| \le \varepsilon$ uniformly for $t \in$ compact subsets of R^+. □

We now turn to the proof of Lemma 3. Following Yamada let $\bar{Y}_n(t)$ denote the storage process with $\bar{r}_n(x) = \bar{r}_n$, $x > 0$ and $\bar{r}_n(0) = 0$. Since $\bar{r}_n(x) \ge r_n(x)$ it is clear that $\bar{Y}_n(t) \ge Y(t)$ and in particular

$$E_x\left(\int_0^t I_{[0,\delta]}(Y_n(s))ds\right) \le E_x\left(\int_0^t I_{[0,\delta]}(\bar{Y}_n(s))ds\right) .$$

Thus to prove Lemma 3 it suffices to prove that

$$\limsup_{n\to\infty} E_x\left(\int_0^t I_{[0,\delta]}(\bar{Y}_n(s))ds\right) \le \varepsilon . \tag{2.22}$$

It is convenient to split the proof into two parts:

$$\lim_{n\to\infty} E_x\left(\int_0^t I_{[0]}(\bar{Y}_n(s))ds\right) = 0 \tag{2.23}$$

$$\limsup_{n\to\infty} E_x\left(\int_0^t I_{(0,\delta]}(\bar{Y}_n(s))ds\right) \le \varepsilon . \tag{2.24}$$

PROOF OF (2.23): The infinitesimal generator G_n' of $\bar{Y}_n(t)$ is

$$G_n'f(x) = -(\sqrt{n}/k_n)\bar{r}_n f'(x) + n\lambda_n \int_0^\infty [f(x+y) - f(x)]dH_n(y), \quad x > 0$$

$$G_n' f(0) = n\lambda_n \int_0^\infty [f(y) - f(0)]dH_n(y).$$

Applying Dynkin's formula as in Theorem 3.1 p. 216 of Rosenkrantz (1981), leads to the formula

$$E_x(\bar{Y}_n(t)) = x - (\sqrt{n}/k_n)(\bar{r}_n - \rho_n)t + (\sqrt{n}/k_n)\bar{r}_n E_x\left(\int_0^t I_{[0]}(Y_n(s))ds\right) . \tag{2.25}$$

In the appendix it will be shown that $\sup_{0\le s\le t} E_x(\bar{Y}_n(s)) < \infty$ for every $t > 0$ and hence

(2.26) $$(\sqrt{n}/k_n)\bar{r}_n\, E_x\left(\int_0^t I_{[0]}(Y_n(s))ds\right) = E_x(\bar{Y}_n(t)) - x + (\sqrt{n}/k_n)(\bar{r}_n - \rho_n)t.$$

By (2.7) $\lim_{n\to\infty} (\sqrt{n}/k_n)(\bar{r}_n - \rho_n)t = dt$ so the right hand side of (2.26) is bounded whilst $\lim_{n\to\infty} (\sqrt{n}/k_n)\bar{r}_n = +\infty$, consequently

(2.27) $$E_x\left(\int_0^t I_{[0]}(Y_n(s))ds\right) = 0(n^{-1/2}).$$

Turning now to the proof of (2.24) we must consider separately the case when $d > 0$ and when $d = 0$.

Case 1: $d = \lim_{n\to\infty} (\sqrt{n}/k_n)(\bar{r}_n - \rho_n) > 0.$

LEMMA 4: For every $\alpha > 0$ the function $f_\alpha(x) = [1 - (x/\alpha)]^+$ is in the domain of the weak infinitesimal generator $\tilde{G}'_n$.

PROOF: See Harrison-Resnick (1976). Of course $\tilde{G}'_n$ is an extension of G'_n and $f_\alpha(x)$ is Lipschitz continuous with $|f_\alpha(x + y) - f_\alpha(x)| \leq y \cdot \alpha^{-1}$, $f'_\alpha(x) = -\alpha^{-1}$ on $[0,\alpha]$. Thus,

(2.28) $$\tilde{G}'_n f_\alpha(x) = (\sqrt{n}/k_n)\bar{r}_n\alpha^{-1} + n\lambda_n\int_0^\infty [f_\alpha(x + y) - f_\alpha(x)]dH_n(y),\ 0 < x \leq \alpha,$$

(2.29) $$|\tilde{G}'_n f_\alpha(0)| \leq (\sqrt{n}/k_n)\rho_n \cdot \alpha^{-1},\quad \tilde{G}'_n f_\alpha(x) = 0,\quad x > \alpha.$$ In particular

$$\tilde{G}'_n f_\alpha(x) \geq (\sqrt{n}/k_n)\bar{r}_n \cdot \alpha^{-1} - n\lambda_n\int_0^\infty \alpha^{-1} \cdot y dH_n(y)$$
$$= \alpha^{-1}(\sqrt{n}/k_n)(\bar{r}_n - \rho_n) \quad \text{on } (0,\alpha].$$

Now for large n, $(\sqrt{n}/k_n)(\bar{r}_n - \rho_n) \geq d/2 > 0$ and this implies $\tilde{G}'_n f_\alpha(x) \geq d/2\alpha$ on $(0,\alpha]$ provided n is large enough. Notice that $\|f_\alpha(x)\| \leq 1$ and hence $T_n(t)f_\alpha(x) - f_\alpha(x) = \int_0^t T_n(s)\tilde{G}'_n f_\alpha(x)ds$ implies $\left|E_x\left(\int_0^t \tilde{G}'_n f_\alpha(\bar{Y}_n(s))ds\right)\right| \leq 2$. On the other hand $E_x\left(\int_0^t \tilde{G}'_n f_\alpha(\bar{Y}_n(s))ds\right) = E_x\left(\int_0^t \tilde{G}'_n f_\alpha(\bar{Y}_n(s))I_{[0]}(Y_n(s))ds\right)$ $+ E_x\left(\int_0^t \tilde{G}'_n f_\alpha(\bar{Y}_n(s))I_{(0,\alpha]}(\bar{Y}_n(s))ds\right)$. From (2.27) and (2.29) we see at once that $E_x\left|\left(\int_0^t \tilde{G}'_n f_\alpha(\bar{Y}_n(s))I_{[0]}(\bar{Y}_n(s)ds\right)\right|$ is bounded, by M say, as $n \to \infty$. On the interval $(0,\alpha]$ however, $\tilde{G}'_n f_\alpha(x) \geq (d/2\alpha)$ and therefore $E_x\left(\int_0^t \tilde{G}'_n f_\alpha(\bar{Y}_n(s))I_{(0,\alpha]}(\bar{Y}_n(s)ds\right)$ $\geq (d/2\alpha)E_x\left(\int_0^t I_{(0,\alpha]}(\bar{Y}_n(s)ds\right)$. Therefore as $n \to \infty$ we get

$$\limsup_{n\to\infty} E_x\left(\int_0^t I_{(0,\alpha]}(\bar{Y}_n(s))ds\right) \leq (2 + M)2\alpha/d\ .$$

The proof is now completed by choosing $\alpha \leq \varepsilon d/(4 + 2M)$.

Case 2: $d = 0$. In this case $\lim_{n\to\infty} G_n'f(x) = (1/2)f''(x)$ for every $f \in D(G) = \{f \in C_0^2(R^+): f'(0) = 0\}$ i.e. the limit process in this case is reflecting Brownian motion $|w(t)|$. Thus the original Trotter-Kato theorem itself implies that $\lim_{n\to\infty} \|E_x(f(\bar{Y}_n(t)) - E_x f(|w(t)|)\| = 0$. It is a consequence of a theorem of Aldous (1978) that $\bar{Y}_n(t)$ converges weakly to $|w(t)|$ or if one prefers, the weak convergence may be deduced from a more general result due to Kurtz (1981), Theorem 4.4. It is well known that reflecting Brownian motion has a local time $\alpha(t,y,\omega)$ and therefore $\int_0^t I_{[0,\delta]}(|w(s)|)ds = \int_0^\delta \alpha(t,y,\omega)dy \leq t$, where $\alpha(t,y,\omega)$ is jointly continuous in (t,y) for each ω. By Lebesgue's dominated convergence theorem then we have $\lim_{\delta\to 0} E\left(\int_0^\delta \alpha(t,y,\omega)dy\right) = 0$ and so given any $\varepsilon > 0$ $\exists \delta > 0$ such that $E\left(\int_0^t I_{[0,\delta]}(|w(s)|)ds\right) < \varepsilon$.

Let us denote by P_n and P the measures induced on $D[0,T]$ by the $\bar{Y}_n(t)$ and $|w(t)|$ processes respectively. It is well known that the functional $\omega \longrightarrow \int_0^T I_{[0,\delta]}(\omega(s))ds$, here ω is a path in $D[0,T]$, is continuous almost everywhere with respect to the measure P, cf. Billingsley (1968), pp. 230-231. This fact together with the weak convergence of P_n to P and Theorem (5.2iii) p. 31 of Billingsley, op. cit., imply

$$\lim_{n\to\infty} E_x\left(\int_0^t I_{[0,\delta]}(\bar{Y}_n(s))ds\right) = E_x\left(\int I_{[0,\delta]}(|w(s)|)ds\right) < \varepsilon. \tag{2.30}$$ □

The proof of Theorem 4 is now complete.

APPENDIX

Let $Bf(x) = (1/2)f''(x) + (\gamma/x)f'(x)$, $\gamma > -(1/2)$ denote the Bessel operator acting on the domain $D(B) = \{f \in C_0^2(R^+) : f'(0) = 0\}$. It was shown in Brezis, et al. (1971) that B acting on $D(B)$ generates a positivity preserving, strongly continuous, contraction semi group $T_1(t) : C_0(R^+) \to C_0(R^+)$. The following *apriori* estimate was also obtained (see Theorem (A.1) p. 411 of Brezis et al. (1971), where a more general result is given):

LEMMA: For every $f \in D(B)$ there exists a constant $\beta > 0$, depending only on γ, such that

(A.1) $$\|f''\| \leq \beta\|Bf\| \ .$$

We next observe that the operator $Gf = Bf + Cf$ where B is Bessel operator defined by (2.15) and $Cf = -df'$, i.e., G is a perturbation of the operator B; clearly $D(C) \supset D(B)$.

THEOREM: There exist constants $a > 0$, $0 < b < 1/2$ such that for every $f \in D(B)$ the inequality

(A.2) $$\|Cf\| \leq a\|f\| + b\|Bf\|, \text{ holds.}$$

REMARK: When (A.2) holds the operator C is said to be *relatively bounded with respect to* B - see Kato (1976), p. 190.

PROOF: Let $\|g\|_{[a,b]} = \sup_{a \leq x \leq b} |g(x)|$ and observe that, for $g \in C_0(R^+)$, $\|g\| = \sup_k \|g\|_{[k,k+1]}$ where the sup is taken over all non-negative integers $k = 0,1,2,\ldots$. The proof of inequality (A.3) below is to be found in Kato, op. cit. p. 192, formula (1.13).

(A.3) $$\|f'\|_{[a,b]} \leq [(b-a)/(n+2)]\cdot\|f''\|_{[a,b]} + [2(n+1)/(b-a)]\cdot\|f\|_{[a,b]}$$

for every $f \in C^2[a,b]$ and every $n > 1$.

Specializing (A.3) to the special case $[a,b] = [k,k+1]$ yields

(A.4) $$\|f'\|_{[k,k+1]} \leq (n+2)^{-1}\|f''\|_{[k,k+1]} + 2(n+1)\|f\|_{[k,k+1]}.$$

If now $f \in C_0^2(R^+)$ we have $\|f''\|_{[k,k+1]} \leq \|f''\|$ and $\|f\|_{[k,k+1]} \leq \|f\|$ and hence

(A.5) $$\|f'\|_{[k,k+1]} \leq (n+2)^{-1}\|f''\| + 2(n+1)\|f\| .$$

Consequently for every $f \in D(B)$ we have

$$\|f'\| = \sup_k \|f'\|_{[k,k+1]} \leq (n+2)^{-1}\|f''\| + 2(n+1)\|f\| \quad \text{and in particular}$$

(A.6) $$\|Cf\| \leq d(n+2)^{-1}\|f''\| + 2d(n+1)\|f\| \leq \beta d(n+2)^{-1}\|Bf\| + 2d(n+1)\|f\|$$

where we used (A.1) in the last step.

Thus by choosing $n > 2\beta d - 2$ we have $b = \beta d(n+2)^{-1} < \frac{1}{2}$ and this completes the proof (A.2) with $a = 2d(n+1)$. □

The following *apriori* estimate is also an easy consequence of the above calculation:

(A.7) $$\|f''\| \leq 2\beta\|Gf\| + 4\beta d(n+1)\|f\|.$$

PROOF: Since $Bf = Gf - Cf$ we have from (A.1) and (A.6) that $\|f''\| \leq \beta\|Gf\| + \beta\|Cf\| \leq \beta\|Gf\| + \beta d(n+2)^{-1}\|f''\| + 2d\beta(n+1)\|f\|$. Since $\beta d(n+2)^{-1} < \frac{1}{2}$ we have $(1 - \beta d(n+2)^{-1})\|f''\| \leq \beta\|Gf\| + 2\beta d(n+1)\|f\|$ and hence $\|f''\| \leq 2\beta\|Gf\| + 4\beta d(n+1)\|f\|$. □

Combining all these estimates together with Theorem 2.7 of Kato p. 501 we arrive at the

THEOREM: The operator $G = B + C$ generates a positivity preserving, strongly continuous contraction semi group $T(t):C_0(R^+) \to C_0(R^+)$ with domain $D(G) = D(B) = \{f \in C_0^2(R^+):f'(0) = 0\}$. Moreover for every $f \in D(G)$ we have the following *apriori* estimate: $\|f''\| \leq 2\beta\|Gf\| + 4\beta d(n+1)\|f\|$. In particular if $f \in D(G)$ then $T(s)f \in D(G)$ and therefore

(A.8) $$\begin{aligned}\|(\partial^2/\partial x^2)T(s)f(x)\| &\leq 2\beta\|GT(s)f\| + 4\beta d(n+1)\|f\| \\ &\leq 2\beta\|T(s)Gf\| + 4\beta d(n+1)\|f\| \\ &\leq 2\beta\|Gf\| + 4\beta d(n+1)\|f\|.\end{aligned}$$

We have used the facts that $T(s)$ commutes with its infinitesimal generator G and that $T(s)$ is a contraction. Notice that the right hand is independent of s.

We next turn our attention to deriving the estimate:

(A.9) $$\sup_{0\leq s\leq t} E_x(\bar{Y}_n(s)^2) \leq x^2 + t.$$

This clearly implies $\sup_{0\leq s\leq t} E_x(\bar{Y}_n(s)) < \infty$ which is all we needed to derive (2.27).

PROOF OF (A.9): Let $U(t,x) = x^2 - t$ and observe that

$$G_n'U(t,x) = -2(\sqrt{n}/k_n)\bar{r}_n x + n\lambda_n \int_0^\infty (2xy + y^2)dH_n(y)$$

$$= 1 - (2\sqrt{n}/k_n)x(\bar{r}_n - \rho_n) \leq 1 \quad \text{on} \quad R^+ .$$

Thus $[(\partial U/\partial t) + G_n']U(t,x) = -1 + G_n'U(t,x) \leq 0$; consequently $\bar{Y}_n(t) - t^2$ is a supermartingale. Thus $E_x(\bar{Y}_n(t)^2 - t) \leq x^2$ or $E_x(\bar{Y}_n(t)^2) \leq x^2 + t$. □

ACKNOWLEDGEMENT

Research supported by the U.S. Air Force of Scientific Research under Grant 82-0167.

REFERENCES

[1] David Aldous (1978), Stopping Times and Tightness, Ann. of Prob. Vol. 6, No. 2, pp 335-340.

[2] P. Billingsley (1968), Convergence of Probability Measures, Wiley, New York.

[3] Brezis, Rosenkrantz, Singer, Lax (1971), On a Degenerate Elliptic-Parabolic Equation Occurring in the Theory of Probability, Comm. Pure and Applied Math., Vol. XXIV, pp 395-416.

[4] D. Burman (1979), An Analytic Approach to Diffusion Approximations in Queueing, Thesis, N.Y.U., Courant Institute of Mathematical Sciences.

[5] E. Cinlar, M. Pinsky (1971), A Stochastic Integral in Storage Theory, Zeit. Wahr. verw. Geb. 17, pp. 227-240.

[6] J. M. Harrison and S.I. Resnick (1976), The Stationary Distribution and First Exit Probabilities of a Storage Process with General Release Rule, Math. of Oper. Res. Vol. 1, No. 4 pp 347-358.

[7] N. Ikeda and S. Watanabe (1981), Stochastic Differential Equations and Diffusion Processes, North-Holland Publishing Co., Amsterdam.

[8] S. Karlin, H. Taylor (1975), A First Course in Stochastic Processes, 2nd ed. Academic Press, New York.

[9] T. Kato (1976), Perturbation Theory for Linear Operators, 2nd ed., Springer-Verlag, New York.

[10] J. F. C. Kingman (1962), On Queues in Heavy Traffic, J. Roy. Stat. Soc., ser. B, 24, pp 383-392.

[11] J. F. C. Kingman (1965), The Heavy Traffic Approximation in the Theory of Queues. Proc. Symp. Congestion Theory. pp 137-169, Univ. of North Carolina Press, Chapel Hill.

[12] T. Kurtz (1969), Extensions of Trotter's Operator Semi Group Approximation Theorems, J. Functional Anal., 3, pp 354-375.

[13] T. Kurtz (1981), Approximation of Population Processes, CBMS-NSF Regional Conference series in Appl. Math., Vol. 36, Published by SIAM, Philadelphia, Pennsylvania.

[14] Y. Okabe and A. Shimizu (1975), On the Pathwise Uniqueness of Solutions of Stochastic Differential Equations, J. Math. Kyoto Univ. 15, pp 455-466.

[15] G. Papanicolaou, D. W. Stroock, S.R.S. Varadhan (1977), Martingale Approach to Some Limit Theorems, Duke Turbulence Conference, Duke Univ. Math. Series III.

[16] W. Rosenkrantz (1980), On the Accuracy of Kingman's Heavy Traffic Approximation in the Theory of Queues, Zeit. Wahr. verw. Geb., 51, pp 115-121.

[17] W. Rosenkrantz (1981), Some Martingales Associated with Queueing and Storage Processes, Zeit. Wahr. Verw. Geb., 58, pp 205-222.

[18] D. W. Stroock, S.R.S. Varadhan (1979), Multidimensional Diffusion Processes, Springer Verlag, New York.

[19] W. Whitt (1974), Heavy Traffic Limit Theorems for Queues: A Survey, Mathematical Methods in Queueing Theory, Lecture Notes in Economics and Mathematical Systems, Springer-Verlag.

[20] K. Yamada (1982), Diffusion Approximation for Storage Processes with General Release Rules (preprint), Institute of Information Sciences, Univ. of Tsukuba, Ibaraki, Japan.

VI

STATIONARITY AND ERGODICITY 2

STATIONNARITE ET ERGODICITE 2

ERGODICITY AND STEADY STATE EXISTENCE. CONTINUITY OF STATIONARY DISTRIBUTIONS OF QUEUEING CHARACTERISTICS.

A. Brandt, P. Franken, and B. Lisek
Humboldt-Universität, Sektion Mathematik
1086 Berlin, PSF 1297
German Democratic Republic

1. Introduction

Our main purpose is to discuss the steady state existence and uniqueness for a queueing system with a given traffic and to construct state processes describing the temporal behaviour of this system in steady state. The word "traffic" stands for the sequence $([T_n, S_n])$ of the arrival epochs of customers T_n and their associated service times S_n. (Notice that for some types of queues, such as priority queues, queues with warming-up or queues with reneging, it is necessary to extend S_n in an appropriate way, because the associated service time does not completely describe the "quality" of a customer.)

Furthermore, we are interested in ergodicity properties of the system considered (i.e. in the convergence of the time-dependent state process to the stationary state process) and in model continuity (i.e. in the continuous dependence of the stationary state process on the traffic).

First we restrict our attention to the system behaviour at the arrival epochs only. We imagine $T_0 = 0$ as the arrival epoch of a customer (who receives the number 0). Then the traffic can be given as the random sequence $\Psi = (X_n)$, $X_n = [A_n, S_n]$, where the A_n denote the interarrival times. Using an appropriately chosen space $\mathcal{Z}$ of system states and denoting by Z_n the (random) state of the system observed by the n-th arriving customer, we can describe the evolution of the system by the recursive stochastic equation

$$Z_{n+1} = f(Z_n, X_n) , \quad n \in \Gamma \quad . \tag{1}$$

The form of the function f depends on the type of the system under consideration and on the choice of the state space $\mathcal{Z}$. For example, for the standard single server queue, if we consider the actual waiting time as the state of the system, we have $\mathcal{Z} = R_+$ and $f(z, [a, s]) = \max(0, z-a+s)$.

In section 2 we sketch some general methods for investigating the recursive equation (1). We point out that equations of this type are of interest not only for queueing theory, cf. e.g. Franken and Lisek (1982).

In section 3 we shall give some recent results concerning steady state existence and ergodicity for more or less standard queueing systems with dependent interarrival and service times. This survey is essentially based on Lisek (1979, 1982), Brandt and Lisek (1981), Brandt (1982a,b), and Wirth (1982, 1983).

The analysis of a queueing system requires the investigation of the system behaviour at arrival epochs, in continuous time and at departure epochs. Thus, we have to discuss all the questions mentioned above with respect to several "time axes". In particular, we have to define for a given queueing system in steady state the so-called arrival-, time- and departure-stationary state processes and to investigate the relationships among probabilistic parameters of these processes. For some specific problems we need state processes which are stationary with respect to some other embedded time axes, e.g. busy-cycle-stationary and batch-stationary processes, cf. FKAS (1981), Wirth (1982).

The so-called point process approach, which is based on a few results from the theory of marked point processes, allows a unified treatment of all these kinds of stationary processes associated with a given queueing system in steady state, cf. e.g. FKAS (1981), Franken (1982). Using this approach one can construct the time-stationary, departure-stationary and other stationary state processes starting with the arrival-stationary process, see section 4.

Finally, we discuss the problem of model continuity, i.e. the continuous dependence of stationary solutions of (1) on the traffic.

2. Some Methods for Constructing Solutions of Equation (1)

Let $\mathcal{X}$ and $\mathcal{Z}$ be complete separable metric spaces and f a measurable function from $\mathcal{Z} \times \mathcal{X}$ into $\mathcal{Z}$. We use the notation (Ψ, P) for a sequence $\Psi = (X_n)$ of $\mathcal{X}$-valued r.v.'s and its distribution P (i.e. P is the probability measure generated by Ψ

on the Borel subsets of $\mathcal{X}^{\Gamma}$).

<u>Definition 1.</u> A pair $(\tilde{\Psi}, \tilde{P})$ consisting of a sequence $\tilde{\Psi} = ([\tilde{X}_n, Z_n], n \in \Gamma)$ of $\mathcal{X} \times \mathcal{Z}$ -valued r.v.'s and its distribution $\tilde{P}$ is called a <u>weak solution</u> of equation (1) if

(i) $\tilde{P}((\tilde{X}_n) \in (.)) = P((X_n) \in (.))$

(the sequences $(\tilde{X}_n)$ and (X_n) are stochastically equivalent),

(ii) $\tilde{P}(Z_{n+1} = f(Z_n, \tilde{X}_n), n \in \Gamma) = 1$.

A weak solution $(\tilde{\Psi}, \tilde{P})$ is called a <u>strong solution</u> of (1) if $(\tilde{X}_n) = (X_n)$ and there exists a measurable function h from $\mathcal{X}^{\Gamma}$ into $\mathcal{Z}^{\Gamma}$ satisfying $(Z_n) = h(\Psi)$ $\tilde{P}$-a.s.

In the following we assume that the sequence $\Psi = (X_n)$ is <u>strictly stationary</u> and <u>ergodic</u> (metrically transitive). Under this assumption we shall discuss the following questions:

(i) Under which additional conditions on f and (Ψ, P) do <u>stationary</u> weak (strong) solutions of (1) exist?

(ii) When is the uniqueness of the solution ensured?

(iii) Consider the sequence $z_0(z) = z$, $z_1(z, X_0) = f(z, X_0)$, ..., $z_{n+1}(z, X_0, \ldots, X_n) = f(z_n(z, X_0, \ldots, X_{n-1}), X_n)$, $n \geq 0$, of consecutive system states generated by the "traffic" and the "initial" state z. What will happen with $z_{n+1}(z, X_0, \ldots, X_n)$ for $n \longrightarrow \infty$?

In recent years several approaches have been developed to answer these questions.

<u>2.1.</u> At first we consider an approach based on some <u>monotonicity</u> and <u>contractivity</u> assumptions on the function f. This approach is due to Loynes (1962) and Brandt (1982a). Assume that $\mathcal{Z}$ may be endowed with the metric ϱ and the semiordering relation $\leq$ satisfying the following regularity conditions:

(2.1.1) The metric ϱ makes $\mathcal{Z}$ to a complete metric space with the property that every bounded subset is totally bounded.

(2.1.2) There exists a smallest element $z_0 \in \mathcal{Z}$, i.e. $z_0 \leq z$ for all $z \in \mathcal{Z}$.

(2.1.3) For all $z_1, z_2, z_3 \in \mathcal{Z}$ with $z_1 \leq z_2 \leq z_3$:

$$\varrho(z_1, z_2) \leq \varrho(z_1, z_3) , \qquad \varrho(z_2, z_3) \leq \varrho(z_1, z_3) .$$

(2.1.4) Consider $z, z' \in \mathcal{Z}$, $(z_n) \in \mathcal{Z}^{\Gamma}$ with the properties

$z_n \uparrow z$ and $z_n \leq z'$ for all n. Then $z_n \leq z \leq z'$ holds for all n.

(2.1.5) f(z, x) is non-decreasing and left continuous in z.

The conditions (2.1.1) - (2.1.4) are fulfilled e.g. for $\mathcal{Z} = R_+^k$ endowed with the maximum metric $\rho(u, v) = \max_i |u_i - v_i|$. Under the condition (2.1.5) the sequence

$$z_r(z_0, x_{n-r}, \ldots, x_{n-1}),\ r \geq 1,$$

is non-decreasing for every $\psi = (x_j)$ and every fixed n.

Theorem 1 (Loynes). Let the conditions (2.1.1) - (2.1.5) be fulfilled. Further we assume that for every n

$$\underline{Z}_n = \underline{z}_n(\Psi) = \lim_{r \to \infty} z_r(z_0, X_{n-r}, \ldots, X_{n-1}) \tag{2}$$

is a proper r.v. Then $([X_n, \underline{Z}_n])$ is a stationary strong solution of (1). Moreover, for an arbitrary weak solution $\tilde{\Psi} = ([\tilde{X}_n, Z_n])$ we have

$$Z_n \geq \underline{z}_n((\tilde{X}_n)) \quad \tilde{P}\text{-a.s.}$$

The proof of Theorem 1 is obvious. The crucial point in the application of this theorem to concrete models is the verification of the properness of $\underline{z}_n(\Psi)$.

The sequence $(\underline{z}_n(\Psi))$ - if it exists - defines a minimal solution of (1). We next search for a maximal solution. To do so we need some additional assumptions.

(2.1.6) Consider $z, z' \in \mathcal{Z}$, $(z_n) \in \mathcal{Z}^\Gamma$ with the properties $z_n \downarrow z$ and $z_n \geq z'$ for all n. Then $z_n \geq z \geq z'$ holds for all n.

(2.1.7) For every $z \in \mathcal{Z}$ and $\varepsilon \in R_+$ there is an element $z_{max}(z, \varepsilon)$ of the ε-ball around z satisfying $z' \leq z_{max}(z, \varepsilon)$ for all z' with $\rho(z, z') \leq \varepsilon$.

(2.1.8) f(z, x) is contractive in z.

Consider the following construction. We choose a $\delta \in R_+$ and a $\psi \in \mathcal{X}^\Gamma$ with $\underline{z}_n(\psi) \in \mathcal{Z}$ for all n. For fixed n

$$z_r(z_{max}(\underline{z}_{n-r}(\psi), \delta), x_{n-r}, \ldots, x_{n-1})$$

is non-increasing for $r \longrightarrow \infty$, because of (2.1.7) and (2.1.8).

Thus, the limit

$$z_n^{\delta}(\Psi) = \lim_{r\to\infty} z_r(z_{max}(\underline{z}_{n-r}(\Psi), \delta), x_{n-r}, \ldots, x_{n-1})$$

exists. It is easy to see that under the conditions of Theorem 1 and (2.1.7), (2.1.8) the sequence $(z_n^{\delta}(\Psi))$ defines a stationary strong solution for every fixed δ. For every fixed n, $z_n^{\delta}(\Psi)$ is non-decreasing for $\delta \longrightarrow \infty$.

Theorem 1' (Brandt). Let (2.1.1) - (2.1.8) be fulfilled. Further assume that $\underline{Z}_n = \underline{z}_n(\Psi)$ and

$$Z_n^{max} = z_n^{max}(\Psi) = \lim_{\delta\to\infty} z_n^{\delta}(\Psi)$$

are proper r.v.'s. Then $([X_n, Z_n^{max}])$ is a stationary strong solution of (1). Moreover, for an arbitrary stationary weak solution $\tilde{\Psi} = ([\tilde{X}_n, Z_n])$ we have

$$Z_n \leq z_n^{max}((\tilde{X}_n)) \qquad \tilde{P}\text{-a.s.}$$

for all n.

The first statement immediately follows from the construction. The maximality property is a consequence of (2.1.6). As shown by Brandt (1982a) for the queueing system G/G/m/∞, there is a close relation between the uniqueness of the stationary weak solution of (1) and ergodic properties of the system considered. The arguments remain valid under more general conditions.

Theorem 2 (Brandt). Assume that for all $z_1, z_2 \in \mathbb{Z}$, $\varepsilon \in R_+$ with $z_1 \leq z_2$ and $0 \leq \varepsilon \leq \rho(z_1, z_2)$ there exists a $z \in \mathbb{Z}$ satisfying $z_1 \leq z \leq z_2$, $\rho(z_1, z) = \varepsilon$, and $\rho(z_1, z) + \rho(z, z_2) = \rho(z_1, z_2)$. (This condition is trivially fulfilled e.g. in Banach spaces.) Further, assume (2.1.1) - (2.1.8) and the properness of $\underline{z}_n(\Psi)$. Then the following statements are equivalent:

(i) $([X_n, \underline{z}_n(\Psi)])$ is the unique stationary weak solution of (1).

(ii) For an arbitrary pair $(Y, (\tilde{X}_n))$ such that $(\tilde{X}_n)$ and (X_n) are stochastically equivalent and Y is a $\mathbb{Z}$-valued random variable:

$$(z_{n+k}(Y, \tilde{X}_0, \ldots, \tilde{X}_{n+k-1}), n \geq 1) \xrightarrow[k\to\infty]{\mathcal{D}} (\underline{z}_n(\Psi), n \geq 1).$$

2.2. The method of "renewing epochs" or "renewing events" was introduced by Franken (1970), Borovkov (1972) and others for analyzing special queueing systems. A systematic treatment was given by Borovkov (1978, 1980). The presentation here is somewhat different from that by Borovkov. It enables us better to explain the relations to the other methods, cf. section 2.3.

Definition 2. Let $z \in \mathbb{Z}$, $\Psi = (x_k) \in \mathbb{X}^{\Gamma}$ and $l \geqslant 0$. The number n is called (z, l)-renewing epoch of Ψ if for all $r > l$

$$z_r(z, x_{n-r}, \ldots, x_{n-1}) = \begin{cases} z, & \text{if } l = 0, \\ z_l(z, x_{n-l}, \ldots, x_{n-1}), & \text{otherwise.} \end{cases}$$

This notion can be interpreted as follows: If the system starts with the state z at the epoch $n - r$, $r > l$, its state at the epoch n (and later) does not depend on the "past" $x_{n-r}, \ldots, x_{n-l-1}$ of Ψ.

For fixed z and l the events

$$B_n = \{\Psi: \text{n is a (z, l)-renewing epoch of } \Psi\}$$

are stationary renewing events in the terminology of Borovkov (1978, 1980).

Theorem 3 (Borovkov). If

$$P(B_0) > 0, \tag{3}$$

then there exists a stationary sequence $(\hat{z}_n(z, \Psi), n \in \Gamma)$ such that $\tilde{\Psi}(z) = ([X_n, \hat{z}_n(z, \Psi)])$ is a stationary strong solution of (1) and

$$P(\bigcup_{m=1}^{\infty} \{z_k(z, X_0, \ldots, X_{k-1}) = \hat{z}_k(z, \Psi), k \geqslant m\}) = 1 . \tag{4}$$

Proof. Define $B = \{\Psi$: for every k there are i and j with $i < -k$, $j > k$, and $\Psi \in B_i \cap B_j\}$. Every sample path $\Psi \in B$ possesses an infinite number of (z, l)-renewing epochs left and right from 0. In view of the assumed stationarity and ergodicity of (X_n), condition (3) is equivalent to

$$P(B) = 1 . \tag{5}$$

For a fixed $\Psi \in B$, $n \in \Gamma$, and $i \leqslant n$ with $\Psi \in B_i$ define

$$\hat{z}_n(z, \Psi) = z_{n-i+1}(z, x_{i-1}, \ldots, x_{n-1}) . \tag{6}$$

According to Definition 2 the right side of (6) does not depend on the choice of the (z, 1)-renewing epoch $i \leq n$. Condition (5) ensures that $\hat{z}_n(z, \Psi)$ is a proper r.v. The stationarity of $(\hat{z}_n(z,\Psi))$ is an immediate consequence of the construction. For all

$$\Psi = (x_n) \in \bigcup_{1 \leq j \leq m} B_j$$

we have

$$z_k(z, x_0, \ldots, x_{k-1}) = \hat{z}_k(z, \Psi), \quad k \geq m.$$

Thus, (5) implies (4).■

The statement of Theorem 3 remains valid if one uses a somewhat weakened notion of a renewing epoch.

Definition 3. Let $z \in \mathcal{Z}$, $\Psi = (x_k) \in \mathcal{X}^{\Gamma}$. The number n is called a z-renewing epoch of Ψ if there is a number $l(n, z, \Psi) \geq 0$ such that for all $r > l(n, z, \Psi)$

$$z_r(z,x_{n-r},\ldots,x_{n-1}) = \begin{cases} z, & \text{if } l(n, z, \Psi) = 0, \\ z_{l(n,z,\Psi)}(z,x_{n-l(n,z,\Psi)},\ldots,x_{n-1}), & \text{otherwise.} \end{cases} \tag{7}$$

In the following let $l(n, z, \Psi)$ be the smallest number satisfying (7). Define $C_n = \{\Psi : n \text{ is a z-renewing epoch of } \Psi\}$.

Theorem 4. If

$$P(C_0) > 0, \tag{8}$$

then the statements of Theorem 3 are valid.

Proof. We have

$$C_0 = \bigcup_{j=0}^{\infty} \{\Psi : l(0, z, \Psi) = j\}.$$

In view of (8) there is an l satisfying $P(\{\Psi : l(0, z, \Psi) = l\}) > 0$. Thus, the condition (3) in Theorem 3 holds, which finishes the proof.■

The definitions of (z, 1)- and z-renewing epochs have the disadvantage that the stationary solution $\tilde{\Psi}$ in Theorem 3 and 4, respectively, depends on the fixed state z. In particular, conditions (3) and (8) do not ensure the uniqueness of the stationary weak solution of (1). Thus, the following modification seems to be reasonable.

<u>Definition 4.</u> The number n is called a <u>renewing epoch</u> of $\psi = (x_k)$ if for every $z \in \mathbb{Z}$ n is a z-renewing epoch and for arbitrary z' and z"

$$z_r(z', x_{n-r}, \ldots, x_{n-1}) = z_r(z'', x_{n-r}, \ldots, x_{n-1})$$

for $r > \max(l(n, z', \psi), l(n, z'', \psi))$.

Define $D_n = \{\psi : n \text{ is a renewing epoch of } \psi\}$.

<u>Theorem 5.</u> If there is a measurable subset $D_0' \subseteq D_0$ with the property

$$P(D_0') > 0, \tag{9}$$

then there is a stationary strong solution $\tilde{\Psi} = ([X_n, \hat{z}_n(\Psi)])$ and for every z

$$P(\bigcup_{m=1}^{\infty} \{z_k(z, X_0, \ldots, X_{k-1}) = \hat{z}_k(\Psi), k \geq m) = 1 . \tag{10}$$

Moreover, $\tilde{\Psi}$ is the only stationary weak solution of (1).

<u>Proof.</u> For every fixed z the renewing epochs are also z-renewing epochs. The assumptions of Theorem 4 are satisfied. Thus, for every z there is a stationary strong solution $\tilde{\Psi}_z = ([X_n, z_n(z, \Psi)])$ with the property (4). From the stationarity, ergodicity and (9) we get the existence of a measurable set D' satisfying P(D') = 1 and

$$D' \subseteq \{\psi : \psi \text{ possesses an infinite number of renewing epochs left and right from } 0\}.$$

For a $\psi = (x_j) \in D'$, $n \in \Gamma$, and $i \leq n$ with $(x_{j-i}) \in D_0'$ we can define $\hat{z}_n(z, \psi)$ in the following way:

$$\hat{z}_n(z, \psi) = z_{n-i+l(i,z,\psi)}(z, x_{i-l(i,z,\psi)}, \ldots, x_{n-1}) .$$

Using Definition 4 we obtain that the solution $\tilde{\Psi}_z =$

$([X_n, \hat{z}_n(z, \Psi)])$ does not depend on z. Denote this solution by $\tilde{\Psi} = ([X_n, \hat{z}_n(\Psi)])$. For an arbitrary z' formula (4) becomes

$$P\left(\bigcup_{m=1}^{\infty}\left\{ z_k(z', X_0, \ldots, X_{k-1}) = \hat{z}_k(\Psi), k \geq m\right\}\right) = 1 . \qquad (11)$$

The uniqueness statement of the theorem follows from (11) in the standard way, cf. FKAS (1981), Chapter 2. ■

2.3. The third approach to equations of type (1) is due to Lisek (1979, 1982). Conditions on (Ψ, P) will be given which ensure the existence of a finite number of strong solutions of (1). These solutions are not necessarily stationary. However, an appropriate mixture of their distributions defines a stationary weak solution.

Theorem 6. Let M be a measurable subset of $\mathcal{X}^{\Gamma}$ with the following properties:

(a) $P(M) = 1$.

(b) There exists a system $\{A(n, \psi), n \in \Gamma, \psi \in M\}$ of non-empty subsets of $\mathbb{Z}$ satisfying the conditions:

(b1) For every fixed $n \in \Gamma$ there exist measurable mappings $\psi \longmapsto z_i(n, \psi)$ defined on $\{\psi: \psi \in M, i \leq \#A(n,\psi)<\infty\}$, $i \geq 1$, such that $A(n, \psi) = \{z_1(n, \psi), z_2(n, \psi), \ldots, z_{\#A(n,\psi)}(n, \psi)\}$. (In other words, the elements of $A(n, \psi)$ with $\#A(n, \psi) < \infty$ are numbered and denoted by $z_i(n, \psi)$.)

(b2) $f(A(n, \psi), x_n) \subseteq A(n+1, \psi)$, $\psi \in M$, $n \in \Gamma$.

(b3) $A(n, (x_j)) = A(n-1, (x_{j+1}))$, $(x_j) \in M$, $n \in \Gamma$.

(b4) $P(\{\psi : \#A(0, \psi) < \infty\}) = 1$.

Then there are a finite number $q = q(P)$, a measurable set $H \subseteq M$ with $P(H) = 1$ and measurable functions $z_{n,1}(\psi), \ldots, z_{n,q}(\psi)$, defined on H, $n \in \Gamma$, satisfying $z_{n,i}(\psi) \in A(n, \psi)$, $i = 1, \ldots, q$, and $z_{n+1,i}(\psi) = f(z_{n,i}(\psi), x_n)$, $n \in \Gamma$, $i = 1, \ldots, q$. These functions define strong solutions $(\tilde{\Psi}_i, \tilde{P}_i)$, $\tilde{\Psi}_i = ([X_n, z_{n,i}(\Psi)])$, with the property that

$$\tilde{P}(.) = \frac{1}{q}\sum_{i=1}^{q} \tilde{P}_i(.) \qquad (12)$$

is the distribution of a stationary weak solution of (1).

Remark. For a given P the number $q = q(P)$ may depend on the choice of the system $\{A(n, \psi)\}$. However, for a fixed system

$\{A(n, \Psi)\}$, $q = q(P)$ is uniquely determined. For finite state space $\mathbb{Z}$ the conditions (a) - (b4) are always satisfied for $A(n, \Psi) = \mathbb{Z}$.

Concerning the uniqueness of the weak solution $(\tilde{\Psi}, \tilde{P})$ defined by (12) we have the following result.

Theorem 7. Under the conditions of Theorem 6 $(\tilde{\Psi}, \tilde{P})$ is the only stationary weak solution of (1) iff $\tilde{\Psi}$ is ergodic.

For proofs of the Theorems 6 and 7 we refer to Lisek (1982).

Assume the conditions of Theorem 6 to be fulfilled. Then there is a close relation between the existence of renewing epochs and the property $q = 1$ (for a fixed system $\{A(n, \Psi)\}$).

Theorem 8. Let $q = 1$ and suppose that for fixed z

$$P(\{\Psi : \text{there is an } l \text{ with } z_j(z, x_{-j}, \dots, x_{-1}) \in A(0, \Psi) \text{ for all } j \geq l\}) = 1 . \qquad (13)$$

Then $P(C_0) = 1$.

Proof. As shown in Lisek (1982), for a.e. Ψ there is a finite number $t(0, \Psi) \leq 0$ with the property

$$\#\{z_{-t(0,\Psi)}(A(t(0, \Psi), \Psi), x_{t(0,\Psi)}, \dots, x_{-1})\} = q . \qquad (14)$$

From (13) and the property (b3) in Theorem 6 we obtain the existence of a finite number $l_z(\Psi)$ satisfying

$$z_j(z, x_{t(0,\Psi)-j}, \dots, x_{t(0,\Psi)-1}) \in A(t(0, \Psi), \Psi), \ j \geq l_z(\Psi) \qquad (15)$$

for a.e. Ψ. For every Ψ with the properties (14) and (15) the value

$$l(0, z, \Psi) = -t(0, \Psi) + l_z((x_{j+t(0,\Psi)})) \qquad (16)$$

satisfies the conditions of Definition 3. Thus, $P(C_0) = 1$ is shown. ∎

Theorem 9. Assume $q = 1$ and the validity of the following condition: There is a measurable subset $M' \subseteq M$ with the properties $P(M') = 1$ and

$$M' \subseteq \{\Psi : \text{for every } z \text{ there is an } l \text{ with } z_j(z, x_{-j}, \dots, x_{-1}) \in A(0, \Psi) \text{ for all } j \geq l\} . \qquad (17)$$

Then there exists a measurable subset $D_0' \subseteq D_0$ with $P(D_0') = 1$.

The proof is similar to that of Theorem 8.

Remark. The Theorems 8 and 9 say that the assumptions $q = 1$ and (13) ($q = 1$ and (17)) ensure the validity of all statements in Theorem 4 (Theorem 5).

<u>Theorem 10.</u> Under the assumption of Theorem 5 it holds that $q = 1$ for every system $\{A(n, \psi)\}$ satisfying the conditions of Theorem 6.

<u>Proof.</u> In Lisek (1982) is shown that for all $k > 0$ and almost all $\psi = (x_n)$

$$z_k(\{z_{0,1}(\psi), \ldots, z_{0,q}(\psi)\}, x_0, \ldots, x_{k-1})$$
$$= \{z_{0,1}((x_{n+k})), \ldots, z_{0,q}((x_{n+k}))\} . \tag{18}$$

Assume $q > 1$. Then $z_{0,1}(\psi) \neq z_{0,2}(\psi)$. From Theorem 5 we obtain that there is a number k with

$$z_k(z_{0,1}(\psi), x_0, \ldots, x_{k-1}) = z_k(z_{0,2}(\psi), x_0, \ldots, x_{k-1}) ,$$

which contradicts (18). ■

<u>3. Arrival-stationary Queueing Processes</u>

As mentioned above, equations of type (1) describe the evolution of queueing systems observed at the arrival epochs only. Here and below we use a modified Kendall notation. By G/G/m/r we denote an m-server queue with r waiting places, FCFS queueing discipline and a stationary and ergodic traffic $\Psi = (X_n)$, $X_n = [A_n, S_n]$ for $n \in \Gamma$. (The ergodicity of Ψ can be relaxed for some queueing systems.) We assume $m_a = E\,A_0 < \infty$, $m_s = E\,S_0 < \infty$. The arrival epochs of customers are

$$T_0 = 0; \quad T_n = \sum_{j=0}^{n-1} A_j \ , \ n \geq 0; \quad T_n = -\sum_{j=n}^{-1} A_j \ , \ n < 0 . \tag{19}$$

Since we do not assume $P(A_n > 0, n \in \Gamma) = 1$, batch arrivals are allowed. However, since the service times S_n are indexed, it follows that the customers in a batch are also numbered. (This assumption is not restrictive in the treatment of batch arrival queues, cf. Wirth (1982).)

No assumptions on the independence in the traffic Ψ are made in general. However, in some special cases the sequence Ψ will satisfy some of the following conditions:

(i) (A_n) and (S_n) are independent sequences.

(ii) The interarrival times A_n are i.i.d.

(iii) The service times S_n are i.i.d.

If we have the properties (i) and (ii), we shall call this a GI/G/m/r queue; if we have (i) and (iii), this is a G/GI/m/r; in the case of the properties (i) - (iii), this is a GI/GI/m/r (notice the latter is GI/G/m/r in Kendall's notation).

The aim of this section is to define the state space $\mathcal{Z}$ and the function f for some special queueing systems and to discuss the questions formulated at the beginning of the previous section. We point out that for a stationary weak solution $\tilde{\Psi}$ of (1) the sequence (Z_n) describes the steady state behaviour of the system at the arrival epochs. The standard queueing characteristics such as actual waiting time, queue size, etc. can be calculated from the weak solution $\tilde{\Psi}$.

3.1. First we consider the system G/G/m/∞. As the state of the system at the arrival epoch of the n-th customer we consider the work load vector Z_n. We have $\mathcal{Z} = \{[z_1, \ldots, z_m]: 0 \leq z_1 \ldots \leq z_m\}$ and

$$Z_{n+1} = f_1(Z_n, [A_n, S_n]) = (\hat{R}(Z_n + S_n J - A_n I))_+ , \tag{20}$$

where $J = [1, 0, \ldots, 0]$, $I = [1, \ldots, 1]$, $[x_1, \ldots, x_m]_+ = [\max(0, x_1), \ldots, \max(0, x_n)]$, and $\hat{R}(x)$ is the vector formed by the components of x rearranged in ascending order.

Endowed with the maximum metric $\rho(x, y) = \max_i |x_i - y_i|$ the space $\mathcal{Z}$ satisfies (2.1.1) - (2.1.5), where $z_0 = 0$. The additional condition

$$m_s/m_a < m \tag{21}$$

ensures the properness of $\underline{z}_n(\Psi)$ in (2), cf. Loynes (1962). Thus, if the traffic Ψ satisfies (21), Theorem 1 determines the minimal solution $([[A_n, S_n], \underline{z}_n(\Psi)])$. However, for $m > 1$ this is not the only stationary solution of (20) in general, cf. Loynes (1962) for counter-examples.

In the case of $m = 1$, $([[A_n, S_n], \underline{z}_n(\Psi)])$ is the only stationary solution of (20). Moreover, for an arbitrary $z \in \mathbb{Z}$

$$P(\bigcup_{m=1}^{\infty} \{z_k(z, X_0, \dots, X_{k-1}) = \underline{z}_k(\Psi), k \geq m\}) = 1 , \tag{22}$$

where $X_j = [A_j, S_j]$. The proof is similar to that of Theorem 5, cf. e.g. FKAS (1981), Chapter 2. The following fact is used: Consider $I_n = (A_n - \underline{z}_n(\Psi) - S_n)_+$ - the time during which the system is empty between the arrival of the n-th and (n+1)st customer. From (21) we obtain

$$P(\sum_{n=1}^{\infty} I_{-n} = \sum_{n=0}^{\infty} I_n = \infty) = 1 . \tag{23}$$

We can also apply Theorem 5 directly, because by using (23) it is easy to show that every $n \in \Gamma$ is a renewing epoch in the sense of Definition 4.

Remark. Arguing in the same way one obtains the existence of the unique solution $\underline{z}_k(\Psi)$ of (20) and the validity of the ergodic property (22) for a single server queue satisfying (21) with an arbitrary conservative queueing discipline, cf. FKAS (1981), Chapter 2.

The conditions (2.1.6) - (2.1.8) are also fulfilled in the case of $G/G/m/\infty$, the condition (21) ensures the properness of $z_n^{max}(\Psi)$, $n \in \Gamma$, cf. Brandt (1982a). Thus, Theorem 1' provides the maximal solution $([[A_n, S_n], z_n^{max}])$. Moreover, the conditions of Theorem 2 are fulfilled. Thus, for $G/G/m/\infty$ queues satisfying (21) the minimal solution $([[A_n, S_n], \underline{z}_n(\Psi)])$ is the only stationary weak solution of (20) iff the ergodic property (ii) in Theorem 2 holds with $X_n = [A_n, S_n]$, $n \in \Gamma$.

Remark. A similar result was proved by Voss (private communication; the paper is to appear in Siberian Math. J. 1983).

Recently, Brandt (1982b) has proved the following

<u>Theorem 11.</u> Consider a $G/GI/m/\infty$ queue satisfying the condition (21). Then the minimal solution $([[A_n, S_n], \underline{z}_n(\Psi)])$ is the only stationary weak solution of (20). Thus, the ergodic statement (ii) in Theorem 2 is true for a $G/GI/m/\infty$ queue.

The proof is based on structural properties of $G/GI/m/\infty$ queues and cannot be derived from the general theory. Theorem 11 generalizes the analogous well-known result by Kiefer and Wolfowitz (1955) for $GI/GI/m/\infty$ queues.

Concerning the treatment of $G/G/m/\infty$ queues by means of renewing events we refer to Borovkov (1978, 1980).

3.2. For the loss system G/G/m/0 we have again $\mathbf{Z} = \{[z_1, \ldots, z_m], 0 \le z_1 \le \ldots \le z_m\}$, where $z_1, \ldots, z_m$ are the residual service times arranged in ascending order. Then Equation (1) has the form

$$Z_{n+1} = f_2(Z_n, [A_n, S_n]) = (\hat{R}(Z_n + S_n 1_{\{Z_{n,1}=0\}} J - A_n I))_+ \quad ,(24)$$

where $Z_n = [Z_{n,1}, \ldots, Z_{n,m}]$.

When investigating a G/G/m/0 queue, it is useful to consider it as a part of the $G/G/\infty$ queue (infinitely many servers) with the same traffic Ψ, cf. Franken (1970), Borovkov (1972, 1978, 1980), FKAS (1981). It can easily be shown that the specification of the equation (1) for $G/G/\infty$ has a unique stationary strong solution and that the ergodic property (10) is valid. In particular, the number of busy servers at the arrival epoch T_n

$$l_n(\Psi) = \#\{j: j < n, T_j + S_j > T_n\}$$

is a proper r.v. for every $n \in \Gamma$.

Consider a sample path $\psi = ([a_n, s_n])$ of the traffic Ψ. A number n is a renewing epoch of ψ for G/G/m/0 in the sense of Definition 4 if there is a number $d = d(n, \psi)$, $0 \le d < \infty$, such that

$$\sum_{j<n-d} 1_{\{t_j+s_j > t_{n-d}\}} < m, \ldots, \sum_{j<n-1} 1_{\{t_j+s_j > t_{n-1}\}} < m ;$$

$$\sum_{j<n-d} 1_{\{t_j+s_j > t_n\}} = 0 .$$

For a given state $z = [z_1, \ldots, z_m]$ we get

$$l(n, z, \psi) = \min\{k: t_{n-d} - t_{n-d-k} \ge z_m\} + d(n, \psi) .$$

In particular, if $d(n, \psi) = 0$, the system $G/G/\infty$ (and thus, G/G/m/0) is empty at the arrival epoch of the n-th customer.

For a G/GI/m/0 queue the condition

$$P(S_0 < A_0) > 0 \tag{25}$$

is sufficient for the validity of $P(1_0(\Psi) = 0) > 0$, cf. Franken (1970). This means that (25) implies (9). For a GI/GI/m/0 queue the condition

$$P(S_0 < m\, A_0) > 0 \tag{26}$$

is sufficient for (9), cf. Borovkov (1978, 1980). Thus, under the conditions (25) and (26), respectively, the statements of Theorem 5 are valid.

Now we apply the method given in section 2.3 to G/G/m/0 queues. If the k-th customer is in the system at t_n, his residual service time is $t_k + s_k - t_n$. Hence the components of the state of the system G/G/m/0 at t_n have to be selected from the a.s. finite set

$$\{t_k + s_k - t_n : k < n,\ t_k + s_k > t_n\}.$$

The assumptions of Theorem 6 are fulfilled and we obtain, cf. Lisek (1979),

<u>Theorem 12.</u> Consider a G/G/m/0 queue with the traffic (Ψ, P). Then there is a finite number $q = q(P)$ and there are measurable functions $z_{n,1}(\psi), \ldots, z_{n,q}(\psi)$, $n \in \Gamma$ (defined for a.e. ψ) such that $(\tilde{\Psi}_i, \tilde{P}_i)$, $\tilde{\Psi}_i = ([[A_n, S_n], z_{n,i}(\Psi)])$ are strong solutions of (24) and

$$\tilde{P}(.) = \frac{1}{q} \sum_{i=1}^{q} \tilde{P}_i(.) \tag{27}$$

is the distribution of a stationary weak solution.

Thus, there always exists a stationary weak solution of the equation (24). It is easy to see that (27) is not the only stationary solution in general.

<u>Example 1.</u> Consider the two server loss system with constant interarrival times $A_n = 1$ and constant service times $S_n = 7/2$. Then there are infinitely many stationary weak solutions.

<u>Example 2.</u> Consider a single server loss system with $A_n = 1$, $n \in \Gamma$, and (S_n) being a stationary Markov chain with the states

5/4 and 7/4, and transition probabilities $p_{12} = p_{21} = 1$. Then we have infinitely many stationary weak solutions.

The uniqueness criterion of Theorem 7 is hard to apply. But, for GI/GI/1/0 queues we have the following result.

Theorem 13. For the GI/GI/1/0 queue there always exists exactly one stationary weak solution.

Proof. According to Theorem 12 there exists a stationary weak solution. Now we consider such a solution (Ψ', P'), $\Psi' = ([[A_n', S_n'], Z_n'])$. Obviously, P'(there are infinitely many n with $Z_n' = 0) = 1$ and thus, the conditional distribution $P'(.|Z_0' = 0)$ is well-defined. From the proof of Theorem 6 one observes that the condition $\{Z_0' = 0\}$ does not depend on the "future" $([[A_n', S_n'], Z_n'], n \geq 0)$ of Ψ', cf. Lisek (1983) for details. Thus,

$$\hat{P}(.) = P'(([[A_n', S_n'], Z_n'], n \geq 0) \in (.) \,/\, Z_0' = 0)$$

$$= P(([[A_n, S_n], z_n(0,[A_0, S_0],\dots,[A_{n-1}, S_{n-1}])], n \geq 0) \in (.)).$$

In other words, the distribution $\hat{P}$ is the same for all stationary weak solutions (Ψ', P'). It is known that the distribution P' is uniquely determined by the conditional distribution $\hat{P}$, cf. e.g. Rolski (1981), Wirth (1982).■

3.3. Now we consider a single server queue with warming-up, i.e. a single server queue in which the first customer of each busy period receives an exceptional service. The investigation of such queues can be reduced to that of a corresponding standard $G/G/1/\infty$ system and a suitably chosen single server loss system, cf. Wirth (1983).

The traffic of the system considered is given as a stationary and ergodic sequence $\Psi = ([A_n, S_n, U_n])$, where A_n, S_n are of the same meaning as above and U_n denotes the potential warm up time of the n-th customer. If Z_n denotes the actual waiting time of a customer, equation (1) becomes

$$Z_{n+1} = \begin{cases} Z_n + S_n - A_n, & \text{if } Z_n + S_n - A_n > 0, \\ U_{n+1}, & \text{otherwise.} \end{cases} \qquad (28)$$

(The waiting time of a customer initiating a busy period is defined as his own warm up time.)

Consider the G/G/1/∞ queue with the traffic $([A_n, S_n])$. If the condition $m_s/m_a < 1$ is satisfied, then there exists a unique stationary and ergodic strong solution $([[A_n, S_n], W_n])$ of (20) satisfying $P(W_0 = 0) > 0$, cf. (23). Thus, the conditional distribution $\hat{P}(.) = P(. \mid W_0 = 0)$ is well-defined. Consider the random sequence $([\hat{A}_n, \hat{S}_n, \hat{U}_n, \hat{W}_n])$ distributed according to $\hat{P}$, where all symbols have the usual meaning. Define the random indices

$$\hat{\vartheta}_0 = 0 \;;$$
$$\hat{\vartheta}_n = \min\left\{k:\ k > \hat{\vartheta}_{n-1},\ \hat{W}_k = 0\right\},\ n \geq 1;$$
$$\hat{\vartheta}_n = \max\left\{k:\ k < \hat{\vartheta}_{n+1},\ \hat{W}_k = 0\right\},\ n \leq -1.$$

Then

$$\hat{\xi}_n = - \sum_{i=\hat{\vartheta}_n}^{\hat{\vartheta}_{n+1}-1} (\hat{S}_i - \hat{A}_i) \geq 0 \qquad \hat{P}\text{-a.s.}$$

is just the length of the n-th idle period conditioned by $W_0 = 0$. Define $\hat{\eta}_n = \hat{U}_{\hat{\vartheta}_n}$. The distribution of the random sequence $\hat{\Psi} = ([\hat{\xi}_n, \hat{\eta}_n])$ is completely determined by $\hat{P}$; $\hat{\Psi}$ is again stationary and ergodic. It holds

$$E_{\hat{P}}\ \hat{\xi}_n = (P(W_0 = 0))^{-1}(m_a - m_s)\ , \tag{29}$$

cf. FKAS (1981), Franken and Lisek (1982), Wirth (1982).

Using the fact that there is a one-to-one correspondence between P and $\hat{P}$, cf. Rolski (1981), Wirth (1982), one can show that there is a one-to-one correspondence between weak solutions of a G/G/1/0 system with the traffic $\hat{\Psi} = ([\hat{\xi}_n, \hat{\eta}_n])$ and weak solutions of (28). Using the criterion (25) and formula (29) we get the following

<u>Theorem 14.</u> If the sequence of potential warm up times (U_n) is independent of $([A_n, S_n])$, the U_n are i.i.d. and the condition

$$P(U_0 < (m_a - m_s)(P(W_0 = 0))^{-1}) > 0 \tag{30}$$

holds, where W_0 denotes the stationary actual waiting time of the G/G/1/∞ queue with the traffic $([A_n, S_n])$, then there exists a unique stationary weak solution of (28), which is strong and which satisfies (10).

Theorem 13 leads to the following result:

Theorem 15. If (U_n) and $([A_n, S_n])$ are independent, U_n, $n \in \Gamma$, are i.i.d. and $A_n - S_n$, $n \in \Gamma$, are i.i.d., then there exists a unique stationary weak solution of (28).

4. Time-stationary Queueing Processes

For a given queue consider a stationary weak solution $\tilde{\Psi} = ([[\tilde{A}_n, \tilde{S}_n], Z_n])$ of the corresponding equation (1). The $\mathbb{Z}$-valued random variables Z_n, $n \in \Gamma$, describe the temporal behaviour of the queue in steady state (at the arrival epochs only). Define

$$Z(t) = f(Z_n, [t - T_n, S_n]), \quad T_n < t \leq T_{n+1}, \tag{31}$$

where f is the same as in (1). For all common queueing systems the temporal behaviour of the system in continuous time given the states at the arrival epochs is described by the stochastic process $(Z(t), t \in R)$. Notice that $(Z(t))$ is non-stationary in general.

Our aim is to define a stationary state process $(\overline{Z}(t))$ describing the temporal behaviour of the queue considered in continuous time. Define a distribution $\overline{P}$ of a random sequence $([\overline{T}_n, [\overline{S}_n, \overline{Z}_n]])$ by

$$\overline{P}(.) = (E_{\tilde{P}}\,\tilde{A}_0)^{-1} \int_0^{\infty} \tilde{P}(\tilde{A}_0 > t, ([\tilde{T}_n - t, [\tilde{S}_n, Z_n]]) \in (.))dt, \tag{32}$$

where the $\tilde{T}_n$ are derived from the $\tilde{A}_n$ via (19). From (32) we have

$$\overline{P}(\overline{Z}_{n+1} = f(\overline{Z}_n, [\overline{T}_{n+1} - \overline{T}_n, \overline{S}_n]) = 1. \tag{33}$$

The stochastic process $(\overline{Z}(t))$ defined by

$$\overline{Z}(t) = f(\overline{Z}_n, [t - \overline{T}_n, \overline{S}_n]) \quad \text{for } \overline{T}_n < t \leq \overline{T}_{n+1} \tag{34}$$

is strictly stationary. We call $(\overline{Z}(t))$ the time-stationary state process associated with the stationary weak solution $\tilde{\Psi}$.

Moreover, we have the following ergodic properties, cf. Nawrotzki (1978), FKAS (1981), Franken (1982),

$$\lim_{t\to\infty} \frac{1}{t} \int_0^t \tilde{P}(Z(u) \in (.))du = \overline{P}(\overline{Z}(0) \in (.)) , \tag{35}$$

$$\lim_{n\to\infty} \frac{1}{n} \sum_{j=1}^{n} \overline{P}(\overline{Z}_j \in (.)) = \tilde{P}(Z_0 \in (.)) . \tag{36}$$

Using (35) one can obtain results similar to (4) in continuous time. The relations (33) - (36) show that $(\overline{Z}(t))$ is the appropriate model for the steady state behaviour of the queue considered in continuous time. For further details see e.g. FKAS (1981) and Franken (1982).

Now we construct a stationary sequence describing the states of the queue immediately after departure epochs. Starting with the process $(\overline{Z}(t))$ and the epochs $(\overline{T}_n)$ one can define the departure epochs $\overline{T}_n^*$ and the system states $\overline{Z}_n^*$ immediately after $\overline{T}_n^*$, $n \in \Gamma$. Let be $\overline{A}_n^* = \overline{T}_{n+1}^* - \overline{T}_n^*$, $n \in \Gamma$. Then the formula

$$P^*(.) = E_{\tilde{P}} \tilde{A}_0 \sum_{j=1}^{\infty} \overline{P}(\overline{T}_j^* < 1, ([\overline{A}_{n+j}^*, \overline{Z}_{n+j}^*], n \in \Gamma) \in (.)) \tag{37}$$

defines the distribution P^* of a stationary sequence $([A_n^*, Z_n^*])$ with the desired properties, cf. FKAS (1981), Franken (1982). We call (Z_n^*) the departure-stationary queueing process.

5. Model Continuity

It is a question of practical and theoretical importance whether the distributions $\tilde{P}$, $\overline{P}$, and P^* describing the steady state behaviour of a queueing system at the arrival epochs, in continuous time and at departure epochs, respectively, depend continuously on the traffic distribution P. We treat this problem in terms of weak convergence of probability measures. The usual weak convergence for traffic distributions is based on the standard metric on $\mathcal{X}^\Gamma$. Unfortunately, the most interesting functions such as

$$\Psi = ([a_j, s_j]) \longmapsto z_{n,i}(\Psi), \ i = 1, \ldots, q(P), \ n \in \Gamma , \tag{38}$$

in the case of loss systems, cf. Theorem 6, Theorem 12, and

$$\Psi = ([a_j, s_j]) \longmapsto l_0(\Psi) = \#\{j: j<0, s_j + t_j > 0\} , \tag{39}$$

cf. section 3.2., are discontinuous for almost all $\psi \in \mathcal{X}^{\Gamma}$ with respect to this metric, cf. Lisek (1981). In a similar way as was done in Lisek (1981) for point processes, one can construct a measurable subset $M \subseteq \mathcal{X}^{\Gamma}$ such that P(M) = 1 for every stationary distribution P and a metric ν on M with the properties:

(5.1) The mappings (38) are continuous for all ψ with $\#\{j: j < n,\ t_j + s_j = t_n\} = 0$.

(5.2) The mapping (39) is continuous for all ψ with $\#\{j: j < 0,\ t_j + s_j = 0\} = 0$.

(5.3) Denote by ρ the Prochorov-metric based on ν and by $\xrightarrow[\rho]{}$ the weak convergence generated by ρ. Then there is the following relation between $\xrightarrow[\rho]{}$ and the usual weak convergence $\Longrightarrow$: Let (Ψ, P) and (Ψ_k, P_k), $k \geq 1$, be stationary traffics, and $m_a^{(k)} = E_{P_k} A_0^{(k)}$, $m_s^{(k)} = E_{P_k} S_0^{(k)}$. If

$m_a^{(k)} \longrightarrow m_a$, $m_s^{(k)} \longrightarrow m_s$, then $P_k \xrightarrow[\rho]{} P$ iff $P_k \Longrightarrow P$.

Now we are able to apply the Continuous Mapping Theorem. For the quantity $l_0(\Psi)$ defined by (39) (the number of busy servers in a G/G/∞ queue) we get the following continuity statement.

<u>Theorem 16.</u> Consider the traffics (Ψ, P), (Ψ_k, P_k), $k \geq 1$. Under the conditions

$$P_k \Longrightarrow P \ ; \quad m_a^{(k)} \longrightarrow m_a \ ; \quad m_s^{(k)} \longrightarrow m_s \ ; \tag{40}$$

$$P(\{\psi: \text{there is a } j < 0 \text{ with } t_j + s_j = 0\}) = 0 \tag{41}$$

it holds that

$$P_k(l_0(\Psi_k) = i) \longrightarrow P(l_0(\Psi) = i)\ , \quad i \geq 0.$$

Using (38) and (39) we obtain

<u>Theorem 17.</u> Consider a loss system G/G/m/0. Under the conditions (40), $q(P_k) \longrightarrow q(P)$, and (41) it holds that $\tilde{P}^k \Longrightarrow \tilde{P}$, where $\tilde{P}^k$ and $\tilde{P}$ are the stationary weak solutions given by Theorem 12.

With different methods (and under the additional condition (9) in the G/G/m/0 case) Theorems 16 and 17 were proved in Borovkov (1972, 1978, 1980) and FKAS (1981). Recently Brandt and Lisek (1982) proved the following continuity result for the G/GI/m/∞

queue.

Theorem 18. Consider a $G/GI/m/\infty$ queue. Under the condition (40) it holds that

$$P_k((\underline{z}_n(\Psi_k)) \in (.)) \Longrightarrow P((\underline{z}_n(\Psi)) \in (.)) ,$$

where $([[A_n^{(k)}, S_n^{(k)}], \underline{z}_n(\Psi_k)])$ and $([[A_n, S_n], \underline{z}_n(\Psi)])$ denote the unique stationary weak solution of (20) for the traffics Ψ_k and Ψ, respectively, cf. Theorem 11.

The continuous dependence of the distributions $\overline{P}$ and P^* on P can be proved using the continuity property of the mapping $P \longmapsto \overline{P}$, cf. (32); for details see FKAS (1981), Brandt and Lisek (1982).

References

Borovkov A.A. (1972), Continuity theorems for multi-server loss systems, Teor. Veroyat. Primenen. 17, 458-468 (in Russian).

Borovkov A.A. (1978), Ergodic and stability theorems for one class of stochastic equations and their applications, Teor. Veroyat. Primenen. 23, 241-262 (in Russian).

Borovkov A.A. (1980), Asymptotic Methods in Queueing Theory (in Russian), Nauka, Moscow.

Brandt A. (1982a), On stationary waiting times and limiting behaviour of queues with many servers I. The general $G/G/m/\infty$ case (to appear in EIK).

Brandt A. (1982b), On stationary waiting times and limiting behaviour of queues with many servers II. The $G/GI/m/\infty$ case (to appear in EIK).

Brandt A. and Lisek B. (1981), On the continuity of G/GI/m queues (to appear in Math. Operationsforsch. Statist., Ser. Statistics).

FKAS (Franken P., König D., Arndt U. and Schmidt V.) (1981), Queues and Point Processes, Akademie-Verlag, Berlin (J. Wiley, 1983).

Franken P. (1970), Ein Stetigkeitssatz für Verlustsysteme, Operationsforschung und Math. Statistik II, 9-23, Akademie-Verlag, Berlin.

Franken P. (1982), The Point Process Approach to Queueing Theory and Related Topics, Seminarbericht Nr. 43, Sektion Mathematik, Humboldt-Universität zu Berlin, G.D.R.

Franken P. and Lisek B. (1982), On Wald's identity for dependent variables, Z. Wahrscheinlichkeitstheorie verw. Geb. 60, 143-150.

Kiefer J. and Wolfowitz J. (1955), On the theory of queues with many servers, Trans. Amer. Math. Soc. 78, 1-18.

Lisek B. (1979), Construction of stationary state distributions for loss systems, Math. Operationsforsch. Statist., Ser.

Statistics 10, 561-581.

Lisek B. (1981), Stability theorems for queueing systems without delay, Elektron. Informationsverarb. Kybernetik 17, 259-278.

Lisek B. (1982), A method for solving a class of recursive stochastic equations, Z. Wahrscheinlichkeitstheorie verw. Geb. 60, 151-161.

Lisek B. (1983), A method for solving a class of recursive stochastic equations II. (paper in preparation).

Loynes R.M. (1962), The stability of a queue with non-independent inter-arrival and service times, Proc. Cambridge Philos. Soc. 58, 497-520.

Nawrotzki K. (1978), Einige Bemerkungen zur Verwendung der Palmschen Verteilung in der Bedienungstheorie, Math. Operationsforsch. Statist., Ser. Optimization 9, 241-253.

Rolski T. (1981), Stationary Random Processes Associated with Point Processes, Lecture Notes in Statistics 5, Springer-Verlag, New York, Heidelberg, Berlin.

Wirth K.-D. (1982), On stationary queues with batch arrivals, Elektron. Informationsverarb. Kybernetik 18, 603-619.

Wirth K.-D. (1983), A new approach to queues with warm up times (submitted to Elektron. Informationsverarb. Kybernetik).

ERGODICITY ASPECTS OF MULTIDIMENSIONAL MARKOV CHAINS WITH APPLICATION TO COMPUTER COMMUNICATION SYSTEM ANALYSIS

Wojciech Szpankowski
Technical University of Gdansk
80-952 Gdansk, ul. Majakowskiego 11/12
POLAND

ABSTRACT

Ergodicity and nonergodicity of multidimensional Markov chains are discussed. We investigate ergodic and nonergodic Markov chains by the means of Lyapunov function method and so called comparison tests method. In the former case two theorems being a generalization of Pakes and Kaplan theorems are presented. Moreover, for multidimensional Markov chains we propose an explicite form for a simple Lyapunov function. On the other hand, comparison tests investigate ergodicity/nonergodicity of M-dimensional Markov chains by the means of appropriate ergodic/nonergodic K-dimensional Markov chains, where K is smaller than M. These results are applied to analysis of some computer communication systems.

1. INTRODUCTION

When computer networks become larger and more sophisticated new problems arise. Packet switching has offered many advantages over conventional circuit-switching techniques, however, under some circumstances unpleasant behaviour may occur as congestions, deadlocks, bistability and hysteresis, divergent behaviour and others. Moreover large complex systems as computer networks with complicated feedback mechanisms pose highly non-trivial design and analysis problems. Fortunately, in many cases a system may be discribed by a multidimensional Markov chain. If such a chain is aperiodic and irreducible, then the convergent/divergent behaviour may be explored as a chain classification problem considering its ergodicity or nonergodicity / transient or null recurrent /. The paper deals with the ergodicity/nonergodicity of multidimensional Markov chains and theirs applications to analysis of computer communication systems.

Queueing theory methods which found a wide applications to the analysis of computer communication systems may be unsuitable in the investigation of pathological behaviour of such systems. Therefore, other methods as qualitative analysis was lately rapidly developed [OLD], [STO], [KAM], [KAL]. Differential and dynamic topology, Lyapunov theory, global analysis, catastrophe theory and stochastic ordering find more and more applications in the analysis of computer networks [OLD], [STO]. In this paper Lyapunov functions and stochastic ordering will be applied to explore ergodicity and nonergodicity of multidimensional Markov chains.

The ergodicity and nonergodicity conditions are quite well discussed in [FOS], [CHU], [KEM], however, the results are not suitable for computer network analysis. An alternative is to use the advantage of Lyapunov functions. In order to enhance the Lyapunov function application we shall prove two theorems which are a generalization of Pakes [PAK] and Kaplan [KAP] lemmas. These theorems will be used to explore properties of multidimensional Markov chains. Nevertheless, Lyapunov functions for the case of multidimensional Markov chains might have a limited application. Therefore, two other lemmas called comparison tests which show that the ergodicity/nonergodicity of M-dimensional Markov chain may be investigated on the basis of the ergodicity/nonergodicity of K-dimensional Markov chains, where K is smaller than M, will be proved. In order to determine the K-dimensional Markov chains needed in the discussed lemmas stochastic ordering will be used. Finally, we shall apply the above theorems to explore the convergent/divergent behaviour of some computer communication systems as: unbuffered ALOHA systems, buffered heterogeneous ALOHA systems and network of queues.

2. BASIC RESULTS

Let $N(t)$ be a Markov chain whose state space C is a countable set of nonnegative integers and t is a discrete time, $t=0,1,\ldots$. Throughout the paper it will be assumed / if not stated otherwise / that $N(t)$ is an irreducible and aperiodic Markov chain [CHU]. In such a case $N(t)$ is called ergodic [CHU], [KEM] if for each $i,j \in C$ the limit $\lim \Pr\{N(t)=j \mid N(0)=i\}$ as t tends to infinity exists, has a positive value and does not depend on i. Otherwise, the chain is nonergodic however, a nonergodic chain may be further classified as null recurrent or transient [CHU]. In this paper we discuss mainly ergodic and nonergodic chains since these are the most important from the practi- point of view. In the first part of this section we shall investigate ergodicity and nonergodicity of a Markov chain by the means of Lyapunov functions. Then, using stochastic ordering comparison tests will be presented.

2.1 LYAPUNOV FUNCTIONS

Let $N(t)$ be a Markov chain as discussed above. In order to investigate its property we introduce a Lyapunov function $V: C \to R$ / C is a state space, while R is a set of real numbers / such that for each $k \in C$ the function is downward bounded, i.e., $V(k) \geq c > -\infty$, $c \in R$. We shall denote by $AV(k)$ an operator defined as follows:

$$AV(k) = E\{V[N(t+1)] - V[N(t)] \mid N(t)=k\} = E\{V[N(t+1) \mid N(t)=k\} - V(k) \quad (1)$$

assuming that the operator exists and for each t and $k \in C$

$$|AV(k)| < \infty \quad (2)$$

The operator $AV(k)$ has an interesting property. To explore it let us notice that

$$V[N(\tau)] = V[N(0)] + \sum_{0 \leq t < \tau} \{V[N(t+1)] - V[N(t)] \quad (3)$$

where τ is a Markov moment [WON]. Averaging both sides of Eq.(3) under condition $N(0)=k$ and denoting $E_k V[N(t)] = E\{V[N(t)] \mid N(0)=k\}$ one obtains [KAL]

$$E_k V[N(\tau)] = E\{V[N(\tau)] \mid N(0)=k\} = V(k) + E_k \sum_{0 \leq t < \tau} AV[N(t)] \quad (4)$$

Then, the following theorem may be proved:

<u>THEOREM 1</u>. Let $N(t)$ be an aperiodic irreducible Markov chain. If there exist such a Lyapunov function $V(k)$, $k \in C$, and a positive constant $\varepsilon > 0$, that for each $k \in C-H$, where H is a finite set of states, the following holds

$$AV(k) < -\varepsilon \quad (5a)$$

while for each $k \in C$

$$|AV(k)| < \infty \quad (5b)$$

then $N(t)$ is ergodic.

PROOF. For simplicity of the considerations let us assume that the set H consists of exactly one state k_o, i.e., $H=\{k_o\}$. We denote by

τ_{k_0} the first return time to state k_0 given the chain starts in k_0 at time zero. In order to prove that N(t) is ergodic it is sufficient to show that the mean value of τ_{k_0} is finite. Moreover, let $\tau = \min\{\tau_{k_0}, T\}$, $T>0$. Then, Eqs.(2),(4),(5) imply that

$$c \leq E_{k_0} V[N(\tau)] = V(k_0) + E_{k_0} AV[N(0)] + E_{k_0} \sum_{0<t<\tau} AV[N(t)] < E_{k_0} V[N(1)] - \varepsilon E_{k_0}\tau$$

hence,

$$E_{k_0}\tau = E_{k_0} \min\{\tau_{k_0}, T\} < \{E_{k_0} V[N(1)] - c\}/\varepsilon < \infty \qquad (6)$$

Let now H be any finite subset of the state space C and by τ_{k_0H}, $k_0 \in H$, we denote the first return time to a state of H given the chain starts in k_0 at time zero. Then, repeating (6) for τ_{k_0H} one shows that $E\tau_{k_0H} < \infty$, which implies that N(t) is ergodic. ∎

In fact Theorem 1 is another formulation of the theorem proved by Foster [FOS] and Pakes [PAK]. Moreover, as a conclusion of the theorem one obtains Pakes Lemma [PAK], that is:

COROLLARY 1 / Pakes 1969 /. Let N(t) be an aperiodic irreducible Markov chain. Then N(t) is ergodic if

$$\begin{aligned} &|E\{N(t+1) - N(t) \mid N(t)=k\}| < \infty, \quad k \in C \\ &\limsup_{k\to\infty} E\{N(t+1) - N(t) \mid N(t)=k\} < 0 \end{aligned} \qquad (7)$$

PROOF. It is enough to assume that $V(k)=k$ for each $k \in C$ in Theorem 1 and take into account the properties of the upper limit. ∎

In many applications it is also important to know when a Markov chain is not ergodic [FAY],[SAA]. Therefore, we shall present a theorem for nonergodicity of a Markov chain. Let us denote by $V_x(k)$, $k \in C$, a parametric Lyapunov function, where x belongs to a set [0,1]. Like in Eq.(1) we define the operator $AV_x(k)$. We shall denote by $A'V_x(k)$ the derivative of $AV_x(k)$ with respect to x assuming it exists. Then the following theorem may be formulated :

<u>THEOREM 2</u>. Let a nonnegative function $V_x(k)$ satisfies the conditions:

1) $\sum_{k\in C} V_x(k) < \infty$, $x \in [0,1]$
2) for any transition probabilities $\{P_{ij}\}$ $i,j \in C$
$$A'V_x(k) = \sum_{j\in C} P_{kj} V'_x(j) - V'_x(k), \quad k \in C, \ x \in [0,1)$$
3) $V_x(k)$ is a nonincreasing function with respect to $k \in C$ for $x \in [0,1]$ and for $x=1$ $V_1(k) =$ const for every $k \in C$.

4) $V'_1(k)$ / the derivative of $V_x(k)$ at the point $x=1$ / is nondecreasing function of $k \in C$

Then $N(t)$ is not ergodic if there exist such constants $\varepsilon > 0$, $d \geq 0$ that for each $x \in [0,1]$ and each $k \in C-H$, where H is a finite subset of C, the following is satisfied:

$$-AV_x(k) \geq d(1-x) \tag{8a}$$

$$A'V_1(k) \geq \varepsilon \tag{8b}$$

PROOF. Let $j_o \in C-H$ be such a state that for $j \in H$ $j \leq j_o$. We introduce a modified Markov chain $N'(t)$ with the state space C-H and transition probabilities $\{P'_{ij}\}$, $i,j \in C-H$ given by

$$P'_{ij} = \begin{cases} P_{ij} & i \in C-H\ ,\ j \in C-H-\{j_o\} \\ \sum_{l \in H} P_{il} & i \in C-H\ ,\ j=\{j_o\} \\ 0 & \text{for other cases} \end{cases}$$

In the other words, the chain $N'(t)$ omits all states of the set H. Let τ_{j_o} and τ'_{j_o} stand for the first return time to j_o given the chains $N(t)$ and $N'(t)$ start in j_o at time zero, respectively. Naturally, τ_{j_o} is stochastically greater than τ'_{j_o} [STO], [KAM], that is, in order to prove that $N(t)$ is not ergodic it is enough to prove that $N'(t)$ is not ergodic. Moreover, it is easy to show, taking into account conditions 2)-4), that the inequalities (8) are satisfied for the chain $N'(t)$ if they hold for the chain $N(t)$. Therefore, we must prove that $N'(t)$ is not ergodic. Let us assume the contrary. Then there exist steady state probabilities $\{\pi_j\}_{j \in C-H}$ satisfying

$$\pi_j = \sum_{i \in C-H} \pi_i P'_{ij}$$

After some algebra, using condition 1), one immediately obtains

$$\sum_{i \in C-H} \pi_i AV_x(i) = 0$$

Then using Fatou's lemma / we need for that the inequality (8a) / and the inequality (8b) we find

$$0 = \lim_{x \to 1^-} \sum_{i \in C-H} \pi_i [-AV_x(i)]/(1-x) \geq \sum_{i \in C-H} \pi_i \lim_{x \to 1} [-AV_x(i)]/(1-x) =$$

$$= \sum_{i \in C-H} \pi_i A'V_1(i) \geq \varepsilon$$

This is a contradition, what finishes the proof. ■

This theorem is a generalization of Kaplan's conditions [KAP] for nonergodicity of a Markov chain and it may be further generalized [SZP2]. However, Kaplan noted that whenever downward transitions are uniformly bounded / $P_{ij}=0$ for $j < i-k$, $k > 0$ / the condition (8a) is satisfied. Therefore:

COROLLARY 2. Let $N(t)$ be a Markov chain with uniformly bounded downward transitions. Then $N(t)$ is not ergodic if the following is satisfied:

$$\liminf_{k \to \infty} E\{N(t+1) - N(t) \mid N(t)=k\} > 0 \tag{9}$$

PROOF. One takes in Theorem 2 $V_x(k) = x^k$, $k \in C$, and take into account properties of the lower limit as well as the fact that the chain is downward uniformly bounded. ■

Let us notice that in both, Kaplan and Pakes, conditions in order to answer if a chain is ergodic or not it is sufficient to check the sign of the average drift $a(k) = E\{N(t+1) - N(t) \mid N(t)=k\}$ for sufficiently large k.

Complications arise when we try to apply the above stated theorems to multidimensional Markov chains $N^M(t) = (N_1(t), \ldots, N_M(t))$. Among others, one should notice that this time there are M average drifts $a_i(k_1,\ldots,k_M) = E\{N_i(t+1) - N_i(t) \mid N^M(t) = (k_1,\ldots,k_M)\}$, $i=1,\ldots,M$, and the above corollaries cannot be directly applied. However, using Theorem 1 and 2 we shall present the conditions on $a_i(k)$, $i=1,\ldots,M$, $k=(k_1,\ldots,k_M)$, which assure ergodicity or nonergodicity of a multidimensional Markov chain.

First of all, we shall present sufficient and necessary conditions for classification of a special two-dimensional Markov chain $N^2(t) = (N_1(t), N_2(t))$ given by Malyshev [MAL]. He has considered a two-dimensional Markov chain of constant average drifts $a_i(k_1,k_2)$ for $k_1 > 0$, $k_2 > 0$, $i=1,2$ and equal a_1, a_2, respectively. On the other hand, for $k_1=0$ or $k_2=0$ the drifts are equal to $a_i' = a_i(k_1,0)$, $k_1 > 0$, $i=1,2$ and $a_i'' = a_i(0,k_2)$, $k_2 > 0$, $i=1,2$. Then Malyshev proved that [MAL]:

<u>LEMMA 1</u>. A. Let $a_1 < 0$, $a_2 < 0$, then a Markov chain $N^2(t)$ is ergodic if and only if

$$a_1a_2' - a_2a_1' < 0 \quad , \quad a_2a_1'' - a_1a_2'' < 0 \ ,$$

recurrent if and only if

$$a_1a_2' - a_2a_1' \leq 0 \quad , \quad a_2a_1'' - a_1a_2'' \leq 0 \ ,$$

transient for other cases.

B. Let $a_1 \geq 0$, $a_2 < 0$, then $N^2(t)$ is ergodic if and only if

$$a_1a_2' - a_2a_1' < 0$$

recurrent if and only if

$$a_1a_2' - a_2a_1' \leq 0$$

transient for other cases.

C. Symmetry to B. ∎

Let us now consider a general M-dimensional Markov chain $N^M(t)$ and let, as before, $V(k)$ and $AV(k)$, $k=(k_1,\dots,k_M) \in C$ be respectively a Lyapunov function and the operator of $N^M(t)$. In order to apply Theorem 1 we introduce the following Lyapunov function:

$$V(k) = \sum_{i=1}^{M} c_i k_i \quad , \quad k=(k_1,\dots,k_M)$$

where c_i are nonnegative constants for $i=1,\dots,M$. Then

$$AV(k) = E\{\sum_{i=1}^{M} c_i N_i(t+1) - \sum_{i=1}^{M} c_i N_i(t) \mid N^M(t)=(k_1,\dots,k_M)\} =$$

$$= \sum_{i=1}^{M} c_i a_i(k_1,\dots,k_M) \tag{10}$$

Using Theorem 1 one may immediately prove the following corollary:

COROLLARY 3. A Markov chain $N^M(t)$ is ergodic if for all $k \in C-H$, where H is a finite set of states, the following simultaneously inequalities are satisfied

$$\sum_{i=1}^{M} c_i a_i(k) < 0 \quad , \quad k=(k_1,\dots,k_M) \in C-H \tag{11a}$$

and for each $k \in C$ and $i=1,\dots,M$

$$|a_i(k)| < \infty \tag{11b}$$

∎

A similar conclusion about nonergodicity of a Markov chain $N^M(t)$ may be stated if one applies Theorem 2. Indeed, putting $V_x(k) = x^{\sum c_i k_i}$, $x \in [0,1]$ and assuming that $N^M(t)$ is downward uniformly bounded one shows:

COROLLARY 4. A downward uniformly bounded Markov chain $N^M(t)$ is not ergodic if for all $k \in C-H$, where H is a finite set of states, the following simultaneously inequalities are satisfied

$$\sum_{i=1}^{M} c_i a_i(k) > 0 \quad , \quad k=(k_1,\dots,k_M) \in C-H \tag{12}$$

∎

The Lyapunov function method presented in this section is very simple and elegant - what should be calculated as an advantage of the method - however, some limitations arise. The theorems do not give us any idea how to construct appropriate Lyapunov functions. We have been only able to present explicite form for two simple Lyapunov functions. Moreover, one should expect that for multidimensional Markov chains sufficient conditions presented in the theorems "are far away" from necessary conditions. In the other words, applying Corollaries 3 and 4 we are able to determine a subset of the ergodicity region, that is, a set of parameters for which a chain is ergodic. Therefore, another method called comparison tests will be studied below.

2.2 COMPARISON TESTS

Let $N^M(t)$ be an M-dimensional Markov chain, while $\tilde{N}^K(t) = \tilde{N}_{m_1,\dots,m_K}(t) = (\tilde{N}_{m_1}(t),\dots,\tilde{N}_{m_K}(t))$ and $\underset{\sim}{N}^K(t) = \underset{\sim}{N}_{m_1,\dots,m_K}(t) = (\underset{\sim}{N}_{m_1}(t),\dots,\underset{\sim}{N}_{m_K}(t))$ be such K-dimensional Markov chains, where the indices $m_i \in \{1,\dots,M\}$, $i=1,\dots,K$, $m_i \neq m_j$ for $i \neq j$, that $N_{m_i}(t)$, $\tilde{N}_{m_i}(t)$ and $\underset{\sim}{N}_{m_i}(t)$ have the same state space. Moreover, the following inequalities hold:

$$\begin{aligned} &\Pr\{\tilde{N}_{m_1}(t) < x_{m_1},\dots, \tilde{N}_{m_K}(t) < x_{m_K} \mid \tilde{N}^K(0) = (k_{m_1},\dots,k_{m_K})\} \leqslant \\ &\Pr\{N_{m_1}(t) < x_{m_1},\dots, N_{m_K}(t) < x_{m_K} \mid N^M(0) = (k_1,k_2,\dots,k_M)\} \end{aligned} \tag{13}$$

and

$$\begin{aligned} &\Pr\{\underset{\sim}{N}_{m_1}(t) < x_{m_1},\dots, \underset{\sim}{N}_{m_K}(t) < x_{m_K} \mid \underset{\sim}{N}^K(0) = (k_{m_1},\dots,k_{m_K})\} \geqslant \\ &\Pr\{N_{m_1}(t) < x_{m_1},\dots, N_{m_K}(t) < x_{m_K} \mid N^M(0) = (k_1,k_2,\dots,k_M) \end{aligned} \tag{14}$$

for each $t \geqslant 0$ and $x=(x_{m_1},\dots,x_{m_K})$, and every $k=(k_1,\dots,k_M)$. Then, the following lemmas called comparison tests will be proved:

<u>LEMMA 2</u>. Let $N^M(t) = (N_1(t),\dots, N_M(t))$ be M-dimensional aperiodic, irreducible Markov chain, while $\tilde{N}^K(t) = (\tilde{N}_{m_1}(t),\dots,\tilde{N}_{m_K}(t))$ be K-dimensional aperiodic irreducible Markov chains for each K-tuple $(m_1,\dots, m_K)$, $m_i \in \{1,\dots,M\}$ $m_i \neq m_j$ for $i \neq j$. If for each $t \geqslant 0$, each $x=(x_{m_1},\dots, x_{m_K})$ and each $k=(k_1,\dots,k_M)$ the inequality (13) holds for every K-tuple $(m_1,\dots,m_K)$, and all K-dimensional Markov chains $(\tilde{N}_{m_1}(t), \dots, \tilde{N}_{m_K}(t))$ are ergodic, then the M-dimensional Markov chain is ergodic.

<u>LEMMA 3</u>. Let $N^M(t) = (N_1(t),\dots,N_M(t))$ is M-dimensional Markov chain and let exist such K-dimensional Markov chain $\underset{\sim}{N}_{n_1}(t),\dots,\underset{\sim}{N}_{n_K}(t))$ that the inequality (14) holds for $t \geqslant 0$, each $(x_{n_1},\dots,x_{n_K})$ and every

$k=(k_1,\dots,k_M)$. Then the M-dimensional Markov chain $N^M(t)$ is not ergodic if the K-dimensional Markov chain $(\underset{\sim}{N}_{n_1}(t),\dots,\underset{\sim}{N}_{n_K}(t))$ is not ergodic.

PROOF OF LEMMA 2. If each of the K-dimensional Markov chain is ergodic, then for any $\varepsilon>0$ there exist such an $x=(x_{m_1},\dots,x_{m_K})$ that for each $k=(k_{m_1},\dots,k_{m_K})$ and each $(m_1,\dots,m_K)$ [CHU]

$$\lim_{t\to\infty} \Pr\{\tilde{N}_{m_1}(t)\geqslant x_{m_1} \text{ or } ..\text{or } \tilde{N}_{m_K}(t)\geqslant x_{m_K} \mid \tilde{N}^K(0)=(k_{m_1},\dots,k_{m_K})\} < \varepsilon \binom{M}{K} \quad (15)$$

Then using (13) and (15) one obtains

$$\lim_{t\to\infty} \Pr\{N_i(t) < x_i,\ i=1,\dots,M \mid N^M(0)=(k_1,\dots,k_M)\} =$$

$$1-\lim_{t\to\infty} \Pr\{N_1(t)\geqslant x_1 \text{ or } \dots \text{ or } N_M(t)\geqslant x_M \mid N^M(0)=(k_1,\dots,k_M)\} \geqslant$$

$$1-\lim_{t\to} \sum_{(m_1,\dots,m_K)} \Pr\{N_{m_1}(t)\geqslant x_{m_1} \text{ or } .. \text{ or } N_{m_K}(t)\geqslant x_{m_K} \mid N^M(0)=(k_1,\dots,k_M)\}\geqslant$$

$$1-\lim_{t\to\infty} \sum_{(m_1,\dots,m_K)} \Pr\{\tilde{N}_{m_1}(t)\geqslant x_{m_1} \text{ or } .. \text{ or } \tilde{N}_{m_K}(t)\geqslant x_{m_K} \mid \tilde{N}^K(0)=(k_{m_1},\dots,k_{m_K})\}$$

$$\geqslant 1-\varepsilon \quad (16)$$

If the M-dimensional Markov chain had been nonergodic, then for any finite set A and any $N^M(0)=(k_1,\dots,k_M)$ the limit $\lim \Pr\{N^M(t)\in A \mid N^M(0)=k\}$ would have tended to zero / as t tends to infinity /, what is imposible because of (16). Thus, the chain $N^M(t)$ is ergodic.

PROOF OF LEMMA 3. Since the K-dimensional Markov chain $(\underset{\sim}{N}_{n_1}(t),\dots,\underset{\sim}{N}_{n_K}(t))$ is not ergodic, therefore, for each $x=(x_{n_1},\dots,x_{n_K})$ and each $(k_{n_1},\dots,k_{n_K})$ the following is satisfied:

$$\lim_{t\to\infty} \Pr\{\underset{\sim}{N}_{n_1}(t) < x_{n_1},\ \dots,\underset{\sim}{N}_{n_K}(t) < x_{n_K} \mid \underset{\sim}{N}^K(0)=(k_{n_1},\dots,k_{n_K})\}=0 \quad (17)$$

Using now (14) and (17) one gets

$$0 \leqslant \lim_{t\to\infty} \Pr\{N_i(t) < x_i,\ i=1,\dots,M \mid N^M(0)=(k_1,\dots,k_M)\}\leqslant$$

$$\lim_{t\to\infty} \Pr\{N_{n_i}(t) < x_{n_i},\ i=1,\dots,K \mid N^M(0)=(k_1,\dots,k_M)\}\leqslant$$

$$\lim_{t\to\infty} \Pr\{\underset{\sim}{N}_{n_i}(t) < x_{n_i},\ i=1,\dots,K \mid \underset{\sim}{N}^K(0)=(k_{n_1},\dots,k_{n_K})\} = 0 \quad (18)$$

Hence, for each finite set A $\lim \Pr\{N^M(t)\in A \mid N^M(0)=k\}=0$, t tends to infinity, what proves that $N^M(t)$ cannot be ergodic.

■

REMARK

Let us outline some remarks about the use of Lemmas 2 and 3, in particular, how to find such K-dimensional Markov chains $\tilde{N}^K(t)$ or $\underset{\sim}{N}^K(t)$

that the inequality (13) or (14) is satisfied. Let us recall, following the work of Kamea, Krangel and O'Brien [KAM] that a stochastic process $\{X(t), t \geq 0\}$ is stochstically smaller or equal to a process $\{Y(t), t \geq 0\}$ if $f(X(t), t \geq 0) \leq f(Y(t), t \geq 0)$ for all nondecreasing real-valued function $f(\cdot)$ defined on the sample paths of $X(t)$ and $Y(t)$. In [KAM] it was also proved that if $\{X(t), t \geq 0\} \leq_{st} \{Y(t), t \geq 0\}$ then a strong sample path comparison holds, that is, it is possible to construct stochastic processes $\{\bar{X}(t), t \geq 0\}$ and $\{\bar{Y}(t), t \geq 0\}$ on the common probability space such that these processes have the same distributions as $\{X(t), t \geq 0\}$ and $\{Y(t), t \geq 0\}$, respectively, and every path of $\{\bar{X}(t), t \geq 0\}$ lies below corresponding sample path of $\{\bar{Y}(t), t \geq 0\}$ [KAM]. Moreover, for queueing processes the stochastic order is equivalent to stochastic order for all finite-dimensional joint distributions, but it is not equivalent to appropriate inequalities for finite-dimensional distribution functions [STO], however, stochstic order implies appropriate inequalities on distribution functions, as it is required in Lemmas 2 and 3. Concluding out, in order to find such K-dimensional Markov chains $\tilde{N}^K(t)$ and $\underset{\sim}{N}^K(t)$ that (13) and (14) are satisfied it is enough to prove that

$$\{N_{m_1}(t), \ldots, N_{m_K}(t), t \geq 0\} \leq_{st} \{\tilde{N}_{m_1}(t), \ldots, \tilde{N}_{m_K}(t), t \geq 0\} \tag{19}$$

$$\{N_{m_1}(t), \ldots, N_{m_K}(t), t \geq 0\} \geq_{st} \{\underset{\sim}{N}_{m_1}(t), \ldots, \underset{\sim}{N}_{m_K}(t), t \geq 0\} \tag{20}$$

under the condition $N_{m_i}(0) = \tilde{N}_{m_i}(0) = \underset{\sim}{N}_{m_i}(0)$, $i=1, \ldots, K$. On the other hand, for proving (19) or (20) a strong sample path comparison may be applied. For instance, the inequality (19) holds if one constructs such stochstic processes $\{\mathcal{N}_{m_1}(t), \ldots, \mathcal{N}_{m_K}(t), t \geq 0\}$ and $\{\tilde{\mathcal{N}}_{m_1}(t), \ldots, \tilde{\mathcal{N}}_{m_K}(t), t \geq 0\}$ on the common probability space with the same distributions as $(N_{m_1}(t), \ldots, N_{m_K}(t), t \geq 0\}$ and $(\tilde{N}_{m_1}(t), \ldots, \tilde{N}_{m_K}(t), t \geq 0\}$, respectively, and sample path of the first process lies below sample path of the second process. In the further part of this paper in order to prove (13) or (14) we shall use the conclusion of these remarks, in particular, we shall prove (19) or (20).

3. APPLICATIONS

We shall apply the theorems presented before to explore convergent/divergent behaviour of some computer communication systems as: homogeneous multiaccess systems, buffered heterogeneous ALOHA system and network of queues. We restrict our considerations to multidimensional Markov chains $N^M(t)=(N_1(t),\dots,N_M(t))$ which satisfy the following stochastic equations:

$$\begin{aligned} N_1(t+1) &= N_1(t) + X_1(t) - Y_1(t) \\ N_2(t+1) &= N_2(t) + X_2(t) - Y_2(t) \\ &\dots\dots\dots\dots\dots\dots \\ N_M(t+1) &= N_M(t) + X_M(t) - Y_M(t) \end{aligned} \tag{21}$$

where $X_i(t)$, $Y_i(t)$, $i=1,\dots,M$ are such random variables that $N^M(t)$ is a Markov chain. The random variables $X_i(t)$, $Y_i(t)$, $i=1,\dots,M$ may be considered as input traffics and output traffics, respectively. Then the Markov chain $N^M(t)$ represents queue lengths in M buffers. Therefore, further we call $N_i(t)$, $i=1,\dots,M$ the queue length in the i-th buffer. Let us denote by $S^i_{in}(n_1,\dots,n_M) = \lim E\{X_i(t)|N^M(t)=(n_1,\dots,n_M)\}$ / as t tends to infinity / a conditional input rate / of packets, messages and so on / and by $S^i_o(n_1,\dots,n_M) = \lim E\{Y_i(t) | N^M(t)=(n_1,\dots,n_M)\}$ / as t tends to infinity / a conditional throughput, $i=1,\dots,M$. Then, it is easily to notice that the average drifts $a_i(n_1,\dots,n_M)= E\{N_i(t+1) - N_i(t) | N^M(t)=(n_1,\dots,n_M)\}$ are expressed as:

$$a_i(n_1,\dots,n_M) = S^i_{in}(n_1,\dots,n_M) - S^i_o(n_1,\dots,n_M) \tag{22}$$

In the analysis of queueing systems it is important to determine so called ergodicity region, that is, the values of system parameters for which the system is stable. Let us denote by λ_k an average input rate to the k-th buffer, i.e., $\lambda_k = \lim EX_k(t)$ / t tends to infinity / $k=1,\dots,M$. For simplicity of the analysis we shall assume that the conditional input rates $S^k_{in}(n_1,\dots,n_M)$ do not depend on $(n_1,\dots,n_M)$ and, therefore, $\lambda_k = S^k_{in}(n_1,\dots,n_M)$ for each $(n_1,\dots,n_M)$. Then, an ergodicity region E will be defined as

$$E = \{ \lambda = (\lambda_1,\dots,\lambda_M) : N^M(t) \text{ is ergodic} \} \tag{23a}$$

while T defined as below

$$T = \{ \lambda = (\lambda_1,\dots,\lambda_M) : N^M(t) \text{ is not ergodic} \} \tag{23b}$$

will denote a nonergodicity region.

3.1 UNBUFFERED HOMOGENEOUS MULTIACCESS SYSTEMS - [KLE], [TOB], [FAY]

Let us consider a multiaccess system with an infinite number of users.

Each user has a buffer of one packet capacity. Users have random access to a broadcast slotted channel, that is, each user randomly selects the slot to transmit a packet. For simplicity, we assume that each user acts in the same way no matter if it has a packet for the first transmission or for retransmissions / random access without retransmission discrimination discipline [SZP1] / and that each user has the same probability of transmitting a packet / homogeneous case/. Let us denote this probability by $r(n)$, where n is the number of active users. In such a case the state of the system is described by a one-dimensional Markov chain $N(t)$; $N(t)$ is the number of active users [KLE],[SZP1]. Naturally, $N(t)$ satisfies Eq.(21). Let $\lambda = EX$ denote the total average input rate. Then the conditional throughput may be expressed as:

$$S_0(n) = nr(n)[1 - r(n)]^{n-1}$$

Let us consider some examples. First we assume that $r(n)=r$ for all $n \geq 0$. Then $\lim S_0(n)=0$ as $n \to \infty$ and one immediately proves, using Corollary 1, that the system is not ergodic for all values of λ . In the other words, the ergodicity region is an empty set.

In the second example, let $r(n)=1/n$ $n>0$ as in [FAY], then as $n \to \infty$ $\lim(1-1/n)^{n-1}=1/e$ and the ergodicity region is formed by all of λ smaller then $1/e=0.367$. Similar conclusions may be drawn out for the other random access protocols / CSMA, S-ALOHA with retransmission discrimination discipline, etc [TOB] /, however, the calculations do not present something new from our point of view. The results discussed in the examples were proved also in [FAY], however, the calculations were much more complex.

The last example deals with similar system as before, however, we consider the aspects presented in [YEM]. Let us denote by n the number of active users, by $r_i(n)$ the probability of transmitting a packet / access rights / by the i-th user, where i is an active user, and by $A(n)$ a set of active users. As before, the system is described by a one-dimansional Markov chain / we are not interested here how to construct such a multiaccess protocol which could recognize the set $A(n)$ /. Then the conditional throughput is equal to:

$$S_0(n) = \sum_{i \in A(n)} r_i(n) \prod_{j \in A(n) - \{i\}} (1 - r_j(n)) \tag{24}$$

To find the ergodicity region one has to take the limit of Eq.(24) as n tends to infinity. In order to determine the biggest ergodicity region with respect to λ, let us assume that as n tends to infinity only one user, let us say the i^*-the user, has full access rights, i.e., $r_{i^*}(n)=1$, while for the others $r_i(n)=0$, $i=1,\dots,M$, $i\neq i^*$. Then, it is evident that as $n\to\infty$ $\lim S_o(n)=1$ and the ergodicity region is a set $\{\lambda:\ \lambda<1\}$. The URN scheme [YEM] realizes such a protocol.

3.2 BUFFERED HETEROGENEOUS ALOHA SYSTEM - [KMA], [SAA], [TSY]

Let us consider a slotted ALOHA multiple access scheme with finite number of buffered users with infinite capacities [KMA], [SAA], [TSY]. Like before users have random access to a common channel, however, the i-th user transmits a packet with probability r_i, $i=1,\dots,M$ and the average input rates per slot equal $\lambda_i=EX_i$, $i=1,\dots,M$ / heterogeneous multiaccess system /. Naturally, the queue lengths in all buffers are described by M-dimensional Markov chain / where M is the number of users / $N^M(t)=(N_1(t),\dots,N_M(t))$ satisfying the stochstic Eqs.(21). To explore the ergodicity of $N^M(t)$ one may use Corollaries 3 and 4 or Lemmas 2 and 3.

Let us first consider Corollary 3, hence, we have to calculate the average drifts $a_k(n_1,\dots,n_M)$. Since the average input rates are equal to λ_k, therefore, according to Eq.(22) we have to determine the conditional throughputs. Basing on the idea of random access protocol one gets immediately

$$S_o^k(n_1,\dots,n_M)=\chi(n_k)\,r_k\prod_{\substack{j=1\\ j\neq k}}^{M}(1-r_j)^{\chi(n_j)}\ ,\quad k=1,\dots,M \tag{25}$$

where $\chi(n_k)=0$ for $n_k=0$, while $\chi(n_k)=1$ for $n_k\geqslant 1$. It is obvious that the conditional throughputs display different values only in a set $\{(n_1,\dots,n_M):\ n_i\leqslant 1,\ i=1,\dots,M\}$ and one can restrict the considerations only to that set. Let us now assume that the set H needed in Corollaries 3 and 4 consists of one point, namely, $H=\{(0,0,\dots,0)\}$ and let all coefficients c_i are equal to one. Then, the following conclusions may be obtained:

a) For the ergodicity of the Markov chain $N^M(t)$ it is sufficient that the 2^{M-1} following inequalities are simultaneously satisfied

$$\sum_{k=1}^{M}\lambda_k<\sum_{k=1}^{M}\chi(n_k)\,r_k\prod_{\substack{j=1\\ j\neq k}}^{M}(1-r_j)^{\chi(n_j)} \tag{26}$$

for each $n_i \leqslant 1$, $i=1,\dots,M$ and $n \neq (0,0,\dots,0)$.

b For the nonergodicity of the Markov chain $N^M(t)$ it is sufficient that the 2^{M-1} following inequalities are simultaneously satisfied

$$\sum_{k=1}^{M} \lambda_k > \sum_{k=1}^{M} \chi(n_k) r_k \prod_{\substack{j=1 \\ j \neq k}}^{M} (1-r_j)^{\chi(n_j)} \tag{27}$$

for each $n_i \leqslant 1$, $i=1,\dots,M$ and $n \neq (0,0,\dots,0)$.

Conditions (26) was obtained also in [FAL], however, using more complex calculations.

Conditions (26) and (27) determine subsets of the ergodicity and nonergodicity regions. Let us denote by E_1 and T_1 the sets of $\lambda = (\lambda_1,\dots,\lambda_M)$ which satisfy the inequalities (26) and (27) , respectively. A simple calculation shows that there exist such values of λ which belong neither to E_1 nor to T_1. Therefore, Lemmas 2 and 3 will be applied.

In order to use Lemma 2 one needs to determine $\binom{M}{K}$ K-dimensional Markov chains $\tilde{N}^K(t)=(\tilde{N}_{m_1}(t),\dots,\tilde{N}_{m_K}(t))$ satisfying (13) or (19). Since we can obtain the ergodicity conditions only for K=1 and 2 / Corollary 1 and Lemma 1 / we apply Lemma 2 only to these cases. For K=1 one has to find M one-dimensional Markov chains $\tilde{N}_m(t)$ satisfying (13) or (19). Let us define $\tilde{N}_m(t)$ as the queue length in the m-th buffer under the condition that all other buffers are never empty, that is, for each $t \geqslant 0$ $N_k(t) > 0$, $k=1,\dots,M$, $k \neq m$. It is " the worst case " since the greatest competition between the m-th users and the others takes place in such a situation. More precisely, the transition probabilities for the one-dimensional Markov chain $\tilde{N}_m(t)$ can be calculated as:

$$\Pr\{\tilde{N}_m(t+1)=n_m \mid \tilde{N}_m(t)=k_m\}=\Pr\{N_m(t+1)=n_m \mid N^M(t)=(k_1,\dots,k_M), k_i > 0,\ i \neq m\} \tag{28}$$

There are two ways of showing that the inequality (13) holds. We can find the probabilities of both sides of (13) and prove / after some complex calculations / that (13) is satisfied, as it was done in [TSY] or we can take the advantage of the stochastic inequality (19). The latter case seems to be simpler and we shall apply it to our considerations. Therefore, besides the analysed system called further basic system we shall consider also a hypothetical system characterized by:

1) for each $t \geqslant 0$ $N_j(t) > 0$, $j \neq m$, that is, all buffers except the

m-th buffer are nonempty

2) input arrival processes to the basic and the hypothetical systems are identical

3 the action of the j-th transmitter in both of the systems is the same, that is, the j-th transmitter in the hypothetical system gets access to the channel if and only if the j-th transmitter in the basic system has an access rights to send a packet.

Moreover, let $\tilde{N}_m(t)$ be the queue length in the m-th buffer in the hypothetical system. It is obvious, ommitting details of proof which are very similar to the ones presented in [WIT], that (19) holds, i.e,

$$\{N_m(t),\ t>0\} \underset{st}{\leq} \{\tilde{N}_m(t),\ t>0\} \quad , \quad N_m(0)=\tilde{N}_m(0)$$

and the distribution function of $\tilde{N}_m(t)$ is exactly the same as for $\tilde{N}_m(t)$. Therefore, to investigate the ergodicity of the multidimensional Markov chain $N^M(t)$ it is sufficient to explore the ergodicity of M one-dimensional Markov chains $\tilde{N}_m(t)$, m=1,...,M, according to Lemma 2. Nevertheless, according to Corollary 1 for the ergodicity of the chain $\tilde{N}_m(t)$ it is sufficient that the following condition is satisfied:

$$\lambda_m < \lim_{k\to\infty} S_o^m(k) \tag{29}$$

where

$$S_o^m(k) = \chi(k)\, r_m \prod_{\substack{j=1\\ j\neq m}}^{M} (1-r_j)$$

that is,

$$\lambda_m < r_m \prod_{\substack{j=1\\ j\neq m}}^{M} (1-r_j) \tag{30}$$

On the other hand, the multidimensional Markov chain $N^M(t)$ is ergodic if the inequalities (30) are satisfied for each m=1,...,M. Therefore, a subset E_2 of the ergodicity region E may be defined as

$$E_2 = \{ (\lambda_1,\ldots,\lambda_M) : \lambda_i < r_i \prod_{\substack{j=1\\ j\neq i}}^{M} (1-r_j) \ ,\ i=1,\ldots,M\} \tag{31}$$

Naturally, $E_2 \subset E$. In order to enlarge E_2, i.e., to find such a set E_3 that $E_2 \subset E_3 \subset E$, one needs to consider Lemma 2 for K=2 and take advantage of Lemma 1. Therefore, we have to find such a two-dimensional Markov chains $\tilde{N}_{nm}(t)$ n,m=1,...,M, n≠m, / or shortly $\tilde{N}^2(t)$ / satisfying (13) for each n and m. The concept of constructing the $\tilde{N}^2(t)$ is the same as in the case K=1. We define $\tilde{N}_{nm}(t)$ as such a two-dimensional Markov chain for which the transition probabilities satisfy the following relationship

$$\Pr\{\tilde{N}_n(t+1)=s_n,\ \tilde{N}_m(t+1)=s_m \mid \tilde{N}_{nm}(t)=(k_n,k_m)\} =$$
$$\Pr\{N_n(t+1)=s_n, N_m(t+1)=s_m \mid N^M(t)=(k_1,..,k_M),\ k_i>0,\ i\neq n,m\} \quad (32)$$

In order to use the two-dimensional Markov chains $\tilde{N}^2(t)$ in Lemma 2 one must prove that (13) or (19) is satisfied. We shall show that (19) takes place. Let $\tilde{N}_{nm}(t)$ represents the queue lengths in the n-th and m-th buffers in a hypothetical system defined similar to the case K=1, however, this time for each $t \geq 0$ $N_j(t) > 0$ for $j\neq n,m$, that is, all buffers except the n-th and the m-th are never empty. Then (19) is satisfied for all n and m. Thus, Lemma 2 may be applied if one finds ergodicity conditions for the two-dimensional Markov chains $\tilde{N}_{nm}(t)$. Here, Lemma 1 might be a help since the chain $\tilde{N}_{nm}(t)$ satisfies the assumptions of the lemma. After some calculations we find that $\tilde{N}_{nm}(t)$ is ergodic if and only if the following conditions hold:

1) if $\lambda_n < r_n(1-r_m)q_{nm}$ then $\lambda_m < r_m[q_{nm} - \lambda_n/(1-r_m)]$

2 if $\lambda_n \geq r_n(1-r_m)q_{nm}$ then $\lambda_m < (1-r_n)(q_{nm} - \lambda_n/r_n)$ (33)

where

$$q_{nm} = \prod_{\substack{j=1\\ j\neq n,m}}^{M} (1-r_j)$$

According to Lemma 2 the M-dimensional Markov chain is ergodic if all input rates λ_i, i=1,...,M, satisfy conditions (33). Let Λ_{nm} be a set defined as:

$$\Lambda_{nm} = \{(\lambda_1,\dots,\lambda_M):\ \lambda_n \text{ and } \lambda_m \text{ satisfy (33)},\ \lambda_i>0, i=1,..,M\}$$

Then the set

$$E_3 = \bigcap_{n=1}^{M}\bigcap_{m=1}^{M} \Lambda_{nm} \quad (34)$$

is a subset of the ergodicity region E such that $E_2 \subset E_3 \subset E$.

Unfortunately, the set E_3 is still only a subset of the ergodicity region. Therefore, let us also determine the values of input rates for which the Markov chain $N^M(t)$ is not ergodic. We use Lemma 3 only in the case K=2. Let $\underset{\sim}{N}_{nm}(t)$ be a two-dimensional Markov chain with the following transition probabilities:

$$\Pr\{\underset{\sim}{N}_n(t+1)=s_n,\ \underset{\sim}{N}_m(t+1)=s_m \mid \underset{\sim}{N}_{nm}(t)=(k_n,k_m)\} =$$
$$\Pr\{N_n(t+1)=s_n,\ N_m(t+1)=s_m \mid N^M(t)=(k_1,..,k_M),\ k_i=0,\ i\neq n,m\}$$

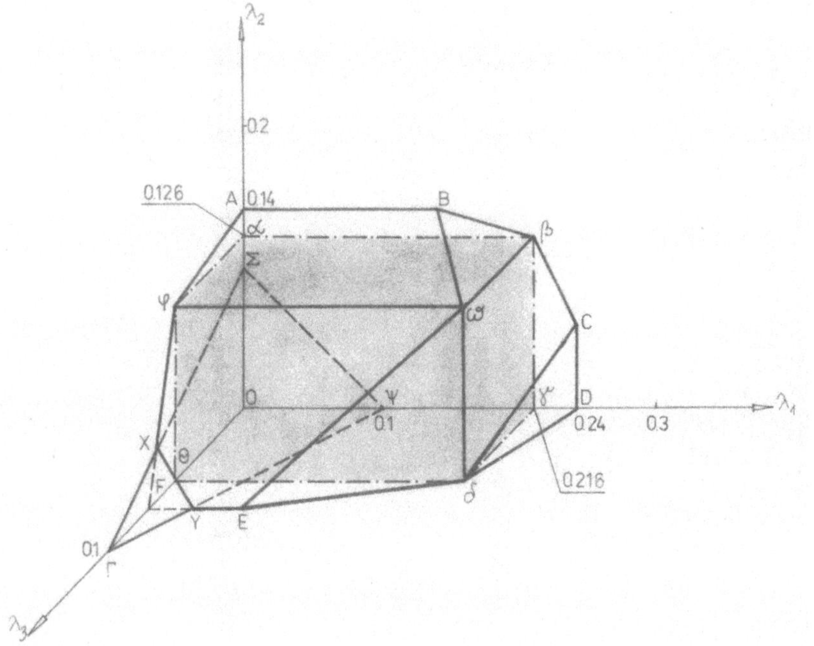

Fig.1 Subsets of the ergodicity region for buffered ALOHA system / $M=3$, $r_1=0.3$, $r_2=0.2$, $r_3=0.1$ /

In other words, the chain $\underset{\sim}{N}_{nm}(t)$ represents the queue lengths in the n-th and the m-th buffers under the condition that all other buffers are always empty, that is, for each $t \geq 0$ $N_j(t)=0$ for $j=1,\ldots,M$, $j \neq n$, m. Using the same explanations as before one quickly shows that the inequality (20) holds, hence, Lemma 3 may be applied. Moreover, according to Lemma 1 the Markov chain $\underset{\sim}{N}_{nm}(t)$ is not ergodic if and only if the input rates λ_n and λ_m satisfy the following conditions

$$\begin{array}{llll} 1 & \text{if} \quad \lambda_n < r_n(1-r_m) & \text{then} & \lambda_m \geq r_m[1-\lambda_n/(1-r_m)] \\ 2 & \text{if} \quad \lambda_n \geq r_n(1-r_m) & \text{then} & \lambda_m \geq (1-r_n)(1-\lambda_n/r_n) \end{array} \tag{35}$$

Let $\underset{\sim}{\Lambda}_{nm}$ be a set

$$\underset{\sim}{\Lambda}_{nm} = \{ (\lambda_1,\ldots,\lambda_M) : \lambda_n \text{ and } \lambda_m \text{ satisfy (35)}, \ \lambda_i > 0, \ i=1,.,M \}$$

Then

$$T_3 = \bigcup_{n=1}^{M} \bigcup_{m=1}^{M} \underset{\sim}{\Lambda}_{nm}$$

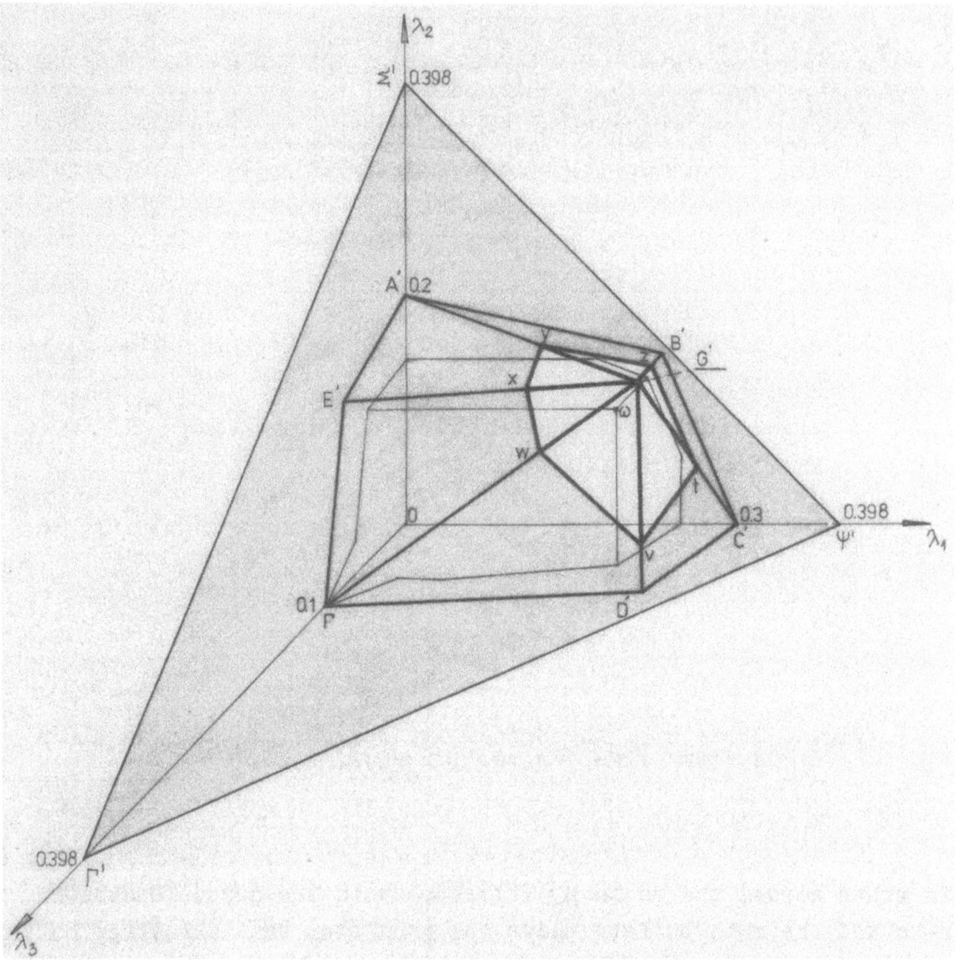

Fig.2 Subsets of the nonergodicity region for buffered ALOHA system / $M=3$, $r_1=0.3$, $r_2=0.2$, $r_3=0.1$ /

To show the relationship between the sets E_1, E_2 and E_3 or between T_1 and T_2 we consider a multiaccess system with random access for $M=3$ and $r_1=0.3$, $r_2=0.2$ and $r_3=0.1$. In Fig.1 the sets E_1, E_2 and E_3 are presented: the set E_1 consists of all points λ lying in the pyramid $O\Gamma\Psi\Sigma$, the set E_2 consists of all points lying in the cuboid $O\alpha\beta\gamma\delta\theta\varphi\omega$, while the set E_3 consists of points inside the surface $OAB\beta C\delta EF\varphi\omega$. It is easy to notice that $E_2 \subset E_3$, however, there are points λ of E_1 which are not included in E_3, namely points of the pyramid FXYΓ.

In Fig.2 the nonergodicity subsets T_1 and T_3 are presented. T_1 consists of all points λ outside the pyramid $O\Gamma'\Psi'\Sigma'$ while T_3 consists of points outside the surface $O'A'B'C'D'E'\Gamma G'$. Let us notice that points belonging to Gxyztuw lie in T_1, but not in T_3. Moreover, only the point $\omega \in T_3$ / Fig.1 and Fig.2 / is common for $\bar{E}_3$ and $\bar{T}_3$ / $\bar{A}$ denotes closure of a set A /. Unfortunately, there are points λ which belong neither to E_1, E_3 nor to T_1, T_3, that is, such points for which the methods presented in this paper cannot be used for ergodic investigation of the chain $N^M(t)$.

3.3 NETWORK OF QUEUES

We shall present one more example showing the fact that even one cannot determine exactly the ergodicity region, using the presented methods / but there are other methods which enable to do it / the methods considered above show some advantages over the exact methods. Let us consider M nodes with infinite capacities, where the i-th node consists of expotential server with parameter μ_i and receives arrivals from outside the system in the form of Poisson process with rate λ_i. Upon leaving the i-th node a customer proceeds to the j-th node with probability p_{ij}, $i,j=1,\dots,M$ and leaves the system with probability $p_{i,M+1}$. It was proved [JAC] that the M-dimensional Markov chain representing the queue lengths in each node is ergodic if quantities γ_n determined as a solution of the following set of M linear equations

$$\gamma_n = \lambda_n + \sum_{j=1}^{M} p_{jn}\gamma_j \tag{36a}$$

satisfy the inequalities

$$\gamma_n < \mu_n \quad , n=1,\dots,M \tag{36b}$$

To determine the ergodicity region one must first solve Eqs.(36a) with λ_n, $n=1,\dots,M$, as parameters and then solve the next M inequalities (36b) to determine the bounds on λ_n. This might be difficult. Instead of determining the ergodicity region by using (36) we might use Lemmas 2 and 3 to simplify the search, however, only a subset of the ergodicity region will be obtained. Let us consider Lemma 2 for K=1. Using the same explanation as in section 3.2 one immediately obtains the following conditions for the ergodicity of the Markov chain

$$\lambda_n < \mu_n - \sum_{j=1}^{M} p_{jn}\mu_j \quad , \qquad n=1,\ldots,M \tag{37}$$

Moreover, using Lemma 2 for K=2 and Lemma 1 one may enlarge the set of $\lambda = (\lambda_1,\ldots,\lambda_M)$ determined in (37). As before, we should consider any two queues, for intance the n-th and the m-th queues, assuming that other buffers are nonempty. Then, using Lemma 1, after some algebra, one obtains the following conditions for ergodicity of the n-th and the m-th queues:

$$\text{if } \lambda_n < \mu_n - \sum_{j=1}^{M} p_{jn}\mu_j \ , \text{ then } \lambda_m < \mu_m - \sum_{j=1}^{M} \mu_j(p_{jn}p_{nm}+p_{jm}) + p_{nm}(\mu_n - \lambda_n)$$

$$\text{if } \lambda_n \geq \mu_n - \sum_{j=1}^{M} p_{jn}\mu_j \ , \text{ then } \lambda_m < \mu_m + [\mu_n - \sum_{j=1}^{M} \mu_j(p_{mn}p_{jm}+p_{jn}) - \lambda_n]/p_{nm} \tag{38}$$

Then, a subset of the ergodicity region for the M-dimensional Markov chain may be determined as in the previous section / Eq.(34) /. The simplicity of (37) and (38) in comparison with (36) is obvious. Naturally, the methods presented above may be applied to a more general case.

4. CONCLUSIONS

In this paper we dealt with the classification of multidimensional Markov chains. Two theorems which generalized Pakes and Kaplan conditions were proved. Then comparison tests were stated / Lemma 2 and 3 / which enables to explore ergodicity and **nonergodicity** of an M-dimensional Markov chain by the means of ergodicity and nonergodicity of K-dimensional Markov chains, where K is smaller than M. Then, we applied those theorems to analyse some computer communication systems. We considered a homogeneous unbuffered multiaccess systems, and heterogeneous buffered ALOHA system as one out of current problem of great interest. Three different subsets of the ergodicity region have been found and compared. Finally, network of queues was considered.

The methods presented in this paper have advantages and drawbacks. The simplicity, generality and utility of these methods were shown. However, these methods, in general, cannot determine the ergodicity region, but only its subset / we have established only sufficient, but not necessary conditions for ergodicity of a Markov chain /. More investigations are needed. Further researches of new Lyapunov functions are one of the direction that should be followed.

REFERENCES

[CHU] Chung, K., Markov chains with stationary transition probabbilities, Springer-Verlag, 1960

[FAL] Falin, G., About ergodicity of a multiaccess system, Tehnischeskaja Kibernetika, 4, 1981, pp.126-131 / in Russian /

[FAY] Fayolle, G., Gelenbe, E., Labetoulle, J., Stability and optimal control of the packet switching broadcast channel, J. ACM, vol.24, No 3, July 1977, pp.375-386

[FOS] Foster, F., On the stochastic matrices associated with certain queueing processes, Ann. Math. Statist., 24, 1953, pp. 355-360

[JAC] Jacson, J., Networks of waiting times, Oper. Res., No 5, 1957 pp. 518-521

[KAL] Kalashnikov, W., Qualitative analysis of complex systems by Lyapunov functions, Moscow 1978 / in Russian /

[KAM] Kamae, T., Krengel, U., O'Brien, G., Stochastic inequalities on partially ordered spaces, Ann. Probab., 5, No.6 December 1977, pp.899-912

[KAP] Kaplan, M., A sufficient condition for nonergodicity of a Markov chain, IEEE Trans. on Information Theory, vol.IT-25, No 4, July 1979, pp.470-471

[KEM] Kemmeny, J., Snell, J., Knapp, A., Denumerable Markov chains, D. Van Nostrand Company, 1966

[KLE] Kleinrock, L., Lam, S., Packet switching in a multiaccess broadcast channel: Performance evaluation, IEEE Trans. on Commun., COM-23, No 4, April 1975, pp.410-422

[KAM] Kamal, S., Mahmoud, S., A study of user's buffer variations in a random access satellite channels, IEEE Trans. on Commun., vol. COM-27, No 6, June 1976, pp.857-868

[MAL] Malyshev, W., A classification of two-dimensional Markov chains and pice-linear martingals, Doklady Akademii Nauk USSR, 3, 1972, pp.526-528 / in Russian /

[OLD] Older, B., Qualitative analysis of congestions - Sensitive routing, Int. Symp. on Flow Control in Comp. Networks, Versailles 1979, pp.131-154

[PAK] Pakes, A., Some conditions for ergodicity and recurrence of Markov chains, Oper. Res., 17, 1969, pp.1058-1061

[SAA] Saadawi, T., Ephremides, A., Analysis, stability and optimization of slotted ALOHA with a finite number of buffered users, IEEE Trans. on Autom. Control, vol.AC-26, No 3, June 1981, pp.680-689

[STO] Stoyan, D., Qualitative Eigenschaften und Abschatzungen Stochastischer Modelle, Akademie-Verlag-Berlin 1977

[SZP1] Szpankowski, W., Analysis and stability considerations in a reservation system, IEEE Trans. on Commun, vol.COM-31, No 5, May 1983

[SZP2] Szpankowski, W., Some sufficient conditions for nonergodicity of a Markov chain, to be published

[TOB] Tobagi, F., Multiaccess protocols in packet communication systems, IEEE Trans on Commun., vol.COM-28, No 4, April 1980, pp. 486-488

[TSY] Tsybakov, B., Mikhaikov, W., Ergodicity of slotted ALOHA system, Problemy Peredachii Informatsii, vol.15, No 4, 1979, pp.73-87 / in Russian /

[WIT] Whitt, W., Comparing counting processes and queues, Adv. Appl. Prob., 13, 1981, pp.207-220

[WON] Wong, E., Stochastic processes in information and dynamical systems, McGraw-Hill 1971

[YEM] Yemini, Y., Kleinrock, L., On general rule for access control or, silence is golden... , Proc. of Int. Symp. on Flow Control in Comp. Networks, Versailles 1979, pp.335-347

ERGODIC CONDITIONS FOR A CLASS OF STATE DEPENDENT QUEUEING SYSTEMS

J. Izydorczyk

Polish Academy of Sciences
Department of Complex Control Systems

44-100 Gliwice, Zwyciestwa 21
Poland

Abstract

Queueing system stability criteria are frequently formulated in the form of testing the sign of queue length (or waiting time) mean increments for heavy load conditions. In this paper we deal with queueing systems which behaviour can be represented by Markov chains on R^N. As stability of such systems is equivalent to ergodicity of proper Markov chains, it must be strongly noticed that criteria of this form are valid only together with some additional set of conditions imposed on the considered chain. We shall suggest four comparatively easy to verify conditions and prove that this conditions are sufficient for considered criteria to be valid. We shall also show that suggested conditions are essential in the sense that skipping one of them turns the criteria to be false and apply the obtained results to check ergodicity of some important class of queueing systems, namely queues with waiting time dependent service times.

Introduction

This paper deals with the stability analysis of state dependent queueing systems. Behaviour of such systems can be usually described (c.f.[BORO76]) by time discrete stochastic processes of the type

$$X_{n+1}= [X_n+ a_n(X_n)]^+, \quad n= 1,2,3,..$$

where $[x]^+= [\max(0,x^{(1)}),.. \max(0,x^{(N)})]$, $x= [x^{(1)},.. x^{(N)}]$ and $a_n(X_n)$ is an N-dimensional random variable dependent on X_n. We shall assume that $\{X_n\}$ is a Markov chain with a state space $\mathcal{H}$ being a subset of the vector space R^N. Such a chain is defined by transition probabilities

$$P(x,A)= \mathrm{Prob}\{[x+a_x]^+\in A\}.$$

Thus in this paper we shall understand stability of some queueing system as ergodicity of a proper Markov chain.

Frequently (c.f. [PUJO78]) testing the sign of $\gamma(x)=\Sigma_{i=1}^{N} E(a_x^{(i)})$, $e_x=[a_x^{(1)},.. a_x^{(N)}]$ for x large enough is approved as criterion of ergodicity. It must however be strongly noticed that this criterion is valid only together with some additional set of conditions imposed on the considered chain. In this paper we shall suggest four comparatively easy to verify conditions and prove that this conditions are sufficient for considered criteria to be valid. The proof is based on earlier results due to Tweedie ([TWEE76]) which main points have been summarized at the beginning of our considerations. We shall also show that suggested conditions are essential in the sense that skipping one of them turns the criteria to be false and apply the obtained results to check ergodicity of some class of queueing systems with waiting time dependent service times.

Preliminaries

In this section some definitions and criteria for classifying general Markov chains will be remainded. Their more complete discussion is given in [TWEE76].

Let $(\mathcal{X},\mathcal{F})$ be an arbitrary measurable space. Suppose $\{X_n\}$ is a temporally homogeneous Markov chain on $(\mathcal{X},\mathcal{F})$ with transition probabilities

$$P(x,A)=\text{Prob}\{X_{n+1}\in A \mid X_n=x\}, \qquad x\in\mathcal{X},\ A\in\mathcal{F}$$

and n-step transitions

$$P^1(x,A)=P(x,A),$$

$$P^{n+1}(x,A)=\int_{\mathcal{X}} P^n(x,dy)P(y,A), \quad n=1,2,3...$$

The chain $\{X_n\}$ is assumed to be irreducible in terms of an auxiliary irreducibility measure φ on $\mathcal{F}$.

Definition 1.
Suppose φ is a σ-finite, non trivial measure on $\mathcal{F}$. Then $\{X_n\}$ is called φ-irreducible if $\Sigma_{n=1}^{\infty} P^n(x\ A)>0$ whenever $\varphi(A)>0$ for every $x\in\mathcal{X}$.

Let φ_0 be a measure on $\mathcal{F}$ such that $\varphi_0(\{x_0\})>0$ and $\varphi_0(\mathcal{X}\setminus\{x_0\})=0$ for some $x_0\in\mathcal{X}$. Further in this paper we shall assume that $\{X_n\}$ is φ_0-irreducible. This might seem to make the irreducibility condition too trivial: however, this is not the case, as the φ-irreducibility implies M-irreducibility.

Definition 2.
If $\{X_n\}$ is φ-irreducible we denote by M a maximal irreducibility measure satisfying:

(1) $\{X_n\}$ is M-irreducible,

(2) if $\{X_n\}$ is Ψ-irreducible for some Ψ then $M(A)=0$ implies $\Psi(A)=0$,

(3) $M(A)=0$ implies $M\{x: \Sigma_{n=1}^{\infty} P^n(x\ A)>0\}=0$,

and we write $\mathcal{F}^+=\{A\in\mathcal{F}: M(A)>0\}$.

The following definitions of transience, recurrence, nullity and ergodicity make it possible to classify φ-irreducible chains in similar manner as in the case of countable state space. Namely: every φ-irreducible chain is either transient or recurrent, every recurrent chain is either null or ergodic.

Definition 3.
We call a φ-irreducible chain:

(1) transient if there exists a sequence $\{A(j)\}$, $A(j)\in\mathcal{F}$, $\bigcup_{j=0}^{\infty} A(j)=\mathcal{X}$ such that $\Sigma_{n=0}^{\infty} P^n(x,A(j))<\infty$ for each $A(j)$ and $x\in\mathcal{X}$,

(2) recurrent if $\Sigma_{n=0}^{\infty} P^n(x,A)=\infty$ for every $A\in\mathcal{F}^+$ and $x\in\mathcal{X}$,

(3) null if there exists a sequence $\{B(j)\}$, $B(j)\in\mathcal{F}$, $\bigcup_{j=0}^{\infty} B(j)=\mathcal{X}$ such that $\lim_{n\to\infty} P^n(x,B(j))=0$ for each $B(j)$ and $x\in\mathcal{X}$,

(4) ergodic if $\limsup_{n\to\infty} P^n(x,B)>0$ for every $B\in\mathcal{F}^+$ and $x\in\mathcal{X}$.

Let us notice that if a φ-irreducible chain is not ergodic then it is null.

Definition 4.
A measure Q non trivial and σ-finite is called invariant for $\{X_n\}$ if

$$Q(A)=\int_{\mathcal{X}} Q(dy)P(y,A) \quad \text{for every } A\in\mathcal{F}.$$

If $\{X_n\}$ is recurrent then the measure Q always exists, is unique up to a constant multiple and is equivalent to the maximal irreducibility measure M, ($M(A)=0 \iff Q(A)=0$). If $\{X_n\}$ is ergodic the unique invariant probability measure π is given by

$$\pi(A)=\lim_{n\to\infty} n^{-1}\Sigma_{m=1}^{n} P^m(x,A) \quad \text{for every } A\in\mathcal{F}^+.$$

In next sections we shall use the notions of a mean drift and a status set. Let R denote the set of all real numbers, while R_+ denotes the set of all nonnegative reals and $g: \mathcal{H} \to R_+$ is a measurable function.

Definition 5.
A function $\mu_g: \mathcal{H} \to R$ defined by $\mu_g(x) = \int_{\mathcal{H}} P(x,dy)g(y) - g(x)$ for every $x \in \mathcal{H}$, will be called a mean drift of $\{X_n\}$.

Definition 6.
Let Θ be a class of chains $\{X_n\}$ on $(\mathcal{H}, \mathcal{F})$, each of which is φ-irreducible for some φ. A set $A \in \mathcal{F}^+$ is called a status set for Θ if, for each $\{X_n\} \in \Theta$ with transition law $P(x,A)$:

(1) $\sum_{n=1}^{\infty} P^n(x,A) < \infty$ for φ-a.a. $x \in \mathcal{H}$ when $\{X_n\}$ is transient,

(2) $\lim_{n \to \infty} P^n(x,A) = 0$ for φ-a.a. $x \in \mathcal{H}$ when $\{X_n\}$ is null.

We shall abuse this notation somewhat by calling A a status set for $\{X_n\}$ if A is a status set for Θ and $\{X_n\} \in \Theta$.

The following theorem shows that notions of a status set and a mean drift can be used to verify ergodicity.

Conditions 1. Suppose $\{X_n\}$ is φ-irreducible and let A be a status set for $\{X_n\}$ in $\mathcal{F}^+$ with $A^c \in \mathcal{F}^+$. Let g be a nonnegative function satisfying for all $x \in \mathcal{H}$ and some $\delta < \infty$ the following inequalities:

C1.1 $\int_{\mathcal{H}} P(x,dy)|g(y)-g(x)| \leq \delta$,
C1.2 $g(x) > \sup_{y \in A} g(y)$, $x \in A^c$.

Theorem 1. [TWEE76]
Any Markov chain $\{X_n\}$ satisfying Conditions 1 is ergodic if and only if for some $B \in \mathcal{F}^+$:

$$\liminf_{n \to \infty} n^{-1} \sum_{m=1}^{n} P^m(x,dy) \mu_g(y) < 0 \quad \text{for each } x \in B.$$

There occure essential difficulties in either infering about ergodicity or even deciding that a considered set is a status set basing directly on the given above definitions of this concepts. Now we give simple criteria for a set A to be status and for a chain $\{X_n\}$ to be ergodic or null in the case of a general state space.

Criterion S1. [TWEE76]

The point set $\{x_0\}$ is a status set for the collection of φ_0-irreducible chains as is each of the sets

$$A_{n,j} = \{y: P^n(y,\{x_0\}) > j^{-1}\}$$

for which $M(A_{n,j}) > 0$.

Criterion S2. [TWEE76]

Suppose $\mathcal{H}$ has a topology $\mathcal{T}$. Every relatively compact set in $\mathcal{T}^+$ is a status set for Θ_s:

$\Theta_s = \{\{X_n\}: \{X_n\}$ is φ-irreducible and $P(x,A)$ is continuous function of x for every $A \in \mathcal{T}\}$.

Let $\{X_n\}$ be a Markov chain satisfying Conditions 1.

Criterion E1.

If there exists $\varepsilon > 0$ such that $\mu_g(x) \leq -\varepsilon$ for every $x \in A^c$ then $\{X_n\}$ is ergodic.

Criterion E2.

If $\mu_g(x) \geq 0$ for every $x \in A^c$ then $\{X_n\}$ is null.

Criteria E1 and E2 are a simple consequence of the Theorem 1.

Modified criteria for a chain on R^N

As our study was motivated by the stability analysis of state dependent queueing systems we shall constrain ourselves to the case when a state space is a subset of the N-dimensional vector space R^N and a chain is defined in terms of state dependent increments a_x.

Let us introduce the following notation:

$\mathcal{B}^N$ -the Borel σ-field in R^N,

$x, y \in R^N$, $\quad x = [x^{(1)} \ldots x^{(N)}]$,

$x^{(i)}, y^{(i)}, \alpha, \beta, \delta, \varepsilon, \nu \in R$.

$[x]^+ = [\max(0, x^{(1)}), \ldots \max(0, x^{(N)})]$.

$I_\beta = \{x \in \mathcal{H}: \Sigma_{i=1}^N x^{(i)} \leq \beta\}$, $\quad I_\beta^c = \{x \in \mathcal{H}: \Sigma_{i=1}^N x^{(i)} > \beta\}$,

$\tilde{I}_\alpha = \{x \in \mathcal{H}: x^{(i)} > \alpha, i=1,\ldots N\}$.

In further considerations we shall deal with a Markov chain $\{X_n\}$ on $(\mathcal{H}, \mathcal{F})$ such that $\mathcal{H} = \{A \in \mathcal{B}^N : A \subset R_+^N\}$, $\mathcal{F} = \{A \in \mathcal{B}^N : A \subset \mathcal{H}\}$, while transitions $P(x, A)$ of $\{X_n\}$ are defined by

$$P(x, A) = \text{Prob}^*\{[x + a_x]^+ \in A\}$$

and increment a_x for each $x \in \mathcal{H}$ is such a random variable that $[x + a_x]^+$ is a random variable on $(\mathcal{H}, \mathcal{F})$.

The following criteria can be used for infering about either ergodicity or nullity of $\{X_n\}$.

Criterion E3.
If there exists $\alpha > 0$ such that $\gamma^+(\alpha) < 0$ then a Markov chain $\{X_n\}$ is ergodic and $\pi(\{0\}) > 0$, (π is the invariant measure of $\{X_n\}$).

Criterion E4.
If there exists $\alpha > 0$ such that $\gamma^-(\alpha) \geqslant 0$ then a Markov chain $\{X_n\}$ is null.

Where: $\gamma^-(\alpha) = \inf_{x \in \widetilde{I}_\alpha} \gamma(x)$, $\gamma^+(\alpha) = \sup_{x \in \widetilde{I}_\alpha} \gamma(x)$,

$\gamma(x) = \sum_{i=1}^{N} E(a_x^{(i)})$.

It must however be stróngly noticed that these criteria are valid together with some additional set of conditions imposed on the considered Markov chain. Such a set of conditions may be chosen in many different ways. We shall suggest four conditions being comparatively easy to verify and prove, using the earlier given criteria, that these cnditions are sufficient for the criteria E3 and E4 to be valid.

Conditions 2. There exist $v > 0$, $\xi > 0$, $\delta > 0$ such that:
    C2.1 $A_x(-v) > \xi$ for every $x \in \mathcal{H}$.
    C2.2 $B_x(v) > \xi$ for every $x \in \mathcal{H}$.
    C2.3 $\eta(x) < \delta$ for every $x \in \mathcal{H}$.
    C2.4 for every $\alpha > 0$ there exists $\beta > 0$ such that $I_\beta^c \subset \widetilde{I}_\alpha$,

where $A_x(v) = \text{Prob}\{a_x^{(i)} \leqslant v, \; i = 1 \ldots N\}$,

$B_x(v) = \text{Prob}\{\sum_{i=1}^{N} a_x^{(i)} > v\}$,

$\eta(x) = \sum_{i=1}^{N} E(|a_x^{(i)}|)$.

Theorem 2.
If $\{X_n\}$ is a Markov chain satisfying Conditions 2 then Criteria E3 and E4 are valid.

Proof
Let x be an arbitrarily chosen point in $\mathcal{X}$ and υ, ξ be the constants from C2.1. We have

$$P^n(x,\{0\}) > \xi^n \quad \text{for} \quad n > x^{max}/\upsilon,$$

where $x^{max} = \max\{x^{(1)} \ldots x^{(N)}\}$, then from Definition 1 $\{X_n\}$ is φ_0-irreducible and from Criterion S1 every set I_β is a status set for $\{X_n\}$, ($M(I_\beta)>0$ as $M(\{0\})>0$ from Definition 2(2)).

The condition C2.1 implies that I_β^c belongs to $\mathcal{F}^+$ for every $\beta>0$ as $P^n(0,I_\beta^c) \geq \xi^n$ for $n>\beta/\upsilon$, while (c.f. Definition 2(3)) $M(\{0\})>0$ implies $M(I_\beta^c)>0$.

Suppose $g(x) = \Sigma_{i=1}^N x^{(i)}$; then $g(x) > \sup_{y\in I_\beta} g(y) = \beta$ for every $x \in I_\beta^c$ and

$$\int_{\mathcal{X}} P(x,dy)|g(y)-g(x)| = E(|\Sigma_{i=1}^N \max(0, x^{(i)}-a_x^{(i)})-x^{(i)})$$

$$\leq E(\Sigma_{i=1}^N |a_x^{(i)}|) - \eta(x) < \delta \quad \text{for every} \quad x\in\mathcal{X},$$

thus $\{X_n\}$ together with a test function $g(x) = \Sigma_{i=1}^N x^{(i)}$ and with a set I_β satisfy Conditions 1 needed for utilisation of Criteria E1 and E2.

Notice, that

$$\mu_g(x) = E(\Sigma_{i=1}^N \max(0, x^{(i)}-a_x^{(i)}) - x^{(i)})$$

$$= \gamma(x) + E(\Sigma_{i=1}^N \max(0,-x^{(i)}-a_x^{(i)})). \qquad (3.1)$$

Proof of the Criterion E3
The existence of $\alpha>0$ such that $\gamma^+(\alpha)<0$ implies existence of $\varepsilon_1>0$ such that $\gamma(x)<-\varepsilon_1$ for every $x\in\tilde{I}_\alpha$. On the other hand if C2.3 is satisfied then there exist constants $\alpha_2>0$ and $\varepsilon_2<\varepsilon_1$ such that $E(\Sigma_{i=1}^N \max(0,-x^{(i)}-a_x^{(i)})) < \varepsilon_2$ for every $x\in\tilde{I}_{\alpha_2}$.
Thus $\mu_g(x) < -\varepsilon$ for all $x\in\tilde{I}_{\alpha_1}$ where $\alpha_1 = \max(\alpha_2, \alpha)$ and $\varepsilon = \varepsilon_1 - \varepsilon_2$. But (c.f. the condition C2.4) we have $\beta>0$ such that $I_\beta^c \subset \tilde{I}_{\alpha_1}$ then $\mu_g(x) < -\varepsilon$ for every $x\in I_\beta^c$, so by Criterion E1 the

chain $\{X_n\}$ is ergodic.

Additionally $\pi(\{0\}) > 0$ as the measure M is equivalent to the invariant measure π and $M(\{0\})>0$ in the case when $\{X_n\}$ is φ_0-irreducible.

Proof of the Criterion E4
There exists $\alpha>0$ such that $\gamma^-(\alpha)= \inf_{x\in\tilde{I}_\alpha} \gamma(x)>0$ and by (3.1) $\mu_g(x)>0$ for every $x\in\tilde{I}_\alpha$. But because of C2.4 there exists $\beta>0$ such that $\mu_g(x)>0$ for every $x\in I^c_\beta$ so Criterion E2 gives nullity for $\{X_n\}$. ∎

Let us now discuss briefly to what extend the above given criteria make it possible to decide about ergodicity or nullity of the Markov chain $\{X_n\}$, assuming that Conditions 2 are satisfied.

1. Suppose there exist γ_∞ and α such that $\gamma(x)=\gamma_\infty$ for every $x\in\tilde{I}_\alpha$. The chain is ergodic if and only if $\gamma_\infty<0$.

2. The approach suggested in [TWEE76] and developped further in this paper does not seem to be appropriate in the case when $\gamma^+(\alpha)>0$ and $\gamma^-(\alpha)<0$ for every $\alpha>\alpha^{min}$.

3. If $\lim_{\alpha\to\infty}\gamma^-(\alpha) = \lim_{\alpha\to\infty}\gamma^+(\alpha) = 0$ it is possible to check ergodicity only in special cases (e.g. Proposition 8.1 of [TWEE76] for $\{X_n\}$ null and Proposition 7.3 of [TWEE76] for $\{X_n\}$ ergodic).

Some of the conditions C2.1-C2.4 may seem to be superflous (especially C2.4). We shall present some exemples proving that skipping one of them turns Theorem 2 to be false.

Example 1 (the condition C2.1 omitted)
Let $\text{Prob}\{a_x=2\}= p_1$, $\text{Prob}\{a_x=-2\}= 1-p_1$ for x=0,2,3, ... and $\text{Prob}\{a_x=2\} =1$ for x=1.
The chain $\{X_n\}$ on $\mathcal{X}=\{0,1,2,..\}$ with $P(x,A)= \text{Prob}\{[x+a_x]^+\in A\}$ is not irreducible thus not ergodic (and not null) even for $p_1<0.5$ when $\gamma(x)<0$ for $x>2$.

Example 2 (the condition C2.1 omitted)
Let $\text{Prob}\{a_x=1\}= p_1$, $\text{Prob}\{a_x=-1\}=1-p_1$ for x=0,1,2... N-1, N+1,.. and $\text{Prob}\{a_x=-1\}= 1$ for x= N.

The chain $\{X_n\}$ on $\mathcal{X}=\{0,1,2,..\}$ with $P(x,A)=\text{Prob}\{[x+a_x]^+\in A\}$ is ergodic for all $p_1\in(0,1)$ despite that $\gamma(x)>0$ for $x=N+1,N+2,..$ when $p_1>0.5$.

Example 3 (the condition C2.3 omitted)
Let $\text{Prob}\{a_x=x\}=p_1$, $\text{Prob}\{a_x=-x\}=1-p_1$ for $x=1,2,3,4,..$ and $\text{Prob}\{a_x=1\}=1$ for $x=0$.
The chain $\{X_n\}$ on $\mathcal{X}=\{0,1,2,...\}$ with $P(x,A)=\text{Prob}\{[x+a_x]^+\in A\}$ is ergodic for all $p_1\in(0.1)$ despite that $\gamma(x)=x(2p_1-1)$ so $\gamma(x)>0$ for all $x=0,1,2,3,...$ when $p_1>0.5$.

Example 4 (the condition C2.4 omitted)
Let $\{X_n\}$ be a chain on $\mathcal{X}=\{x:\ x=[x^{(1)},x^{(2)}],\ x^{(1)},x^{(2)}\in\{0,1,2,..\}\}$ while a_x are defined by

$$\text{Prob}\{a_x=[1.1]\}=p_1p_2\ ,$$
$$\text{Prob}\{a_x=[1.-1]\}=p_1(1-p_2)\ ,$$
$$\text{Prob}\{a_x=[-1.1]\}=(1-p_1)p_2\ ,$$
$$\text{Prob}\{a_x=[-1.-1]\}=(1-p_1)(1-p_2)\quad\text{for all}\quad x\in\mathcal{X}.$$

The chain $\{X_n\}$ is ergodic if and only if $p_1<0.5$ and $p_2<0.5$ despite that $\gamma(x)=p_1+p_2-1$ for all $x\in\mathcal{X}$, so when $p_1=0.1$ and $p_2=0.6$ it is null and $\gamma(x)=-0.3$.

The most strange condition for Theorem 2 is C2.4. In several situations C2.4, calling for a special property of a state space in multidimensional case, seems to be technical in nature. It is however rather difficult or even impossible to replace this condition by weaker one when the criteria E1, E2 and S1 or S2 are used for the proof of E3 or E4.
When we check Conditions 1 we must take a set A such that:

1. A is a status set,
2. $g(x)>\sup_{y\in A} g(y)$ for all $x\in A^c$.

From S2 or S1 when C2.3 is satisfied we can check that a set A is status if A is bounded and for $g(x)=\sum_{i=1}^{N} x^{(i)}$ from 2. a set A is I_β so it is also bounded. But $\mu_g(x)$ for $x\in A^c$ in E1 or E2 must be replaced by $\gamma(x)$ for $x\in\tilde{I}_\alpha$, thus $\tilde{I}^c_\alpha=\mathcal{X}\setminus\tilde{I}_\alpha$ must be a bounded set. If C2.4 is not fulfilled a set $\tilde{I}^c_\alpha$ is not bounded for any $\alpha>\alpha_{min}$.

Queues with waiting time dependent service times

Now we apply Theorem 2 to stability analysis of some important class of queueing systems. In further considerations we shall deal with service systems having the following features:

1. Demands arrive in time epochs $t_0, t_1, t_2, \dots$ and time periods $c^n = t_{n+1} - t_n$ are independent random variables distributed identically as some random variable c.

2. Demands are served individually under FIFO discipline on N identical, independently operating parallel servers.

3. Service time b^n of the n-th demand which spends time w waiting in the queue is a random variable distributed identically as some random variable b_w.

Let us denote by $X^{(i)}(t)$ the occupation time of the i-th server at the instant t, i.e. $X^{(i)}(t)$ is the time which elapses from t until the server becomes idle for the first time if no demands arrive after time t, and let $X_n^{(i)} = X^{(i)}(t_n - 0)$.

It is easy to see that the process $\{X_n\} = \{[X_n^{(1)} \dots X_n^{(N)}]\}$ is a Markov chain and the transitions $P(x,A)$ of $\{X_n\}$ are defined by $P(x,A) = \text{Prob}\,\{[x+a_x]^+ \in A\}$ with $a_x^{(i)} = -c$ for $i \neq j$ and $a_x^{(j)} = b_w - c$ where $w = x^{(j)} = \min_{i=1\dots N} \{x^{(i)}\}$.

Theorem 3.
Let $\{X_n\}$ be a chain defined by a_x such that:
C3.1 There exist $\xi>0$ and $\upsilon>0$ such that $\text{Prob}\{b_w - c < -\upsilon\} > \xi$ and $\text{Prob}\{b_w - Nc > \upsilon\} > \xi$ for every $w>0$.
C3.2 $0<E(c)<\infty$.
C3.3 There exists b_{max} such that $\text{Prob}\{b_w < b_{max}\} = 1$, $w>0$ for the case when $N>1$ or $E(b_w) < b_{max}$, $w>0$ for the case when $N=1$.

If there exist $\varepsilon>0$ and $\alpha>0$ such that $E(b_w) < NE(c) - \varepsilon$ for every $w>\alpha$ then the chain $\{X_n\}$ is ergodic.

If there exists $\alpha>0$ such that $E(b_w) > NE(c)$ for every $w>\alpha$ then the chain $\{X_n\}$ is null.

Proof

For the proof of this theorem we shall use Theorem 2, thus we must show that Conditions 2 are satisfied. The condition C3.1 immediately implies C2.1 and C2.2. From C3.2 and C3.3 we have $\eta(x) < b_{max} + NE(c)$ so C2.3 is fulfilled. In the case when N=1 the condition C2.4 is trivial, but when N>1 it yields from C3.3 that $Prob\{X_n^{(i)} - X_n^{(j)} > b_{max}\} = 0$ so we can take as a state space of $\{X_n\}$ the set

$$\mathcal{H} = \{x \in R_+^N : x^{(i)} - x^{(j)} \leqslant b_{max}\}.$$

In this case $\Sigma_{i=1}^N x^{(i)} > \beta$ implies $x^{(j)} > \beta - (N-1)b_{max}$ for every j=1,.N and C2.4 is fulfilled. Finally we note that $\gamma(x) = E(b_w) - NE(c)$ where $w = \min_{i=1...N} \{x^{(i)}\}$. ∎

Theorem 3 is an extension of results given in [CALL73],[TWEE75], [LASL78] for the single channel case and in [SUGA65] for the multichannel case with deterministic dependence of service time upon waiting time.

Final remarks

We have suggested a set of additional conditions which are sufficient for the criteria E3 and E4 to hold. Although we have demonstrated that none of this conditions can be skipped, it seems that this is not a set of nessesary conditions. For example applying the results of Theorem 3 to the classical M/G/n queue we can prove ergodicity only under superflous condition that service times are bounded by some finite value.

As presented in comments to Theorem 2 there is a little chance for improving this situation following the approach developed by Tweedie until we will not be able to prove that an unbounded set is a status set.

Acknowledgements

I wish to thank Dr A. Wolisz for his comments and help in preparation of the manuscript.

References

[BORO76] Borovkov A.A. "Stochastic processes in queueing theory," Springer Verlag, 1976.

[CALL73] Callahan J.R. "A queue with waiting time dependent service times," Naval Res. Logist. Quart. 20, p.321, 1973.

[LASL78] Laslett G.M., Pollard D.B., Tweedie R.L. "Techniques for establishing ergodic and recurrence properties of continuous valued Markov chains," Naval Res. Logist. Quart. 25, p.455, 1978.

[PUJO78] Pujolle G. "Applications of some Markov chains results to computer systems modelling," IRIA Rapport de Recherche 289, 1978.

[SUGA65] Sugavara S., Takahashi M. "On some queues occuring in an integrated iron and steel works," J. of O.R. Japan 8, p.16, 1965.

[TWEE75] Tweedie R.L. "Sufficient conditions for ergodicity and recurrence of Markov chains on a general state space," Stoch. Proc. Appl. 3, p.385, 1975.

[TWEE76] Tweedie R.L. "Criteria for classifying general Markov chains" Adv. Appl. Prob. 8, p.737, 1976.

VII

QUEUES AND NETWORKS 3

FILES D'ATTENTE ET RESEAUX 3

A Bottleneck-driven Scheduler

Giuseppe Serazzi

Istituto di Analisi Numerica - CNR, Pavia
Dip. Informatica e Sist. - Università di Pavia - Italia

ABSTRACT

An adaptive scheduling algorithm which controls the input to the system in order to maximize a given performance criterion, i.e., the *system throughput rate*, is presented. The system load is adjusted depending on the characteristics of the mix of jobs in execution and of the mix of jobs submitted to the system and waiting in the input queue. The performance objective is achieved by maximizing the utilization factor of all the resources through the best match between the resource capacities and the workload demand pattern. The evaluation of the adaptive scheduling algorithm is performed through simulation experiments using data collected from real workloads.

1. The Problem

It is known that computer system performances are highly dependent on how the system reacts to workload fluctuations. The unpredictable composition of the program mixes together with the peculiar characteristics of the jobs generate on the system resources peaks of load having highly variable intensities and durations. The nonstationarity of workload characteristics causes dynamic migrations from one system's resource to another of the congestion due to service requests. It seems therefore evident that to take care of the fluctuations of the resource demands, that may change dramatically over short-time intervals, automatic control mechanisms are required.

The possibility of controlling in real time the utilization of the resources of a computer system has already been studied for over a decade. Several papers deal with the problem of dynamic control of a given specific function by means of local control mechanisms. These mechanisms are local in the sense that they do not control the performances of the global system, but rather that of one, or a few, of its resources.

The principle behind this approach is that an optimal use of the controlled resource permits a better utilization of the global system. Indeed, we may find many papers that investigate the optimization of CPU utilization by means of adaptive algorithms that control CPU scheduling and quantum length (see e.g. [2,3,4,10,13,16]). Local controls may be very dangerous if they are not exerted within a more general scheme which controls the performances of the entire system. Indeed, local control does not permit the prediction of bottleneck migrations and might worsen global performances by optimizing a *non-critical* resource. A higher viewpoint is usually applied in the dynamic control mechanisms concerning the performances of virtual memory systems.

* Partial support for this research has been provided by the HUSPI Project (Honeywell-Università di Pavia).

The overall performances of virtual memory systems are often controlled via the degree of multiprogramming (see e.g. [2,7,8]). The basic principle is that the load of the system is left free to increase while a suitable constraint holds. The constraint depends on the criterion used to estimate the optimal load, e.g., the knee-criterion, the L=S criterion, the 50% criterion. In this paper an approach which takes care of the time-variant situations of load on the various resources is described.

An *adaptive scheduling algorithm* that will regulate the load of the system in order to optimize a given performance measure is proposed. Even though the approach presented can be applied to the optimal control of various performance indexes, e.g., the system's response time, the utilization of a particular resource , the main memory occupancy, and so on, here we shall be concerned only with the optimization of *system throughput*. Other objectives can be achieved by applying minor modifications to the approach described below.
Section 2 deals with the basic assumptions concerning the dynamic behavior of the jobs during their executions. In Section 3 the properties of the system model used for the transient analysis are reviewed. The optimization objectives and the controller functions of the adaptive control system are described in Section 4 and 5. The results of the simulation experiments conducted on two real workloads are presented and evaluated in Section 6.

2. Basic Assumptions

Workload fluctuations are mainly due to the following factors:

- each job has its own characteristics that are, in general, statistically different from those of the other jobs;
- the amount of unsatisfied service requests for the various resources varies dynamically in an unpredictable way depending on the resource contentions;
- the statistical characterization of each job's resource request pattern varies with the state of its execution.

In such a variable situation the use of statistical techniques, like clustering, factor analysis, and so on, to characterize the workload is not adequate. In fact these statistical techniques ignore the individuality of each job's requests and are lacking in the ability to model the interactions between time and space requests of each job. To study the resource contentions it is fundamental to know how a job's resource demands are distributed among the resources during its execution. This knowledge would allow us to predict resource demands in the near future. The availability of these data makes it possible the design of an adaptive scheduling policy to bring about a *balance* between resource demands and unused processing power. The working set model proposed by Denning [5] is the first, and perhaps the most important, effort in the study of job behavior. Denning points out that the resource requests that a job makes during its execution are predictable with high probability.

Empirical analyses of the reference patterns of many jobs in a paged memory system indicate that most jobs can be characterized by their locality, [6,9]. These jobs have the property of concentrating references within relatively small regions of the virtual space over relatively long intervals of time. It is the presence of loops, arrays or other data structures on which the job operates for long periods of time that causes the phenomenon of locality to appear. The execution of the same sequence of instructions for a large number of times has as a consequence that the same resource demand pattern is maintained for a long period of time.

Therefore, a correspondence between a given locality and a particular resource demand pattern can be established. Since the execution time of a single sequence of instructions characterizing a locality is very short in comparison to the job's execution time, the resource demand pattern of the corresponding job tends to remain nearly constant during the most part of its execution. As a consequence, on each resource the fractions of residual resource demands, i.e., the proportions of remaining service, of a job with one locality remain nearly constant throughout its execution.

Let us remark that the fractions of residual resource demands of a given job can be considered also as fractions of resource busy time, i.e., the resource utilizations, caused by the execution of that job alone.

Let us consider a job i having an average number V_j^i of requests, or visits, for the j-th resource. Its minimum turnaround time, or ***minimum response time***, $R_{\min}^i$ obtainable when no resource contention occurs, e.g., when the job is executed in monoprogramming mode, is:

$$R_{\min}^i = \sum_{j=1}^{n} (V_j^i / \mu_j) \qquad (1)$$

where $1/\mu_j$ is the mean service time per visit of the j-th resource, n being the number of resources in the system.

The fractions τ_{ij} 's of $R_{\min}^i$ given by

$$\tau_{ij} = V_j^i / (\mu_j R_{\min}^i) \qquad (j = 1,\ldots,n) \qquad (2)$$

characterize the resource demand pattern of job i , that is, the fractions of $R_{\min}^i$ that job i spends on each resource. For the above considerations, the τ_{ij} 's are nearly constant throughout the execution of a one-locality job and are slowly varying in time for a several locality job.

If job i is executed in monoprogramming mode, τ_{ij} is the fraction of time resource j is busy during the job's execution. The various jobs of the workload may be grouped into *classes* using as classifying parameters their $R_{\min}$ values and their fractions of resource busy time τ_{ij} 's. The jobs of a class i have same (or very similar) $R_{\min}^i$ and τ_{ij} values. In the sequel we will consider *only* jobs having one *locality*, i.e., for which the τ_{ij} values are constant.

If the job has more than one locality the new τ_{ij} values may be estimated applying a forecasting technique that use the recent-past τ_{ij} values to predict the near-future τ_{ij} values.

For example, at a given instant t the workload of the system may be constituted by the jobs of the classes described in Table 1.

Moreover, our system will be supposed to have a limited amount of main memory or a bounded number of simultaneously executing jobs.

3. The Model

In order to motivate the method proposed we shall describe the model adopted for the simulation of the dynamics of the system we wish to control.

In a multiprogrammed computer system the jobs in execution compete against each other for the utilization of the resources. The resource contention has been modeled assuming the scheduling discipline of *all* system resources to be of the *processor sharing* (PS) type. According to the PS concept, [11], fractions of the resource capacity are assigned to all the request pending on it. Thus, a job during its execution utilizes fractions of the resources capacities on a full-time basis. The behavior of a job 's execution is described by the following assumptions:

a) all the resources for which the job has service requests start executing its requests at the same instant

b) the fractions of the resource capacities assigned to the job are such that the execution of its requests terminate on all the resources at the same instant.

The rationale behind these assumptions is the fine intermixing of job requests for the various resources, the repetitivity of similar sequences of operations (within each job), and the scheduling algorithms of some resources (e.g., the Round Robin algorithm of the CPU).

Let us consider, for example, a job i that spends 1/3 of its time at the CPU and 2/3 of its time at an I/O device, i.e., $\tau_{i1} = 1/3$ and $\tau_{i2} = 2/3$. By assumptions a) and b), the fraction of CPU capacity actually used by job i will be *always* half of the I/O capacity used by this same job i , independently of the dynamic variations of the system load.

The errors introduced by these assumptions are usually very small since most of the jobs behave as described in the previous section, and their requests for the different resources are thoroughly intermixed. For example, before and after an I/O operation an amount of CPU time is required. Also, before and after heavy CPU computations, I/O operations are usually performed. Furthermore, the assumptions introduced seem reasonable if one considers the *macro-level* on which the control algorithm is exerted.

According to these assumptions, an τ_{ij} value may be considered as the fraction of the capacity of resource j used for $R^i_{\min}$ time units by a class-i job executed in monoprogramming mode. In a multiprogramming environment the resource contentions will result in a response time R^i higher than $R^i_{\min}$, thus making the fractions of resource capacity actually obtained by class-i jobs smaller than the τ_{ij} 's .

The fractions of resource capacities used by the jobs will depend on the characteristics of the *mix* of jobs in execution at the instant of time considered. To determine the actual fractions of resource capacity assigned to a job at each instant we have to analyze the strategy according to which resource capacities are distributed among the jobs.

If the jobs have the same priority, we assume that the fractions of resource capacities are assigned according to an *egalitarian* policy. The basic principle is that no preference is given to any class of jobs and, on the other hand, the system takes care of the resource contention by processing the requests in an optimal way following assumptions a) and b) . A formal description of this optimality of processing will be given in the sequel.

The state of the system is described by the vector $\mathbf{N}(t) = (N_1(t), \ldots, N_m(t))$ whose components represent the number of jobs of each class in execution at time t. The global number of jobs in executions, i.e., the multiprogramming level, is $N(t) = \sum_{i=1}^{m} N_i(t)$. The summation

$$\sum_{i=1}^{m} N_i(t)\, \tau_{ij} \qquad (j = 1,\ldots,n) \tag{3}$$

represents the fraction of capacity of resource j requested by the jobs in execution at instant t. Whenever summation (3) exceeds one, only a part of the requested capacity can be allocated to the jobs.

To each fraction of the resource capacity assigned to a job there corresponds a service rate, i.e., a *processing speed*, that, due to our assumptions, is independent of the resources but depends only on the class of the job. Indeed, according to these assumptions, the simultaneous progress of a job's execution must have the same processing speed on all the resources and the fractions of resource capacities must be computed in such a way that a job has requests pending on the resources on a full-time basis during its execution.

Let $\mathbf{s}(t) = \{ s_1(t),\ldots, s_m(t) \}$ be the vector whose components represent the fraction of the requested resource capacity actually allocated to the jobs of each class. The $s_i(t)$ may be regarded as the instantaneous *processing speed* of each class-i job. Clearly it will be $0 \le s_i(t) \le 1$, $(i = 1,\ldots,m)$. The value $s_i(t) = 1$ corresponds to the case in which job i is executed in monoprogramming mode. Due to our assumptions, the $s_i(t)$ of a class-i job will never be equal to zero while the job i is in execution (work conservative policy, [11]).

Let us now compute the $\mathbf{s}(t)$ values. The most requested resource, i.e., the *bottleneck of the system*, at instant t is the resource (denoted in the sequel by index b), such that:

$$\sum_{i=1}^{m} N_i(t)\, \tau_{ib} = \max_{1 \le j \le n} \sum_{i=1}^{m} N_i(t)\, \tau_{ij} \tag{4}$$

and it is known that this is the resource that limits the system's overall performance. Without loss of generality, we assume that Eq. (4) is satisfied by only *one* resource.
If the right-hand side of Eq. (4) is less than or equal to one, the requests will be processed by all the resources at the maximum speed, i.e., $s_i(t) = 1$, $(i = 1,\ldots,m)$. Whereas, if the right-hand side of Eq. (4) is greater than one the requests, at least on resource b, are greater than the resource capacity. In this case the processing speed of the jobs requiring service on resource b will be lower than 1 since only a fraction of the requested capacity will be assigned to each class. The maximum processing speed for the jobs is given by

$$s_b(t) = \min \left[1, \frac{1}{\sum_{i=1}^{m} N_i(t)\, \tau_{ib}} \right] \tag{5}$$

At each instant of time the following inequalities must hold:

$$\sum_{i=1}^{m} N_i\,(t)\,\tau_{ij}\,s_i\,(t) \le 1 \qquad (j = 1,\ldots,n) \tag{6}$$

where each term on the left-hand side represents the fraction of the resource j's capacity (normalized to one multiplying both terms by $1/\mu_j$) assigned to class-i jobs, making their processing speed equal to $s_i\,(t)$.

Setting $\mathbf{s}\,(t) = \{\, s_i\,(t) = s_b\,(t)\,\}_{i=1,\ldots,m}$, we may see that the left-hand side of inequality (6) is equal to one only for resource b whereas is less than one for all the other resources.

Let us consider the system whose workload is described by the 3 × 4 matrix $\boldsymbol{\tau}$ and by the vector $\mathbf{R}_{min}$ reported in Table 1. If in a time interval the system state is $\mathbf{N} = \{\, 1,1,1\,\}$, the bottleneck is resource 3. Indeed from Eq. (5) we obtain the value 1.1 and the processing speed s_b is 0.909 for all the jobs. If in another time interval 12 jobs are in execution, that is $\mathbf{N} = \{\, 10,1,1\,\}$, the bottleneck is resource 1; from Eq. (5) we obtain the value 5.5, and s_b is 0.181 for all the jobs.

classes of jobs $[i]$	1 CPU	2 I/O disk	3 I/O drum	4 I/O tape	R^i_{min} $[sec]$
	resource $[j]$				
1	0.5	0.1	0.4	0.	8
2	0.4	0.	0.6	0.	12
3	0.1	0.7	0.1	0.1	20

Table 1- Characterization of the jobs in execution at a given instant t by means of their τ_{ij} and R^i_{min} values.

If in another time interval the system state is $\mathbf{N} = \{\, 1,1,10\,\}$, the bottleneck is resource 2; from Eq. (5) we obtain the value 7.1 . The processing speed s_b is 0.1408 for jobs of classes 1 and 3 while the job of class 2 may have a higher processing speed because it does not request service from the bottleneck. To calculate $s_2(t)$ we may act in the following way: determine the capacity of resources 1 and 3, available for the job of class 2, that is, $1 - (\,1.5\times 0.1408\,) = 0.7888$ for resource 1 and $1 - (\,1.4\times 0.1408\,) = 0.8028$ for resource 3; since the fraction of capacity required from the job of class 2 to the more requested resource from that class, i.e., resource 3, is lower than the available capacity ($0.6 < 0.8028$) , $s_2(t)$ may be set to 1. Thus, the vector of the processing speeds is: $\mathbf{s}\,(\,t\,) = \{\, 0.1408, 1., 0.1408\,\}$.

The dynamic behavior of the computer system is known if we may determine at each instant of time the vector $\mathbf{s}(t)$. The egalitarian principle of the system is reflected in the assignment of the fractions of resource capacities to the jobs, according to which no job must be favored or penalized.

In an informal way, we may say that the strategy applied by the system in determining $\mathbf{s}(t)$ is optimal in the sense that it cannot be further improved by increasing the processing speed of an arbitrary class of jobs without decreasing the processing speed of those classes of jobs that are executed more slowly.
Indeed, as seen in the example considered, the vector $\mathbf{s}(t) = \{ s_i(t) = s_b(t) \}_{i=1,\dots,m}$ may be further improved for the jobs that do not request the bottleneck without violating the basic assumptions of the system's operation.
Applying iteratively the operations described in Eqs. (4), (5), (6), and updating at each iteration the values of the available resources capacity, it is possible to determine the hierarchy of the bottlenecks.

A complete description of the mathematical model is given in [1] while an algorithm for the identification of the bottleneck hierarchy is described in [15].

4. Optimization Objectives

The main objective we want to achieve through dynamic load optimization is the *maximization* of the *system's throughput*. This objective may be stated as an optimal control problem that may be exerted at different levels. At a *global level* the problem is:

given M jobs, that may have different characteristics, determine their sequence of execution such that the time T_M required to complete their execution will be minimized;

This problem of global optimal control may be solved, at least from a theoretical point of view, applying the approach of dynamic modeling described in [1] and outlined in the previous section, that requires only macro-level parameters. Even if its analytical solution may be found, [1], the practical applicability of such a global optimal control is severely limited. Major drawbacks are related to the complexity of the analytical solution and to the input data required. The computations required for the solution of the problem may cause a *high overhead* in the system, that would introduce serious perturbations on its functioning. Furthermore, the global optimal control requires the *knowledge* of the job characteristics for *all* the M jobs that will be executed in the observation interval.
The input data of the controller are related on both the jobs already arrived to the system and on the jobs not yet submitted. To be practical, a controller must avoid the high overhead of hunting for the maximum of the global control function; it may utilize supplemental variables whose values indicate the most desirable direction for the adjustment of load to achieve the objective of global optimal control. Instead of the knowledge of the characteristics of *all* the M jobs to be executed in the observation interval we assume to know *only* the characteristics of the jobs already arrived to the system, and eligible for execution, but whose execution has not yet started, i.e., of the M_q jobs in the *system's input queue* Q.

This knowledge may be based on prior executions of the same jobs or on the user declarations in the job-control statements. Thus, we solve the same optimal control problem as stated before but with horizon limited only to the input queue Q. This approach, at a *near-global* level, seems more realistic than that at the global level since it does not require information on jobs that have not yet been submitted to the system. Furthermore, its ability in adapting itself in real time to some unpredictable load fluctuations is high since its "horizon" is limited only on the input queue characteristics.

The minimization of the execution time T_q of the M_q jobs will be achieved by maximizing the utilization factors of *all* the resources through the best match between the resource capacities and the workload demand pattern.

By the definition of the *utilization factor* of a resource we may see that the left-hand side of inequality (6) is the utilization factor of resource j, that we repeat here for convenience:

$$U_j(t) = \sum_{i=1}^{m} N_i(t)\, \tau_{ij}\, s_i(t) \qquad (7)$$

When several jobs share the resources, the sum of all the utilization factors U_j's usually exceeds one due to the parallelism of the resource operations, and the following inequality holds:

$$\sum_{j=1}^{n} U_j(t) \le n \qquad (8)$$

n being the number of the system 's resources. The function

$$P(t) = \sum_{j=1}^{n} U_j(t) \qquad (9)$$

is the *level of parallelism* of the system at instant t , and will be used as our *criterion* for dynamic throughput optimization.

Our objective is the definition of a *job execution sequence* (JS) such that the execution time T_q of the M_q jobs will be minimized, that is, the throughput rate in the interval $[0, T_q]$, will be maximized.

The throughput rate is directly related to the fractions of the resource capacities used, i.e., to the utilization factor of the resources, that will depend on JS . As a consequence, the level of parallelism and the utilization factors are functions of both t and JS . At a near-global level the problem can be stated in the following way:

given M_q jobs in the input queue Q, find a JS such that

$$\max_{JS} \left\{ \frac{\int_0^{T_q} P(t, JS)\, dt}{T_q} \right\} \qquad (10)$$

will be attained.

The solution of this near-global optimal control problem depends, see Eq. (9), on the $U_j(t, JS)$'s. Thus, we will concentrate our efforts on the maximization of the resource utilizations $U_j(t, JS)$. The $U_j(t, JS)$ may be maximized with respect to the $s_i(t)$'s, the τ_{ij}'s, and the $N_i(t)$'s.

The basic idea is that, knowing at each instant the unused fractions of resource capacities, we may find the way for their "best" utilization.
Indeed, from Eq. (5) we may see that it is possible to increase the processing speed of the jobs that do not require services from the bottleneck without decreasing the $s_b(t)$ until the bottleneck does not *switch* to another resource.
In this work we maximize the $U_j(t,JS)$'s acting only on the $N_i(t)$ values. Other possibilities, like that of acting on the execution priorities, are not examined here.
This objective is achieved by starting the execution of those jobs that have resource demand patterns (τ_{ij}' s) which "best" fit the unused fractions of resource capacities, given by $[1 - U_j(t,JS)]$ ($j = 1,\ldots,n$). If among the jobs waiting for execution in the input queue there is one, say job i, that does not require service from the bottleneck, i.e., for which is $V_b^i = 0$, it must be selected for the next execution. In this case, if its requests do not cause a switch of the bottleneck, $s_b(t)$ does not change but $P(t,JS)$ increases since some of the terms $N_i(t)\,\tau_{ij}$ are increased (see Eq. (7)). Thus, the throughput of the system will increase without any increase in the system 's response time.

5. The Scheduler

The assumptions concerning the flow of the jobs in input to the system are the followings. Each job arrived to the system waits in an external queue until it becomes "eligible" for execution, that is, until the resources it requires are available. When it is eligible, a job departs the external queue and joins the *input queue* Q on which the *scheduler* acts.
In our case the scheduler consists of an adaptive scheduling algorithm (AA) which, using measurements concerning the input and the output traffic of jobs, controls the access of jobs to the system resources so as to maximize the given performance criterion. In this paper the performance criterion considered is the maximization of system throughput.

Ruschitzka and Fabry reported in [14] a classification scheme of the scheduling algorithms based on: (1) the decision mode, (2) the priority function, and (3) the arbitration rule. The *decision mode* of the implemented adaptive scheduling algorithm is *nonpreemptive*. The controller is invoked only when a job terminates its execution or when a job in the input queue finds the required resources available. Each time the controller is invoked we solve a near-global problem with a different $\mathbf{M}_q$ (Eq. (10)).
The *priority function* of the AA is a function of both the job characteristics and the system load that tries to maximize the criterion described.
To avoid, or at least to reduce, the instability problems due to the overhead and to the communication delay of the typical feedback control mechanisms, the implemented AA rely only on information at a *macroscopic level*. In other words, it does not use information that must be computed at each instant of time such as the amount of service already received by each job in execution. Indeed, to accomplish its task, the AA must know only the state of the system $\mathbf{N}(t) = \{N_1(t),\ldots,N_m(t)\}$, defined as the number of jobs of each class in execution at time t, and the state of the input queue $\mathbf{M}_q(t) = \{M_{q_1}(t), \cdots, M_{q_m}(t)\}$, defined as the number of jobs of each class in the input queue Q.

Note that the relevant characteristics, in terms of the τ_{ij}'s and the $R^i_{\min}$'s, of the jobs either in execution or in the input queue are assumed to be known or at least estimated and then adjusted during the execution of each job.

When the scheduler is invoked the priority function updates the following variables:

- number of jobs, per class, in execution (the vector $\mathbf{N}(t)$)
- number of jobs, per class, in the input queue (the vector $\mathbf{M}_q(t)$)

and, by Eqs. (4) and (5) applied to $\mathbf{N}(t)$ and $\mathbf{M}_q(t)$, the following operations must be performed:

- computation of the *actual* bottleneck (indicated with the subscript ba) due to the jobs in execution $\mathbf{N}(t)$
- computation of the *forecast* bottleneck (indicated with the subscript bq) that would be generated by the simultaneous execution of all the M_q jobs in the input queue.

In the theoretical case in which no resource interferences are generated by the execution of the M_q jobs, that is, with the *optimal JS*, the minimum execution time T_{theor} is given by:

$$T_{theor} = \sum_{i=1}^{m} M_{q_i} \, \tau_{i,bq} \, R^i_{\min} = \sum_{i=1}^{m} (M_{q_i} \, V^i_{bq} \, / \, \mu_{bq}) \tag{11}$$

The strategy of the AA is that of adjusting the load of the system in such a way that the actual bottleneck coincides as much as possible during the observation interval with the forecast bottleneck. The loading on the resource bottleneck bq must be the minimum sufficient to satisfy Eq. (4) and in the same time the loadings on other resources must be kept at the maximum possible.

The operations performed by the priority function are:

step 1 if $ba \neq bq$, select job i having $\max_i \tau_{i,bq}$, i.e., assign the highest priority to job i

step 2 if $ba \equiv bq$, select job i having $V^i_{bq} = 0$ (if it exists) provided that it does not induce a bottleneck migration

step 2.1 if job i does not exist, select job h having $\min_h \tau_{h,bf}$, provided that it does not induce a bottleneck migration

step 2.2 if job h does not exist, select job k among the remaining having $\min_k \tau_{k,bf}$ provided that it does not induce a bottleneck migration

step 2.3 if job k does not exist, select job z having $\min_z R^z_{\min}$

The priority function of the heuristic AA proposed will produce the best choice for the job to be selected subject to the constraints given by the unmodifiable current load $\mathbf{N}(t)$ and by the unmodifiable input queue Q status $\mathbf{M}_q(t)$. Given an existing loading situation, the AA identifies the job to be selected so that the mean response time will have the minimum increase with respect to its optimal value T_{theor} (see Eq. (11)). The optimal choice is that performed by *step* 2. Indeed, in this case the load on resource *bq* does not change while the load on other resources increases. The choices performed by *steps* 2.1 to 2.3 result in increasing degradations of the system response time with respect to the one obtained with T_{theor}.

Actions performed by *steps* 2.1 and 2.2 tend to minimize the overload on the bottleneck *bq* selecting jobs with $\min_i \tau_{i,bq}$ and, in the same time, increasing more than proportionally the load on other resources since it must always be $\sum_{j=1}^{n} \tau_{ij} = 1 \quad (i = 1,...,m)$. When the $\mathbf{M}_q$ characteristics are such that the previous steps cannot be executed successfully, actions performed by *step* 2.3 tend to minimize the period of time during which the load of the system is out of the optimal condition.

The *arbitration rule* of the AA that must resolve conflicts among jobs with equal highest priority is described by the operations performed in *steps* 2.1,2.2, and 2.3.

6. Evaluation of the Adaptive Algorithm

The computer system has been modeled according to the basic assumptions reported in Sec. 3. The processor-sharing policy is adopted by all the servers, with the constraints previously described. The workload of the model consists of *m* classes of jobs. A detailed description of the simulator used can be found in [1].

The empirical data used in the evaluation were gathered on a Honeywell DPS8/44 (workload W_1) and on a Univac 1100/81 (workload W_2) of two university computing centers. The traces include data regarding the jobs characteristics and their arrival time at the system. The jobs executed have been characterized through a resource-oriented approach, and six classes have been identified in both the workloads W_1 and W_2 by applying the *k-means* clustering algorithm. The resource demands of the jobs of each class have been considered to be the same as those of the centroid of the class. During its execution each job behave according to the assumptions reported in Sec. 2.

The length of each trace was limited to 1000 jobs that we judged sufficient to reproduce the effects of the various scheduling sequences. To evaluate the effectiveness of the AA, the total real time necessary to complete the executions of all the 1000 jobs of each trace was measured. The minimum value of this time, T_{theor}, is given by Eq. (11).

To avoid the interferences on the performance of the AA produced by memory contention, the memory requirements of all jobs have been assumed equal. Different simulation runs were performed for various lengths M_q of the input queue Q seen by the AA. When M_q = 1000 the AA operates in its 'best' condition, i.e., the control is exerted at the global level, while with M_q = 1 the AA operates as a FCFS scheduling algorithm. In order to verify the correctness of the simulator as well as that of the implemented AA, a *test* workload W_t was initially executed. W_t consists of 1000 jobs of two classes C_1 and C_2, whose characteristics were artificially defined to obtain the maximum effectiveness from the AA.

The system model consists of 2 servers and the service requirements of the two classes are $\tau_{1,j} = \{1,0\}_{j=1,2}$ and $\tau_{2,j} = \{0,1\}_{j=1,2}$. The R_{min} of each job is 1 time unit for both classes. The input sequence *JS* of 1000 jobs of W_t has been constructed to simulate the "worst" way to schedule a multiprogramming system, that is, in our case all class-1 jobs followed by all class-2 jobs. Indeed, the system has a bottleneck on resource 1 for the first half of its processing time while resource 2 is not utilized and the dual situation is verified during the second half of its processing time. The multiprogramming level was arbitrarily set equal to eight.

Figure 1 shows the percentage increase of the global execution time T_{W_t} of workload W_t of 1000 jobs with respect to the minimum time T_{theor} for input queue lengths M_q ranging from 1 (FCFS scheduling) to 1000. The execution time increase is $100*(T_{W_t} - T_{theor}) / T_{W_{theor}}$.

T_{theor} was 500 time units. The simulation results have the form expected by the application of the AA. Indeed the global execution time when $M_q \geq 500$ is halved with respect $M_q = 1$ and similarly the throughput is doubled. In Table 2 the resource utilizations and the throughput rates of the two classes of jobs of W_t for $M_q = 1, 10, 500$, and 1000 are reported. Table 3 illustrates the centroid characteristics of the job classes of workload W_1 in terms of resource demand patterns τ_{ij}'s (given by Eq. (2)), and minimum response times R^i_{min}'s

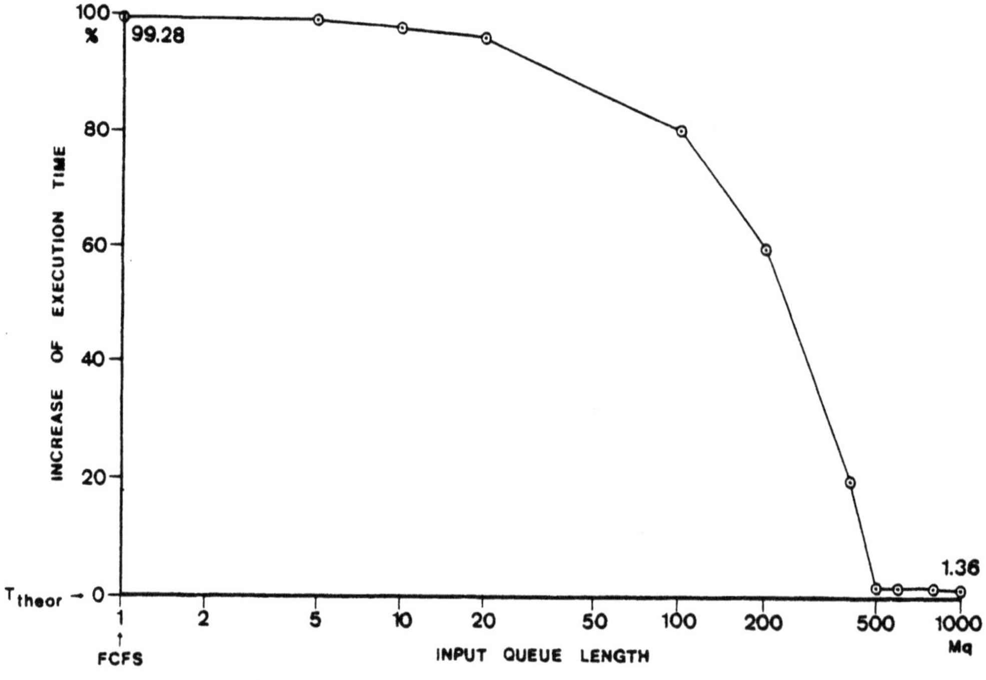

Figure 1 - Increase of the global execution time T_{W_t} of test workload W_t with respect to its minimum theoretical value T_{theor}.

Q lengths	U_1	U_2	X_1	X_2
M_q = 1 (FCFS)	0.5027	0.5028	0.1272	0.1276
M_q = 10	0.5101	0.5101	0.1288	0.1298
M_q = 500	0.9994	0.9992	0.2567	0.2564
M_q = 1000	0.9994	0.9992	0.2567	0.2564

Table 2- Resource utilizations and throughput rates of test workload W_t.

(given by Eq. (1)). The percentage of the 1000 jobs of W_1 each class represents are also reported. Only the three resources more heavily requested in the system have been considered, namely, the CPU and two I/O resources. The multiprogramming level was set to eight to agree with the value obtained from the empirical data. The minimum theoretical execution time (given by Eq. (11)) was T_{theor} = 8619.9 sec. (about 2*h* and 23 min).

In Fig. 2 the increments of T_{W_1} with respect to T_{theor} for the various values of M_q from 1 to 1000 are reported. The resource utilizations corresponding to several M_q values are shown in Table 4. It should be noted that the loads on the resources produced by the execution of W_1 with a FCFS algorithm are unbalanced and generate a bottleneck on resource 3, whose utilization was very high, U_3 = 0.9619 . Clearly, with these characteristics of the load, the improvement that may be achieved is quite limited. However, with M_q = 1000 the utilization of resource 3 is increased up to U_3 = 0.9997 . From Fig. 2 it may be seen that a substantial decrease of T_{W_1} was obtained when M_q changed its value from 1 to 5 . For M_q = 5 the value of T_{W_1} is just 0.81% greater than T_{theor}. The second experiment was concerned with the execution of two different traces collected during corresponding observation intervals of two consecutive days. The basic characteristics of this new workload W_2 , are reported in Table 5. With this workload only two resources, namely, the CPU and an I/O resource, were heavily loaded while the others are very low utilized.

The main simulation results of the two traces, referred as W_2' and W_2'' , are reported in Tables 4, 5 and in Fig. 2. The corresponding T_{theor} was 8826.99sec (about 2*h* and 27 min). For both W_2' and W_2'' the general considerations made above for W_1 on the high utilization, with M_q=1 , of one resource, can be repeated. Indeed, in Table 4 we see that with a FCFS

Class	τ_{ij}			R^i_{min}	Jobs per	X_i [prog/sec]	
(i)	1	2	3	[sec]	class [%]	M_q = 1	M_q = 1000
1	0.192	0.	0.808	9.94	20.3	0.021	0.017
2	0.115	0.084	0.801	28.70	29.6	0.007	0.008
3	0.069	0.906	0.025	5.63	22.1	0.039	0.044
4	0.087	0.850	0.063	26.22	9.2	0.008	0.009
5	0.107	0.893	0.	18.26	2.2	0.024	0.018
6	0.541	0.459	0.	7.40	16.6	0.074	0.053

Table 3- Characteristics of the job classes of workload W_1 and throughput rates obtained by simulation.

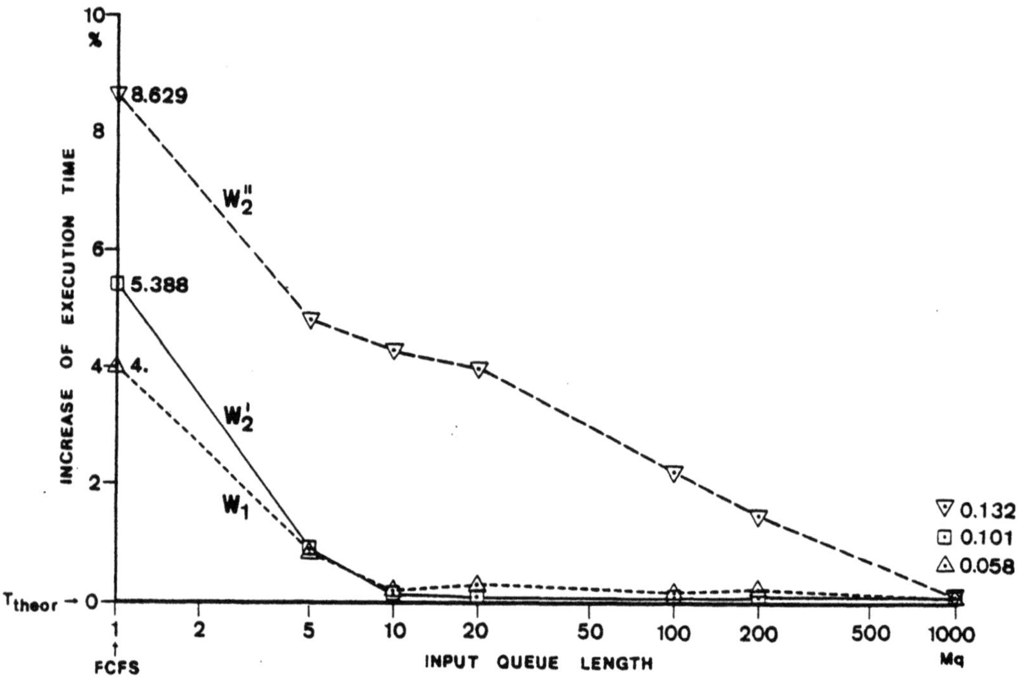

Figure 2 - Increase of the global execution time of workloads W_1, W_2', and W_2'' for various M_q values with respect to T_{theor}.

Q lengths	W_1			W_2'		W_2''	
	U_1	U_2	U_3	U_1	U_2	U_1	U_2
M_q = 1(FCFS)	0.2643	0.5373	0.9619	0.9492	0.8221	0.9209	0.7975
M_q = 10	0.2745	0.5578	0.9986	0.9988	0.8650	0.9592	0.8307
M_q = 1000	0.2747	0.5582	0.9997	0.9994	0.8657	0.9991	0.8654

Table 4- Resource utilizations for workloads W_1, W_2', W_2''.

Classes id.	Resources (γ_{ij}) 1	2	R^i_{min} [sec]	Jobs per class [%]	$X_i\{W_2'\}$ M_q = 1	M_q = 1000	$X_i\{W_2''\}$ M_q = 1	M_q = 1000
1	0.401	0.599	22.2	4.90	0.0101	0.0112	0.0097	0.0112
2	0.425	0.575	4.96	28.5	0.0449	0.0497	0.0441	0.0502
3	0.536	0.464	2.67	10.8	0.0840	0.0925	0.0841	0.0923
4	0.307	0.693	13.7	37.1	0.0165	0.0180	0.0160	0.0170
5	0.764	0.236	45.3	16.4	0.0048	0.0049	0.0046	0.0050
6	0.344	0.656	50.8	2.30	0.0045	0.0047	0.0044	0.0048

Table 5- Characteristics of the job classes of W_2 and throughput rates obtained with two different traces, W_2' and W_2''.

scheduling we have $U_1 = 0.9492$ with W_2' and $U_1 = 0.9209$ with W_2''. In spite of these "bad" initial conditions, the AA with $M_q = 1000$ achieved an reduction of 5.27% for $T_{W_2'}$ and of 8.49% for $T_{W_2''}$ with respect to the corresponding values obtained with FCFS scheduling. The system throughput rate was increased by the same percentages.

7. Summary

An adaptive algorithm (AA) which regulates the load to the system in order to maximize a given performance criterion, i.e., the *system throughput rate*, has been presented. The AA adjusts the load on the system depending on the characteristics of both the actual mix of jobs in execution and the mix of jobs submitted to the system and waiting in input queue.

The evaluation of the AA was performed used traces collected from two real workloads, allowing the algorithm to know the characteristics of a variable number of jobs, that is, with various input queue lengths.

A characterization of the dynamic behavior of the jobs has been introduced in Sec. 2 and the model used in the simulation was briefly described in Sec. 3.

Figure 2 shows the execution time increase over the minimum theoretical execution time T_{theor} of the workloads considered for input queue lengths from 1 to 1000. The 'best' results correspond to $M_q = 1000$. In this case, the optimal control is exerted at a global level and the corresponding mathematical problem has a theoretical optimal solution, [1]. It was shown that the heuristic near-global AA implemented leads to results close to the global AA even for quite limited input queue lengths, e.g., from 5 to 20.

The results of the simulation experiments conducted on two real workloads have proved the ability of the AA to adapt its decisions to the time-varying characteristics of a load.

Acknowledgements

The author would like to thank C. Baiocchi for many stimulating discussions concerning the topic of this paper and G. Balbo for the helpful comments on an earlier draft of the paper.

References

[1] Baiocchi, C., Capelo, A., Comincioli, V., and Serazzi, G. A mathematical model for transient analysis of computer systems. *Performance Evaluation* (to appear), 1983.

[2] Badel, M., Gelenbe, E., Leroudier, J., and Potier, D. Adaptive optimization of a time-sharing system's performance. *Proc. IEEE (Special Issue on Interactive Computer Systems)*, vol. 63, 924-939, June 1975.

[3] Bhat, U.N., and Nance, R.E. An evaluation of CPU efficiency under dynamic quantum allocation. *J. ACM* 26 , 4, 761-778, Oct. 1979.

[4] Blevins, P.R., and Ramamoorthy, C.V. Aspects of a dynamically adaptive operating system. *IEEE Trans. on Comp.* , C-25, 713-725, July 1976.

[5] Denning, P.J. The working set model for program behavior. *Comm. ACM* 11 , 5, 323-333, May 1968.

[6] Denning, P.J., and Kahn, K.C. A study for program locality and lifetime function. *Proc. 5th ACM Symp. on Operating System Principles* , 207-216, Nov. 1975.

[7] Denning, P.J., and Kahn, K.C. An L=S criterion for optimal multiprogramming. *ACM SIGMETRICS* , Cambridge, MA, 219-229, March 1976.

[8] Denning, P.J., Kahn, K.C., Leroudier, J., and Suri, R. Optimal multiprogramming. *Acta Informatica* , 7, 197-216, 1976.

[9] Denning, P.J., and Schwartz, S.C. Properties of the working set model. *Comm. ACM* 15 , 3, 191-198, March 1972.

[10] Jain, R.K. Control-theoretic approach to computer systems performance improvement. *Proc. CPEUG XIV* , Boston, 93-100, 1978.

[11] Kleinrock, L. *Queueing Systems* , *Volume II* ; *Computer Applications*, J. Wiley, 1976.

[12] Leroudier, J., and Potier, D. Principles of optimality for multiprogramming. *ACM SIGMETRICS* , Cambridge, MA, 211-218, March 1976.

[13] Potier, D., Gelenbe, E., and Lenfant, J. Adaptive allocation of central processing unit quanta. *J. ACM* 23 , 1, 97-102, Jan. 1976.

[14] Ruschitzka, M., and Fabry, R.S. A unifying approach to scheduling. *Comm. ACM* 20 , 7, 469-477, 1977.

[15] Serazzi, G. The dynamic behavior of computer systems. in Ferrari, D. and Spadoni, M. (eds.): *Experimental Computer Performance and Evaluation*, North Holland, 127-163, 1981.

[16] Sherman, S., Baskett III, F., and Browne, J.C. Trace-driven modeling and analysis of CPU scheduling in a multiprogramming system. *Comm. ACM* 15 , 12, 1063-1069, 1972.

RESULTS FOR DUAL RESOURCE QUEUES

A.R. Unwin
Trinity College Dublin

ABSTRACT

In dual resource queues jobs not only require the resource of a server but some other resource as well. The amount of the second resource required is variable. Computer jobs requiring both CPU and core are an example. This paper derives stability conditions for these models and properties of their solutions when service times are exponential.

1. Introduction

Models with product-form solutions have been applied successfully many times. Nevertheless there are features of computer systems which they do not model well. An example of this is a queue where jobs require two resources at the same time, such as at the CPU where a job must be in core to receive service. If it can be assumed that the computer's memory is infinite, the resource of core is irrelevant to performance. If not, then using a product-form model could lead to large errors.

There has been research on limited memory models, mostly concerned with particular computer installations. Brown et al. |1| studied the CDC6400. They basically used a decomposition and approximation approach and assumed there was always a queue for memory, Brandwajn |2| and Bard |3| are others who have modelled computer systems with the limited memory size restriction and who have found approximate solutions for their models. More recent work is reported in Sauer |5|.

It is important to state that a restriction on memory size does not necessarily have the same effect as a restriction on multiprogramming level. Jacobson et al. |6| have described an iterative method of solution for systems with a fixed number of memory partitions. Their method would not apply to a model in which jobs require different amounts of core and the restriction is not on the number of jobs but on the sum of their memory requirements.

A queue where jobs require not only a server but a variable amount of another resource as well will be called a dual resource queue in this paper. The terms "simultaneous resource possession" (Jacobson et al. |6|) and "passive resources" (Keller |7|) have both been used for models in which all jobs require the same amount of the second resource.

Rather than looking for approximate solutions for networks which include dual resource queues this paper considers exact properties of dual resource queues in isolation. This shows more clearly the effects of a restriction on memory size and should be useful for suggesting and evaluating approximations.

Dual resource queues may also arise in satellite communications. Messages require different amounts of bandwidth for different amounts of time (Aein |8|). If insufficient bandwidth is available for an incoming message it cannot be sent and is lost. Unlike the corresponding computer system model there is therefore no queueing. These 'loss' models have been extensively analysed in |9|.

2. Example of a Dual Resource Queue

The simplest dual resource model may be described as a two-server model. There are two classes of customer, one class needs one server (and either one of the two will do) while the other class needs both servers. If the queue discipline is first-come-first-served the situation can arise where one server is idle although jobs requiring just one server are in the queue. This happens when only one server is busy and a job requiring both servers is first in the queue. Diagrams showing possible states are shown below. Clearly the servers are not being used to their full potential and this is what makes models of this kind interesting.

When this model is used in the remainder of the paper it will be assumed that the rate of service is $i\mu$ when i jobs are in service.

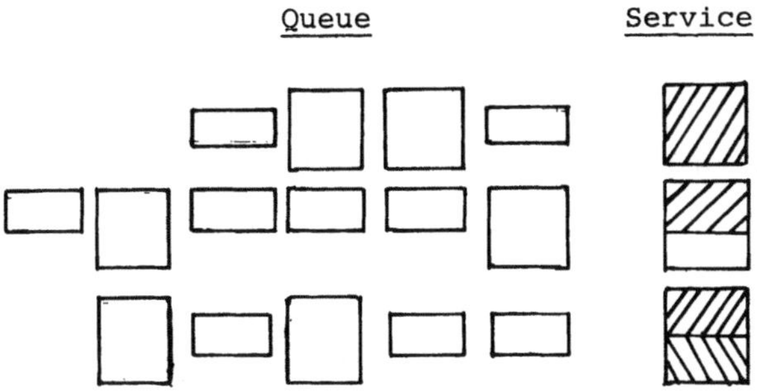

Figure 1

3. Structure of Dual Resource Queues

There are N job classes. Jobs of class i are all of size i (in the sense that they require i units of service capacity) and arrive at the service system in a Poisson stream of rate $\lambda_i = \lambda z_i$. The N arrival streams are independent and the total arrival rate is

$$\lambda = \sum_{i=1}^{N} \lambda_i$$

The arrivals form one queue which is served in FCFS order. The service system has a capacity N units. It can serve any combination of jobs whose total capacity requirement does not exceed N. Its service rate may differ for each combination but the probability distribution of the time to the next departure is always exponential. Let the number of service states (possible combinations of jobs) be m. The states of the model will be of the form

$$(n,i) \quad n \geq 1 \quad 1 \leq i \leq m$$

where n is the number of jobs present in the system and i specifies the service state. There is also the empty state, 0. If there are not job classes of every size there will be fewer service states and there may be more states of the model like the empty state which cannot have a queue.

The difficulties in solving any dual resource model are directly related to the number of service states, that is the number of different combinations of jobs that may be in service at any one time.

For N=2 the number of service states is 3. For N=3 it would be 6. For N=6 it would be 29. For general N it is the sum of the numbers of partitions of 1, 2, 3, ... and N. This assumes that there are job classes of all N possible sizes, which is unlikely. For instance, with N=6 there might only be classes of sizes 1, 2 and 4 and then the number of service states would be 16. This excludes service states such as one of class one being served on its own because there could never be a queue with that service state. No matter what class of job was at the head of the queue it could always be accepted for service. This was not the case for the model with N=6 and job classes of all sizes. If a job of size 6 were at the head of the queue it could not be accepted for service even if only 1 of the 6 units of service capacity were in use. If the queue discipline were not FCFS but some pre-emptive discipline there would not be so many service states. On the other hand much more information would have to be given about the state of the queue than merely the number present.

Omahen and Marathe |10| studied a wide range of queue disciplines for the N=2 model given as an example earlier. Their results suggest interesting possibilities but many of the results would not generalise to other dual resource models as with N=2 there are never jobs of both classes in service together.

The more customers vary in their requirements the more it seems likely that it would be worthwhile introducing special queue disciplines. Special disciplines often mean extra overheads and the cost of running the system in terms of time and money increases. The advantages of sophisticated and costly disciplines have to be assessed in relation to the performance of the system under the simple first-come-first-served discipline.

Dynamic scheduling disciplines which try to get the best fit of customers at any one time would have most potential. This is recognised by Omahen |11| in his analysis of the optimal throughput of multi-resource systems (i.e. systems where customers may require several resources simultaneously). He shows how to formulate the problem of determining a system's optimal throughput as a linear program. He suggests that scheduling might be performed by restricting the combination of jobs worked on to optimal patterns given by the LP solution.

In a dual resource model the service system serves all jobs which fit into its capacity simultaneously. The rate at which it serves jobs may vary according to which other jobs are present, so that a job may be served at several different rates during its period in service. As exponential distributions are assumed these complications raise no particular analytic difficulties.

The range of service models included in this formulation is large. All jobs may be served at the same fixed rate independently of the number of jobs present. Jobs of different classes may be served at different rates. For particular combinations of jobs all the server's efforts may be directed to one class only. Although the queue itself is FCFS, all manner of priorities may be introduced into the service system.

4. Model Equations

The steady-state equations for a dual resource queue have the following general form for $n>N$:

$$(\lambda+\mu_i)p_{ni} = \lambda p_{n-1i} + \sum_{j=1}^{m} \mu_j \gamma_{ji} p_{n+1j} \qquad (4.1)$$

p_{ni} = probability of state (n,i)

μ_i = rate of service in state i

γ_{ji} = probability that a transition from service state j leads to service state i when there are queues present.

A typical γ_{ji} term would be the product of the probability that a job of a certain size leaves state j, the probability of other jobs being at the front of the queue and entering to change the state to i, and the probability that the job left at the head of the queue is too large to enter.

$\Gamma = (\gamma_{ij})$ is, of course, a stochastic matrix.

For N=2 (our earlier example) $\Gamma = \begin{pmatrix} z_2 & z_1 z_2 & z_1^2 \\ 1 & 0 & 0 \\ 0 & z_2 & z_1 \end{pmatrix}$

For $n \leq N$ the equations are similar but do not satisfy a general form. This is because some jobs may enter service immediately on arrival for states with $n<N$.

The structure of the model is best seen by initially ignoring the equations for $n \leq N$ and defining m generating functions

$$P_i(s) = \sum_{n=N+1}^{\infty} p_{ni} s^n \qquad 1 \leq i \leq m$$

After multiplying each balance equation by s^{n+1} and adding over n we may write the equations as:

$$\underline{p}'A = \underline{b}' \tag{4.2}$$

where $\underline{p}' = (P_1(s), \ldots, P_m(s))$

$$A = -\lambda I s^2 + (\lambda + \underline{\mu}')Is - \underline{\mu}'\Gamma$$

$$\underline{\mu}' = (\mu_1, \ldots, \mu_m)$$

$$\underline{b}' = s^{N+1}(\lambda s \underline{p}'_N - \underline{p}'_{N+1}(\mu'\Gamma))$$

$$\underline{p}'_n = (p_{n1}, \ldots, p_{nm})$$

Using Cramer's rule the solution of (4.2) is

$$p_i(s) = \frac{|A_i|}{|A|} \tag{4.3}$$

where $\underline{b}'$ replaces the ith row of A in A_i.

Before considering what can be done with (4.3) we shall derive the stability condition under which a steady-state solution is possible.

5. Stability Condition

The denominator of (4.3) is a polynomial of degree 2m. One of the roots of the polynomial is s=1, as can easily be seen by summing the rows of $|A|$:

$$\sum_{j=1}^{m} a_{ij} = \mu_i(s-1)-\lambda s(s-1) \qquad i=1, \ldots, m$$

Define the matrix $A^*(s)$ by $A^*(s)_{i1} = \mu_i-\lambda s$, $A^*(s)_{ij} = A(s)_{ij}$ $j\neq 1$
$(s-1)|A^*(s)| = |A(s)|$

Theorem 1

The stability condition for the model is given by

$$|A^*(1)| > 0$$

and may be written

$$\lambda < \frac{|F|}{|G|}$$

where the matrices F and G have elements

$$F_{i1} = 1 \quad G_{i1} = \mu_i^{-1}$$

$$F_{jj} = G_{jj} = 1-\gamma_{jj} \qquad j\neq 1$$

$$F_{ij} = G_{ij} = -\gamma_{ij} \qquad i\neq j,\ j\neq 1$$

For the N=2 example this gives the following expression for the stability condition

$$\lambda < \frac{\begin{vmatrix} 1 & -z_1 z_2 & -z_1^2 \\ 1 & 1 & 0 \\ 1 & -z_2 & 1-z_1 \end{vmatrix}}{\begin{vmatrix} \frac{1}{\mu} & -z_1 z_2 & -z_1^2 \\ \frac{1}{\mu} & 1 & 0 \\ \frac{1}{2\mu} & -z_2 & 1-z_1 \end{vmatrix}} = \mu \frac{1}{1-\frac{1}{2}z_1^2}$$

Proof:

To find the stability condition we find the output rate, λ', of the system when there is always an infinite queue of jobs to be served, (Lavenberg |12|). The stability condition is then

$$\lambda < \lambda'.$$

The state of the queue can be ignored if it is always infinite and only the m service states need be considered. Transition points between the m states form a Markov Chain whose matrix Γ has elements γ_{ij}. Let the equilibrium distribution of Γ be

$$\pi_i \qquad 1 \leq i \leq m.$$

Since each transition leads to one departure, the average time between departures is

$$\Sigma \frac{\pi_i}{\mu_i}$$

and the output rate is

$$\lambda' = (\Sigma \frac{\pi_i}{\mu_i})^{-1} \tag{5.1}$$

By definition

$$\pi' \Gamma = \pi \tag{5.2}$$

and $\Sigma \pi_i = 1$ (5.3)

(5.2) is equivalent to

$$\pi'(I-\Gamma) = 0$$

One of these equations is redundant so replace the first one by (5.3). Solving for π_i:

$$\pi_i = |F|^{-1} |G_i|$$

where F is defined in the statement of the theorem and G_i = G for all but its first column:

$$(G_i)_{j1} = \delta_{ij}$$

Using (5.1) we find

$$\lambda' = \frac{|F|}{|G|}$$

and so the stability condition is

$$\lambda < \frac{|F|}{|G|}$$ as required.

To prove the first part of the theorem recall that

$$A^*(s)_{i1} = \mu_i - \lambda s$$

and $A^*(s)_{ij} = A(s)_{ij} \qquad j \neq 1$

Setting s=1 removes all λ coefficients barring those in the first column and we may write:

$$|A^*(1)| = (\sum_{j=1}^{m} \mu_j)(|F| - \lambda |G|)$$

after extracting the service rates, μ_j, from each row. It is now obvious that

$$|A^*(1)| > 0 \iff \lambda < \frac{|F|}{|G|}$$

For one important class of models we may write down the stability condition in a more specific form.

Theorem 2

If the service system of the model is such that all jobs in service at any one time get equal attention from the server, the stability condition is

$$\lambda < \left(\Sigma \frac{\pi_j}{\mu_j}\right)^{-1}$$

where μ_j is the overall service rate in state j and

$$\pi_j = \frac{\left(\sum_{k=1}^{N} i_k\right)!}{\prod_{k=1}^{N} (i_k!)} \left(\prod_{k=1}^{N} z_k^{\;i_k}\right) \sum_{\ell=X}^{N} z_\ell$$

where i_k = the number of class k jobs present in state j and

$$X = N+1 - \sum_{k=1}^{N} k i_k$$

$\sum_{\ell=X}^{N} z_\ell$ = proportion of jobs which are too large to fit into the space capacity of state j.

A related result has been proved by Schoute |13| by substitution. The following proof is given because it emphasises the combinatorial nature of the models.

Proof:

(i) Suppose there are no jobs in service but an infinite queue is waiting. When the server decides to start work he lets in as many jobs from the front of the queue as he can. The resulting initial state is of the form

$$(i_1 \ldots i_k \ldots i_N)\ell$$

where ℓ is the size of the job now at the front of the queue and the probability distribution over these states is given by

$$P|(i_1 \ldots i_j \ldots i_N)\ell| = \frac{\left(\sum_{j=1}^{N} i_j\right)!}{\prod_{j=1}^{N} (i_j)!} \left(\prod_{j=1}^{N} z_j^{\,i_j}\right) z_\ell \tag{5.4}$$

because P(state S) = P(job sizes in S) X number of different orderings of that set of job sizes.

(ii) Suppose the model is like (i) but the server refuses to serve whoever is first in the queue. He must leave. The probability distribution of the initial state after this is obviously the same as in (i).

(iii) Suppose the model is like (ii) with one difference. Instead of refusing to serve the first in the queue, the server waits until he has let in as many jobs as possible before picking one of those at random and refusing to serve him. That one must leave, other jobs are then let in depending on how much service space is now free and on the order of the queue. As the ordering of jobs accepted for service is irrelevant (iii) is equivalent to (ii) and the resulting initial state probability distribution is the same. So (i), (ii) and (iii) all have the same distribution.

The transition from (i) to (iii), removing a job at random and letting in as many new jobs as possible, may be repeated. Clearly this will give the same state probability distribution and this is a stationary distribution for transitions of that kind.

The transition probabilities are only dependent on the current state so that the states $|(i_1 \ldots i_N)\ell|$ form a Markov Chain under that transition rule. As all states communicate the chain is irreducible. All states are positive recurrent (for the chain is finite) and therefore the stationary distribution given in (5.4) is the unique equilibrium distribution. Returning to the original service system whose probabilities are to be found, the assumption that all jobs in service get equal attention ensures that at a departure the job chosen to depart is picked at random from those being served. Hence the probability distribution at departures must be the same as that for the Markov Chain.

To obtain the form given in the theorem, sum over all possible job sizes ℓ that could be first in the queue when the jobs in service are given by $(i_1 \ldots i_j \ldots i_N)$.

The formula for π_j is like a multinomial based on the numbers of different classes of jobs in service. It differs in its final factor, $\sum_{\ell=X}^{N} z_\ell$, which relates to the job at the head of the group. If state j uses up all but 3 units of capacity the job at the head of the queue must be of size 4 or larger. If state j uses up all available capacity any job can be first in the queue and then X=1 and $\sum_{\ell=X}^{N} z_\ell = 1$.

The key assumption for this theorem is that if 10 jobs are present each gets $\frac{1}{10}$th of the server. The rate of service in the different states can vary as much as before, only how it is distributed between the jobs in service is restricted.

To get an idea of the importance of the memory size restriction consider again the simple two class model given as an example. Instead of specifying memory size to be 2 let it be N. Theorem 2 enables us to write down a formula for the stability condition for general N for this model but it is more revealing to consider the stability limit as a percentage of the optimal throughput possible as N increases. Optimal throughput is found by discarding the FCFS queueing discipline assumption. For the two class model it is

$\mu\frac{N-1}{2z_2}$ (N odd and $z_1 < \frac{2z_2}{N-1}$)

$\mu\frac{N}{2z_2+z_1}$ (otherwise)

With $\mu = 1$, $z_1 = 0.2$, $Z_2 = 0.8$ the following results are obtained:

N	Stability Limit on λ	% of Optimal Throughput
2	1.02	92
3	1.22	98
4	2.07	93
5	2.40	96
6	3.15	94
7	3.56	95
8	4.23	95
9	4.70	94
10	5.33	96

Table 1

The lower limits for odd values of N arise because class 2 jobs do not fit so well in such cases. As N increases the optimal throughput increases by more than the stability limit for N odd and by less than the stability limit for N even. These changes are also attributable to the fitting effect. However, once N exceeds $(1 + \frac{2z_2}{z_1})$, 9 for the model here, the % increases steadily for all N as the fitting effect becomes progressively less important. It is clear that by N=10 the absolute difference between the exact limit and the approximation is tiny as we would anticipate. For lower N the differences may also seem small but it is useful to bear the next graph in mind before jumping to conclusions. This shows how a slight difference in stability conditions can mean a very large difference in expected waiting time if the arrival rate is close to the limit.

Both systems A and B are M/M/1 queues but the service rate in B is faster so that the stability limit is higher. Other queueing models manifest similar behaviour for high arrival rates.

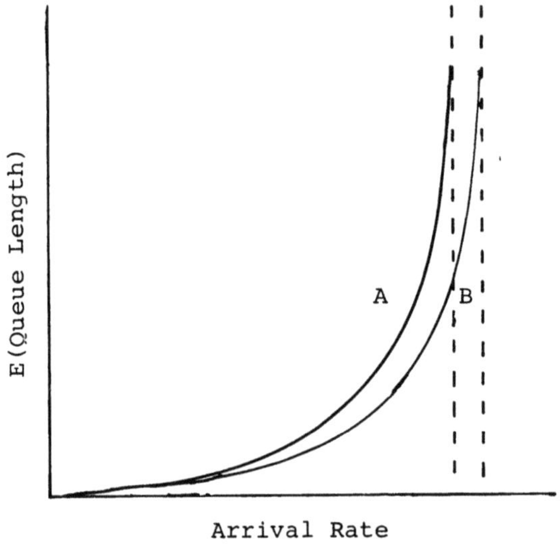

Figure 2

6. Solution of Dual Resource Queues

The solution given at the end of section 4 is incomplete because the vector b contains unknown probabilities. These can be found numerically using analytic properties of the generating function which is an infinite polynomial whose coefficients are probabilities. The generating function must be finite for all points within the unit circle ($|s|\leq 1$) and the denominator in equation (4.3) has zeros in this region. For the function to remain finite such zeros of the denominator must be zeros of the numerator and this gives a set of equations for the unknown probabilities in b. This is an approach which has been suggested for many other underdetermined queueing models. It is known to work in practice although no complete theoretical justification appears to have been published. Most articles prove that the number of zeros equals the number of unknowns but do not prove that the equations resulting from setting the numerator equal to zero are sufficient for determining the unknowns.

Counting the number of zeros of an arbitrary polynomial within the unit circle can be difficult. The following heuristic argument uses the queueing structure of the model to show that the denominator polynomials for dual resource queues have a simple property: exactly half of the zeros lie outside the unit circle.

The models being discussed may have unlimited queues. It is instructive to restrict the number permitted to join the queue and we shall examine models where the number in the system is limited to M. Any jobs arising when M are already present are lost. The balance equations for a finite system of this kind will be the same as for the infinite system up to n=M-1. The equations for n=M will be

$$\mu_i p_{Mi} = \lambda p_{M-1i} \qquad 1 \leq i \leq M$$

This assumes that M is large enough so that all service states i are possible when M-1 are present. If service capacity were 1,000 and jobs of size 1 were possible, M would have to be bigger than 1,000. Otherwise the equations at n=M would take a different form.

The solution to the finite model is

$$W_i(s) = \frac{|A_i| - \lambda s^{M+1}(s-1)\,|B_i|}{|A|} \qquad (6.1)$$

where

$$W_i(s) = \sum_{n=N+1}^{M} p_{ni} s^n,$$

$|A_i|$ and $|A|$ are as in (4.3) and $|B_i|$ equals $|A|$ but for the ith row which is replaced by $\underline{p}'_M$, the vector with components p_{MJ}.

Equation (6.1) is derived in the same way as (4.3). The right-hand-side vector, $\underline{b}$, in the unlimited system has become $\underline{b} - \lambda s^{M+1}(s-1)\underline{p}_M$ in the finite system. Otherwise there is no change.

Unlike $P_i(s)$, which is an infinite order polynomial, $W_i(s)$ has a highest power M. This means that it must be finite for all values of s in the complex plane, not just for values within the unit circle.

Hence all zeros of $|A|$ must be zeros of $|A_i| - \lambda s^{M+1}(s-1)|B_i|$. If these zeros are to provide sufficient equations to solve the finite system there must be exactly m more than for the infinite system because there are m more unknowns (the p_{Mj}'s). The extra zeros are those outside the unit circle and so this reasoning suggests that $|A(s)| = 0$ has m roots outside the unit circle. From its definition, $|A(s)|$ is a polynomial of order 2m and we have already shown that s=1 is always a single root. There are therefore m-1 other roots within the unit circle. For dual resource models it is common that one of these roots is s=0. It is not always a root (consider M/M/1 as a dual resource model with only one class) and sometimes it is a multiple root (when there are different job classes of the same job size). The other roots within the unit circle may be real or complex.

Using the roots within the unit circle the infinite queue model may be solved as those zeros of the denominator are also zeros of the numerator. The resulting set of equations eliminates the unknowns in the numerator. For the two class model with N=2 the problem reduces to finding the root within the unit circle of a quartic.

Finding roots of polynomials, particularly if they are close together, is not easy, so it is useful to observe that the roots themselves are not what we want. To solve our system we need one of the two polynomials of order m which multiplied together give $|A(s)|$. One of the polynomials has all its roots outside the unit circle, the other has all but one of its roots within the unit circle and the last root equal to 1. Knowing one polynomial we could obviously get the other. To eliminate our numerator unknowns we would divide the polynomial with roots inside the unit circle into the numerator and use the fact that there must be no remainder to obtain sufficient equations to solve for the unknowns. A related idea is used by Grassmann and Chaudhry |15| in their solution of some bulk queueing models.

To obtain an expression for R(s), the P.G.F. of the number in the system, we need to take account not only of the generating functions $\{P_i(s)\}$ but also of the probabilities of states not included in the generating functions. Letting their contribution to R(s) be r(s) gives

$$R(s) = \sum_{i=1}^{m} P_i(s) + r(s)$$

$$= P(s) + r(s)$$

For the N = 2 example:

$$r(s) = p_o + (p_{12}+p_{13})s + (p_{21}+p_{22}+p_{23})s^2$$

The unknown probabilities in r(s) can be found by using the balance equations not included in (4.2).

P(s) may be obtained from the set of equations

$$P_i(s) = \frac{|A_i(s)|}{|A(s)|} \qquad 1 \leq i \leq m$$

$$\text{as } P(s) = \frac{|\bar{A}(s)|}{|A(s)|}$$

where $\bar{a}_{i1} = b_i$ and $\bar{a}_{ij} = a_{ij} - a_{i1}$ $j \neq 1$ and the unknowns in $\underline{b}$ are found as described above.

The form of the solution is interesting. After removing the unknowns, the denominator is an mth degree polynomial and may be written as

$$C \prod_{j=1}^{m} (1-\omega_j s)$$

where (ω_j^{-1}) are the roots of $|A(s)| = 0$ outside the unit circle. The denominator is the same for all generating functions, $\{P_i(s)\}$, and for any combination of them such as $P(s) = \sum_{i=1}^{m} P_i(s)$. If we were dealing with P(s), say, we could expand the solution in partial fractions. Provided that there were no multiple roots we would have a sum of m geometric generating functions. This suggests the name 'summation-form' for the shape of the solution in contrast to 'product-form', the name given to solutions which are products of geometric factors. The most important difference between summation-form and product-form is that the factors of the latter are simple functions of the system's parameters. With summation-form solutions the factors, $\{\omega_i\}$, are by no means simple functions of the parameters. There is a way in which we can describe the solutions to dual resource models as 'product-form', although without the advantage of simple factors.

Theorem 3

For n>N the dual resource model's solution may be written

$$\underline{p}_n = R\underline{p}_{n-1}$$

where $\underline{p}_n$ is a vector with components p_{ni} and R is an mxm matrix. The proof is straightforward.

The result of this theorem is the starting point of another approach which may be used to solve dual resource models. Instead of finding roots of polynomials the idea is to find the matrix R directly by an iterative process. Neuts has described the application of this method to a variety of queueing models in a series of papers and a book |16|. He does not appear to have discussed dual resource models but they could be solved by a variant of his approach.

7. Comments

The difficulty of solving dual resource queues in isolation confirms the need for approximations in network applications. Table 1 illustrates how model properties may vary with changed resource limits. In certain circumstances changing the job size distribution may have more effect on throughput than increasing resource limits.

The usual approach to underdetermined queueing models for which roots of polynomials have to be found includes the application of Rouché's theorem to confirm that there are the required number of roots in the unit circle. In this paper the close connections between infinite and finite models have been used to sidestep Rouché. It is satisfying to exploit queueing properties of the model rather than analytic ones and similar arguments would apply to any other model in which only one queue can become infinite.

Throughout this paper all service and interarrival distributions have been assumed to be exponential. Models with distributions which are combinations of exponentials could theoretically also be solved but the state space and consequently the scale of the problem would be increased enormously. Models in which all jobs require the same deterministic service time have been solved |9| and will be the subject of another paper.

The stability condition formula is not affected if the Poisson assumption is dropped provided that the arrival process is still a recurrent renewal process. What happens to the stability condition when service time assumptions are relaxed is unclear. It seems plausible that the condition should be stricter for service distributions with greater variance but even for the very simplest dual resource model used as an example in earlier sections this has not been shown for general distributions.

REFERENCES

|1| Brown, R.M., Browne, J.C. and Chandy, K.M., *Memory Management and Response Time*, Comm ACM, 1977 20, p. 153-166.

|2| Brandwajn, A., *A Queueing Model of Multiprogrammed Computer Systems under Full Load Conditions*, JACM 1977 24, p. 222-240.

|3| Bard, Y., *The Modelling of some Scheduling Strategies for an Interactive Computer System*, p. 113-137 of Chandy and Reiser |4|.

|4| Chandy, K.M. and Reiser, M. (eds), *Computer Performance*, North Holland, 1977.

|5| Sauer, C., *Approximate Solution of Queueing Networks with Simultaneous Resource Possession*, IBM Journal of R & D 1981, p. 894-903.

|6| Jacobson, P.A. and Lazowska, E.D., *Analysing Queueing Networks with Simultaneous Resource Possession*, CACM 1982, 25 p. 142-151.

|7| Keller, T.W., *Computer Systèms Models with Passive Resources*, Ph.D. Thesis, University of Texas, 1976.

|8| Aein, J.M., *A Multi-User-Class, Blocked-Calls-Cleared, Demand Access Model*, IEEE Trans. Comm. 1978 26, p. 378-385.

|9| Unwin, A.R., *Dual Resource Queueing Models*, Ph.D. Thesis, Trinity College Dublin, 1982.

|10| Omahen, K.J. and Marathe, V., *A Queueing Model for a Multiprocessor System with Partitioned Memory*, Tech. Report CSD-TR 132 Purdue, U. 1975.

|11| Omahen, K.J. *Capacity Bounds for Multiresource Queues*, JACM 1977 24, p. 646-663.

|12| Lavenberg, S.S., *Stability and Maximum Departure Rate of Certain Open Queueing Networks having Finite Capacity Constraints*, R.A.I.R.O. Informatique 1978 12, p. 353-370.

|13| Schoute, A.L., *Comparison of Global Memory Management Strategies in Visual Memory Systems with Two Classes of Processes*, p. 389-414 of Gelenbe |14|.

|14| Gelenbe, E. (ed.), *Modelling and Performance Evaluation of Computer Systems*, North Holland, 1976.

|15| Grassmann, W.K. and Chaudhry, M.L., *A New Method to Solve Steady State Queueing Equations*, Naval Research Logistics Quarterly 1982, 29, 461-473.

|16| Neuts, M.F., *Matrix-Geometric Solutions in Stochastic Models - An Algorithmic Approach*, The John Hopkins University Press, Baltimore 1981.

EXACT ANALYSIS OF A PRIORITY QUEUE WITH FINITE SOURCE

Michel VERAN

Centre de Recherches CII HONEYWELL BULL
BP 53X / 38041 GRENOBLE Cedex / FRANCE

ABSTRACT

This paper presents a method to evaluate the exact stationary solution of a single queue with preemptive-resume priorities and finite source. In the case of an infinite source (Poisson arrivals) the solution is given by a well-known simple formula. Models requiring finite customer populations are frequently used for computer performance evaluation (i.e. a processor shared by real time applications), for which the infinite source hypothesis may cause unacceptable errors.

Such a model has already been fully analysed by Jaiswal, but this may yield to very complex computations when applied to real case problems. The same limitation occurs when using a numerical approach. Various approximate solutions have also been proposed.

An exact expression of the utilization factor for each customer population is first proved. A simple iterative algorithm is derived and a PASCAL implementation is described. Several examples are presented.

1 - INTRODUCTION

We consider a simple closed multi-class queueing model, constitued by a service station (with a single server and a preemptive-resume scheduling discipline), connected to an infinite server (delay).

There are R customer classes, with different priority levels, strictly decreasing from class 1 to class R. Each class contains a finite number of customers (N_r for class r).

A customer is either in the service station (waiting or in service), or is in the external delay. The service of a class r customer is interrupted upon arrival of a class s>r customer. This service will be resumed when no higher priority customer is in the station.

Service times and external delays are assumed to be independant and exponentially distributed.

The model parameters are summurized as follows :

- R : number of classes,
- N_r : number of class r customers,
- $Z_r=1/\lambda_r$: mean external delay for class r,
- $S_r=1/\mu_r$: mean service time for class r.

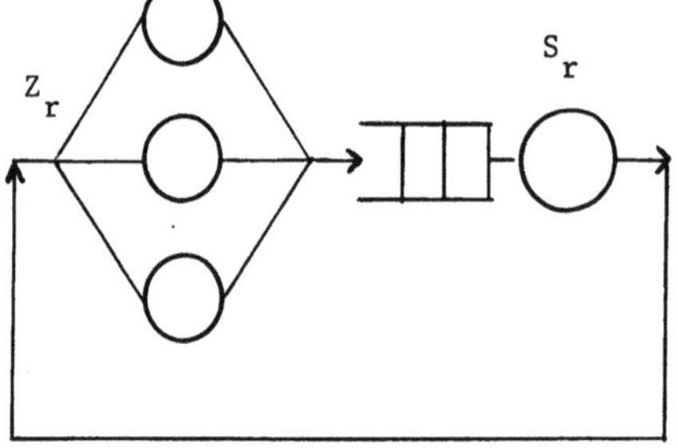

In the case of Poisson arrivals (i.e. when $N_r \to \infty$ and $\lambda_r \to 0$, with $N_r \cdot \lambda_r \to \lambda_r^*$ finite) and exponential services, we have the well known expressions for steady state :

. utilization of the server for class r :

$$U_r = \frac{\lambda_r^*}{\mu_r} \qquad \text{with} \quad \sum_{r=1}^{R} U_r < 1$$

. mean response time for class r (waiting time + service time) :

$$R_r = \frac{1}{1 - U^{[r-1]}} \left(S_r + \frac{\sum_{s=1}^{r} U_s \cdot S_s}{1 - U^{[r]}} \right)$$

with

$$U^{[r]} = \sum_{s=1}^{r} U_s$$

(sum of the utilization factors of classes having priority greater or equal to class r)

The use of such an infinite source model may induce important errors when the classes contain few customers (for example a multiprogram ming system for which the influence of program priorities is to be studied).

In the finite source case, the exact expression of utilization U_r may be obtained using Jaiswal's general results [1], but in practice this approach is quickly limited by complexity when the number of classes and customers is increased.

More recently We-Min Chow [10] proposed an exact analysis of the M/M/1 priority queues system with state dependent arrival and service rates. This study includes the finite source case, but is limited to 2 classes.

For general applications the finite source model is solvable :

- either by numerical resolution of the global balance equations (this facility is provided by some queueing network solution packages [2] [3]). This method is drastically limited by the state number (equal to $(N_1 + 1) . (N_2 + 1) \dots (N_R + 1)$). Some difficult convergence problem may also be encountered when very different service times are to be handled (see example 4.2);

- or by an approximate method [4] [5] [6], whose accuracy is often difficult to estimate.

We describe hereafter an exact mathematical analysis of the finite source model under exponential assumptions. This analysis is based on the underlying Markov chain and yields a simple recursive formula for the utilization factors.

The particular case with only one customer in each class is first considered (§2), then generalized to any customer population. An algorithm is described (§3) which provides an efficient way to calculate the utilization factors. Response times and throughputs are easily derived using Little's formula. It is worth noting that this method avoids having to evaluate the equilibrium state probabilities.

A PASCAL implementation of this algorithm is given in the annex. Various examples are presented (§4).

2 - SYSTEM ANALYSIS

. System with 2 customers

In the very simple case of 2 classes and one customer per class, the system is described by a continuous time Markov chain with 4 states.

Assuming that class 1 has the highest priority, these states are defined as follows :

0 : the service station is empty,

1 : the class 1 customer is alone in the station,

2 : the class 2 customer is alone in the station,

1,2 : both customers are in the station (only the class 1 customer is in service).

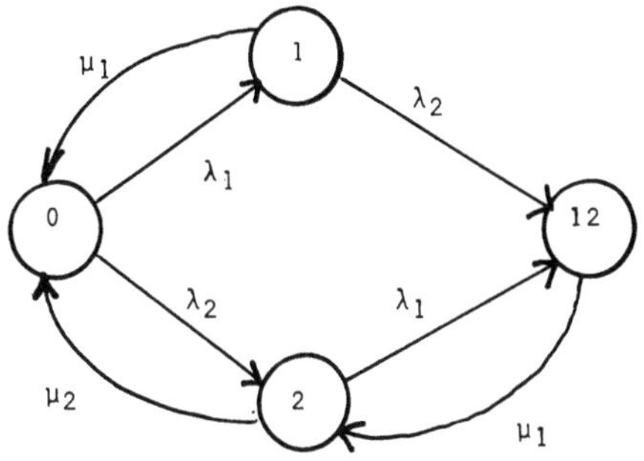

The global balance equations are :

$$p(0).(\lambda_1+\lambda_2) = p(1).\mu_1 + p(2).\mu_2 \qquad (1)$$

$$p(1).(\lambda_2+\mu_1) = p(0).\lambda_1 \qquad (2)$$

$$p(2).(\lambda_1+\mu_2) = p(0).\lambda_2 + p(1,2).\mu_1 \qquad (3)$$

$$p(1,2).\mu_1 = p(1).\lambda_2 + p(2).\lambda_1 \qquad (4)$$

The utilization factors are defined by :

$$U_1 = p(1) + p(1,2)$$
$$U_2 = p(2)$$

and $p(0) = 1 - U_1 - U_2$

(2) and (4) yield :

$$(\ p(1) + p(1,2)\ ).\mu_1 = (\ p(0) + p(2)\ ).\lambda_1$$

and $$U_1 = \frac{\lambda_1}{\lambda_1 + \mu_1} \qquad (5)$$

(3) and (4) give :

$$p(2).\mu_2 = (\ p(0) + p(1)\ ).\lambda_2$$

and $$U_2 = p(0)\ .\ \frac{\alpha_2(\lambda_2)}{\mu_2} \qquad (6)$$

with $$\alpha_2\ (\lambda_2) = \lambda_2\ .\ \frac{\lambda_1 + \lambda_2 + \mu_1}{\lambda_2 + \mu_1} \qquad (7)$$

and $$p(0) = 1 - U_1 - U_2 = \frac{\mu_1}{\lambda_1 + \mu_1}\ .\ \frac{\mu_2}{\alpha_2(\lambda_2) + \mu_2} \qquad (8)$$

System with R classes and one customer in each class

We consider the system limited to the n first classes ($n<R$).

The states of the system are defined by :

$$\underline{e} = (e_{r_1}, e_{r_2}, \ldots, e_{r_p})$$

with $1 \leq r_1 < r_2 < \ldots < r_p \leq n \leq R$

where e_r means that the class r customer is in the service station.

For a system limited to n classes, let $p_n(\underline{e})$ be the stationary probability of state $\underline{e}$, and $p_n(\underline{0})$ the stationary probability that the system be empty.

Due to the preemptive priority disicipline and the external delay, introducing the class n+1 does not modify the behaviour of higher priority classes 1, 2, ..., n.

It is clear that the utilization factor of class n can be derived from the model limited to the n first classes.

Thus $$p_n(\underline{0}) = 1 - U_1 - U_2 - \ldots - U_n = p_{n-1}(\underline{0}) - U_n \quad (9)$$

$$U_n = p_n(e_n) = p_{n-1}(\underline{0}) - p_n(\underline{0}) \quad (10)$$

(5) and (8) imply that

$$p_0(\underline{0}) = 1$$

$$p_1(\underline{0}) = \frac{\mu_1}{\lambda_1 + \mu_1}$$

$$p_2(\underline{0}) = p_1(\underline{0}) \cdot \frac{\mu_2}{\alpha_2(\lambda_2) + \mu_2}$$

<u>Theorem</u> :

The utilization factor of class n and the stationary probability of the system being empty verify the following relations :

$$U_n = p_n(e_n) = p_n(\underline{0}) \cdot \frac{\alpha_n(\lambda_n)}{\mu_n} \quad (11)$$

$$p_n(\underline{0}) = p_{n-1}(\underline{0}) \cdot \frac{\mu_n}{\alpha_n(\lambda_n) + \mu_n} \quad (12)$$

where the coefficients $\alpha_n(\lambda)$ are defined as follows :

$$\alpha_n(\lambda) = \alpha_{n-1}(\lambda) \cdot \frac{\alpha_{n-1}(\lambda_{n-1}+\lambda) + \mu_{n-1}}{\alpha_{n-1}(\lambda) + \mu_{n-1}} \qquad n>1 \qquad (13)$$

and $\alpha_1(\lambda) = \lambda$

Demonstration

The global balance equations for the n classes system can be written as follows :

$$p_n(\underline{0}) \sum_{r=1}^{n} \lambda_r = \sum_{r=1}^{n} p_n(e_r) \cdot \mu_r \qquad (14)$$

$$p_n(\underline{e}) \left(\sum_{\{r \neq r_i\}} \lambda_r + \mu_{r1} \right) = \sum_{\{r<r_1\}} p_n(\underline{e}+e_r) \cdot \mu_r + \sum_{\{r=r_i\}} p_n(\underline{e}-e_r) \cdot \lambda_r \qquad (15)$$

$\underline{e}=(e_{r1}, e_{r2}, \ldots, e_{rp})$
$0 < r_1 < \ldots < r_p < n$

We suppose now that the utilization factors have been computed for classes 1, 2, ..., n-1, with :

$$U_{n-1} = p_{n-1}(e_{n-1}) = p_{n-1}(\underline{0}) \cdot \frac{\alpha_{n-1}(\lambda_{n-1})}{\mu_{n-1}} \qquad (16)$$

By summing all states which do not contain e_n, we get :

$$\sum_{\substack{\{\underline{e}\} \\ e_n \notin \underline{e}}} p_n(\underline{e}) \cdot \lambda_n = p_n(e_n) \cdot \mu_n \qquad (17)$$

and for the system limited to the n-1 first classes, using (16) :

$$\sum_{\substack{\{\underline{e}\} \\ e_{n-1} \notin \underline{e}}} p_{n-1}(\underline{e}) \cdot \lambda_{n-1} = p_{n-1}(e_{n-1}) \cdot \mu_{n-1}$$

$$= p_{n-1}(\underline{0}) \cdot \alpha_{n-1}(\lambda_{n-1}) \qquad (18)$$

We remark that the right part of the general relation (15) associated with a state $\underline{e} = (e_{r1}, e_{r2}, \ldots, e_{rp})$ does not depend on any state containing e_r such that $r > r_p$.

We consider now the set of relations (15) issued from states which contain neither e_{n-1} nor e_n. These equations are identical to the equations derived from the system with n-1 classes, by simply substituting λ_{n-1} by $\lambda_{n-1}+\lambda_n$ in the left parts.

We apply (18), and derive :

$$\sum_{\substack{\{\underline{e}\} \\ e_{n-1} \notin \underline{e} \text{ and } e_n \notin \underline{e}}} p_n(\underline{e}) = p_n(\underline{0}) \cdot \frac{\alpha_{n-1}(\lambda_{n-1}+\lambda_n)}{\lambda_{n-1}+\lambda_n} \qquad (19)$$

We sum up equations (15) issued from states $\underline{e}$ which contain neither e_{n-1} nor e_n, with equation (16). We get :

$$\sum_{\substack{\{\underline{e}\} \\ e_{n-1} \notin \underline{e} \text{ and } e_n \notin \underline{e}}} p_n(\underline{e}) \cdot (\lambda_{n-1}+\lambda_n) = p_n(e_{n-1}) \cdot \mu_{n-1} + p_n(e_n) \cdot \mu_n \qquad (20)$$

Now we define :

$$P_n(\underline{0}) = p_n(\underline{0}) + p_n(e_{n-1})$$

$$\underset{\substack{\{\underline{e}\} \\ e_{n-1} \notin \underline{e} \text{ and } e_n \notin \underline{e}}}{P_n(\underline{e})} = p_n(\underline{e}) + p_n(\underline{e}+e_{n-1})$$

From (14) and (15), we obtain :

$$p_n(\underline{0}) \cdot \left(\lambda_n + \sum_{r=1}^{n-2} \lambda_r \right) = \sum_{r=1}^{n-2} p_n(e_r) \cdot \mu_r + p_n(e_n) \cdot \mu_n$$

$$\underset{\{\underline{e}\}}{p_n(\underline{e})} \cdot \left(\lambda_n + \sum_{\{r \neq r_i\}} \lambda_r \right) = \sum_{\{r < r_1\}} p_n(\underline{e}+e_r) \cdot \mu_r$$

$$\underline{e} = (e_{r1}, \ldots, e_{rp})$$
$$0 < r_1 < \ldots < r_p < n-1 \qquad + \sum_{\{r=r_i\}} p_n(\underline{e}-e_r) \cdot \lambda_r$$

These equations are identical to equations derived from the n-1 classes system limited to states $\underline{e}$ without class n-1, after substitution of λ_{n-1} by λ_n, $p_{n-1}(e_{n-1}) \cdot \mu_{n-1}$ by $p_n(e_n \cdot \mu_n$, and $p_{n-1}(\underline{e})$ by $p_n(\underline{0})$.

Applying relation (16), it follows that :

$$p_n(e_n) = p_n(\underline{0}) \cdot \frac{\alpha_{n-1}(\lambda_n)}{\mu_n} = \left(p_n(\underline{0}) + p_n(e_{n-1}) \right) \cdot \frac{\alpha_{n-1}(\lambda_n)}{\mu_n}$$

From (19), (21) and (20), we obtain :

$$U_n = p_n(e_n) = p_n(\underline{0}) \cdot \frac{\alpha_n(\lambda_n)}{\mu_n}$$

where $$\alpha_n(\lambda_n) = \alpha_{n-1}(\lambda_n) \; \frac{\alpha_{n-1}(\lambda_{n-1}+\lambda_n) + \mu_{n-1}}{\alpha_{n-1}(\lambda_n) + \mu_{n-1}}$$

and $$p_n(\underline{0}) = p_{n-1}(\underline{0}) - p_n(e_n) = p_{n-1}(\underline{0}) \cdot \frac{\mu_n}{\alpha_n(\lambda_n)+\mu_n}$$

. System with R classes and several customers in each class.

To generalize the previous results to systems in which each class r contains $N_r > 1$ customers, we have simply to observe that for a given class the server utilization factor is independant of the scheduling discipline (FIFO or priority within the class).

The class n utilization can be derived from $p_n(\underline{0})$, issued from a system with N_1 customers of class 1, N_2 customers of class 2, ..., N_n customers of class n, considered as a system with $N_1+N_2+ ...+N_n$ one customer classes.

The global utilization factor for class n is still given by :

$$U_n = p_{n-1}(\underline{0}) - p_n(\underline{0})$$

where $p_n(\underline{0})$ is the stationary probability of the system being empty when the system is limited to the first n classes. This can be written :

$$p_n(\underline{0}) = p_{n-1}(\underline{0}) \cdot \frac{\mu_n}{\alpha_{n,1}(\lambda_n)+\mu_n} \cdot \frac{\mu_n}{\alpha_{n,2}(\lambda_n)+\mu_n} \cdots \frac{\mu_n}{\alpha_{n,N_n}(\lambda_n)+\mu_n}$$

The coefficients $\alpha_{n,i}(\lambda)$ are given by :

$$\alpha_{n,1}(\lambda) = \alpha_{n-1,N_{n-1}}(\lambda) \cdot \frac{\alpha_{n-1,N_{n-1}}(\lambda+\lambda_{n-1}) + \mu_{n-1}}{\alpha_{n-1,N_{n-1}}(\lambda) + \mu_{n-1}}$$

$$\alpha_{1,1}(\lambda) = \lambda$$

$$\alpha_{n,i}(\lambda) = \alpha_{n,i-1}(\lambda) \cdot \frac{\alpha_{n,i-1}(\lambda+\lambda_n) + \mu_n}{\alpha_{n,i-1}(\lambda) + \mu_n} \qquad i=2, \ldots, N_n$$

The coefficient $\alpha_{n,i}(\lambda)$ corresponds to the coefficient $\alpha_{N_1+N_2+\ldots+N_{n-1}+i}(\lambda)$ in the equivalent one customer per class model.

3 - COMPUTING METHOD

We calculate the series of terms $p_n(\underline{0})$ by a recurrence, starting with a system restricted to one class.

The term $p_1(\underline{0})$ corresponds to class 1 (highest priority) alone in the system. It can be computed by the well-known recurrence ("repairman" model) :

$$p_{1,0}(\underline{0}) = 1$$

$$p_{1,i}(\underline{0}) = \frac{\mu_1 \cdot p_{1,i-1}(\underline{0})}{i.\lambda_1 + \mu_1 . p_{1,i-1}(\underline{0})} \qquad i=1,\ 2,\ \ldots,\ N_1$$

and $p_1(\underline{0}) = p_{1,N_1}(\underline{0})$

$$U_1 = 1 - p_1(\underline{0})$$

The general term $p_n(\underline{0})$ can be derived from $p_{n-1}(\underline{0})$ by using the N_n coefficients $\alpha_{n,i}(\lambda_n)$, $i=1,\ 2,\ \ldots,\ N_n$.

The coefficient $\alpha_{n,N_n}(\lambda_n)$ implies the following coefficients ...

- at level n :

$$\alpha_{n,N_n-1}(\lambda_n)\ ;\ \alpha_{n,N_n-1}(2\lambda_n),$$

$$\alpha_{n,N_n-2}(\lambda_n)\ ;\ \alpha_{n,N_n-2}(2\lambda_n)\ ;\ \alpha_{n,N_n-2}(3\lambda_n),$$

i.e. $\alpha_{n,i}(k.\lambda_n)$ with $i=1,\ 2,\ \ldots,\ N_n$.
$k=1,\ 2,\ \ldots,\ N_n-i+1$

- at level n-1 :

$\alpha_{n-1,i}(k.\lambda_{n-1}+k_n.\lambda_n)$ with $i=1,2,\ldots,N_{n-1}$
$k=0,1,\ldots,N_{n-1}-i+1$
$k_n=1,2,\ldots,N_n$

- down to level 1 :

$\alpha_{1,i}(k.\lambda_1+\sum_{j=2}^{n} k_j.\lambda_j)$ with $i=1,2,\ldots,N_1$
$k=0,1,\ldots,N_1-i+1$
$k_j=0,1,\ldots,N_j$
$k_n=1,2,\ldots N_n$

It thus implies :

$$\frac{N_n \cdot (N_n+1)}{2} \quad \text{coefficients } \alpha_{n,i} ,$$

$$N_n \left(\frac{N_{n-1} \cdot (N_{n-1}+3)}{2} \right) \quad \text{coefficients } \alpha_{n-1,i} ,$$

$$N_n \cdot (N_{n-1}+1) \ldots (N_2+1) \left(\frac{N_1 \cdot (N_1+3)}{2} \right) \text{coefficients } \alpha_{1,i}.$$

The following algorithm iteratively calculates $p_n(\underline{0})$. At level n (as the class n is introduced), we have first to calculate all the necessary coefficients $\alpha_{1,1}$, equal to the terms $k_1.\lambda_1 + k_2.\lambda_2 + \ldots k_n.\lambda_n$, with $k_i=0,1,\ldots,N_i$, for $i=1,2,\ldots,n-1$, and $k_n=1,2,\ldots,N_n$.

These coefficients are paired to calculate coefficients $\alpha_{1,2}, \ldots$ up to α_{1,N_1}. We calculate then coefficients $\alpha_{2,1}, \ldots,\alpha_{2,N_2}, \ldots$, up to coefficient α_{n,N_n}, which yields the utilization U_n.

The resolution of level n needs $(N_1+1).(N_2+1)\ldots(N_{n-1}+1).N_n$ memory words for storing coefficients $\alpha_{1,1}$.

To calculate the next class, all coefficients have to be computed again.

With R classes, the algorithm needs at most $\prod^{R-1} (N_r+1).N_R$ memory words.

Example 3.1 exhibits the iterative calculation of coefficients α in the case of a 3 class system.

The annex contains a PASCAL implementation of this algorithm.

- <u>Example 3.1</u> : 3 classes with $N_1=1,\ N_2=2,\ N_3=3$.

Class 1 :

$$p_1(0)=\frac{\mu_1}{\lambda_1+\mu_1} \qquad U_1 = 1 - p_1(0)$$

Class 2 :

$$p_2(0)=\frac{\mu_2}{\alpha_{2,1}(\lambda_2)+\mu_2}\cdot\frac{\mu_2}{\alpha_{2,2}(\lambda_2)+\mu_2}\cdot p_1(0)$$

$\alpha_{1,1}$: λ_2, $2\lambda_2$, $\lambda_1+\lambda_2$, $\lambda_1+2\lambda_2$ $\xrightarrow{\mu_1}$ $\alpha_{2,1}$: $\alpha_{2,1}(\lambda_2)$, $\alpha_{2,1}(2\lambda_2)$ $\xrightarrow{\mu_2}$ $\alpha_{2,2}$: $\alpha_{2,2}(\lambda_2)$

$$U_2 = p_1(0) - p_2(0)$$

Class 3 :

$$p_3(0)=p_2(0)\ \frac{\mu_3}{\alpha_{3,1}(\lambda_3)+\mu_3}\cdot\frac{\mu_3}{\alpha_{3,2}(\lambda_3)+\mu_3}\cdot\frac{\mu_3}{\alpha_{3,3}(\lambda_3)+\mu_3}$$

$$U_3 = p_2(0)-p_3(0)$$

$\alpha_{1,1}$: λ_3, $2\lambda_3$, $3\lambda_3$, $\lambda_2+\lambda_3$, $\lambda_2+2\lambda_3$, $\lambda_2+3\lambda_3$, $2\lambda_2+\lambda_3$, $2\lambda_2+2\lambda_3$, $2\lambda_2+3\lambda_3$, $\lambda_1+\lambda_3$, $\lambda_1+2\lambda_3$, $\lambda_1+3\lambda_3$, $\lambda_1+\lambda_2+\lambda_3$, $\lambda_1+\lambda_2+2\lambda_3$, $\lambda_1+\lambda_2+3\lambda_3$, $\lambda_1+2\lambda_2+\lambda_3$, $\lambda_1+2\lambda_2+2\lambda_3$, $\lambda_1+2\lambda_2+3\lambda_3$

$\xrightarrow{\mu_1}$ $\alpha_{2,1}(\ldots)$: λ_3, $2\lambda_3$, $3\lambda_3$, $\lambda_2+\lambda_3$, $\lambda_2+2\lambda_3$, $\lambda_2+3\lambda_3$, $2\lambda_2+\lambda_3$, $2\lambda_2+2\lambda_3$, $2\lambda_2+3\lambda_3$

$\xrightarrow{\mu_2}$ $\alpha_{2,2}(\ldots)$: λ_3, $2\lambda_3$, $3\lambda_3$, $\lambda_2+\lambda_3$, $\lambda_2+2\lambda_3$, $\lambda_2+3\lambda_3$

$\xrightarrow{\mu_2}$ $\alpha_{3,1}(\ldots)$: λ_3, $2\lambda_3$, $3\lambda_3$

$\xrightarrow{\mu_3}$ $\alpha_{3,2}(\ldots)$: λ_3, $2\lambda_3$

$\xrightarrow{\mu_3}$ $\alpha_{3,3}(\ldots)$: λ_3

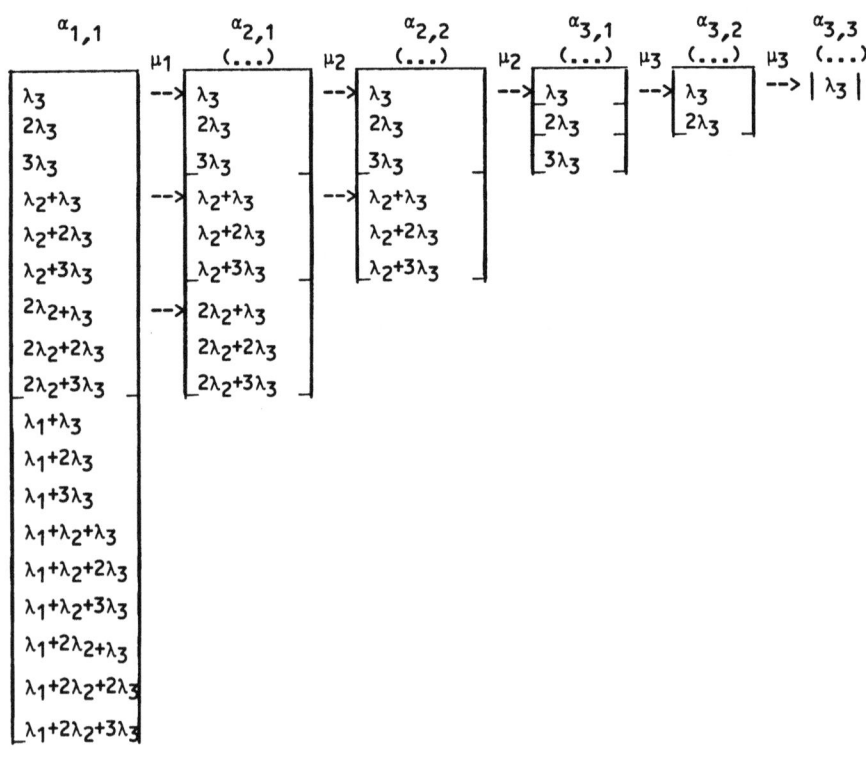

4 - EXAMPLES

. Example 4.1

A processor is shared by real-time tasks.

These tasks are executed on 4 different priority levels, with 5 tasks on each level. Tasks related to a same level are supposed to have the same statistical behaviour (service times to the processor, external delays before a new execution demand).

$$S_1 = 10ms, \quad S_2 = 5ms, \quad S_3 = 10ms, \quad S_4 = 5ms$$
$$Z_1 = 400ms, \quad Z_2 = 300ms, \quad Z_3 = 200ms, \quad Z_4 = 100ms$$

The processor utiization factors and the task mean response times are to be evaluated.

A numerical solution of the Markov chain (1296 states) needs about 3mn of computing time (on a DPS/68) using the QNAP package [2].

The algorithm described in this paper yields the exact solution in a few seconds.

The following results have been obtained :

S1= 10.0	Z1= 400.0	U1= 0.12164	R1= 11.04851
S2= 5.0	Z2= 300.0	U2= 0.08126	R2= 7.66817
S3= 10.0	Z3= 200.0	U3= 0.22857	R3= 18.75234
S4= 5.0	Z4= 100.0	U4= 0.20454	R4= 22.22440

. <u>Example 4.2</u>

This short example illustrates a situation where approximate methods may give unacceptable errors.

The system to be studied contains 4 customers with a high priority (Z_1=500s, S_1=100s), and one customer with a low priority (Z_2=1s, $S_2=10^{-3}$s). The response time of the low priority custome is to be evaluated (the service of this customer is interrupted by long but infrequent services of high priority customers).

We get easily for class 1 :

$$U_1 = 0.6016$$

$$R_1 = 4*100/0.606 - 500 = 164.83 \text{ s}$$

For class 2 we can write :

$$U_2 = (1 - U_1) \cdot \frac{\alpha_2(\lambda_2)}{\alpha_2(\lambda_2)+\mu_2}$$

The computation of the successive coefficients α_i is shown below :

$\alpha_{1,1}$	μ_1	$\alpha_{1,2}$	μ_1	$\alpha_{1,3}$	μ_1	$\alpha_{1,4}$	μ_2	α_2
1	--->	1.0019802	--->	1.0039604	--->	1.0059409	--->	1.0079212
1.002		1.0039802		1.0059606		1.0079409		
1.004		1.0059803		1.0079606				
1.006		1.0079803						
1.008								

So we get $U2 = 4.011\ 10^{-4}$

$$R_2 = 10^{-3}/4.011\ 10^{-4} - 1 = 1.493 \text{ s}$$

It is interesting to notice that in this particular case we can get solution bounds. It is easy to demontrate that coefficients $\alpha_n(\lambda)$ increase with n and λ.

We derive simply : $\lambda_n < \alpha_n(\lambda_n) < \lambda_1+\lambda_2+ \dots + \lambda_n$

and in this example :

$$(1 - U_1) \frac{\lambda_2}{\lambda_2+\mu_2} < U_2 < (1 - U_1) \frac{4.\lambda_1+\lambda_2}{4.\lambda_1+\lambda_2+\mu_2}$$

which yields : $1.492 \text{ s} < R_2 < 1.513 \text{ s}$

In this simple case the "Shadow CPU" approximate method as proposed in [4] consists in changing the service time of the low priority class μ_2 by $\mu^* = \mu_2 \,.\, (1 - U_2)$ - the server is "slowed down" due to high priority services- .

We get a considerable underestimate for response time :

$$R_2 = 10^{-3} \;/\; (1-U_1) = 2.5\ 10^{-3}\ \text{s}$$

5 - CONCLUSION

A single server station systeme with a preemptive-resume discipline and a finite source (M/M/1/N/ Priority) can be solved by using a simple recursion formula (as far as mean response times and utilization factors are concerned) for any number of classes and customers. A practical algorithm has been described with its PASCAL implementation.

This algorithm can be used in an approximate iterative method for solving general queueing networks with preemptive priority stations. In this method the priority stations are replaced by equivalent delays (as in [8]), these delays being calculated at a lower level by the exact algorithm. This method is described in [9]. Its accuracy has been found to be fairly good in all studied cases. Further intensive testing is currently being caried out.

- REFERENCES

[1] Jaiswal, N. "Priority queues", New-York : Academic Press, 1968.

[2] Merle D.; Potier D.; Véran M. "QNAP : a software tool for computer system evaluation", Int. Conf. on Performance of Computer Installations, North-Holland, 1978.

[3] Wallace, V.L.; Rosenberg, R.S. "Markovian models and numerical analysis of computer system behavior", Proc. 1966 AFIPS Spring Jt. Conf., Vol. 28.

[4] Sevcik, K.C. "Priority scheduling disciplines in queueing network models of computer systems", Proc. IFIP Congress 77, North-Holland Publ. Co., Amsterdam.

[5] Sauer, C.H.; Chandy, K.M. "Approximate analysis of central server models", IBM J. Res. Dev. 19, May 75.

[6] Bard, Y. "Some extensions to multiclass queueing network analysis", 4th Int. Symp. on modelling and performance evaluation of computer systems, Vienna, Feb. 79.

[7] Buzen, J.P.; Goldberg, P.S. "Guidelines for the user of infinite source queueing models in the analysis of computer system performance", Proc. of the NCC, pg. 371-374, 1974.

[8] Jacobson, P.A., Lazowska E.D. "The Method of Surrogate Delays : Simultaneous Resouce Possession in Analytic Models of Computer Systems"; Proc. ACM SIGMETRICS Conference on Measurement and Modelling of Computer Systems, Las Vegas, Sept. 1981.

[9] Véran M. " Résolution d'un réseau de files d'attente avec partage de ressource"; Rapport de Recherche, Centre de Recherches CII-HB, Décembre 81.

[10] Chow We-Min "M/M/1 Priority Queues with State Dependent Arrival and Service Rates", RC8570, IBM, Thomas J. Watson Research Center, Yorktown Heights, NY.

- <u>ANNEX</u> PASCAL implementation of the algorithm ...

```
program PRIOR(output);
(* resolution of a service station with preemptive-resume
   priorities and finite sources*)
(*this version is limited to 10 classes; no check on input data
  is provided*)

var R : integer;                          (*number of classes <10*)
    LD : array [1..10] of real;           (*arrival rates*)
    NBR : array [1..10] of integer;       (*customer numbers*)
    MU : array [0..10] of real;           (*service rates*)
    L : array [0..2000] of real;          (*working area*)
    IR, IR1, IL, IL1, I, J, N : integer;
    PO, POP : real;
    U : array [1..10] of real;             (*utilizations*)
begin (*initialization*)
R:=4; (*example with 4 classes and 5 customers per class*)
LD[1]:=1.0/400.0; MU[1]:=1.0/10.0;   NBR[1]:=5;
LD[2]:=1.0/300.0; MU[2]:=1.0/5;      NBR[2]:=5;
LD[3]:=1.0/200.0; MU[3]:=1.0/10.0;   NBR[3]:=5;
LD[4]:=1.0/100.0; MU[4]:=1.0/5.0;    NBR[4]:=5;
(*computation of U[1] : highest priority class utilization*)
PO:=1.0;
for I:=1 to NBR[1] do PO:=MU[1]*PO/(I*LD[1]+MU[1]*PO);
U[1]:=1.0-PO;
(*iteration on classes*)
for IR:=2 to R
do begin
   L[0]:=0.0;       (*initialization of the working stack*)
   for I:=1 to NBR[IR] do L[I]:=L[I-1] + LD[IR];
   IL:=NBR[IR]; IL1:=IL;
   for IR1:=IR-1 downto 1
   do begin
      for N:=1 to NBR[IR1]
      do for J:=1 to IL1
         do begin IL:=IL+1; L[IL]:=L[IL-IL1]+LD[IR1]; end;
      IL1:=IL;
      end;
```

```
   for IR1:=1 to IR-1 (*working stack reduction*)
   do begin
      IL1:=IL div (NBR[IR1]+1);
      for N:=1 to NBR[IR1]
      do begin
         IL:+IL-IL1;
         for I:=1 to IL
         do  L[I]:=L[I]*(L[I+IL1]+MU[IR1])/(L[I]+MU[IR1]);
         end;
      end;
   POP:=P0; (*last class*)
   for N:=1 to NBR[IR]
   do begin
      IL:=IL-1;
      P0:=P0*MU[IR]/(L[1]+MU[IR]);
      for I:=1 to IL
      do L[I]:=L[I]*(L[I+1]+MU[IR])/(L[I]+MU[IR]);
      end;
   U[IR]:=POP-P0;
   end;
(*output of utilization factors and of response times*)
for IR:=1 to R
do writeln(' S',IR :1,'=',1.0/MU[IR] :5:1,
           ' Z',IR :1,'=',1.0/LD[IR] :5:1,
           ' U',IR :1,'=',U[IR]     :10:5,
           ' R',IR :1,'=',NBR[IR]/( U[IR]*MU[IR] ) - 1.0/LD[IR]
                                    :10:5 );
end.
```

VIII

COMPUTER SYSTEMS EVALUATION 1

EVALUATION DE SYSTEMES INFORMATIQUES 1

PERFORMANCE EVALUATION OF A DATA BASE MANAGEMENT SYSTEM

Alfred SAAL, Otto SPANIOL
Fachbereich Informatik
Universität Frankfurt
P.O. Box 11 19 32
D - 6000 Frankfurt

Abstract

Transaction behaviour, locking policies and multiprogramming environment are most important factors for the performance of data base management system. In this paper we deal with a simple model in order to derive quantitative results concerning the tradeoff between several system parameters. The analysis is performed by means of mathematical methods and results are presented either in closed form or as approximations which are easily evaluated by numerical methods. The simple structure of our formulae can be used to get insight into the influence of system parameters like number of I/O-servers, CPU processing rate, number of terminals, transaction load, number of data base granules etc. The application of our results is demonstrated by some examples.

1. INTRODUCTION: DESCRIPTION OF THE DATABASE MANAGEMENT SYSTEM MODEL

In this paper we deal with a management system for a centralized database which is based on a model used by Potier and Leblanc [PL] ; the system works as follows:

- the database is assumed to consist of a set of cells (elementary units of storage). A fixed number of such cells forms a logical object ('granule'). The nature of granules is unimportant for our purpose.

- a transaction (TA) is composed of a series of database operations (queries and updates of database objects) which require a certain number of processing steps, i.e. accesses to CPU and storage units;

- transactions are issued from a finite number, N, of terminals. After the generation of a TA the corresponding terminal becomes inactive until the TA is terminated. At that time the terminal begins to produce a new transaction (thinking time);

- in order to preserve database consistency [DA] the locking of database objects becomes necessary; in this paper we consider exclusive locks only. Furthermore we restrict on a static locking policy where a central mechanism tests (in negligible time) after TA generation whether all granules the TA will refer to during its lifetime are available. If a granule is actually locked then the TA becomes blocked; none of the locks will be granted. After finishing of other TA (which implies the release of the corresponding locks) one or more of the blocked TA are examined again. This simple locking mechanism avoids deadlocks; a transaction will be eventually finished when all locks have been granted;

- if all locks for TA may be granted the TA becomes running: After a certain number of CPU and I/O-requests the transaction will finish; locks will be released and the transaction returns to the originating terminal, i.e. to the thinking state.

The database management system may be described by a closed queueing network with 3 stations (see figure 1) where the stations refer to the different states of a TA: thinking, blocked or running.

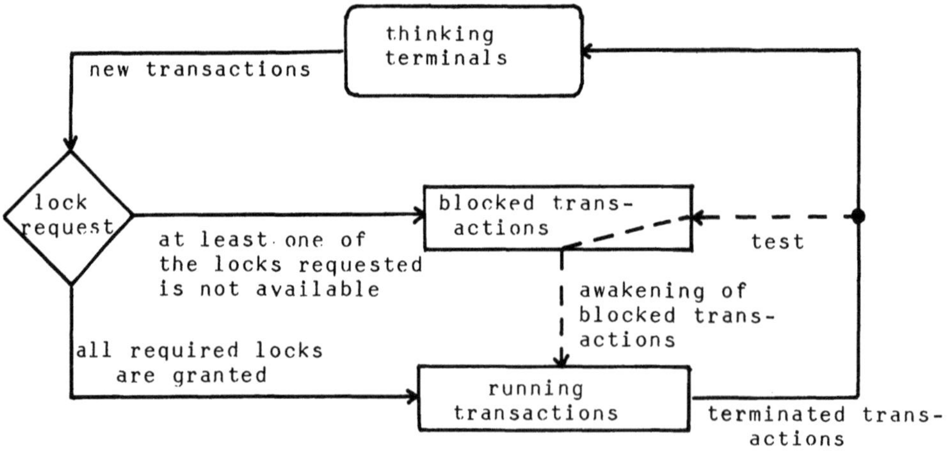

Figure 1: Database queueieng network

In this paper total transaction throughput will be used as the measure of database performance. In order to evaluate this parameter the following network components have to be determined:

a. Probability that locks are granted for a newly arriving or a previously blocked transaction.

b. Behaviour of running transactions (number and duration of CPU as well as of I/O-accesses; concurrency between simultaneously running transactions; service time distribution of running transactions).

2. MODEL ANALYSIS

2.1. TA ACCEPTANCE PROBABILITY

The probability that a TA may become running (i.e. that all locks may be granted) obviously depends on the number of TA which are actually in the runnning state.

The following assumptions are sufficient to develop a simple expression for that probability.

Assumptions:

A 1. the number of database granules is given by m

A 2. the number, n, of exclusive locks needed by a transaction is a constant

A 3. locked objects are uniformly distributed (without distribution) over the set of granules.

If these assumptions are satisfied the probability of lock granting is given by:

q_r: = Pr (locks for another transaction may be granted supposed that r transaction are already running)

$$= \frac{\binom{m-rn}{n}}{\binom{m}{n}} = \begin{cases} \prod_{i=0}^{n-1} \frac{m-rn-i}{m-i} & \text{if } (r+1)n \leq m \\ 0 & \text{else} \end{cases}$$

Remarks: Assumptions A 2 and A 3 should be considered as a crude approximation for real databases. A more refined analysis would result in modified probabilities q_r^* which might be used in the subsequent formulae instead of our approximative values q_r.

2.2. CONDITIONAL SYSTEM THROUGHPUT

If all locks have been granted a transaction is processed, i.e. it performs a certain number of alternative CPU and I/O-accesses. Transaction service time depends on the number of simultaneously running transactions due to the concurrency with regard to CPU and I/O-units.

For the subsequent analysis we assume that the following assumptions are valid for transaction behaviour:

A 4: transactions are served by a single CPU and D identical I/O-units (disks).

A 5: CPU- und I/O-accesses alternate with each other (central server model). The number of each kind of access per transaction is geometrically distributed:

Pr (exactly i CPU-accesses per transaction) = $(1-p_F)^{i-1} \cdot p_F$ (i=1,2...)

p_F is the probability that current CPU- (or I/O-) access is the last one.

The average number of CPU-accesses per transaction is given by p_F^{-1}.

A 6: Disks are accessed independent of each other with probability 1/D.

A 7: The CPU is operating with processor sharing discipline whereas disks are served according to FIFO. Service times (per access) are assumed to be independent exponentially distributed random variables with parameters μ_{CPU} and μ_{DISK} respectively.

Remark: If the CPU would operate with a FIFO strategy instead as with processor sharing then due to exponential service time assumptions the same stationary state probabilities are obtained in both cases.

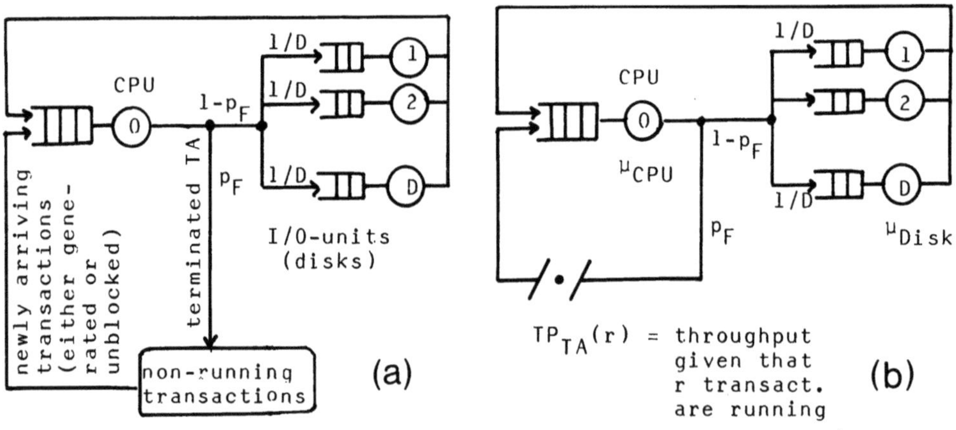

Figure 2: (a) processing of running transactions
(b) closed network representation with r customers

TA service time and TA throughput when r TA are simultaneously running are obtained from the analysis of a closed queueing network where each departing TA is immediately replaced by a new one, see figure 2.

The stationary state probabilities for the states $(r_o, r_1, \ldots r_D)$ of the closed queueing network are obtained in closed form from a BCMP-analysis ([BCMP], [GM]); see [Sa] for more detail:

$$\Pr(r_o, \ldots, r_D) = \gamma_D^{-1}(r) \cdot \left(\frac{1}{2\cdot\mu_{CPU}}\right)^{r_o} \cdot \left(\frac{1}{2D\cdot\mu_{DISK}}\right)^{r-r_o} \quad \text{where } r = \sum_{i=o}^{D} r_i .$$

If $\alpha := \frac{\mu_{CPU}}{D\cdot\mu_{DISK}}$ then the normalizing constant $\gamma_D(r)$ is given by:

$$\gamma_D^{-1}(r) = \left(\frac{1}{2\mu_{CPU}}\right)^r \cdot \binom{r+D}{D} \qquad \text{if } \alpha = 1$$

$$\gamma_D^{-1}(r) = \left(\frac{1}{2\mu_{CPU}}\right)^r \cdot \left[\frac{1}{(1-\alpha)^D} + \alpha^r \cdot \sum_{j=1}^{D} \frac{\binom{r+D-j}{r}}{(j-1)!\cdot(1-\alpha)^{j-1}\cdot(1-\alpha^{-1})}\right] \qquad \text{if } \alpha \neq 1$$

Remark: $\alpha = 1$ corresponds to a 'balanced' system, i.e. to a central server network where the single CPU works as fast as the community of I/O-units.

CPU throughput is given by

$$TP_{CPU}(r) = \mu_{CPU} \cdot \Pr(\text{CPU busy})$$

$$= \mu_{CPU} \cdot \left[1 - \sum_{\substack{(r_o,\ldots,r_D) \\ r_o=0}} \Pr(r_o, \ldots, r_D)\right]$$

$$= \mu_{CPU} \cdot \left[1 - \gamma_D^{-1}(r) \cdot \binom{r+D-1}{r} \cdot \left(\frac{1}{2D\mu_{DISK}}\right)^r\right]$$

Thus transaction throughput (under the condition that r TA are actually running) is obtained from:

$$TP_{TA}(r) = TP_{CPU}(r) \cdot p_F$$

where p_F is the probability that the TA will not require another CPU access.

$TP_{TA}(r)$ is upperbounded by:

$$TP_{TA}(r) \leq \mu_{CPU} \cdot p_F \overset{\text{def}}{=\!=} TP_{TA}^{(max)}$$

with equality if the CPU is always busy.

It will be seen from the numerical examples of section 3 that the conditional throughput approaches its upper limit very fast if the number of simultaneously running TA is increasing.

In the following discussion station 3 (running transactions) of figure 1 will be replaced by an exponential FIFO-server with processing rate $TP_{TA}(r)$.
We are now ready for the analysis of the gobal database model.

2.3. ANALYSIS OF TOTAL TRANSACTION THROUGHPUT

In order to analyze the queueing network of figure 1 we introduce another assumption:

A 8: The thinking time per terminal is an exponentially distributed random variable with parameter λ (per terminal).

Thus if u terminals are actually thinking then new transactions are generated with rate $\lambda \cdot u$.

Due to the results of the previous section the database model now looks as follows (see figure 3):

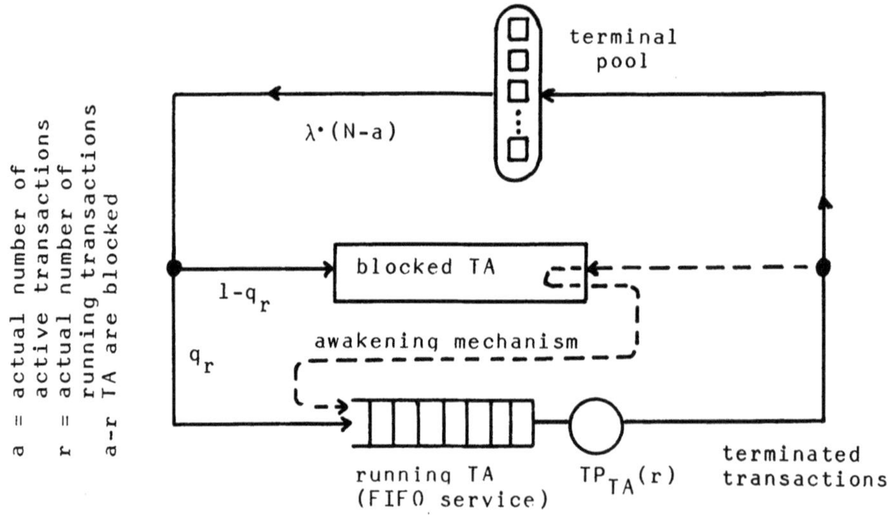

Figure 3: Database network revisited

Whereas the 'thinking' station (infinite server, exponential) and the 'running' station (FIFO, exponential) fit very well into the usual network philosophy the behaviour of the 'blocked' station is more complicated: if a runnning TA becomes terminated (and releases its locks) then possibly one or even more of the blocked stations may become running. Seeking for such candidates will be called the 'awakening strategy'. Good strategies may considerably improve the relation r/a between the number of running and non-thinking transactions respectively. On the other hand the time necessary for 'optimal' strategies may become significant. Furthermore an older TA should be earlier reactivated than a younger one. For this reason the following simple policy will be adopted:

A 9: (awakening strategy): blocked stations are considered to be 'served' according to FIFO; if one of the r running TA becomes terminated the oldest blocked TA is tested whether locks may be granted now or not (due to the model assumptions the probability of lock granting is the same for all blocked transactions). Thus a blocked TA becomes running with probability q_{r-1} (since r-1 TA are still running), with probability $1-q_{r-1}$ the population size of 'running' and 'blocked' stations remains unchanged.

Due to our model assumptions the state of the closed network is described by the number of terminals which are actually 'running' or 'active' (a terminal is called active if the correponding TA is either running or blocked).

The state space is therefore given by:

$$S = \left\{ (r,a) \;\middle|\; \begin{array}{l} r = \text{number of running transactions,} \\ a = \text{number of active transactions} \end{array} \right\} .$$

Obviously the following restrictions have to be satisfied:

1. $0 \leq r \leq a \leq N$
2. $r > 0$ if $a > 0$ (not all active transactions may be simultaneously blocked).

Thus our state space consists of

$$\binom{N+2}{N} - N = \frac{N^2+N+2}{2} \qquad \text{states.}$$

For this reason the determination of state probabilities by solving the set of balance equations is numerically impossible (for values of N which exceed a certain threshold).

In the following, therefore, a simplified - and approximate - solution technique will be proposed:

The 'blocked' and 'running' stations are combined to a single station called 'active'; this leads to the simple closed queueing network of figure 4.

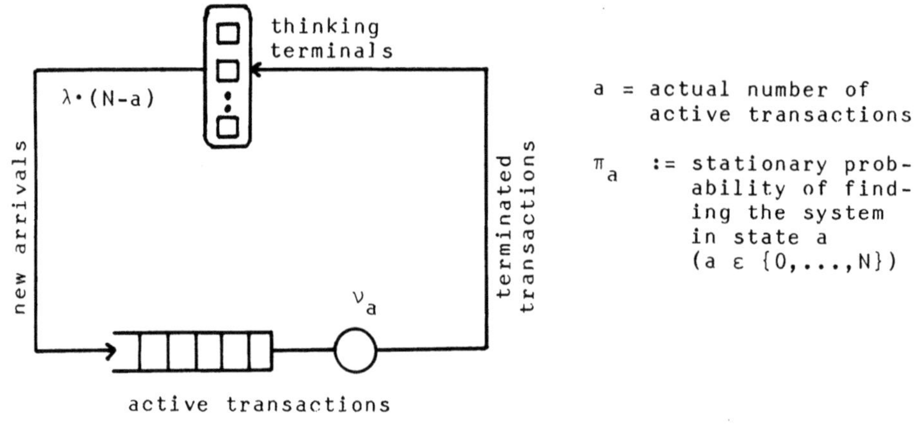

Figure 4: Combination of running and blocked transactions towards 'active' transactions (database state determined by number, a, of active transactions)

In figure 4, ν_a is defined to be the average transaction throughput rate given that N-a terminals are thinking:

$$\nu_a := \sum_{r=1}^{a} Pr\,(r|a) \cdot TP_{TA}\,(r)$$

where

$Pr\,(r|a)$: = probability that r TA are running given that a terminals are active.

The state probabilites of the N+1 states contained within figure 4 are now determined by the following system of balanced equations:

$$\lambda \cdot (N-i+1)\cdot\pi_{i-1}+\nu_{i+1}\cdot\pi_{i+1} = \left[\lambda\cdot(N-i)+\nu_i\right]\cdot\pi_i \qquad (i=0,\ldots,N)$$

$$\sum_{i=0}^{N} \pi_i = 1\,.$$

It is easy to show that these equations may be reduced to the following difference equations:

$$\pi_i = \frac{\nu_{i+1}}{\lambda\cdot(N-i)}\cdot\pi_{i+1} \quad (i=0,\ldots,N-1); \quad \sum_{j=0}^{N}\pi_j = 1.$$

Thus the solution is given by

$$\pi_i = \frac{\gamma_i}{G_N} \qquad \text{where } \gamma_i = \frac{\prod_{j=i+1}^{N}\nu_j}{(N-i)!\,\lambda^{N-i}} \qquad \text{and } G_N := \sum_{r=o}^{N}\gamma_r$$

is normalizing constant.

Transaction throughput TT is now obtained as follows:

$$TT = \sum_{a=1}^{N} \nu_a \cdot \pi_a$$

$$= \sum_{a=1}^{N} \frac{\prod_{j=1}^{N} \nu_a}{(N-a)!\, \lambda^{N-a} \cdot G_N}$$

An upper bound for total throughput is given by:

$$\nu_a = \sum_{r=1}^{a} \Pr(r|a) \cdot TP_{TA}(r)$$

$$\leq \sum_{r=1}^{a} \Pr(r|a) \cdot TP_{TA}^{(max)} = TP_{TA}^{(max)} .$$

$$TT = \sum_{a=1}^{N} \nu_a \cdot \pi_a \leq TP_{TA}^{(max)} \cdot (1-\pi_0) \overset{def}{=} TT^{(max)} .$$

In order to determine TT we have to deal with the (still unknown) probabilities $\Pr(r|a)$ that r $(1 \leq r \leq a)$ out of the a active TA are running. These values are, moreover, influenced by the 'awakening mechanism' which tries to unblock one or more blocked TA when locks are released by a terminated transaction.

'Optimal' strategies will result (on the average) in a higher number of running TA:

$$\sum_{r=1}^{a} r \cdot \Pr(r|a)_{[\text{good strategy}]} \geq \sum_{r=1}^{a} r \cdot \Pr(r|a)_{[\text{bad strategy}]} .$$

Disregarding TA finishing (and thus any reactivation of blocked TA) the following simple relationship is obtained for the parameters in question (by "^" we indicate that the expressions are only approximations since 'awakening strategies' are disregarded):

$\hat{\Pr}(r|a)$ = Pr (r out of a-1 active TA are running; the next TA is not accepted)

+Pr (r-1 out of a-1 active TA are running; the next arriving TA is accepted)

$$= \hat{\Pr}(r|a-1) \cdot (1-q_r) + \hat{\Pr}(r-1|a-1) \cdot q_{r-1}.$$

By using the initial conditions

$$\hat{\Pr}(1|1) = \Pr(1|1) = 1$$
$$\hat{\Pr}(0|c) = \Pr(0|c) = 0 \quad (\text{if } c > 0)$$

the probabilities $\hat{Pr}(r|a)$ are easily evaluated.

Thus finally the following estimation of the system throughput is obtained:

$$TT = TT(N, D, \lambda, m, n, \mu_{CPU}, \mu_{DISK}, P_F)$$

$$\cong \hat{TT} := \sum_{a=1}^{N} \hat{\nu}_a \cdot \hat{\pi}_a$$

where $\hat{\pi}_a := \frac{\hat{\gamma}_a}{\sum_{r=0}^{N} \hat{\gamma}_r}$; $\hat{\gamma}_i = \frac{\prod_{j=i+1}^{N} \hat{\nu}_j}{(N-i)!\lambda^{N-i}}$

$$\hat{\nu}_a = \sum_{r=1}^{a} \hat{Pr}(r|a) \cdot TP_{TA}(r)$$

$$\hat{Pr}(r|a) = \hat{Pr}(r|a-1)\cdot(1-q_r) + \hat{Pr}(r-1|a-1)\cdot q_{r-1}$$

$$q_r = \binom{m-nr}{n} \Big/ \binom{m}{n}$$

$$TP_{TA}(r) = \mu_{CPU} \cdot \left[1-\gamma_D^{-1}(r) \cdot \binom{r+D-1}{r} \cdot \left(\frac{1}{2D\mu_{DISK}}\right)^r \right] \cdot p_F$$

and

$\gamma_D(r)$ is a normalizing constant (see 2.2).

<u>Remark</u> on the accuracy of probabilities $\hat{Pr}(r|a)$:

$\hat{Pr}(r|a)$ reflects the distribution which is obtained if the a active transactions join an 'empty' system without intermediate departures. This distribution should correspond very well with the distribution which includes departures if the awakening mechanism has the following properties:

- only one of the blocked TA (the test candidate) will be tested and possibly unblocked after a TA departure
- the choice of the test candidate is influenced neither by the locks which are needed by the candidate nor by the locks which are released by the departing TA.

A random choice of exactly one candidate would satisfy both conditions. It is not quite clear, however, whether the FIFO selection (assumption A 9) fully satifies the second conditions since if a blocked TA has been tested several times without success it could be more likely or perhaps less likely that the next trial will be successful (with respect to younger blocked TA). Nevertheless any such influence will not be very serious since the number of locks needed is constant and since locks are uniformly distributed over the set of objects; furthermore the dependence will be even less important if new arrivals are observed between successive departures.

The above informal arguments indicate that the probability distribution $Pr(r|a)$ is not very sensitive of the transaction history if the beforementioned conditions are satisfied.

It is true that the closed form expression for transaction system throughput TT has been obtained by means of several very restrictive assumptions some of which will not be satisfied even approximately in real databases.

On the other hand it is well known that analytical results obtained from idealistic modeling asssumptions are often very robust.
In our opinion, the importance of the results derived in this paper come from the fact that the effects of the eight system parameters

- N number of terminals
- D number of I/O-units
- λ TA generation rate per terminal
- m number of database granules
- n number of exclusive locks needed by a TA
- μ_{CPU} CPU processing rate
- μ_{DISK} I/O processing rate
- p_F^{-1} average number of CPU accesses per TA

may be studied very easily for a broad range of parameter values without relying on software packeges or simulations. Some numerical results are shown in the following section.

3. NUMERICAL RESULTS

Due to the limited space available only a few effects are discussed. It should become clear, however, how to deal with other cases.
Some of the database parameters will be fixed in the following:

$\lambda = 2/3$; $D = 3$; $\mu_{DISK} = \mu_{CPU} = 20$ (i.e. the CPU is overloaded with respect to the three disks).

In figure 5 the conditional transaction throughput TP_{TA} (r) is given for different transaction durations.

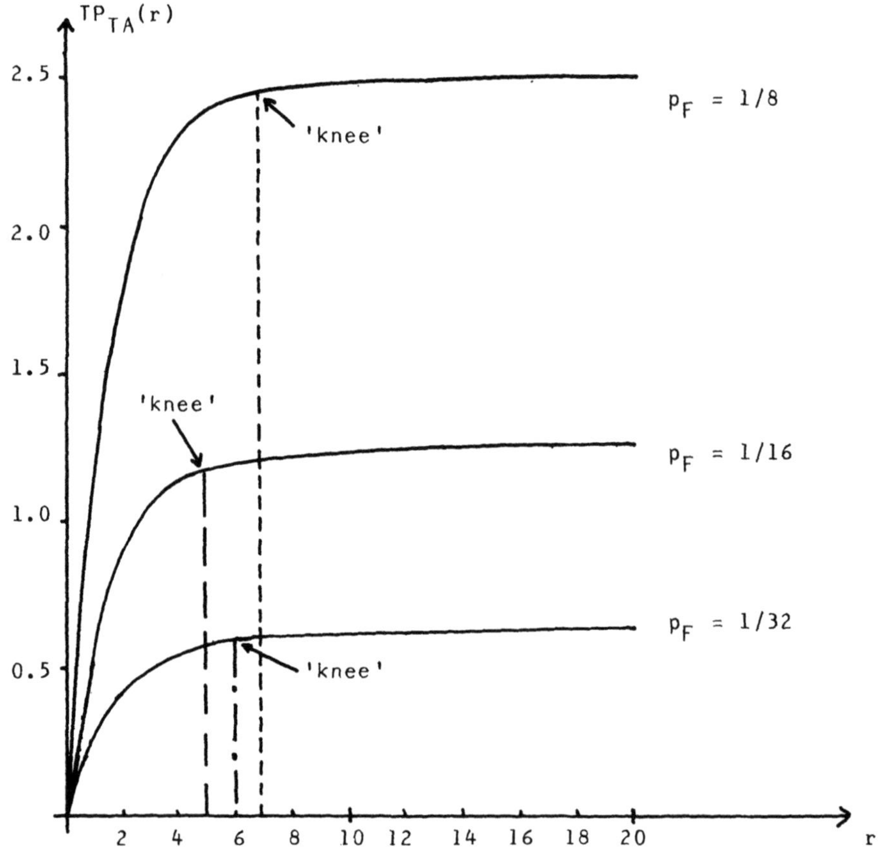

Figure 5: Conditional transaction throughput $TP_{TA}(r)$ as a function of the number, r, of running transactions

The conditional throughput TP_{TA} (r) is at first very sharply increasing with r but then remains almost constant; the latter effect is due to the fact that TP_{TA} rapidly approaches its maximum

$$TP_{TA}^{(max)} = \mu_{CPU} \cdot p_F$$

which could be obtained if the CPU were always busy. If μ_{DISK} were larger when compared to μ_{CPU} then the flattening of the curves would be observed for higher numbers of running transactions.

In order to obtain a high transaction throughput we have to guarantee that the average number of running TA stays above the 'knee' indicated in figure 5. It is of little importance whether the number of running TA is much larger or little larger than the value indicated by the knee.
TA service time, on the other hand, will be increasing if the database is overloaded, i.e. if CPU and/or I/O-resources become a bottleneck of the system.

The tradeoff between total system throughput TT and number, N, of terminals is illustrated in figure 6 in dependence of the number of locked objects, n, per TA.

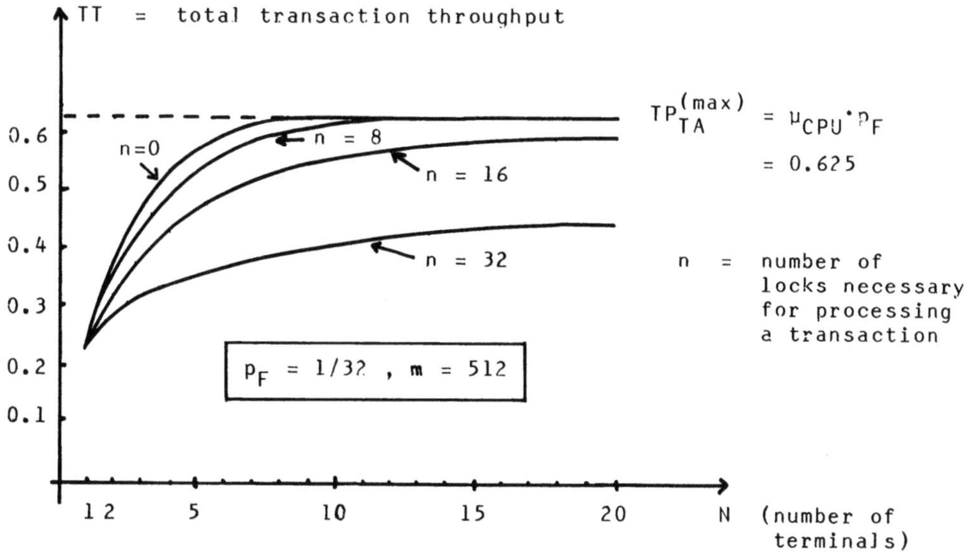

Figure 6: Total transaction throughput, TT, as a function of the number, N, of transaction generating terminals

TT is increasing with N since more terminals bring more load to the system and the resources (CPU and I/O) become better utilized and decreasing with n since higher lock number requests result in more blocking situations; thus the access probabilities

$$q_r = \binom{m-nr}{n} \Big/ \binom{m}{n}$$

become very small even for moderate r.

Therefore the number of running TA becomes so small that the CPU is not sufficiently utilized. This negative influence of parameter n is only observed, however, if the mean number of running transactions stays below the 'knee' given in figure 5.

In figure 7, for example, the mean number of running transactions is greater than 6 (where 6 is to be considered as the knee value for this case) for $n \leq 8$ and $N \geq 10$, and, therefore, total throughput is not significantly reduced in this case with respect to the maximum attainable throughput (see figure 6); conversely, for $n \geq 16$ the knee threshold is not reached for $N \leq 20$ and total throughput remains significantly below its maximum (apart from the trivial case N = 1 where no blocking occurs).

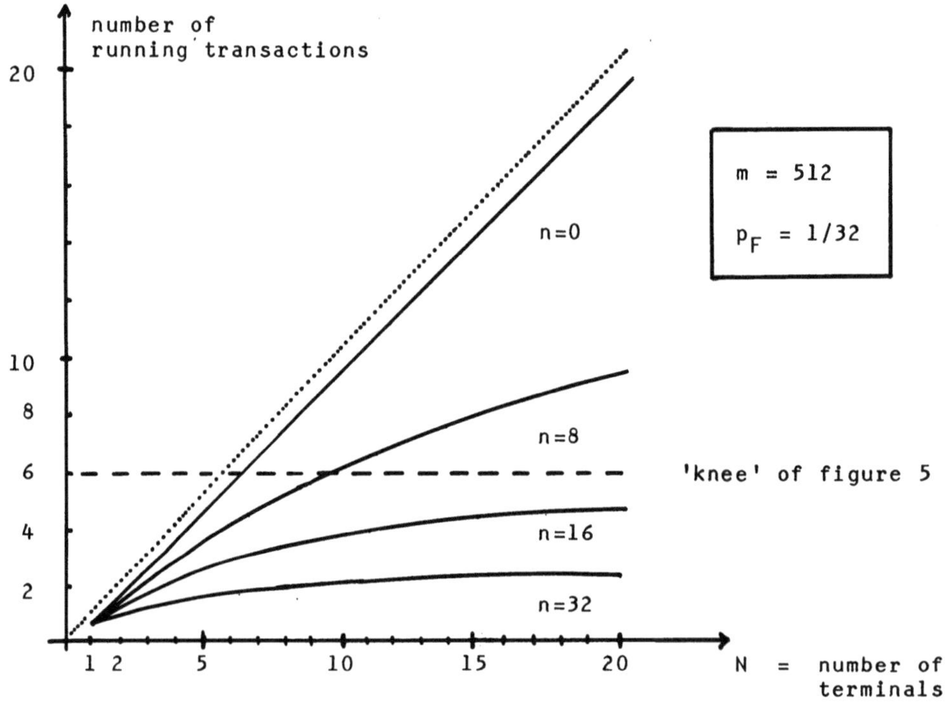

Figure 7: Tradeoff between number of terminals and number of running transactions

Remark: The number of simultaneously running transactions is upper bounded by min $(N,\frac{m}{n})$. The bound $\frac{m}{n}$ (optimal distribution of locks) will not be tightly approached very often, however, even if N becomes very large since in this case most of the terminals become blocked but according to our assumptions only one of the blocked TA is tested and possibly unblocked after finishing and lock release of another transaction. If most objects are locked it will be highly unlikely that all the locks necessary for the test candidate become available from locks released by a departing TA. For this reason TT will not necessarily converge to $TT^{(max)}$ if N goes to infinity.

4. CONCLUSION

A simple database management system has been analyzed by mathematical techniques. Results are obtained in easily evaluated closed form. The effect of eight important system parameters may be derived from our formulae. Several numerical examples have been presented to demonstrate the application of our results.

5. REFERENCES

[BCMP] Baskett, Chandy, Muntz, Palacios: 'Open, closed and mixed networks of queues with different classes of customers'. Journal of the ACM 22; 1975.

[Da] Date: 'An introduction to database systems'. Addison Wesley, 1977.

[GM] Gelenbe, Muntz: 'Probabilistic models of computing systems - Part I'. Acta Informatica 7, 1976.

[PL] Potier, Leblanc: 'Analysis of locking policies in DBMS'. CACM 23, 1980.

[Sa] Saal: 'Analyse eines Transaktionensystems'. Diploma thesis; University of Bonn 1981.

THE OUTPUT PROCESS OF THE SINGLE SERVER QUEUE WITH PERIODIC ARRIVAL PROCESS AND DETERMINISTIC SERVICE TIME

DESCRIPTION DU FLUX DE SORTIE D'UN MULTIPLEXEUR DE CANAUX DETERMINISTES IDENTIQUES

Pierre **BOYER** Alain **DUPUIS** Annie **GRAVEY** Jean-Marc **PITIE**

CENTRE NATIONAL D'ETUDES DES TELECOMMUNICATIONS
LAA/SLC/EVP - BP 40 - 22301 LANNION Cedex
Tél. : (96) 38.24.52
Télex : 730963F

ABSTRACT

We consider n independant identical deterministic packet streams. Packets must share the transmission capacity of a single outgoing trunk.

Through a single server queue model, we show that the output process is periodic and we derive closed expressions for the joint distribution function of successive busy and idle periods and for the waiting time till idleness of the output line.

RESUME

On considère n canaux de transmission indépendants identiques véhiculant chacun des paquets de longueur constante séparés par des blancs de durée également constante. Ces paquets doivent se partager la capacité d'écoulement d'un unique canal de sortie identique aux canaux d'entrée. Le multiplexage des différents flux est assuré par une file d'attente régie par une discipline PAPS.

Le flux s'écoulant sur le canal de sortie est périodique et l'on donne la loi conjointe des périodes d'activité et d'inactivité qui se révèle être une loi produit.

On donne aussi certaines caractéristiques intrinsèques du flux de sortie (nombre de périodes d'activité, nombre de périodes d'activité d'une longueur donnée) ainsi que la distribution de l'attente de libération du canal.

INTRODUCTION

The present study is motivated by the evaluation of AMBRE (see [Cou 81]), a loop-like packet network integrating data, speech and videocommunications. Most of the traffic handled by the ring is speech and videocommunication; moreover, there is no bandwidth compression. We assume that all communications are periodic successions of packets of fixed length.

When AMBRE operates as a concentrating network the output process analysis of every subnetwork can be carried through using a multiplexer model. A multiplexer consists of n identical input lines leading to an infinite buffer and one output line (identical to the input lines) leading out. We assume that each input line delivers a packet of length θ seconds every T seconds. Moreover the n input lines are independant of each other. The service discipline is FIFO.

Through a single server queue model we first show that the traffic on the output line is periodic. We then give the joint distribution of the active periods and idle periods of the output line. Lastly we give the distribution of the waiting time till idleness of the output line.

The model described here has previously been formulated and studied by A. E. ECKBERG [Eck 79] who derived an algorithm for exactly computing the survivor function of packet delay. Under different assumptions concerning the arrival process such a multiplexer has been extensively studied by M. RUBINOVITCH [Rub 73], J. W. COHEN [Coh 74], H. KASPI and M. RUBINOVITCH [KaR 75]. These three studies assume that idle periods on the input lines are exponentially distributed. In [KaR 75] input lines may be different. These three studies yield M/G/1 models for the output process.

CONTENTS

I
THE ARRIVAL PROCESS

I-1 Description of the packet transmission system

Packets of the same length are transmitted over a single trunk, one packet at a time. The transmission time for a packet is θ seconds and the packets are transmitted in the order of their arrival (FCFS discipline).

Packets arrive in a buffer memory, prior to their transmission, over n identical incoming trunks. Every T seconds each incoming trunk delivers a packet. Furthermore we assume that the n arrival streams are independant.

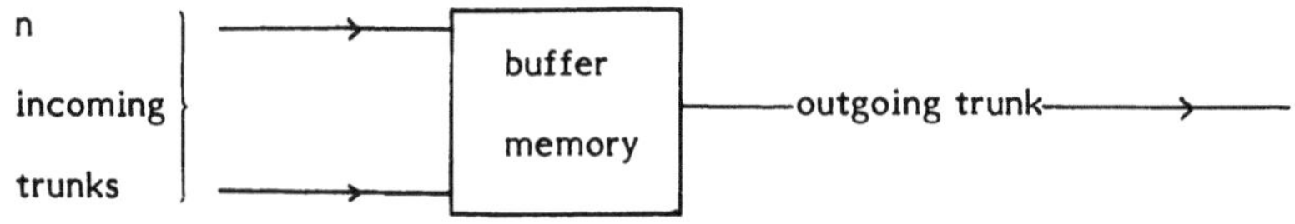

figure 1

We suppose that $T > n\theta$ (I - 1)

Figure 2 represents the busy and idle periods of an outgoing trunk in the case n = 4.

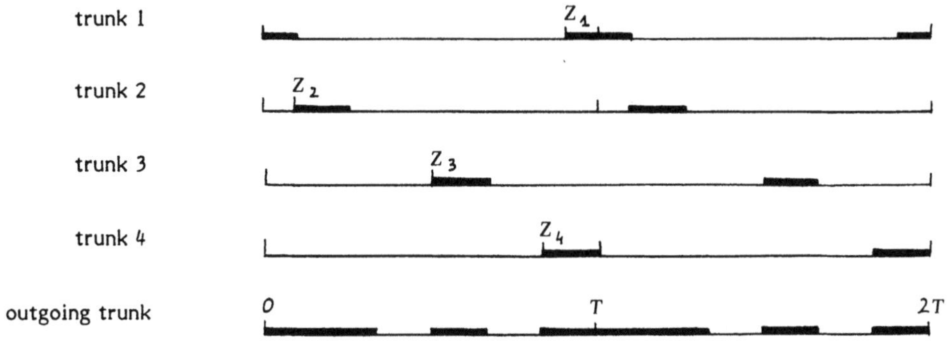

figure 2

I-2 The single server queue model

The study of the transmission system is done through a single server queue model. The date of arrival of a customer in the queuing model is taken to be the date of arrival of the first bit of the corresponding packet in the transmission system. So, the arrival process to the queue is the superposition of n identical independant periodic processes. The service time corresponds to the transmission time of a packet. Hence, it is deterministic of length θ seconds.

Let us call k-customer a customer corresponding to a packet delivered by trunk k. For each k ($k = 1,..,n$), exactly one k-customer will arrive to the queue during a fixed interval $[0 , T[$. Let Z_k be its date of arrival and Y_k ($k = 2,..,n$) be the interarrival time between a one-customer C and the first k-customer to arrive to the queue after C.

We will next prove that the two following assumptions are equivalent.

i) $(Z_1,..,Z_n)$ are n independant random variables, uniformly distributed over $[0,T[$.

ii) $(Z_1,Y_2,...,Y_n)$ are n independant random variables, uniformly distributed over $[0,T[$

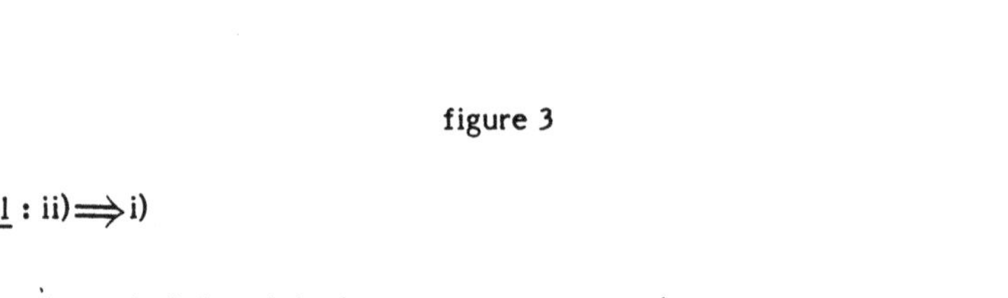

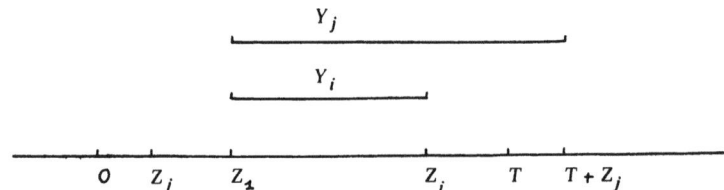

figure 3

Result I-2-1 : ii) $\Longrightarrow$ i)

Proof : using the periodicity of the input streams we see that

$$Z_j = (Y_j+Z_1).1_{\{Y_j<T-Z_1\}} + (Y_j+Z_1-T).1_{\{Y_j\geq T-Z_1\}}$$

Let $\Psi_{Z_j}(t) = Ee^{itZ_j}$ be the characteristic function of Z_j .

Since Z_1 is uniformly distributed over $[0,T[$:

$$\Psi_{Z_1}(t) = \frac{e^{itT}-1}{itT}$$

For $j = 2,\dots,n$ $\quad \Psi_{Z_j}(t) = E\,\psi(t,Z_1,Y_j)$ where

$$\psi(t,z_1,y_j) = 1_{\{y_j<T-z_1\}} \cdot e^{it(z_1+y_j)} + 1_{\{y_j \geq T-z_1\}} \cdot e^{it(z_1+y_j-T)}$$

Since Y_j is uniformly distributed over $[0,T[$ and independant of Z_1:

$$\Psi_{Z_j}(t) = \frac{1}{T^2}\int_0^T dz_1 \int_0^T \psi(t,z_1,y_j)dy_j$$

since :
$$\frac{1}{T}\int_0^T \psi(t,z_1,y_j)dy_j = \frac{e^{itT}-1}{itT}$$

we have :
$$\Psi_{Z_j}(t) = \frac{e^{itT}-1}{itT}$$

Hence Z_j is uniformly distributed over $[0,T[$.

Let $\Psi_{Z_1,\dots,Z_n}$ be the characteristic function of $(Z_1,\dots,Z_n)$

$$\begin{aligned}\Psi_{Z_1,\dots,Z_n}(t_1,\dots,t_n) &= E\left\{e^{i(t_1Z_1+\dots+t_nZ_n)}\right\} \\ &= E\left\{ e^{it_1Z_1} \prod_{j=2}^{n} e^{it_jZ_j} \right\} \\ &= E\left\{ e^{it_1Z_1} \prod_{j=2}^{n} \psi(t_j,Z_1,Y_j) \right\}\end{aligned}$$

Using that $Z_1,Y_2,\dots,Y_n$ are independant :

$$\begin{aligned}\Psi_{Z_1,\dots,Z_n}(t_1,\dots,t_n) &= \frac{1}{T}\int_0^T e^{itz_1}dz_1 \prod_{j=2}^{n} \frac{1}{T}\int_0^T \psi(t_j,y_j,z_1)dy_j \\ &= \frac{1}{T}\int_0^T e^{itz_1} \prod_{j=2}^{n} \frac{e^{it_jT}-1}{it_jT}\,dz_1 \\ &= \prod_{j=1}^{n} \frac{e^{it_jT}-1}{it_jT}\end{aligned}$$

Hence the random variables $Z_1,\dots,Z_n$ are independant.

Theorem I-2-2

Let $Z_1,\dots,Z_n$ be n independant random variables uniformly distributed over $[0,T[$. Let F be a random variable over $[0,T[$, independant of $(Z_1,\dots,Z_n)$. Let $Y_1,\dots,Y_n$ be n random variables defined by :

$$Y_j = 1(Z_j \geq F) \cdot (Z_j - F) + 1(Z_j < F) \cdot (Z_j - F + T)$$

Then $(F,Y_1,\dots,Y_n)$ are $(n+1)$ independant random variables and for each j, Y_j is uniformly distributed over $[0,T[$.

Proof:

Let μ_F be the probability measure for F and $\Psi_{Y_j}(t)$ be the characteristic function of Y_j.

$$\Psi_{Y_j}(t) = E\, e^{itY_j}$$

$$= E\, \psi(t,T-F,Z_j)$$

$$= \int_0^T \mu_F(df) \frac{1}{T}\int_0^T \psi(t,T-f,z_j)dz_j$$

Using that :

$$\frac{1}{T}\int_0^T \psi(t,T-f,z_j)dz_j = \frac{e^{itT}-1}{itT}$$

we see that

$$\Psi_{Y_j}(t) = \frac{e^{itT}-1}{itT}$$

Hence Y_j is uniformly distributed over $[0,T[$.

Let

$$\Psi_{F,Y_1,\dots,Y_n}(t_0,\dots,t_n) = E\, e^{(it_0F+t_1Y_1+\dots+t_nY_n)}$$

$$= \int_0^T e^{it_0f} \mu_F(df) \prod_{j=1}^{n} \frac{1}{T}\int_0^T \psi(t,T-f,z_j)dz_j$$

using that $(F,Z_1,\dots,Z_n)$ are independant. Furthermore we have already seen that the second integral does not depend on f. Hence:

$$\Psi_{F,Y_1,\dots,Y_n}(t_0,\dots,t_n) = \Psi_F(t) \prod_{j=1}^{n} \Psi_{Y_j}(t)$$

$(F,Y_1,\dots,Y_n)$ are independant random variables.

Remarks: 1 - On any fixed interval I of length T, exactly n customers arrive to the queuing system; the dates of arrival are independant and uniformly distributed over I ($I = [F, F+T[$)

2 - Using theorem 1.2.2 with $F = Z_1$ and $(Z_2,\dots,Z_n)$ we see that i)$\Longrightarrow$ii)

From now on we will assume indifferently either i) or ii).

I-3 The arrival process is not a renewal process

Let $[0, T[$ be a fixed interval, Z_k be the date of arrival of the k-customer arriving during $[0, T[$ and $(Z^*_1,\dots,Z^*_n)$ be the n order statistics for $(Z_1,\dots,Z_n)$:

$$0 \leq Z^*_1 < \dots < Z^*_n < T$$

Since $(Z_1,\dots,Z_n)$ are independant and uniformly distributed over $[0, T[$, the joint density for $(Z^*_1,\dots,Z^*_n)$ is :

$$f(z_1,\dots,z_n) = 1_{(0 \leq z_1 < \dots < z_n < T)} \cdot \frac{n!}{T^n}$$

Let $(Z^*_i)_{i \geq 1}$ be the arrival process to the queue. Since for each k the arrival process of the k-customers is periodic:

$$Z^*_{k+nj} \overset{as}{=} Z^*_k + jT \qquad 1 \leq k \leq n,\ j \geq 0$$

If $(L_k)_{k>0}$ is the interarrival process:

$$L_k = Z^*_{k+1} - Z^*_k \qquad k \geq 1$$

We have $$L_{k+nj} \overset{as}{=} L_k \qquad 1 \leq k \leq n,\ j \geq 0$$

moreover $$\sum_{k=i+1}^{n+i} L_k \overset{as}{=} T \qquad i \geq 0$$

Using the expression for $f(z_1,\dots,z_n)$ we can show that $(L_k)_{k \geq 1}$ are not identically distributed and that

$$EL_k = T/(n+1) \qquad 1 \leq k \leq n-1$$

$$EL_n = E(T-Z^*_n)+EZ^*_1 = 2T/(n+1)$$

II
PERIODICITY OF THE OUTPUT PROCESS

II-1 All k-customers experience the same waiting time

Let 0 be a fixed point and W_k be the waiting time experienced by the kth customer arriving to the queue after 0. We assume that W_1 is almost surely bounded by a constant $A \geq 0$.

For $k \geq 1$, we let $\qquad U_k = \theta - L_k = \theta - Z^*_{k+1} + Z^*_k$

Using I-3 we have $\qquad U_{k+nj} \overset{as}{=} U_k \qquad$ (II-1-1)

and $\qquad U_1+...+U_n \overset{as}{=} n\theta - T \qquad$ (II-1-2)

The queue discipline is FIFO . Hence we can apply Lindley's formula to $(W_k)_{k \geq 1}$:

if $k \geq 1 \qquad W_{k+1} = \max(0, W_k + U_k)$

Theorem II-1 :
Let j_A be the smallest integer such that $j_A(n\theta - T) + A < 0$. If $j \geq j_A$ we have for $k = 1,2,...,n$:

$$W_{k+1+jn} = \max(0, U_k, U_k+U_{k-1}, ..., U_k+...+U_1, U_k+...+U_1+U_n, ..., U_k+...+U_1+U_n+...+U_{k+2})$$

and

$$W_{1+jn} = \max(0, U_n, U_n+U_{n-1}, ..., U_n+...+U_2)$$

Lastly, for each k, all k-customers experience then the same waiting time.

Proof: $W_2 = \max(0, W_1+U_1)$
$W_3 = \max(0, U_2, U_2+U_1+W_1)$

.

.

$W_{k+1} = \max(0, U_k, U_k+U_{k-1}, ..., U_k+...+U_1+W_1)$

.

.

$W_n = \max(0, U_{n-1}, ..., U_{n-1}+U_{n-2}+...+U_1+W_1)$
$W_{1+n} = \max(0, U_n, U_n+U_{n-1}, ..., U_n+U_{n-1}+...+U_1+W_1)$

since $U_1+...+U_n \overset{as}{=} n\theta-T<0$, we see that:

$$W_{1+n} = \max(0,U_n,U_n+U_{n-1},...,U_n+...+U_2,n\theta-T+W_1)$$

And, using II-1-1, we see that as long as $j<j_A$

$$W_{1+jn} = \max(0,U_n,U_n+U_{n-1},...,U_n+...+U_2,W_1+j(n\theta-T))$$

and $$W_{1+j_An} = \max(0,U_n,U_n+U_{n-1},...,U_n+...+U_2)$$

Using II-1-1, II-1-2, we prove then easily theorem II-1 by induction.

From now on, we suppose that for each k, all k-customers experience the same waiting time.

II-2 Description of the busy periods of the server.

The state of the server is described by the following function :

$$1_S(t) = \begin{cases} 0 & \text{if the queue is empty} \\ 1 & \text{otherwise} \end{cases}$$

Clearly, if $W_k = 0$ a new busy period begins at Z^*_k and if $W_k \neq 0$ the kth customer is served just after the $(k-1)^{th}$ customer during the same busy period of the server.

Since $$Z^*_{k+n} \overset{as}{=} Z^*_k + T$$

$$W_{k+n} \overset{as}{=} W_k$$

we see that $$1_S(t+T) \overset{as}{=} 1_S(t)$$

Hence, the number of busy periods beginning during an interval of length T (random or fixed) does not depend on that interval. Let X be that number. Let also be X_k the number of busy periods of length $k\theta$ beginning during any interval of length T. It is important to note that neither X nor X_k depend on the interval during which the output process is observed.

III
A PROBABILITY MEASURE
ON THE STATES OF THE OUTPUT PROCESS

Let 0 be a fixed point. We note V the event $\{$the queue is empty at $0^-\}$. In that part we give the distribution of the successive busy and idle periods of the server during the interval $[0, T[$, assuming V.

III-1 Derivation of P(V)

For each k (k = 1,...,n) let Z_k be the date of arrival of the only k-customer arriving during $[0, T[$. We assume that $Z_1,...,Z_n$ are n independant random variables, each uniformly distributed over $[0, T[$.

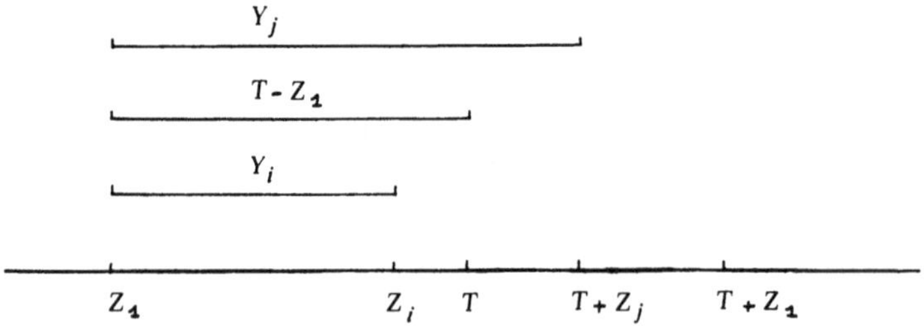

figure 4

If $j \geq 2$, let $Y_j = (T-Z_1+Z_j).1_{(Z_j<Z_1)} + (Z_j-Z_1).1_{(Z_j \geq Z_1)}$

Using theorem I-2-2, we know that $(Z_1,Y_2,...,Y_n)$ are n independant random variables , each uniformly distributed over $[0,T[$.

$$P(V) = E1_{(V)} = E[E[1_{(V)}|Y_2,...,Y_n]]$$

as

$$E[1_{(V)}|Y_2,...,Y_n] = P[V|Y_2,...,Y_n] = 1-n\frac{\theta}{T}$$

Indeed, if we know $Y_2, \ldots, Y_n$, we know the busy periods of the server during $[Z_1, Z_1+T[$. Morever, since Z_1 is uniformly distributed over $[0, T[$, the observation point T is uniformly distributed over $[Z_1, Z_1+T[$. Hence the probability that the server is idle at that point is clearly $1-n\theta/T$ and $P(V) = 1-n\theta/T$.

III-2 Description of the output process conditionally to V

$P^V(.)$ is the conditional probability measure assuming V.

III-2-a Joint distribution of the busy periods and the idle periods.

Let X be the number of busy periods beginning during $[0, T[$. If we assume $\{X = p\}$ let $k_1,\ldots,k_p,t_1,e_2,\ldots,e_p$ be such that:

$k_i\theta$ is the length of the ith busy period.
the first busy period begins at date t_1.
the length of the idle period preceding the ith busy period is e_i.

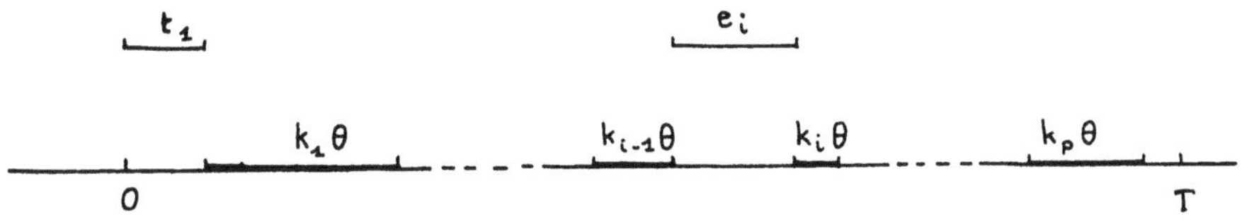

figure 5

We remark that, conditionally to V, these random variables describe the output process of the queuing system.

Clearly, we must have :

$$\begin{cases} k_i \in \{1,2,\ldots,n+1-p\} & 1 \leq i \leq p. \\ e_i > 0 & 2 \leq i \leq p \\ t_1+e_2+\ldots+e_p < T-n\theta \end{cases}$$

Let $A(p,k_1,\dots,k_p,t_1,e_2,\dots,e_p,dt_1,de_2,\dots,de_p)$ be the following event :

$$\left\{\begin{array}{l} \text{the queue is empty at } 0^- \\ X = p \\ \text{the lengths of the p busy periods are } k_1\theta, \dots, k_p\theta \\ \text{the first busy period begins between } t_1 \text{ and } t_1 + dt_1 \\ \text{for } i = 2,\dots,p \text{, the length of the idle period preceding the } i^{th} \text{ busy period is in} \\ \quad [e_i, e_i + de_i[\end{array}\right\}$$

Result III-2-a :

$P^V(A(p,k_1,\dots,k_p,t_1,e_2,\dots,e_p,dt_1,de_2,\dots,de_p))$

$$= \frac{1}{T^n} \frac{1}{1-\frac{n\theta}{T}} \prod_{i=1}^{p} \frac{(k_i\theta)^{k_i-1}}{k_i!} \cdot 1(t_1 \geq 0) \cdot 1(e_2>0) \cdots 1(e_p>0) \cdot 1(t_1+e_2+\dots+e_p<T-n\theta)\, dt_1\, de_2 \dots de_p$$

Proof is given in the appendice

III-2-b Joint distribution of the lengths of the busy periods

Let $A(p,k_1,\dots,k_p)$ be the following event :

$$\left\{\begin{array}{l} \text{the queue is empty at } 0^- \\ X = p \\ \text{the lengths of the p busy periods are } k_1\theta, \dots, k_p\theta \end{array}\right\}$$

Since $$P^V(A(p,k_1,\dots,k_p)) = \int_{\substack{t_1 \geq 0 \\ e_2>0,\dots,e_p>0 \\ t_1+e_2+\dots+e_p<T-n\theta}} P^V(A(p,k_1,\dots,k_p,t_1,e_2,\dots,e_p,dt_1,de_2,\dots,de_p))$$

$$P^V(A(p,k_1,\dots,k_p)) = \frac{n!}{p!} \prod_{i=1}^{p} \frac{k_i^{k_i-1}}{k_i!} \left(\frac{\theta}{T}\right)^{n-p} \left(1-n\frac{\theta}{T}\right)^{p-1}$$

remark: the lengths of the busy periods during $[0, T[$ are not independant.

We see that : $P^V(X=p,V) = P^V(X=p) = \sum_{\substack{k_1+\dots+k_p=n \\ k_i>0}} P^V(A(p,k_1,\dots,k_p))$

Using one of the multinomial Abel identities (see Rior 79 , p 26) :

$$(R) \qquad \sum_{\substack{k_1+\dots+k_p=n \\ k_i>0}} \prod_{i=1}^{p} \frac{k_i^{k_i-1}}{k_i!} = \frac{p}{n} \frac{1}{(n-p)!} n^{n-p} = p \frac{n^{n-p-1}}{(n-p)!}$$

we have:

$$P^V(X=p) = \binom{n-1}{p-1} (1-n\frac{\theta}{T})^{p-1} (n\frac{\theta}{T})^{n-p} \qquad 1 \le p \le n$$

and :

$$E^V X = E(X|V) = n(1-(n-1)\frac{\theta}{T})$$

<u>III-2-c $P^{V,\{X=p\}}(.)$ is a product form probability measure.</u>

$P^{V,\{X=p\}}(.)$ is the conditional probability measure assuming V and $\{X=p\}$.

Let $A(p,t_1,e_2,\dots,e_p,dt_1,de_2,\dots,de_p)$ be the following event :

- the queue is empty at 0^-
- $X = p$
- the first busy period begins between t_1 and t_1+dt_1
- for $i=2,\dots,p$ the length of the idle period preceding the ith busy period is in $[e_i , e_i+de_i[$

<u>Theorem III-2-c</u> : $P^{V,\{X=p\}}$ is a product form probability measure on the space of the states of the output process, that is:

$$P^{V,\{X=p\}}(A(p,k_1,\dots,k_p,t_1,e_2,\dots,e_p,dt_1,de_2,\dots,de_p))$$

$$= P^{V,\{X=p\}}(A(p,k_1,\dots,k_p)) \,.\, P^{V,\{X=p\}}(A(p,t_1,e_2,\dots,e_p,dt_1,de_2,\dots,de_p))$$

with $\displaystyle\sum_{\substack{k_1+\dots+k_p=n \\ k_i>0}} P^{V,\{X=p\}}(A(p,k_1,\dots,k_p)) = 1$

and $\displaystyle\int_{\substack{t_1\ge 0 \\ e_i>0 \quad 2\le i\le p \\ t_1+e_2+\dots+e_p<T-n\theta}} P^{V,\{X=p\}}(A(p,t_1,e_2,\dots,e_p,dt_1,de_2,\dots,de_p)) = 1$

Proof : For any event A included in $\{X=p\}$ $\quad P^{V,\{X=p\}}(A) = P^V(A) / P^V(X=p)$

Hence, using III-2-a

$$P^{V,\{X=p\}}(A(p,k_1,\dots,k_p,t_1,e_2,\dots,e_p,dt_1,de_2,\dots,de_p))$$

$$= \frac{n}{p}\frac{(n-p)!}{n^{n-p}} \prod_{i=1}^{p} \frac{k_i^{k_i-1}}{k_i!} \frac{p!}{(T-n\theta)^p} 1(t_1 \geq 0) \cdot 1(e_2>0) \cdot \dots \cdot 1(e_p>0) \cdot 1(t_1+e_2+\dots+e_p<T-n\theta) dt_1 de_2 \dots de_p$$

Using III-2-b

$$P^{V,\{X=p\}}(A(p,k_1,\dots,k_p)) = \frac{n}{p}\frac{(n-p)!}{n^{n-p}} \prod_{i=1}^{p} \frac{k_i^{k_i-1}}{k_i!}$$

Using (R) we check that this is a probability law on the set

$$\{ (k_1,\dots,k_p),\ k_i \in \mathbb{N}^*,\ k_1+\dots+k_p = n \}$$

Moreover, using (R)

$$P^{V,\{X=p\}}(A(p,t_1,e_2,\dots,e_p,dt_1,de_2,\dots,de_p))$$

$$= \sum_{\substack{k_1+\dots+k_p=n \\ k_i>0}} P^{V,\{X=p\}}(A(p,k_1,\dots,k_p,t_1,e_2,\dots,e_p,dt_1,de_2,\dots,de_p))$$

$$= \frac{p!}{(T-n\theta)^p} 1(t_1 \geq 0) \cdot 1(e_2>0) \cdot 1(e_p>0) \cdot 1(t_1+e_2+\dots+e_p<T-n\theta) dt_1 de_2 \dots de_p \ .$$

Result III-3-c follows immediatly.

Remark : in the same way, we have also studied the output process assuming

$$V_1 = \{ \text{the 1-customers do not wait} \}$$

The results are similar (see [EVP 81]) and $P^{V_1,\{X=p\}}$ is also a product form probability measure on the space of the states of the output process.

IV
DISTRIBUTION OF THE RESIDUAL BUSY PERIOD

In the transmission system, let us consider a $(n+1)^{th}$ trunk. As long as the outgoing trunk is busy (i.e packets of the n incoming trunks are transmitted), the packets delivered by the $(n+1)^{th}$ trunk wait. They are transmitted as soon as the outgoing trunk is free.

If a packet is delivered by the $(n+1)^{th}$ trunk at date u we call W(u) the time it will wait before being transmitted.

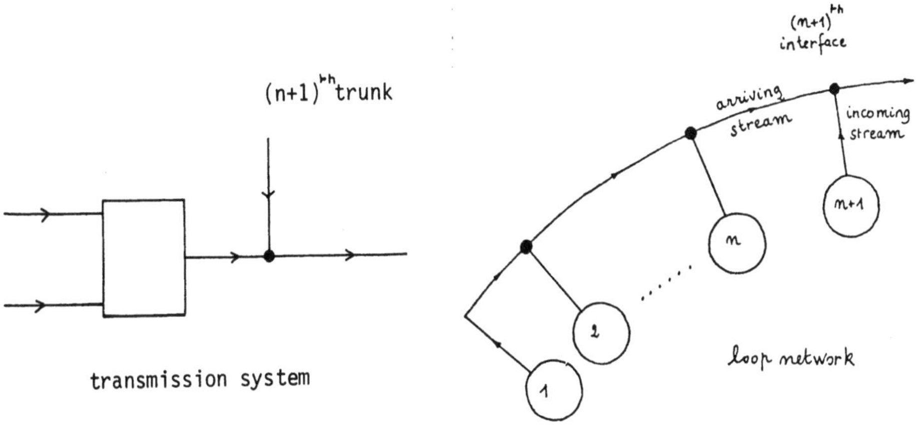

figure 6

Using the single server queue model we see that

W(u) = 0 if the server is idle

= a if the server is busy and if the busy period ends at date u+a

We note also that W(u) is the waiting time experienced by an incoming packet in station n+1 if the loop-like packet network operates as a concentrator. Indeed it has been proved [EVP 83] that the stream of packets arriving in the $(n+1)^{th}$ station is the same than the one departing from the transmission system described in I-1.

In that part, we first give E(X) and ($E(X_k)$, k=1,...,n), mean values of the number of busy periods and of busy periods of length $k\theta$ beginning in any interval of length T. With those results we will then be able to give the distribution of W.

IV-1 E(X), E(X_k)

Let $[0 , T[$ be a fixed interval and t any point in that interval. We have seen in I-2 that on all the intervals $[t , t+T[$ the arrival process has the same distribution.

If V_t = { the queue is empty at t^- } we have already seen in III-1 that

$$P(V_t) = 1 - \frac{n\theta}{T}$$

Let B(t,dt,k) be the following event :

$$\left\{ \begin{array}{l} \text{the queue is empty at } t^- \\ \text{a busy period of length } k\theta \text{ begins during } [t , t+dt[\end{array} \right\}$$

Result IV-1-1

$$P(B(t,dt,k)) = \binom{n}{k} \left(\frac{k\theta}{T}\right)^{k-1} \left(1 - \frac{k\theta}{T}\right)^{n-k-1} \left(1 - \frac{n\theta}{T}\right) \frac{dt}{T}$$

Proof :We will use the results in III-2-a to derive P(B(t,dt,k)) from $P(A(p,k ,\ldots,k_p,0,e_2,\ldots,e_p,dt,de_1,\ldots,de_p))$

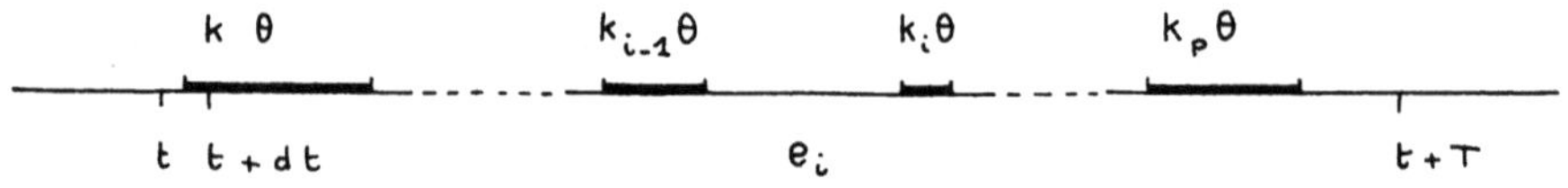

figure 7

if k=n

$$B(t,dt,n) = P(A(1,n,0,dt))$$

$$= \left(\frac{n\theta}{T}\right)^{n-1} \frac{dt}{T}$$

if k<n

$$P(B(t,dt,k)) = \sum_{p=2}^{n-k+1} \sum_{\substack{k_2+\ldots+k_p=n-k \\ k_i>0}} \int_{(e_2,\ldots,e_p)} P(A(p,k,\ldots,k_p,0,e_2,\ldots,e_p,dt,de_2,\ldots,de_p))$$

Since $\int_{(e_2,\ldots,e_p)} 1(e_2>0)\cdots 1(e_p>0)\ 1(e_2+\ldots+e_p<T-n\theta)\ de_2\ldots de_p = \frac{(T-n\theta)^{p-1}}{(p-1)!}$

and $\sum_{\substack{k_2+\ldots+k_p=n-k \\ k_i>0}} \frac{k_2^{k_2-1}\ldots k_p^{k_p-1}}{k_2!\ldots k_p!} = \frac{p-1}{n-k}\,\frac{1}{(n-k-p+1)!}\,(n-k)^{n-k-p+1}$

we find

$$P(B(t,dt,k)) = \left(\frac{k\theta}{T}\right)^{k-1} \binom{n}{k} \left(1-\frac{k\theta}{T}\right)^{n-k} \frac{dt}{T} \sum_{p=2}^{n-k+1} \frac{p-1}{n-k} \binom{n-k}{p-1} \left(\frac{(n-k)\theta}{T-k\theta}\right)^{n-k-p+1} \left(\frac{T-n\theta}{T-k\theta}\right)^{p-1}$$

$$= \left(1-\frac{n\theta}{T}\right)\left(\frac{k\theta}{T}\right)^{k-1}\binom{n}{k}\left(1-\frac{k\theta}{T}\right)^{n-k-1}\frac{dt}{T}$$

If B(t,dt) is the following event:

$$\left\{\begin{array}{l}\text{the queue is empty at } t^- \\ \text{a busy period begins during } [t,t+dt[\end{array}\right\}$$

We see that $B(t,dt) = \sum_{k=1}^{n} B(t,dt,k)$

Result IV-1-2

$$P(B(t,dt)) = \left(1-\frac{(n-1)\theta}{T}\right)\frac{ndt}{T}$$

Proof:

$$P(B(t,dt)) = \sum_{p=1}^{n} \sum_{\substack{k_1+\ldots+k_p=n \\ k_i>0}} \int_{(e_2,\ldots,e_p)} P(A(p,k_1,\ldots,k_p,0,e_2,\ldots,e_p,dt,de_2,\ldots,de_p))$$

$$= \sum_{p=1}^{n} n! \left(\frac{\theta}{T}\right)^{n-p} \sum_{\substack{k_1+\dots+k_p=n \\ k_i>0}} \frac{k_1^{k_1-1}\dots k_p^{k_p-1}}{k_1!\dots k_p!} \int_{\substack{e_2+\dots+e_p<T-n\theta \\ e_i>0}} \frac{de_2}{T}\dots\frac{de_p}{T}\frac{dt}{T}$$

$$= \frac{ndt}{T}\left(1 - \frac{(n-1)\theta}{T}\right)$$

Let X (respectively X_k) be the number of busy periods (respectively of busy periods of length $k\theta$) beginning during $[0, T[$.

Result IV-1-3

$$E(X_k) = \binom{n}{k}\left(\frac{k\theta}{T}\right)^{k-1}\left(1 - \frac{k\theta}{T}\right)^{n-k-1}\left(1 - \frac{n\theta}{T}\right)$$

$$E(X) = n\left(1 - \frac{(n-1)\theta}{T}\right)$$

Proof: Let $X_k(t)$ be the counting process defined by:

$$\{X_k(t) = i\} \Leftrightarrow \{\text{exactly } i \text{ busy periods of length } k\theta \text{ begin during } [0,t[\ \}$$

Clearly $X_k(t+dt) = \begin{cases} X_k(t)+1 & \text{if a busy period of length } k\theta \text{ begins during } [t,t+dt[\\ X_k(t) & \text{otherwise} \end{cases}$

Hence $E(X_k(t+dt)) = E(X_k(t)) + P(B(t,dt,k))$

Using IV-1-2 we have

$$\lim_{dt\to 0} \frac{E(X_k(t+dt)) - E(X_k(t))}{dt} = \binom{n}{k}\left(\frac{k\theta}{T}\right)^{k-1}\left(1 - \frac{k\theta}{T}\right)^{n-k-1}\left(1 - \frac{n\theta}{T}\right)\frac{1}{T}$$

Hence $E(X_k(t))$ admits a constant derivative on $[0, T[$. Since $X_k(0) = 0$, we have as

$$E(X_k) = \binom{n}{k}\left(\frac{k\theta}{T}\right)^{k-1}\left(1 - \frac{k\theta}{T}\right)^{n-k-1}\left(1 - \frac{n\theta}{T}\right)$$

Diagram 1 shows $(E(X_k), k=1,\dots,n)$ for n=10 and different values of $\rho = n\theta/T$.
The derivation of E(X) is quite similar.

Remarks : 1 - $E(X) = \sum_{k=1}^{n} E(X_k)$

2 - $\sum_{k=1}^{n} kE(X_k) = n$

3 - if we observe X and $(X_k, k=1,\dots,n)$ on a random interval $[F,F+T[$ where F is independant of the arrival process, using I-2 , we see that we get the same mean values for X and $(X_k, k=1,\dots,n)$.

4 - it is possible to show that X and X_k are independant from the event V {the queue is empty at 0^-} . Hence $E^V(X) = E(X)$ and $E^V(X_k) = E(X_k)$.

IV-2 Distribution of W

Using the results in I-2 we see that W(u) does not depend on u. From now on, we will suppose that u=0.

Theorem IV-2-1 : the distribution function of W admits a unique discontinuity at 0 of magnitude $1-n\theta/T$. Its continuous part admits a density α on $]0,n\theta]$

$$\alpha(a) = \frac{1}{T}\sum_{k=i}^{n} EX_k \qquad a \in](i-1)\theta, i\theta]$$

Proof: Clearly, W is smaller than the biggest busy period of the server: $W \leq n\theta$ almost surely

Moreover $P(W=0)$ = P(the queue is empty at 0^-)
$= 1 - n\theta/T$

If $(i-1)\theta < a < a+da < i\theta$

$$P(W \in [a,a+da[) = \sum_{k=i}^{n} P(B(a-k\theta, da, k))$$

$$= \frac{da}{T}\sum_{k=i}^{n} EX_k$$

Using IV-2-1 we see that, on the set $](i-1)\theta, i\theta]$

$$P(W \leq a) = 1 - n\frac{\theta}{T} + \sum_{k=1}^{n} \inf(a, k\theta)\, EX_k$$

Diagram 2 shows $P(W \leq a)$ for $n\theta / T = .5$

We will now prove that for each $i \geq 1$

$$E\,W^i = \frac{1}{(i+1)T} \sum_{k=1}^{n} (k\theta)^{i+1}\, EX_k \qquad \text{(IV-2-2)}$$

<u>Proof:</u> $E\,W^i = \int_0^{n\theta} a^i \alpha(a) da$

$$= \sum_{i=1}^{n} \alpha(i\theta) \int_{(i-1)\theta}^{i\theta} a^i\, da$$

$$= \frac{1}{T} \sum_{k=1}^{n} EX_k \int_0^{k\theta} a^i da$$

$$= \frac{1}{(i+1)T} \sum_{k=1}^{n} (k\theta)^{i+1}\, EX_k$$

For $i=1$

$$E(W) = \frac{1}{2T} \sum_{k=1}^{n} (k\theta)^2\, EX_k$$

with a bit of tedious algebra, we get

$$E(W) = \frac{\theta}{2} \sum_{k=1}^{n} k \frac{n!}{(n-k)!} \left(\frac{\theta}{T}\right)^k$$

Diagram 3 shows $E(W)/T$ as a function of T/θ for different values of n (full curves) and of $\rho = n\theta/T$ (dotted curves).

<u>remark :</u> $\lim_{n\theta \to T} E(W) = T/2$

APPENDICE

Derivation of $P(A(p,k_1,\ldots,k_p,t_1,e_2,\ldots,e_p,dt_1,de_2,\ldots,de_p))$

If $(Z_1,\ldots,Z_n)$ are n independant random variables, uniformly distributed over $[0,T[$, let $(Z_1^*,\ldots,Z_n^*)$ be the n order statistics for $(Z_1,\ldots,Z_n)$. The joint density for $(Z_1^*,\ldots,Z_n^*)$ is:

$$f(z_1,\ldots,z_n) = \frac{n!}{T^n} 1_{(0\leq z_1<z_2<\ldots<z_n<T)}$$

Let 0 be a fixed point. We assume that the queuing system is empty at 0^-. We note X the number of busy periods beginning during $[0,T[$, and if $X=p$, $k_1\theta,\ldots,k_p\theta$ the lengths of the p busy periods. Let ξ_1 be the date of the beginning of the first busy period and ϵ_i be the length of the idle period preceding the ith busy period.

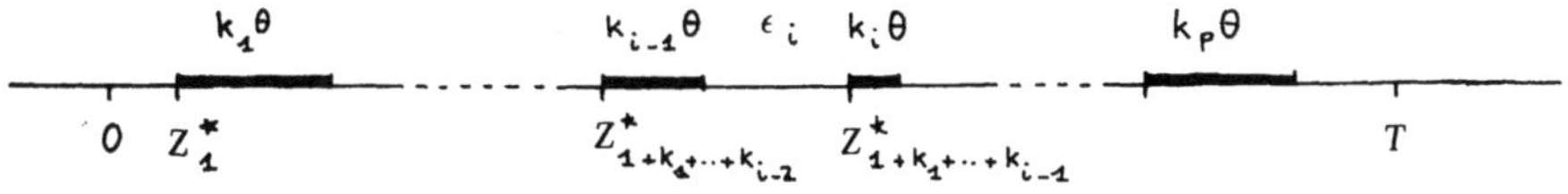

figure 8

Clearly, we have

$$(H)\quad\begin{cases} 1\leq p\leq n \\ 1\leq k_i\leq n-p+1 \qquad (i=1,\ldots,p) \\ k_1+\ldots+k_p=n \\ \\ \xi_1\geq 0 \\ \epsilon_i>0 \qquad (i=2,\ldots,p) \\ \xi_1+\epsilon_2+\ldots+\epsilon_p<T-n\theta \end{cases}$$

Let $A(p,k_1,\ldots,k_p,t_1,e_2,\ldots,e_p,dt_1,de_2,\ldots,de_p)$ be the following event:

$$\left| \begin{array}{l} \text{the queue is empty at } 0^- \\ X = p \\ \text{the lengths of the p busy periods are } k_1\theta ,\ldots, k_p\theta \\ \text{the first busy period begins between } t_1 \text{ and } t_1 + dt_1 \\ \text{for } i = 2,\ldots,p \text{ , the length of the idle period preceding the } i^{th} \text{ busy} \\ \text{period is in } [e_i , e_i + de_i [\end{array} \right.$$

If (H) is true for $(t_1,e_2,\ldots,e_p)$ it is possible to find $dt_1,de_2,\ldots,de_p$ such that (H) is true for $(\xi_1,\epsilon_2,\ldots,\epsilon_p)$ as long as

$$\begin{array}{ll} t_1 \leq \xi_1 < t_1+dt_1 & \\ e_i \leq \epsilon_i < e_i+de_i & (i=2,\ldots,p) \end{array}$$

The first busy period begins during $[t_1,t_1+dt_1[$ if and only if :

$$t_1 \leq Z_1^* < t_1+dt_1$$

Moreover, assuming that $X \geq i$ and that the total length of the (i-1) first busy periods is $k_1\theta +\ldots+k_{i-1}\theta$, the $(i-1)^{th}$ busy period ends at date $k_{i-1}\theta + Z^*_{1+k_1+\ldots+k_{i-2}}$ and the ith busy period begins at date $Z^*_{1+k_1+\ldots+k_{i-1}}$. Hence the length of the ith idle period is

$$\epsilon_i = Z^*_{1+k_1+\ldots+k_{i-1}} - (k_{i-1}\theta + Z^*_{1+k_1+\ldots+k_{i-2}})$$

and we must have

$$e_i \leq Z^*_{1+k_1+\ldots+k_{i-1}} - (k_{i-1}\theta + Z^*_{1+k_1+\ldots+k_{i-2}}) < e_i+de_i$$

For $j=1,\ldots,k_i$, let $Y_j=Z^*_{k_1+\ldots+k_{i-1}+j}$. Assuming that the queuing system is empty at date Y_1^- , a busy period of length $k_1\theta$ begins at Y_1 if and only if

$$Y_{j-1} < Y_j \leq Y_1 + (j-1)\theta \qquad j=2,\ldots,k_i$$

Hence $P(A(p,k_1,\dots,k_p,t_1,e_2,\dots,e_p,dt_1,de_2,\dots,de_p))$

$$= \frac{n!}{T^n}\int_0^T \cdots \int_0^T \Bigg[\Big(1(t_1 \le z_1 < t_1+dt_1) \prod_{j=1}^{k_1-1} 1(z_j < z_{j+1} \le z_1 + j\theta)\Big)$$

$$\prod_{i=2}^{p} \Big(1(e_i \le z_{1+k_1+\dots+k_{i-1}} - (k_{i-1}\theta + z_{1+k_1+\dots+k_{i-2}}) < e_i + de_i)$$

$$\prod_{j=1}^{k_i-1} 1(z_{j+k_1+\dots+k_{i-1}} < z_{j+1+k_1+\dots+k_{i-1}} \le j\theta + z_{1+k_1+\dots+k_{i-1}})\Big)\Bigg] dz_1 \dots dz_n$$

$$= \frac{n!}{T^n}\Bigg[\Big(\int_{t_1}^{t_1+dt_1} dz_1 \int_{z_1}^{z_1+\theta} dz_2 \cdots \int_{z_{k_1-1}}^{z_1+(k_1-1)\theta} dz_{k_1}\Big)$$

$$\prod_{i=2}^{p}\Big(\int_{e_i+k_{i-1}\theta+z_{1+k_1+\dots+k_{i-2}}}^{e_i+de_i+k_{i-1}\theta+z_{1+k_1+\dots+k_{i-2}}} dz_{1+k_1+\dots+k_{i-1}}$$

$$\int_{z_{1+k_1+\dots+k_{i-1}}}^{\theta+z_{1+k_1+\dots+k_{i-1}}} dz_{2+k_1+\dots+k_{i-1}}$$

$$\cdots \int_{z_{k_1+\dots+k_i-1}}^{(k_i-1)\theta+z_{1+k_1+\dots+k_{i-1}}} dz_{k_1+\dots+k_i}\Big)\Bigg]$$

Using the next lemma, we see that

$P(A(p,k_1,\dots,k_p,t_1,e_2,\dots,e_p,dt_1,de_2,\dots,de_p))$

$$= \frac{n!}{T^n}\prod_{i=1}^{p}\frac{(k_i\theta)^{k_i-1}}{k_i!} 1(t_1+e_2+\dots+e_p < T-n\theta)\cdot 1(t_1 \ge 0)\prod_{i=2}^{p} 1(e_i>0)\, dt_1\, de_2 \dots de_p$$

<u>lemma</u> :

$$F_n = \int_a^{a+\theta} dx_1 \int_{x_1}^{a+2\theta} dx_2 \dots \int_{x_{n-1}}^{a+n\theta} dx_n = \frac{[(n+1)\theta]^n}{(n+1)!}$$

<u>Proof</u> : let $y_i = x_i - a$

Then $$F_n = \int_0^{\theta} dy_1 \int_{y_1}^{2\theta} dy_2 \dots \int_{y_{n-1}}^{n\theta} dy_n$$

Let $$F_n(k) = \int_0^{\theta} dy_1 \int_{y_1}^{2\theta} dy_2 \dots \int_{y_{n-1}}^{n\theta} \frac{(y_n)^k}{k!} dy_n .$$

$$F_n(k) = \int_0^{\theta} dy_1 \dots \int_{y_{n-2}}^{(n-1)\theta} \left\{ \frac{(n\theta)^{k+1}}{(k+1)!} - \frac{(y_{n-1})^{k+1}}{(k+1)!} \right\} dy_{n-1}$$

$$= \frac{(n\theta)^{k+1}}{(k+1)!} F_{n-1} - F_{n-1}(k+1)$$

hence $$F_n(0) = \sum_{k=0}^{n-2} (-1)^k \frac{[(n-k)\theta]^{k+1}}{(k+1)!} F_{n-k-1} + (-1)^{n-1} F_1(n-1)$$

since $$F_1(n-1) = \int_0^{\theta} \frac{(y_1)^{n-1}}{(n-1)!} dy_1 = \frac{\theta^n}{n!}$$

We have

$$F_n = F_n(0)$$

$$F_n = \sum_{k=0}^{n-1} (-1)^k \frac{[(n-k)\theta]^{k+1}}{(k+1)!} F_{n-k-1}$$

By induction, we now prove $F_n = \frac{[(n+1)\theta]^n}{(n+1)!}$

Clearly $F_1 = \theta$

if for $k=1,\ldots,n-1$ $F_k = \frac{[(k+1)\theta]^k}{(k+1)!}$

$$F_n = \sum_{k=0}^{n-1} (-1)^k \frac{[(n-k)\theta]^{k+1}}{(k+1)!} F_{n-k-1}$$

$$= \sum_{j=1}^{n} (-1)^{n-j} \frac{(j\theta)^{n-j+1}}{(n-j+1)!} F_{j-1}$$

$$= \theta^n \sum_{j=1}^{n} (-1)^{n-j} \frac{j^n}{j!(n+1-j)!}$$

$$= \frac{\theta^n}{(n+1)!} \left\{ - \sum_{j=0}^{n+1} (-1)^{n+1-j} \binom{n+1}{j} j^n + (n+1)^n \right\}$$

using [Fel 66], pp65 : $\sum_{j=0}^{n+1} (-1)^{n+1-j} \binom{n+1}{j} j^n = 0$

and $F_n = \frac{[(n+1)\theta]^n}{n!}$

CONCLUSION

In this paper, the study of the output process of the multiplexer is conducted. In [Gra 83] further results concerning the distributions of the waiting time and the number of customers in the system are given.

AKNOWLEDGMENTS

The authors thank Jean PELLAUMAIL (INSA-Rennes) for many useful discussions during this work.

BIBLIOGRAPHY

[Coh 74] J.W. COHEN. "Superimposed renewal processes and storage with gradual input". Stochastic Processes and their applications 2 (1974)

[Cou 81] Jean-Pierre COUDREUSE. "Un réseau local expérimental de vidéocommunications en anneau à commutation de paquets : le projet AMBRE". IDATE 3èmes Journées Internationales des Réseaux locaux 10-1981

[Eck 79] A. E. ECKBERG. "The Single Server queue with periodic arrival process and deterministic service time".
IEEE Transactions on communications 3 (1979).

[Fel 66] W. FELLER . "Probability theory and its applications".
Wiley 1966

[KaR 75] H. KASPI M. RUBINOVITCH. "The stochastic behaviour of a buffer with non-identical input lines".
Stochastic Processes and their applications 3 (1975)

[Rior 79] J. RIORDAN. "Combinatorial identities".
Krieger 1979

[Rub 73] M. RUBINOVITCH. "The output of a buffered data communication system".
Stochastic Processes and their applications 1 (1973)

[EVP 81] P.BOYER A.DUPUIS A. GRAVEY JM.PITIE. "Etude de la superposition de canaux déterministes identiques à phases aléatoires".
Technical Report of the Centre National d'Etudes des Télécommunications, NT/LAA/SLC/99 (1983)

[Gra 83] A.GRAVEY. "Temps d'attente et nombre de clients dans une file nD/D/1".
to be published in Annales de l'Institut Henri Poincaré.

[EVP 83] JM.PITIE P.BOYER. "Concentration de flux dans les réseaux arborescents à débit uniforme"
Technical Report of the Centre National d'Etudes des Télécommunications, NT/LAA/SLC/140 (1983)

Diagram 1

$E(X_k)$ n=10

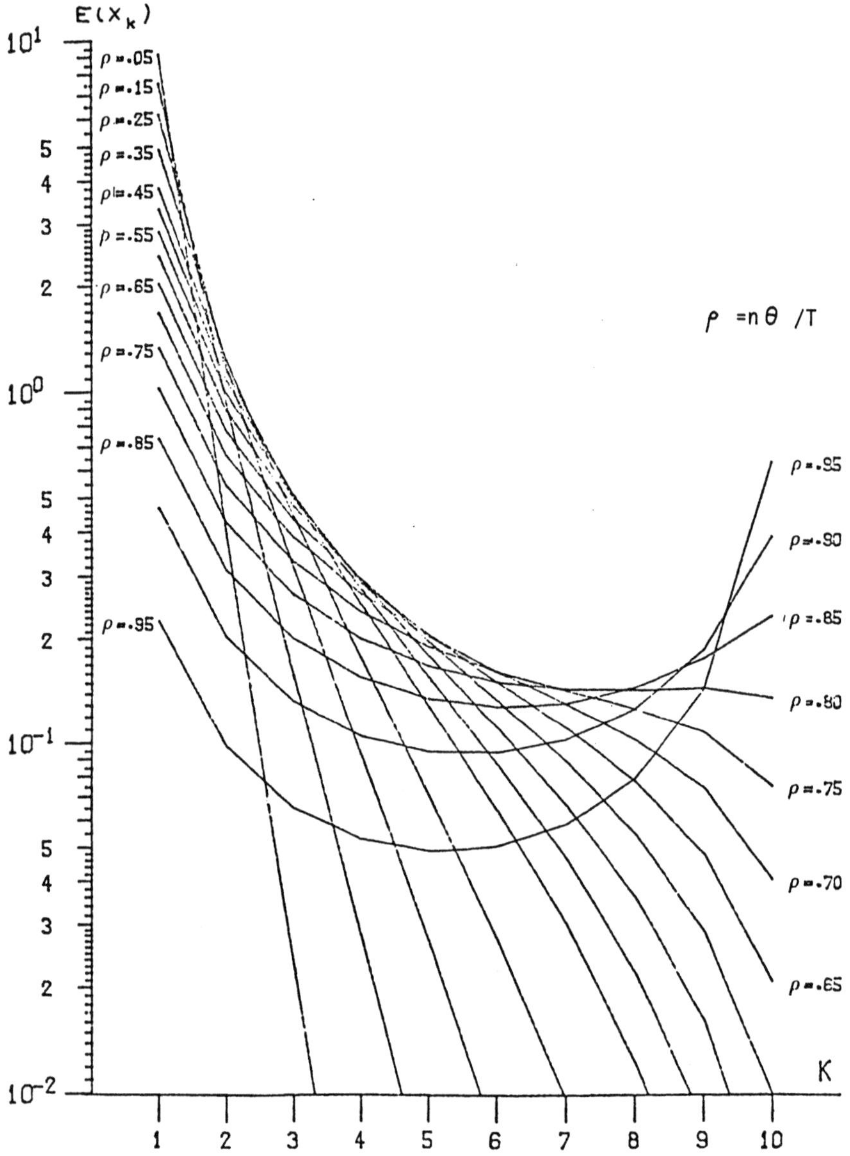

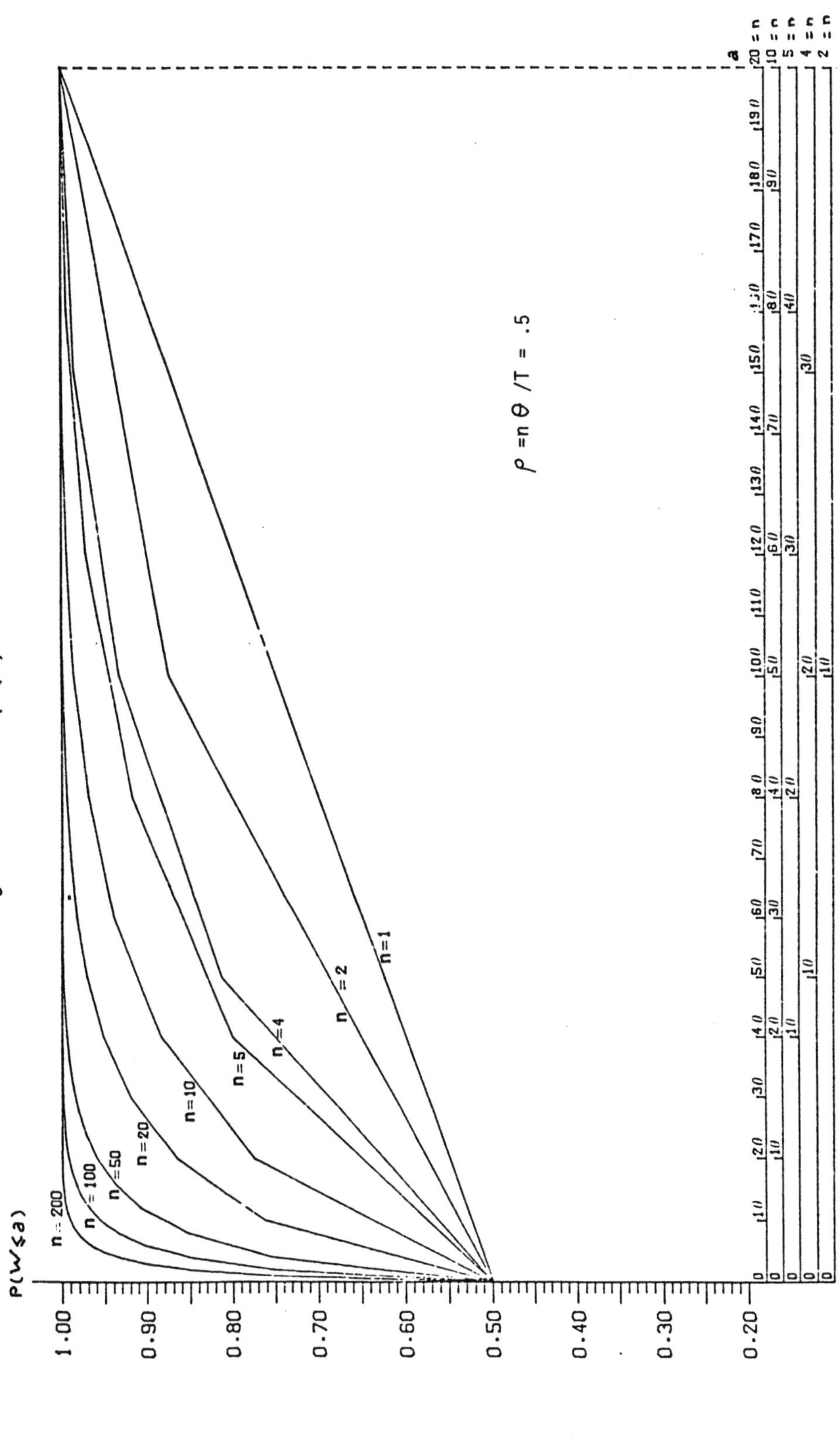

Diagram 2 P(W≤a)

Diagram 3 E(W)/T

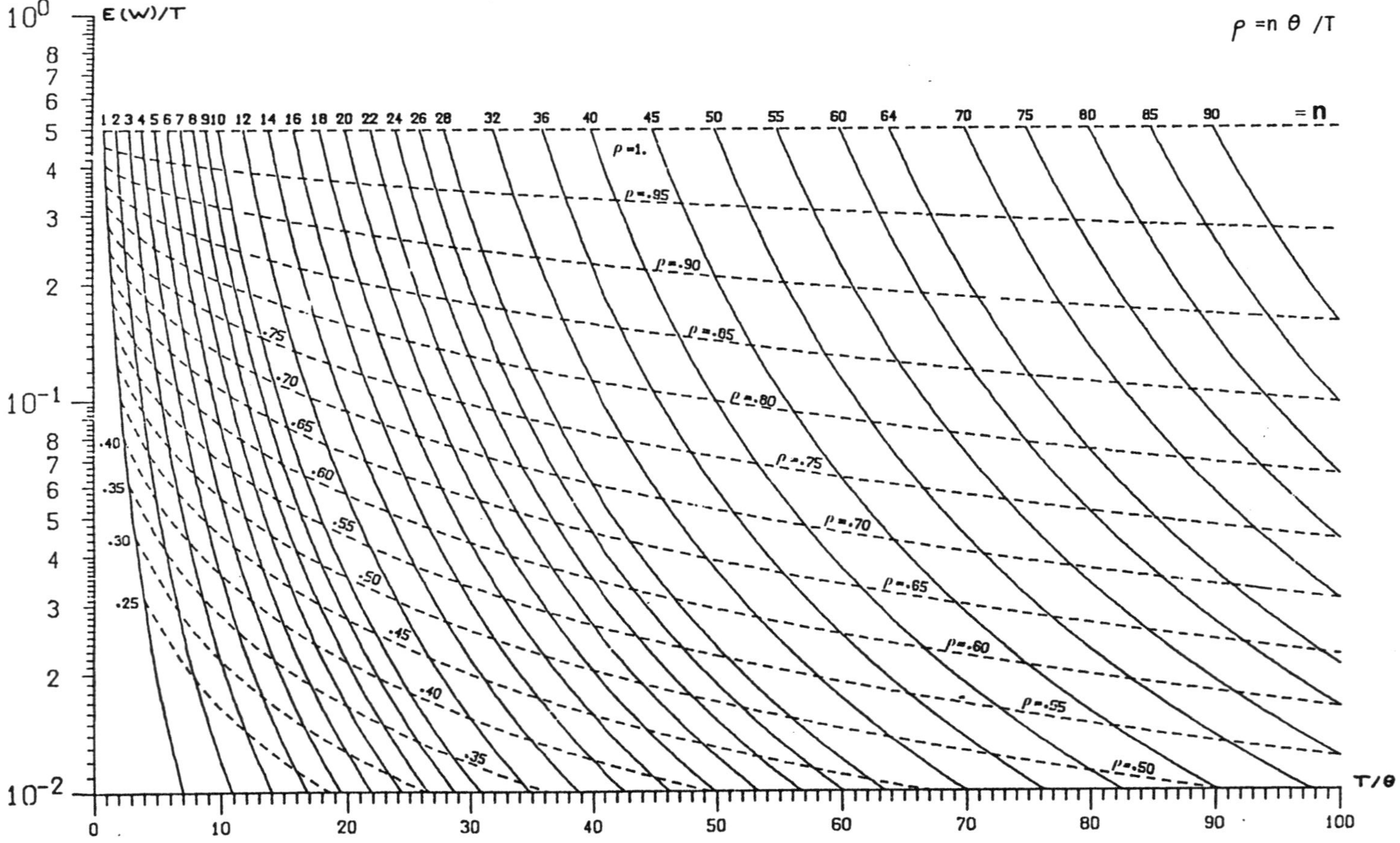

IX

ANALYSIS PROTOCOLS 2

ANALYSE DE PROTOCOLES 2

Performance Modelling for Multiprocessor-Systems, Fundamental Concepts and Tendencies

Ulrich Herzog
Universität Erlangen-Nürnberg

Abstract

This contribution surveys first performance problems and fundamental modelling concepts for multiprocessor computer systems processing independent user tasks. Uniting classical as well as recent results allows to accurately describe their system behavior.

Progressive multiprocessor systems, however, take advantage of the parallelism inherent in many problems, i.e. application programs are decomposed into sets of cooperating subtasks and processed in parallel, when possible. So one may increase not only the throughput of a system: run-times (and therefore response-times) for individual application programs may be reduced significantly, too. Then, however, difficult coordination problems (synchronization between tasks, data- and code-sharing, etc.) may accur and have to be considered in modelling such systems. Measurements also show that system overhead due to interprocessor communication can be significant and has to be taken into account.

1. Motivation

There is a well known theorem in queuing theory that says that best performance, i.e. fastest response time, can be achieved iff all processing power is concentrated in a single server. Some numerical results are shown in figure 1, and we may learn from the lower curve that a uniprocessor with rate $\mu_1=1$ is about five times as fast as an ensemble of ten processors each of which with a processing rate $\mu_{10}=0.1$.

Experiences with progressive multiprocessor systems also show that considerable overhead occurs due to coordination problems, memory interference and process communication. So, at a first glance, multiprocessor computer systems seem to be inferior to uniprocessors.

Little success of industrial and university projects in the past supports this believe.

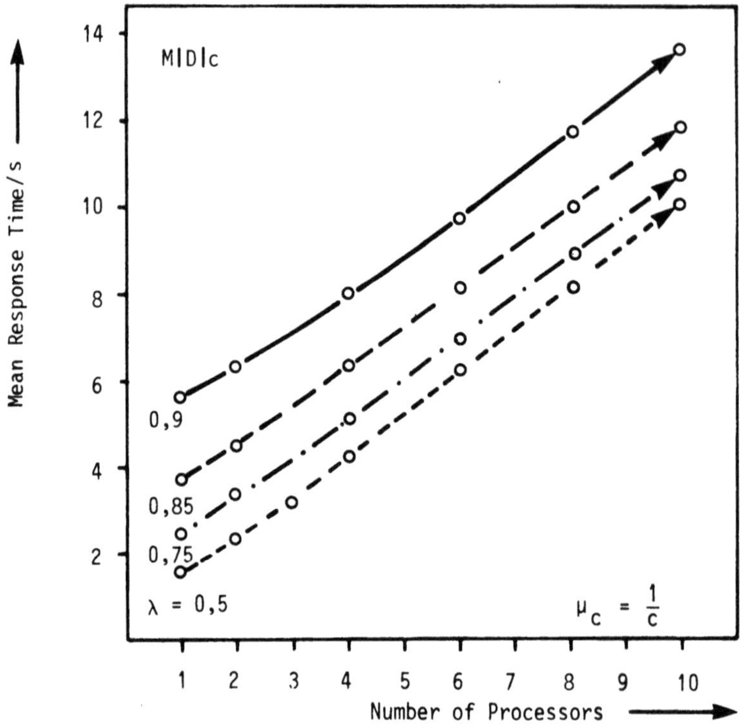

Fig. 1.: Loss in response time when a given processing power is split into c parts. (λ arrival rate, processing rate μ_c).

There are, however, several reasons why research and development of multiprocessor systems are intensified:

- Limited processing power even of the largest mainframes but increasing requirements in modern chemistry, physics, powerplant design and control, pattern recognition, etc. etc.
- Increasing need for reliable and faulttolerant computer systems.
- Enormous advances in microelectronic technology used the best advantage in regular, multiple times repeated structures.

And there are assumptions in our theorem not valid for all computer organizations and application programs:

- User tasks are assumed to be indivisible and independent on each other.
- The sequence of microscopic processing steps is stated to be indentical in both uni - and multiple processor systems.

Consequently, there is an increasing scientific and commercial interest in a modelling technique which includes the particulars of multiprocessor hardware, system software as well as application programs.

2. Modelling Classical Multiprocessor Systems With Independent User Tasks

2.1. Standard Analysis

Multiprocessor systems with two, three or four processing units have been built since many years by the industry: Burroughs, Control Data, IBM, Siemens and many others. The most commonly quoted reasons are robustness and throughput [23]: Given that the workload at an installation is larger than what the largest uniprocessor can handle, there are two possibilities

A) We use several uniprocessors and partition the workload between them; the corresponding queuing model consists of c identical single server systems and performance characteristics may be obtained by classical M/G/1 results.

B) We use a multiprocessor system, i.e. several processing units under a single operating system; accordingly, M/G/c-results may be applied for performance analysis.

The speedup S, defined as the expected response time ratio

$$S = \frac{E[T_R, \text{ several uniprocessors}]}{E[T_R, \text{ multiprocessor}]} ,$$

is shown in figure 2 using classical M/M/c results of A.K. Erlang[3]. In recent years, however, considerable progress has been made in developing approximations for various operating characteristics of the M/G/c queue [2,15]. Some results are available even for systems with interarrival and service times distributed according to phase type distributions [14].

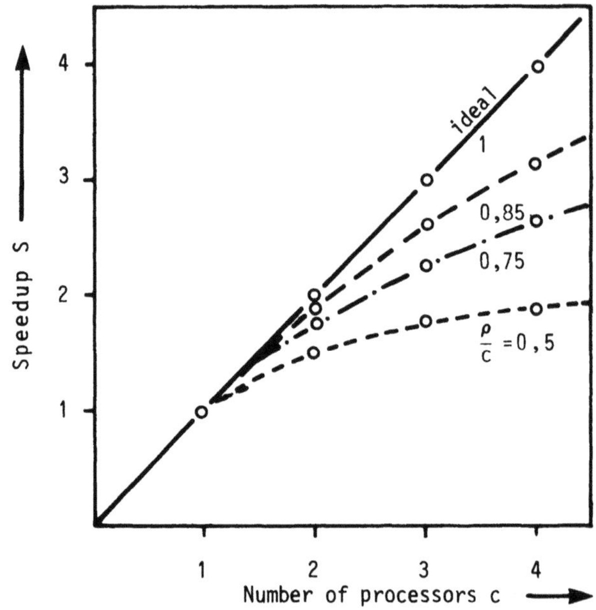

Fig. 2: Speedup for multiprocessor configurations (ρ utilization).

All these results prognosticate asymptotic speedup S. From real system measurements, however, we know that they are too optimistic because of additional multiprocessor overhead. Each processor in a multiprocessor configuration performs worse then when it is a uniprocessor configuration. The principal reasons are

- contention for memory
- lockouts for critical sections
- more complex scheduling
- communication delays.

Taking into account the influence of shared resources is demonstrated next by memory contention models.

2.2. Memory Interference

Memory conflicts may occur whenever two or more processors attempt to gain access to the same memory module simultaneously. The effect of memory conflicts, referred to as memory interference, may decrease the execution rate of each processor significantly. Much attention has been paid to the analysis of this phenomena and standard as well as newest results are summarized in [5-7]. Embedding these results in the M/G/c-queuing model realistic speedup values may be obtained, cf. figures 3 and 4: Dependent on the instruction service time ratio

$$\gamma = \frac{E[\text{Memory Acces Time}]}{E[\text{Processing Time}]}$$

each additional processor contributes relatively less to the overall performance and the curve may even drop again, the famous breakdown phenomena of many multiprocessor projects.

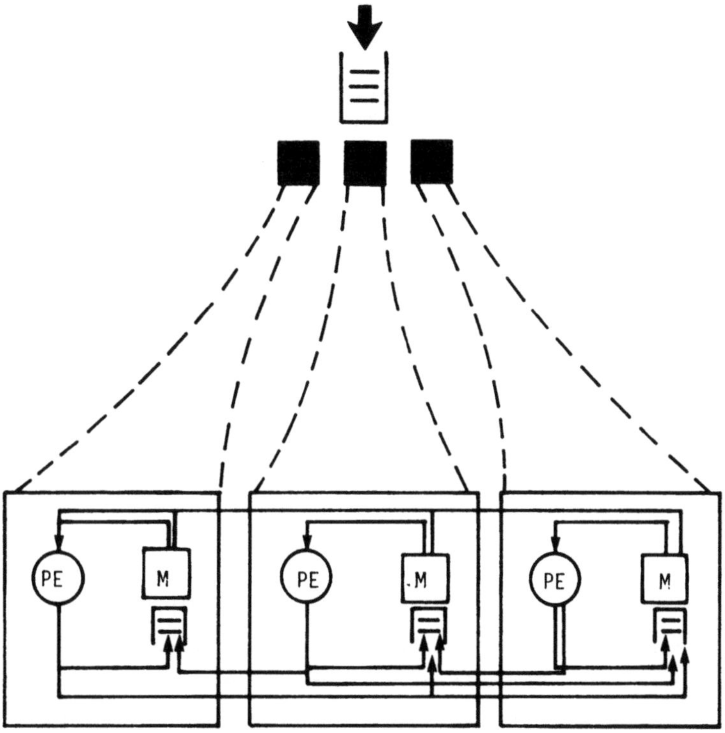

Fig. 3.: Hierarchical modelling technique demonstrated for a three processor configuration (PE: processing element, M: memory unit)

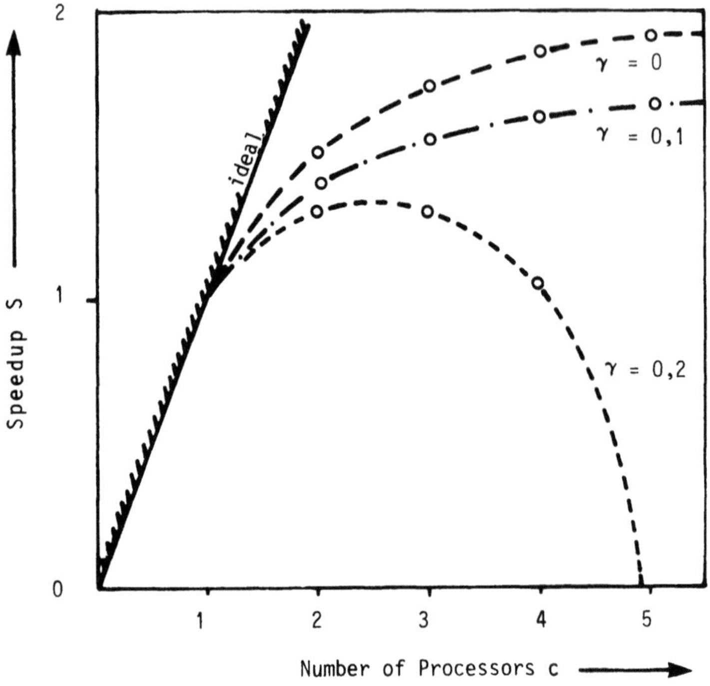

Fig. 4.: Speedup S for multiprocessor configurations taking into account memory conflicts (extreme example where all processors access the same memory block; γ instruction service time ration).

3. Modelling Progressive Multiprocessor Concepts

3.1. General Remarks

In the classical modelling technique one usually assumes concurrent processes to be independent of each other; on the other hand side it is also standard to assume processes, being dependent on each other, - e.g. I/O and CPU phases - take a sequential turn. Little research considers I/O and CPU overlap and recognizes that programs may be decomposed into well defined cooperating subtasks and processed concurrently.

Modern multiprocessor projects such as CM*[19,20],SMS [21] and EGPA [7,18] take advantage of the inherent parallelism of many application programs. Therefore, the response time for each application may be reduced drastically. Then, however, difficult coordination problems may occur and have to be considered in modelling such systems:

- synchronization between tasks and subtasks
- process communication delays
- data and code sharing problems.

We next describe a basic synchronization model and conclude with some most advanced research results.

3.2. The basis synchronization problem

Be given a program structure as shown in figure 5 :

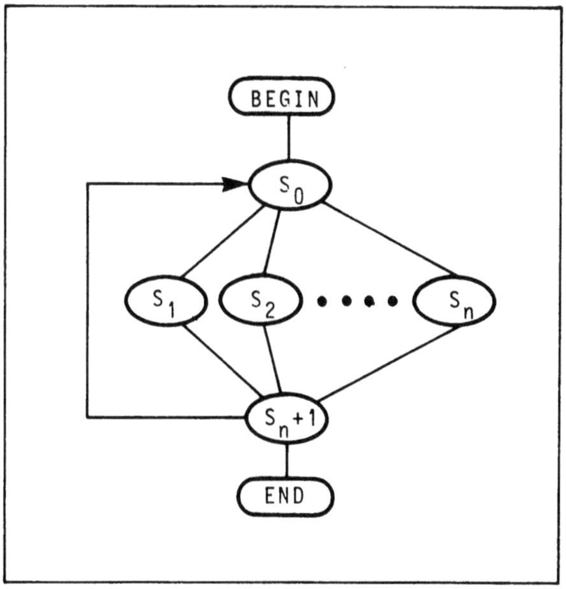

Fig. 5.: Type-1-program structure [9].

Problems are often of this type: algorithm for the solution of linear algebraic or partial differential equation systems, optimization procedures, simulations including subruns for the purpose of estimating confidence intervals, problems of picture processing, etc. etc.

A possible implementation on a hierarchically organized multiprocessor such as SMS or EGPA is illustrated by the timing diagram in figure 6:

- At first the source program is translated, loaded and then started by the coordinating B-processor
- The B-processor initiates the execution of n independent subtasks by the application processors A_1 to A_n.

- Having completed its subtask, each A-processor sends asynchronously a message to the B-processor
- Postprocessing and preparation of a new loop cycle by the B-processor is only possible when all subtasks are completed.

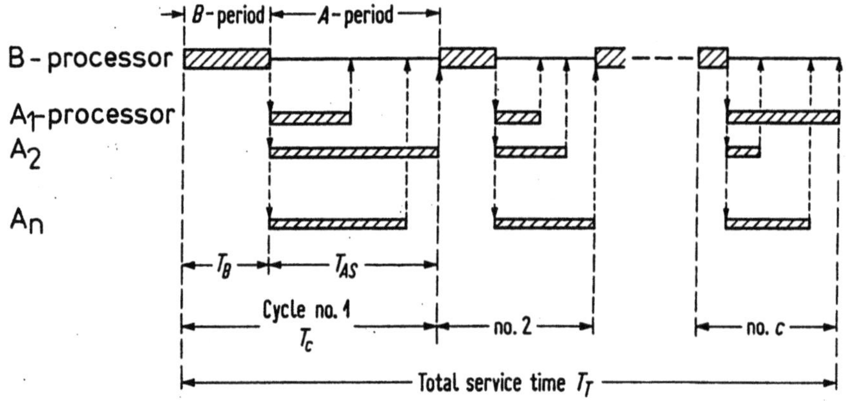

Fig. 6.: Timing diagram

Performance characteristics may be readily obtained by the following procedure:

- Obviously, the A-period is determined by the maximum of n service times T_{Ai}, $i \in \{1,2,\dots,n\}$. Therefore, its d.f. $F_{AS}(t)$ is given by the product of all d.f.s $F_{Ai}(t)$:

$$F_{AS}(t) = P(T_S \leqq t) = P(T_{A_1} \leqq t) \cdot \dots\dots\dots \cdot P(T_{A_n} \leqq t)$$

$$F_{AS}(t) = \prod_{i=1}^{n} F_{A_i}(t)$$

- Example: Suppose, all service intervals T_{Ai} are exponentially distributed with uniform service rates $\epsilon i = \epsilon$. We then obtain

$$F_{AS}(t) = \prod_{i=1}^{n} F_{A_i}(t) = (1 - e^{-\epsilon t})^n$$

$$= \sum_{k=0}^{n} \binom{n}{k} (-1)^k \cdot e^{-k\epsilon t}$$

with mean and variance

$$E[T_{AS}] = \frac{1}{\epsilon} \sum_{k=1}^{n} 1/_k$$

$$VAR[T_{AS}] = \frac{1}{\epsilon^2} \sum_{k=1}^{n} 1/_{k^2} \quad .$$

- Since the cycle time T_C is the sum of the B-period (T_B) and the A-period (T_{AS}), its d.f. is determined by the convolution of these two d.f.s.

$$F_C(t) = F_B(t) * F_{AS}(t)$$

- Example: Suppose the B-period is constant ($T_B = t_b$) and, as before all intervalls T_{Ai} are exponentially distributed.
Then

$$F_C(t) = \begin{cases} o & o \leqq t < t_B \\ \sum_{k=o}^{n} \binom{n}{k} (-1)^k \cdot e^{-k\epsilon(t-t_b)} & t_B \leqq t < \infty \end{cases}$$

with mean and variance

$$E[T_C] = t_B + \frac{1}{\epsilon} \sum_{k=1}^{n} 1/_k$$

$$VAR[T_C] = 1/_{\epsilon^2} \sum_{k=1}^{n} \frac{1}{k^2} \quad .$$

- Finally, the d.f. of the total service time is determined by the c-fold convolution of the cycle time

$$F_T(t) = F_C(t) * F_C(t) * \ldots\ldots\ldots * F_C(t)$$

with mean and variance

$$E[T_T] = c.E[T_C] = c \cdot E[T_{AS}] + c \cdot E[T_B]$$

$$VAR[T_T] = c \cdot VAR[T_C] = c \cdot VAR[T_{AS}] + c \cdot VAR[T_B] \; .$$

3.3. Trends

Many extensions of the above basic results have been obtained taking into consideration various operating modes for multiprocessors, different program structures as well as communication overhead [9,10,12]. And an important generalisation - at first under BCMP-assumptions - has been published only recently by Heidelberger and Trivedi [8].

A totally different approach is chosen by Kleinöder [11]: the task structure is modeled by a modified data flow graph. Combining this model with all implementation constraints, e.g.

- processor/memory topology
- multiprocessor scheduling
- communication overhead

an implementation graph is found which may be evaluated by efficient numerical techniques.

The main objective of performance modelling is not analysis but synthesis, synthesis of optimal structures and operating modes [12]. How do we find a multiprocessor-configuration best suited for a distinct spectrum of applications? How to allocate functions and data to different processors and memory modules? How to schedule tasks and subtasks, globally as well as locally. You may find a summary of related publications as well as interesting new results in Fromm's dissertation [6]. And you may learn that there are still many important and challenging problems for future research.

References:

Modelling and Theory

[1] Agrawala, A.; Herzog, U. (Eds.): Performance Evaluation of Multiple Processor Systems. Special Issue of the IEEE Transactions on Computers, January 1983

[2] Boxma, O.J.; Cohen, J.W.; Huffels, N.: Approximations of the Mean Waiting Time in an M/G/s Queuing System. Operations Research, Vol. 27, No. 6, 1979, pp 1115-1127

[3] Brockmeyer, E.; Halstrøm, H.L.; Jensen, A.: The Life and Works of A.K. Erlang. Kopenhagen: Trans. Dan. Acad. Techn. Sci., No. 2, 1948

[4] Dempster, M.A.; Lenstra, J.K.; Rinnooykan (Eds.): Deterministic and Stochastic Scheduling. Nato Advanced Study Institutes Series, Series C, Reidel Publ. Company, Boston-London, 1982

[5] Fromm, H.J.: Diplomarbeit Zur Modellierung der Speicherinterferenz bei hierarchisch organisierten Multiprozessorsystemen Arbeitsberichte des IMMD, Erlangen, Bd. 13, No 3, 1980

[6] Fromm, H.J.: Dissertation Multiprozessor-Rechenanlagen: Programmstrukturen, Maschinenstrukturen und Zuordnungsprobleme Arbeitsberichte des IMMD,Erlangen, Bd. 15, No 5, 1982

[7] Fromm, H.J.; Hercksen, U.; Herzog, U.; John, K.H.; Klar, R.; Kleinöder, W.: Experiences with Performance Measurement and Modelling of a Processor Array. IEEE-TC-January 1983, p 15-31

[8] Heidelberger, P.; Trivedi, K.: Analytic Queuing Models for Programs with Internal Concurrency. IEEE-TC-January, 1983

[9] Herzog, U.; Hoffmann, W.; Kleinöder, W.: Performance Modeling and Evaluation for Hierarchically Organized Multiprocessor Computer Systems. Proc. Int. Conf. on Parallel Processing, Bellaire, 1979, pp 21-24

[10] Herzog, U.: Performance Characteristics for Hierarchically Organized Multiprocessor Computer Systems with Generally Distributed Processing Times. Archiv für Electronik und Übertragungstechnik, Electronics and Communication, 34, 1980, p 45-51

[11] Kleinöder, W.: Dissertation Stochastische Bewertung von Aufgabenstrukturen für hierarchische Mehrrechnersysteme Arbeitsberichte des IMMD, Erlangen, Bd. 15, No 10, 1982

[12] Kleinöder, W.; Herzog, U.: Einführung in die Methodik der Verkehrstheorie und ihre Anwendung bei Multiprozessor-Rechenanlagen. Computing, Suppl. 3, p 31-64, 1981

[13] Pulkkis, G.: A Comparison of Some Mathematical Models of the Bus Traffic in a Single Bus Multimicroprocessor. Elektron. Informationsverarbeitung und Kybernetik EIK (1982) 1/2, pp 41-55

[14] Takahashi, Y.: Asymptotic Exponentiality of the Tail of the Waiting Time Distribution in a Ph/Ph/c Queue. Adv. in Appl. Probab., to appear

[15] Tijms, H.; van Horn, M.: Approximations for the Steady State Probabilities in the M/G/c Queue. Research Report 49, Vrije Universiteit, Amsterdam

<u>Tables</u>

[16] Hillier, F.S.; Yu, O.S.: Queuing Tables and Graphs. Publication in Operations Research Series, Volume 3, North Holland, New York-Oxford, 1981

[17] Kühn, P.: Tables on Delay Systems. Institut for Switching and Data Techniques, University of Stuttgart, 1976

Multiprocessor Systems

[18] Händler, W.; Hofmann, F.; Schneider, H.J.: A General Purpose Array with a Broad Spectrum of Applications. Computer Architecture (Informatik Fachberichte, Bd. 4) Händler (Ed) p 311-325. Springer 1975

[19] Jones, A.; Chansler, R.; Durham, I.; Feiler, P.; Scelza, D.; Schwans, K.; Vegdahl, S.: Programming Issues Raised by a Multiprocessor. Proc. IEEE, Vol. 66, No 2, February 1978

[20] Jones, A.; Schwarz, P.: Experiences Using Multiprocessor Systems - A Status Report. Computing Surveys, Vol. 12, No 2, June 1980, p 121-165

[21] Kober, R.; Kuznia, C.: SMS 201 - A Powerful Parallel Processor with 128 Microprocessors. Euromicro, 1979. Amsterdam, North-Holland

[22] Recoque, A.: Survey of Main Trends in Computer Hardware Architecture. Proc. IFIP Information Processing 80, S.H. Lavington (Ed.) North Holland Publishing, 1980, pp 115-125

[23] Satayanarayan, M.: A Survey of Multiprocessing Systems IBM Research Report R C 7346, Yorktown Heights, 1978

[24] Schütt, D.: Parallelverarbeitende Maschinen Informatik-Spektrum 3, Springer, 1980, pp 1-8

MODELISATION D'UN RESEAU A COMMUTATION DE PAQUETS ET DE SON CONTROLE DE FLUX DE BOUT EN BOUT

Michel DAO
C.N.E.T.
PAA-ATR-SST
38-40 rue du Général Leclerc
92131 Issy-Les-Moulineaux

__Résumé__ :

Dans cet article nous étudions un réseau à commutation de paquets où un contrôle de flux de bout en bout est implémenté (contrôle de flux par fenêtre). Après avoir défini les caractéristiques des réseaux visés, nous introduisons plusieurs modèles correspon dants à différents états de fonctionnment du réseau. A partir de ces modèles nous calculons les temps mouens de séjour des paquets dans le réseau par la méthode d'analyse des valeurs moyennes. Dans une dernière partie une application de cette étude est présentée.

__Abstract__ :

In this paper we study a store-and-forward computer network using an end-to-end window flow control. We first define the characteristics of the networks under study and then introduce several models corresponding to different working states of the network. We derive from these models the packets mean sojourn time in the network by means of the mean value analysis method. In a last section, we present an application of this study.

1. Introduction

On considère un réseau à commutation de paquets constitué de noeuds de commutation et de lignes de transmission entre ces noeuds. D'autre part des hôtes (terminaux et ordinateurs) sont raccordés à ce réseau afin de communiquer 2 à 2 entre eux. Une des contraintes imposées ici est que les paquets échangés entre 2 hôtes (dans un sens donné) empruntent un chemin fixe (noeuds et liaisons). C'est le cas des circuits virtuels traités dans l'avis X25 ([1]) à la différence que ceux-ci constituent des liaisons bidirectionnelles entre 2 hôtes alors que nous considérons ici des liaisons unidirectionnelles. Par abus de langage nous désignerons une liaison <u>fixe</u> et <u>unidirectionnelle</u> entre 2 hôtes par le terme de <u>circuit virtuel</u> ; il faut souligner que cette étude peut s'appliquer à des réseaux sans circuits virtuels au sens usuel du terme dès l'instant où le routage des paquets allant d'un hôte vers un autre est fixe.

Par ailleurs on perd peu en généralité en considérant des liaisons unidirectionnelles car une liaison bidirectionnelle peut être considérée comme 2 liaisons unidirectionnelles de sens opposés ; nous permettons en outre ici que ces 2 liaisons suivent des chemins différents.

La deuxième caractéristique des réseaux étudiés ici a trait au contrôle de flux. On considère qu'il est ici de type contrôle par fenêtre et s'effectue de bout en bout.

2. Contrôle de flux par fenêtre

Considérons un point A émetteur de paquets et un point B récepteur de paquets associé au point A.

Dans le cas qui nous intéresse les "points" A et B seront des ports d'entrée et de sortie correspondants à <u>un</u> circuit virtuel.

Les 2 fonctions principales du contrôle de flux entre ces 2 points sont les suivantes :

- prévention de la baisse du débit due à une surcharge des équipements permettant l'acheminement des paquets entre A et B.

- adaptation de la vitesse d'émission du point A à la capacité de réception du point B.

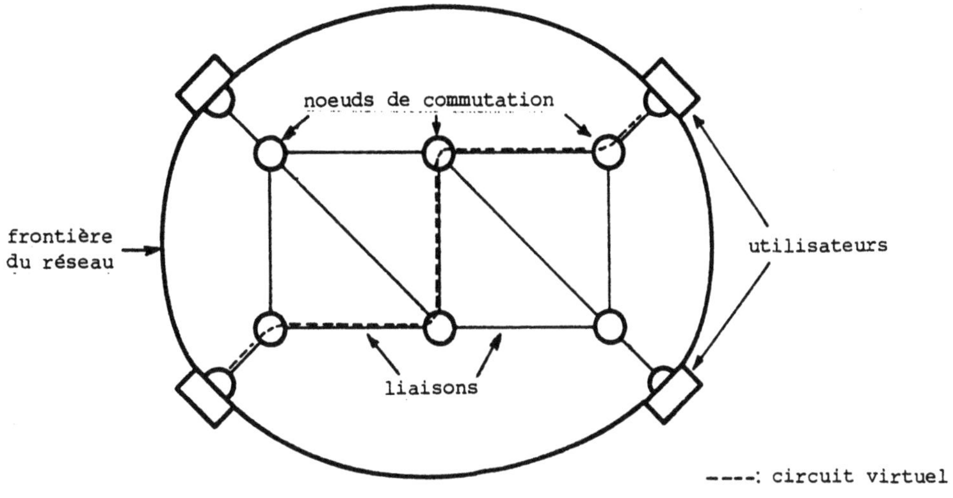

Fig. 1. Schéma d'un réseau

Un type de contrôle de flux couramment employé est le contrôle de flux par fenêtre agissant comme suit.

Pour plus de clarté nous reprendrons ici la terminologie de l'avis X25 bien que ce type de contrôle de flux puisse être mis en oeuvre indépendamment de cet avis.

Les paquets émis par le point A sont numérotés séquentiellement au moyen de P(S) modulo 8 (ou 128). Pour chaque paquet correctement reçu, de numéro P(S), le point B envoie au point A l'information donnant le numéro P(R) (en général égal à P(S) + 1) à partir duquel celui-ci peut envoyer L paquets de numéro P(R), ..., P(R) + L - 1. Le paramètre L s'appelle la taille de la fenêtre utilisée. Cette information (accusé de réception) peut être acheminée soit par des paquets spéciaux ("stand-alone"), soit avec des paquets de données émis dans l'autre sens ("piggyback").

Ce mécanisme implique qu'il y a au maximum L paquets entre les points A et B ; les envois des accusés de réception sont des mises à jour de la fenêtre et permettent au point A d'envoyer de nouveaux paquets vers B.

Ce type de contrôle de flux peut s'appliquer à plusieurs niveaux dans un réseau à commutation de paquets :

1) au niveau transport (niveau 4 de la norme ISO) : entre 2 équipements terminaux de traitement de données (ETTD) ou par l'utilisation de l'option "bout-en-bout" de l'avis X25

2) au niveau accès au réseau : entre un ETTD et un équipement terminal de circuit de données (ETCD) ; par exemple si l'avis X25 est utilisé entre ces 2 points

3) au niveau réseau : entre 2 ETCD correspondants (ceci est en général un protocole interne au réseau)

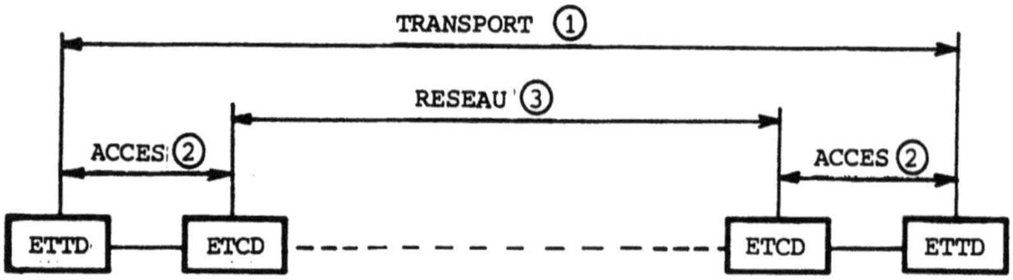

Fig. 2. Niveaux d'implémentation du contrôle de flux

La présente étude peut s'appliquer à des réseaux utilisant les niveaux 1) ou 3) de contrôle de flux, c'est à dire ceux ayant un contrôle de flux de bout en bout.

Par contre les réseaux du type TRANSPAC ne peuvent être l'objet d'applications directes de ce travail car aucun contrôle de flux de bout en bout n'est implémenté.

On fera les hypothèses qu'aucune erreur de transmission ne se produit et, dans une première partie, que les accusés de réception traversant le réseau le font instantanément.

3. Modèle du réseau

Considérons un noeud de commutation et l'ensemble des lignes de transmission qui lui sont associées en sortie (voir Fig. 3.). La partie commutation de cet ensemble sera modélisée comme une file d'attente à serveur exponentiel et on modélisera tout ce qui concerne la transmission des paquets entre 2 noeuds (envoi, gestion du niveau 2 (réémissions, ...)) par une file d'attente à serveur exponentiel.

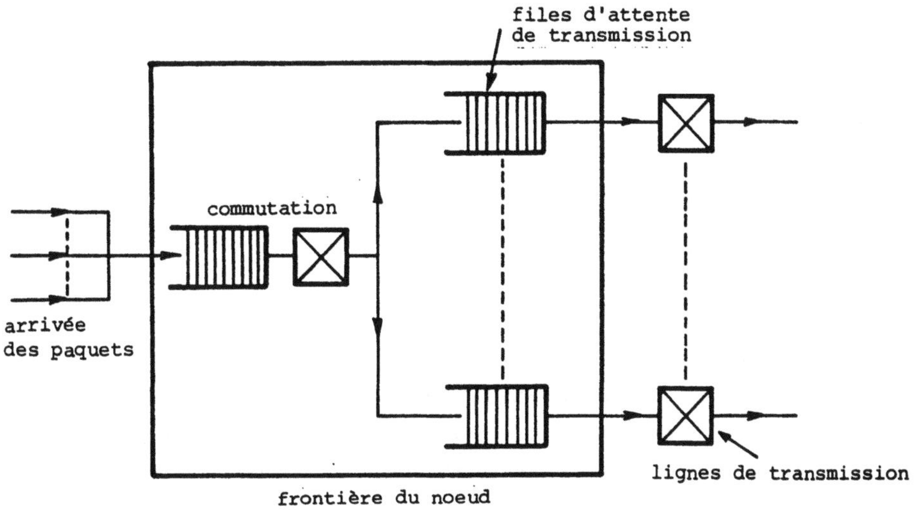

Fig. 3. Schéma d'un noeud de commutation

Globalement le réseau sera donc modélisé comme un réseau de stations, chaque station correspondant à une file d'attente à serveur exponentiel d'un des 2 types définis précedemment (commutation ou transmission).

Il est à noter que chaque station est généralement partagée par plusieurs circuits virtuels.

On peut distinguer 2 sortes de trafic soumis à un tel réseau :

un "trafic de circuits virtuels" et un "trafic de paquets".

Le "trafic de circuits virtuels" correspond à l'établissement et à la rupture des circuits virtuels et le "trafic de paquets" à la circulation des paquets sur les circuits virtuels établis.

Ces 2 types de trafic correspondent grossièrement au trafic des appels et au trafic de conversation dans un réseau téléphonique.

Nous n'étudions ici que le deuxième type de trafic ; c'est à dire que nous supposerons qu'il y a un nombre constant de circuits virtuels actifs dans le réseau. Ceci permet quand même une étude de la réalité car d'une part la durée de l'état transitoire dû à l'établissement et à la rupture des circuits virtuels commutés peut être considéré comme négligeable par rapport au temps passé en état stationnaire et d'autre part la variation de l'état stationnaire aprés de tels changements d'état est très faible. L'évolution de l'état stationnaire du réseau due au premier type de trafic est donc assez lente.

4. Modèles des circuits virtuels

Dans la suite on sera amené à considérer chaque circuit virtuel séparément et on modélisera chacun d'entre eux comme un ensemble de stations traversées par le circuit virtuel et par une file d'attente (F1) modélisant le contrôle de flux (voir Fig. 4.).

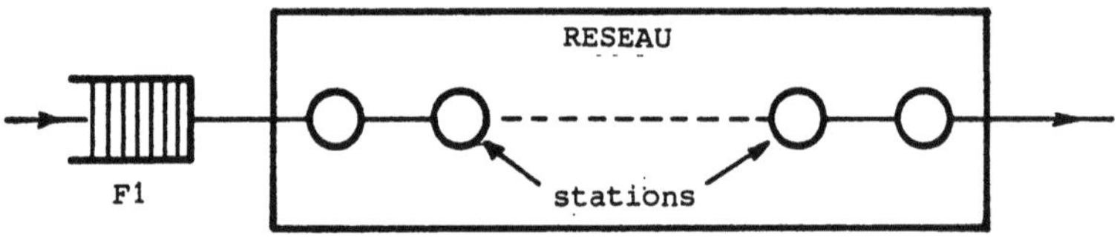

Fig. 4. Modélisation d'un circuit virtuel

Les paquets générés par l'utilisateur sont mis dans la file d'attente F1 et sont envoyés dans le réseau (si la file F1 est non vide) tant que le nombre de paquets sur le circuit virtuel est inférieur à la taille de la fenêtre (L) et sinon à chaque fois qu'un paquet quitte le réseau à l'extrémité destination. Donc à tout instant il se trouve au maximum L paquets du circuit virtuel dans le réseau.

On considère d'autre part que les arrivées à la file F1 d'un circuit virtuel sont poissonniennes.

On examinera 2 classes de problèmes que nous appelons "cas-saturé" et "cas non-saturé" (voir Fig. 5.).

L'état non-saturé correspond au cas dans lequel la file F1 peut se vider : il peut donc y avoir moins de L paquets du circuit virtuel présents dans le réseau. Dans l'état saturé on considère qu'il y a toujours au moins un paquet dans la file F1 : il y a donc toujours L paquets du circuit virtuel présents dans le réseau.

Physiquement l'état saturé correspond au cas où le temps moyen de séjour dans le réseau des paquets d'un circuit virtuel est trop grand par rapport au taux d'arrivée des paquets. Une condition de saturation sera définie au paragraphe 7.4.

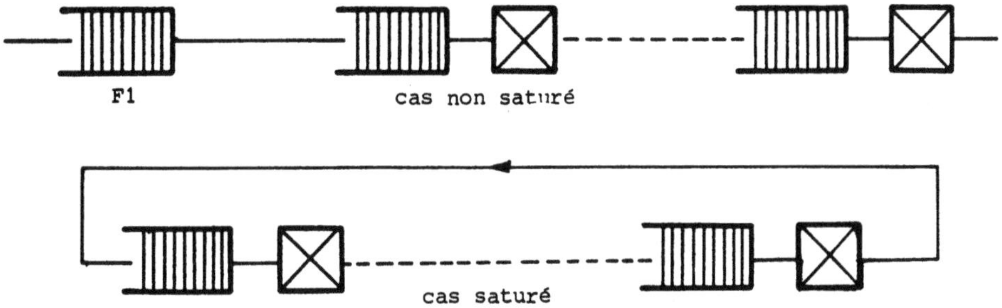

Fig. 5. Les 2 types de circuits virtuels modélisés

L'étude proprement dite sera décomposée en 3 parties correspondant aux 3 cas : tous les circuits virtuels sont non-saturés, tous les circuits virtuels sont saturés et au cas mixte où une partie des circuits virtuels sont saturés et l'autre partie non-saturés.

Il faut souligner que dans l'application de cette étude l'état d'un circuit virtuel (saturé ou non-saturé) sera un résultat et non un paramètre.

5. Notations

On suppose qu'il y a un nombre constant M de circuits virtuels actifs dans le réseau et N stations et on représente le "routage" de ces circuits par une matrice (q_{ij}, $1<i<N$, $1<j<M$) où $q_{ij} = 1$ si le circuit j passe à la station i et 0 sinon .

Soit d'autre part μ_{ij}^{-1} le temps moyen de service à la station i des paquets du circuit virtuel j, pour les circuits virtuels non saturés λ_j le taux d'arrivée des paquets à ce circuit et L_j la taille de fenêtre utilisée sur le circuit virtuel j.

On note W_{ij} le temps de séjour moyen d'un paquet du circuit j à la station i et t_j le temps moyen total passé par un paquet du circuit j dans le réseau :

$$t_j = \sum_{i=1}^{N} q_{ij}.W_{ij}$$

6. Exactitude des solutions

D'après ce qui précède la modélisation choisie représente le réseau comme un réseau de files d'attente.

Lorsque tous les circuits virtuels sont à l'état stationnaire saturé et que les taux de services aux différentes stations sont indépendants des circuits virtuels ($\mu_{ij} = \mu_i \; \forall \; j$) le réseau est un réseau BCMP fermé ; en effet chaque station est constituée d'une file d'attente et d'un serveur exponentiel, on peut considérer chaque circuit virtuel comme une classe d'utilisateurs et il y a un nombre constant de paquets (clients) - égal à $\sum_{j=1}^{M} L_j$ - présents dans le réseau.

Sevcik et Mitrani ont montré ([2]) que dans un réseau BCMP fermé l'état dans lequel un client de la classe j (ici un paquet du circuit virtuel j) voit le système quand il arrive dans une file d'attente est l'état stationnaire du système constitué par la population globale avec un client de moins dans la classe j (K - $\{j\}$).

Les approximations faites ici consistent d'une part à considérer dans tous les cas le réseau comme BCMP (en fait on suppose que l'état dans lequel un client - ici un paquet - voit le réseau en arrivant à une file d'attente est l'état stationnaire) et d'autre part à assimiler la population K - $\{j\}$ à la population globale K ; ces 2 approximations sont en effet légitimes si on considère des réseaux de grande tailles (comprenant un grand nombre de circuits virtuels).

Dans les 2 autres cas (cas où tous les circuits virtuels sont non-saturés et cas mixte) le réseau considéré est ouvert et de plus le contrôle de flux rend les arrivées des paquets au réseau non poissonniennes et on opère les mêmes approximations que dans le cas saturé en supposant que l'erreur faite est du même ordre de grandeur (en considérant toujours une population de grande taille).

7. Calcul des W_{ij}

Soit N_{ik}^{j} le nombre moyen de paquets du circuit virtuel k qu'un client du circuit virtuel j trouve à son arrivée à la station i. On a :

$$W_{ij} = \sum_{k=1}^{M} q_{ik} \cdot \frac{N_{ik}^{j}}{\mu_{ik}} + \frac{1}{\mu_{ij}} \qquad (1)$$

où chaque terme de la somme correspond à l'attente du paquet j due aux paquets du circuit virtuel k le précédant à la station i et $\frac{1}{\mu_{ij}}$ au temps moyen de service du paquet du circuit virtuel j.

Les q_{ik} doivent intervenir pour prendre en compte le routage d'un circuit virtuel. Nous aurions pû les faire intervenir d'une autre manière (par exemple dans N_{ik}^{j}) mais pour plus de clarté il nous a semblé préférable de les introduire dès le début des calculs.

Pour $k \neq j$ N^j_{ik} est égal au nombre moyen de paquets du circuits k présents dans le réseau ($\bar{n}_k$) multiplié par la probabilité qu'un paquet du circuit k se trouve au noeud i sachant qu'il est dans le réseau ($p_i(k)$):

$$N^j_{ik} = \bar{n}_k \cdot p_i(k) \tag{2}$$

et N^j_{ij} est égal au nombre moyen de paquets du circuit j présents dans le réseau sachant qu'il y en a au moins 1 ($\bar{n}'_j$), diminué d'une unité et multiplié par $p_i(j)$:

$$N^j_{ij} = (\bar{n}'_j - 1) \cdot p_i(j) \tag{3}$$

Dans le cas où le circuit j est non-saturé on a :

$$\bar{n}'_j = \frac{\bar{n}_j}{1 - R_j(0)} \tag{4}$$

où $R_j(m)$ est la probabilité d'avoir m paquets du circuit virtuel j présents dans le réseau ; ici $1 - R_j(0)$ représente donc la probabilité qu'il y ait au moins un paquet du circuit virtuel j présent dans le réseau.

Dans le cas où le circuit virtuel j est saturé on a toujours :

$$\bar{n}'_j = \bar{n}_j = L_j \tag{5}$$

puisqu'il y a toujours L_j paquets du circuit virtuel j présents dans le réseau ; donc dans ce cas on a :

$$N^j_{ij} = (L_j - 1) \cdot p_i(j) \tag{6}$$

7.1 Cas non-saturé (tous les circuits virtuels non-saturés)

Dans le cas où le circuit virtuel j est non-saturé on le modélise comme une file $M/M/L_j$ ce qui en constitue une bonne approximation bien que les différents services ne soient pas totalement indépendants et qu'ils se déroulent avec une certaine séquentialité (2 paquets sortiront du circuit dans le même ordre qu'ils y sont entrés).

On a donc en posant $\rho_j = \lambda_j t_j$:

$$R_j(m) = K.\frac{\rho_j^m}{m!} \qquad \text{si } m < L_j$$

et

$$R_j(m) = K.\frac{\rho_j^m}{L_j^{(m-L_j)}.L_j} \qquad \text{si } m > L_j$$

où K représente une constante de normalisation ($K=R_j(0)$).

n_j correspond alors au nombre moyen de serveurs occupés dans une file $M/M/L_j$ et donc en appliquant la formule de Little aux clients en service on obtient :

$$n_j = \rho_j \tag{7}$$

et on a :

$$R_j(0) = \left(\sum_{i=0}^{L_j-1} \frac{\rho_j^i}{i!} + \frac{\rho_j^{L_j}.\ L_j}{L_j!(L_j - \rho_j)} \right)^{-1} \tag{8}$$

D'autre part à l'état stationnaire $p_i(j)$ est égal au rapport du temps moyen passé par un paquet du circuit virtuel j au serveur i sur le temps total moyen passé dans le réseau par ce paquet :

$$p_j(j) = \frac{W_{ij}}{t_j} \tag{9}$$

Donc dans le cas où tous les circuits virtuels sont non-saturés on a d'après (2), (7) et (9) :

$$N_{ik}^{j} = \rho_k . \frac{W_{ik}}{t_k} = \lambda_k W_{ik} \qquad k \neq j \tag{10}$$

et d'après (3), (4) et (7) :

$$N_{ij}^{j} = \left(\frac{\rho_j}{1-R_j(0)} - 1\right). \frac{W_{ij}}{t_j} \tag{11}$$

Donc d'après (1), (10) et (11) :

$$W_{ij} = \sum_{\substack{k=1 \\ k\neq j}}^{M} q_{ik} \frac{\lambda_k W_{ik}}{\mu_{ik}} + \left(\frac{\rho_j}{1-R_j(0)} - 1\right). \frac{W_{ij}}{t_j} . \frac{1}{\mu_{ij}} + \frac{1}{\mu_{ij}}$$

d'où

$$W_{ij} = \sum_{k=1}^{M} q_{ik} \frac{\lambda_k W_{ik}}{\mu_{ik}} + \left(\frac{\rho_j}{1-R_j(0)} - 1\right). \frac{W_{ij}}{\mu_{ij} t_j} + \frac{1}{\mu_{ij}} - \frac{\lambda_j W_{ij}}{\mu_{ij}}$$

soit

$$W_{ij} \left(1 + \frac{\lambda_j}{\mu_{ij}} \left(1 + \frac{1}{\rho_j} . \frac{1}{1-R_j(0)}\right)\right) - \frac{1}{\mu_{ij}} = \sum_{k=1}^{M} q_{ik} \frac{\lambda_k W_{ik}}{\mu_{ik}}$$

Soit en posant :

$$f(j) = 1 + \frac{1}{\rho_j} - \frac{1}{1-R_j(0)}$$

on obtient :

$$W_{ij}\left(1+\frac{\lambda_j f(j)}{\mu_{ij}}\right)-\frac{1}{\mu_{ij}}=\sum_{k=1}^{M} q_{ik}\frac{\lambda_k W_{ik}}{\mu_{ik}} \tag{12}$$

W_{ij} intervenant (par l'intermédiaire de t(j)) de manière non linéaire dans f(j) on ne peut en général obtenir de formule close pour W_{ij}. On peut cependant utiliser cette équation de 2 manières :

1- Soit en calculant W_{ij} à partir de (12) ce qui donne :

$$W_{ij}=\left(\sum_{k=1}^{M} q_{ik}\frac{\lambda_k W_{ik}}{\mu_{ik}}+\frac{1}{\mu_{ij}}\right)\left(1+\frac{\lambda_j f(j)}{\mu_{ij}}\right)^{-1} \tag{13}$$

et on pourra utiliser cette équation pour obtenir W_{ij} grâce à une méthode numérique itérative (voir 9.).

2- Soit remarquer que le membre de droite de l'équation (12) étant indépendant de j on a $\forall$ j et k :

$$W_{ij}\left(1+\frac{\lambda_j f(j)}{\mu_{ij}}\right)-\frac{1}{\mu_{ij}}=W_{ik}\left(1+\frac{\lambda_k f(k)}{\mu_{ik}}\right)-\frac{1}{\mu_{ik}} \tag{14}$$

d'où en calculant les W_{ik} en fonction de W_{ij} et en introduisant ces valeurs dans l'équation (13) on obtient :

$$W_{ij}=\left(1+\frac{\lambda_j f(j)}{\mu_{ij}}\right)^{-1}\left(1-\sum_{k=1}^{M}\frac{q_{ik}\lambda_k}{\mu_{ik}+\lambda_k f(k)}\right)^{-1} \times \tag{15}$$

$$\times\left(\frac{1}{\mu_{ij}}+\sum_{k=1}^{M} q_{ik}\lambda_k\frac{\frac{1}{\mu_{ik}}-\frac{1}{\mu_{ij}}}{\mu_{ik}+\lambda_k f(k)}\right)$$

On obtient facilement les f(j) pour des tailles de fenêtre assez petites (voir calcul en annexe) :

pour $L_j = 1$ $\quad f(j) = 1$

pour $L_j = 2$ $\quad f(j) = \frac{1}{2}$

pour $L_j = 3$ $\quad f(j) = \frac{3}{\rho_j + 6}$

Donc lorsque $L_k \in \{1,2\}$ $\forall$ $k \in [1..M]$ on obtient une formule close pour W_{ij}. Par exemple lorsque $L_k = 1$ et $\mu_{ik} = \mu_i$ $\forall k$, on a d'après (15) :

$$W_{ij} = (\lambda_j + \mu_i)^{-1}(1 - \sum_{k=1}^{M} \frac{q_{ik} \lambda_k}{\mu_i + \lambda_k})$$

résultat qui avait été obtenu par J. Labetoulle et G. Pujolle dans [3].

Sinon on calcul W_{ij} d'une manière numérique itérative (voir 9.)

7.2 Cas saturé (tous les circuits saturés)

Dans ce cas on considère qu'il y a toujours L_k paquets du circuit k présents dans le réseau (et ceci $\forall$ k) donc en prenant pour $p_i(k)$ la même valeur que celle du cas non-saturé (soit $p_i(k) = W_{ik} / t_k$) on a d'après (2), (5) et (9) :

$$N^j_{ik} = L_k . \frac{W_{ik}}{t_k} \quad \text{si } k \neq j \tag{16}$$

et d'après (3), (5) et (9) :

$$N^i_{ij} = (L_j - 1) . \frac{W_{ij}}{t_j} \tag{17}$$

donc d'après (1), (16) et (17) :

$$W_{ij} = \sum_{\substack{k=1 \\ k \neq j}}^{M} q_{ik} \frac{L_k W_{ik}}{t_k \mu_{ik}} + (L_j - 1).\frac{W_{ij}}{t_j \mu_{ij}} + \frac{1}{\mu_{ij}}$$

et on obtient en s'inspirant du calcul pour le cas non-saturé :

$$W_{ij} \left(1 + \frac{1}{\mu_{ij} t_j}\right) - \frac{1}{\mu_{ij}} = \sum_{k=1}^{M} q_{ik} \frac{L_k W_{ik}}{t_k \mu_{ik}} \qquad (18)$$

On peut faire une analogie entre cette formule et celle obtenue dans le cas non-saturé en remarquant que en remplacant f(j) par $1/\lambda_j t_j$ et en posant $\lambda_k = L_k/t_k \quad \forall\, k$ (le taux d'arrivée λ_k est égal au taux de départ L_k/t_k puisque le circuit est saturé) dans (12) on obtient la formule ci-dessus.

De même que dans le cas non-saturé on peut utiliser cette formule de 2 manières :

1- Soit en calculant directement W_{ij} à partir de (18) :

$$W_{ij} = \left(1 + \frac{1}{\mu_{ij} t_j}\right)^{-1} \left(\sum_{k=1}^{M} q_{ik} \frac{L_k W_{ik}}{t_k \mu_{ik}} + \frac{1}{\mu_{ij}} \right) \qquad (19)$$

2- Soit en remarquant que le membre de droite de (18) est indépendant de j et obtenir une formule analogue à celle du cas non-saturé :

$$W_{ij} = \left(1 + \frac{1}{\mu_{ij} t_j}\right)^{-1} \left(1 - \sum_{k=1}^{M} \frac{q_{ik} L_k}{1 + t_k \mu_{ik}}\right)^{-1} \times$$
$$\times \left(\frac{1}{\mu_{ij}} + \sum_{k=1}^{M} q_{ik} L_k \frac{\frac{1}{\mu_{ik}} - \frac{1}{\mu_{ij}}}{1 + \mu_{ik} t_k} \right) \qquad (20)$$

On peut remarquer qu'un circuit virtuel saturé ayant une taille de fenêtre égale à L_k peut être considéré comme L_k

circuits virtuels ayant tous une taille de fenêtre de 1 et possédant les mêmes caractéristiques que le circuit initial : μ_{ik} et q_{ik}.

7.3 Cas mixte

Dans ce cas on considère que l'ensemble $[1..M]$ des circuits virtuels est constitué de 2 sous-ensembles S et $\bar{S}$ en formant une partition : S est l'ensemble des circuits saturés et $\bar{S}$ l'ensemble des circuits non-saturés.

Le calcul de W_{ij} va s'effectuer d'une manière analogue à celle des 2 cas précédents en remplaçant dans (1) N_{ik}^{j} par la valeur obtenue dans (10) si $k \in \bar{S}$ et par celle obtenue dans (16) si $k \in S$ et N_{ij}^{j} par la valeur obtenue dans (11) si $j \in \bar{S}$ et par celle obtenue dans (17) si $j \in S$ et on obtient donc si le circuit j est saturé :

$$W_{ij} = \sum_{k \in \bar{S}} q_{ik} \frac{\lambda_k W_{ik}}{\mu_{ik}} + \sum_{\substack{k \in S \\ k \neq j}} q_{ik} \frac{L_k W_{ik}}{t_k \mu_{ik}} + \frac{1}{\mu_{ij}} \left(1 + (L_j - 1) \frac{W_{ij}}{t_j}\right)$$

et si le circuit j est non-saturé :

$$W_{ij} = \sum_{\substack{k \in \bar{S} \\ k \neq j}} q_{ik} \frac{\lambda_k W_{ik}}{\mu_{ik}} + \sum_{k \in S} q_{ik} \frac{L_k W_{ik}}{t_k \mu_{ik}} + \frac{1}{\mu_{ij}} \left(1 + \left(\frac{P_j}{1 - R_j(0)} - 1\right) \frac{W_{ij}}{t_j}\right)$$

et on obtient en s'inspirant des calculs précédents :

$$W_{ij}\left(1 + \frac{1}{\mu_{ij} t_j}\right) - \frac{1}{\mu_{ij}} = \sum_{k \in \bar{S}} q_{ik} \frac{\lambda_k W_{ik}}{\mu_{ik}} + \sum_{k \in S} q_{ik} \frac{L_k W_{ik}}{t_k \mu_{ik}} \qquad (21)$$

si le circuit j est saturé et :

$$W_{ij}\left(1 + \frac{\lambda_j f(j)}{\mu_{ij}}\right) - \frac{1}{\mu_{ij}} = \sum_{k \in \bar{S}} q_{ik} \frac{\lambda_k W_{ik}}{\mu_{ik}} + \sum_{k \in S} q_{ik} \frac{L_k W_{ik}}{t_k \mu_{ik}} \qquad (22)$$

si le circuit j est non-saturé.

Ces 2 formules correspondent aux formules (18) et (12) où la sommation sur tous les circuits virtuels différents de j (tous saturés dans (18) et tous non-saturés dans (12)) a été scindées en 2 parties correspondant à S et $\bar{S}$ et où l'apport du circuit virtuel j (hors de la sommation) reste le même.

De même que dans 7.1 et 7.2 on peut calculer directement W_{ij} à partir de (21) pour le cas où le circuit virtuel j est saturé ce qui donne :

$$W_{ij} = \left(\sum_{k \in \bar{S}} q_{ik} \frac{\lambda_k W_{ik}}{\mu_{ik}} + \sum_{k \in S} q_{ik} \frac{L_k W_{ik}}{t_k \mu_{ik}} + \frac{1}{\mu_{ij}}\right)\left(1 + \frac{1}{t_j \mu_{ij}}\right)^{-1} \qquad (23)$$

et à partir de (22) dans le cas où le circuit virtuel j est non-saturé ce qui donne :

$$W_{ij} = \left(\sum_{k \in \bar{S}} q_{ik} \frac{\lambda_k W_{ik}}{\mu_{ik}} + \sum_{k \in S} q_{ik} \frac{L_k W_{ik}}{t_k \mu_{ik}} + \frac{1}{\mu_{ij}}\right)\left(1 + \frac{\lambda_j f(j)}{\mu_{ij}}\right)^{-1} \qquad (24)$$

Quand le circuit j est saturé on remarque que :

$$W_{ij}\left(1 + \frac{1}{t_j \mu_{ij}}\right) - \frac{1}{\mu_{ij}} = W_{ik}\left(1 + \frac{1}{t_k \mu_{ik}}\right) - \frac{1}{\mu_{ik}} \quad \forall\, k \in S \qquad (25)$$

$$W_{ij}\left(1 + \frac{1}{t_j \mu_{ij}}\right) - \frac{1}{\mu_{ij}} = W_{ik}\left(1 + \frac{\lambda_k f(k)}{\mu_{ik}}\right) - \frac{1}{\mu_{ik}} \quad \forall\, k \in \bar{S} \qquad (26)$$

d'où en remplaçant W_{ik} dans (21) par les valeurs obtenues à partir de (25) et (26) on obtient :

$$W_{ij} = \left(1 + \frac{1}{t_j \mu_{ij}}\right)^{-1}\left(1 - \sum_{k\varepsilon\bar{S}} \frac{q_{ik} \lambda_k}{\mu_{ik} + \lambda_k f(k)} - \sum_{k\varepsilon S} \frac{q_{ik} L_k}{1 + t_k \mu_{ik}}\right)^{-1} \times$$

$$\times \left(\sum_{k\varepsilon\bar{S}} \frac{q_{ik} \lambda_k \left(\frac{1}{\mu_{ik}} - \frac{1}{\mu_{ij}}\right)}{\mu_{ik} + \lambda_k f(k)} + \sum_{k\varepsilon S} \frac{q_{ik} L_k \left(\frac{1}{\mu_{ik}} - \frac{1}{\mu_{ij}}\right)}{1 + t_k \mu_{ik}} + \frac{1}{\mu_{ij}}\right) \quad (27)$$

Si le circuit j est non-saturé il suffit de remplacer $1+1/t_j\mu_{ij}$ par $1+\lambda_j f(j)/\mu_{ij}$ dans (27) d'où :

$$W_{ij} = \left(1 + \frac{\lambda_j f(j)}{\mu_{ij}}\right)^{-1}\left(1 - \sum_{k\varepsilon\bar{S}} \frac{q_{ik} \lambda_k}{\mu_{ik} + \lambda_k f(k)} - \sum_{k\varepsilon S} \frac{q_{ik} L_k}{1 + t_k \mu_{ik}}\right)^{-1} \times$$

$$\times \left(\sum_{k\varepsilon\bar{S}} \frac{q_{ik} \lambda_k \left(\frac{1}{\mu_{ik}} - \frac{1}{\mu_{ij}}\right)}{\mu_{ik} + \lambda_k f(k)} + \sum_{k\varepsilon S} \frac{q_{ik} L_k \left(\frac{1}{\mu_{ik}} - \frac{1}{\mu_{ij}}\right)}{1 + t_k \mu_{ik}} + \frac{1}{\mu_{ij}}\right) \quad (28)$$

7.4 Condition de saturation

Il est important d'établir une condition permettant de distinguer les circuits virtuels saturés et non-saturés.

Si on considère un circuit virtuel saturé j t_j représente le temps moyen de séjour et t_j/L_j le taux moyen de sortie des paquets. Une condition de non-saturation pour le circuit virtuel j sera alors :

$$\lambda_j t_j / L_j < 1$$

En effet dans ce cas il existe pour le circuit virtuel j des périodes où il n'y a aucun paquet sur ce circuit.

8. Prise en compte des accusés de réception

On considère que les accusés de réception sont acheminés en "stand-alone" c'est à dire qu'ils font l'objet de paquets spéciaux et d'autre part qu'ils sont servis instantanément à chaque station.

Il suffit alors de remplacer dans les calculs t_k par :

$$t_k^* = t_k + \sum_{i=1}^{N} q_{ik}^* W_{ik}^*$$

où q_{ik}^* représente le routage de l'accusé de réception et :

$$W_{ik}^* = W_{ik} - \frac{1}{\mu_{ik}}$$

Le temps moyen de traversée du réseau sera toujours donné par t_k.

9. Application

Afin de tester la validité des méthodes présentées dans cette étude, nous avons d'une part appliqué les résultats obtenus à un réseau test et d'autre part effectué une simulation de ce réseau.

9.1. Présentation du réseau test

Le réseau utilisé à la fois pour l'application des résultats analytiques et pour la simulation est celui représenté dans la Fig. 7. Il s'agit d'un réseau comportant 7 noeuds et 7 hôtes ainsi que 30 liaisons unidirectionnelles reliant les noeuds et les hôtes.

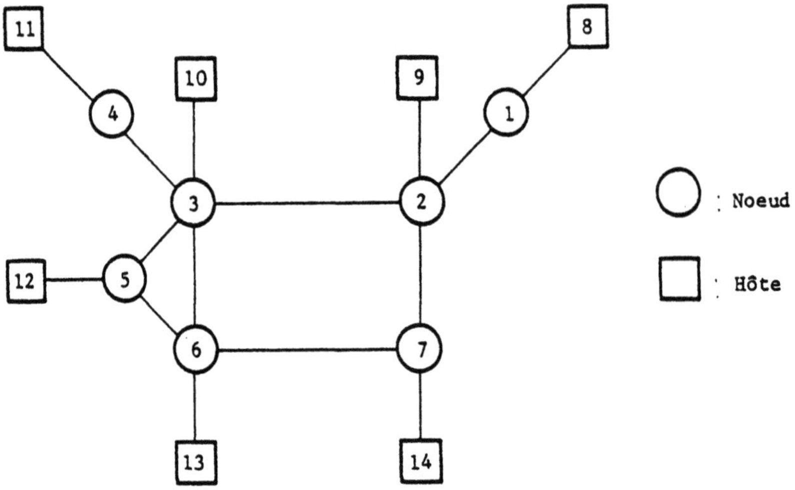

Figure 7 - Schéma du réseau test

Nous avons défini 14 circuits virtuels établis sur le réseau conformément au tableau de routage de la Fig. 8.

La démarche adoptée pour appliquer les résultats obtenus dans la partie théorique de cette étude a été de considérer séparément un des circuits virtuels (en l'occurrence le circuit n° 1) et d'étudier son comportement en faisant varier d'une part les taux d'arrivée aux autres circuits virtuels et d'autre part le taux d'arrivée au circuit n° 1.

Une variable intéressante pour étudier le comportement d'un circuit virtuel est le temps moyen de réponse (délai de bout-en-bout dans le réseau) des paquets d'un circuit virtuel donné. En fixant les taux d'arrivée sur les autres circuits nous obtenons le temps moyen de réponse en fonction du taux d'arrivée et ceci jusqu'à saturation du circuit ce qui permet d'obtenir le début maximal et le temps de réponse maximal sur ce circuit.

N° Circuit Virtuel	Hôte Source	Noeuds Traversés	Hôte Destination
1	8	1 2 7 6	13
2	9	2 7	14
3	10	3 5	12
4	11	4 3 2 1	8
5	12	5 6 7 2	9
6	13	6 3 4	11
7	14	7 6 3	10
8	8	1 2 3 4	11
9	9	2 3 6	13
10	10	3 5	12
11	11	4 3 2 7	14
12	12	5 3 2 1	8
13	13	6 7 2	9
14	14	7 6 3	10

Figure 8 - Routage des Circuits Virtuels.

9.2 Application des résultats analytiques

De plus nous pouvons faire varier la taille de la fenêtre utilisée sur un circuit virtuel.

Un programme de calcul des temps moyens de séjour des paquets d'un circuit virtuel dans le réseau a été écrit, utilisant les résultats obtenus au chapitre 7.

Puisqu'à priori, étant donné les paramètres du réseau et des différents circuits virtuels établis, nous ne pouvons savoir si un circuit virtuel est saturé ou non, nous avons établi un algorithme utilisant les résultats du cas mixte (réseau comportant des circuits saturés et non-saturés - section 7.3) de la façon suivante (les notations sont celles utilisées au chapitre 7) :

1) Initialisation de tous les W_{ij} par :

$$W_{ij} = \sum_{k=1}^{M} q_{ik} L_k/\mu_{ik}$$

On considère initialement que tous les circuits virtuels sont non-saturés.

2) Itération du calcul des W_{ij} grâce aux équations (27) et (28) de la section 7.3 suivant que le circuit est saturé ou non-saturé.

3) Arrêt si pour tous les circuits virtuels soit le calcul a convergé, soit diverge.

4) On teste la condition de saturation (cf. 7.4) pour chaque circuit virtuel non-saturé : tous les circuits virtuels non-saturés ne satisfaisant pas cette condition passent à l'état saturé.

5) Si au moins un circuit virtuel a changé d'état en 4) on retourne en 2), sinon arrêt.

Cet algorithme donne donc comme résultat d'une part les temps moyens de séjour dans chacune des stations du réseau pour les différents circuits virtuels, et d'autre part l'état de chaque circuit virtuel (saturé ou non-saturé).

Les résultats obtenus grâce à cet algorithme sont représentés sur les courbes des Fig. 9 à 15. Les courbes des Fig. 9 à 13 représentant (pour des tailles de fenêtre allant de 1 à 5) les temps moyens de réponse du circuit virtuel 1 en fonction du taux d'arrivée à ce circuit et ceci pour 2 taux d'arrivée aux autres circuits virtuels : 1 et 30 paquets/s.

Lorsque le circuit est saturé nous pouvons en déduire le début maximal du circuit virtuel. En effet dans ce cas si W est la taille de la fenêtre utilisée sur ce circuit virtuel et T le temps moyen de séjour, nous avons comme expression du débit D :

$D = W/T$

puisqu'il y a toujours W paquets du circuit virtuel présents dans le réseau.

Les courbes des Fig. 14 et 15 représentent donc respectivement l'évolution du débit maximal et du temps de traversée maximal sur le circuit virtuel 1 (à saturation) pour des tailles de fenêtre allant de 1 à 5.

Pour des raisons de commodité de comparaison les résultats de simulation sont aussi représentés sur les courbes des Fig. 9 à 15 et les commentaires concernant les résultats analytiques et de simulation seront effectués conjointement au chapitre suivant (11.2.1.).

9.3 Simulation

Pour la modélisation effectuée le point important est la validation des modèles de circuits virtuels. Pour cette raison (ainsi que pour des raisons de durée de simulation), la simulation effectuée n'a portée que sur 1 circuit virtuel. Les autres circuits virtuels utilisant des ressources communes au circuit virtuel simulé sont intervenus comme des flux externes partageant certaines des files modélisant les ressources du circuit virtuel simulé.

En effet on peut considérer que (si le système est ergodique) le nombre de paquets des flux externes passant par les files d'attente du circuit virtuel simulé est peu affecté par le contrôle de flux, l'influence de celui-ci étant de régulariser le trafic.

Cette simulation a été écrite en FORTRAN pour des raisons de rapidité d'exécution : un "RUN" prenant entre 500 et 1 000 s CPU suivant les paramètres.

Une version modifiée de la simulation a été écrite pour l'étude du cas où le circuit virtuel est saturé. En effet, dans ce cas, le système n'étant plus ergodique le nombre de paquets générés excède le nombre de paquets sortant du circuit virtuel et une saturation des tableaux de stockage des paquets survient. La modification effectuée a consisté en la suppression de la file d'attente initiale des paquets : un nouveau paquet est généré et entre dans le réseau à chaque fois que le contrôle de flux le permet.

Nous allons maintenant comparer les résultats obtenus par cette simulation à ceux obtenus par les méthodes analytiques.

9.4 Comparaison des résultats analytiques et de simulation

Les courbes des Fig. 9 à 13 représentent l'évolution du temps moyen de réponse du circuit virtuel 1 pour 2 taux d'arrivée des autres circuits virtuels (1 et 30 paquets par seconde) et pour des tailles de fenêtres allant de 1 à 5.

Ces courbes montrent les résultats obtenus grâce aux méthodes analytiques développées précédemment ainsi que ceux de la simulation.

La première remarque suscitée par ces courbes est le bon accord entre les résultats analytiques et la simulation : l'erreur relative est au maximum égale à 10 % (fenêtre égale à 4 et 5, taux d'arrivée aux autres circuits virtuels égal à 30) et majoritairement inférieure à 5 %. Les résultats sont globalement meilleurs pour une faible charge du réseau.

Une caractéristique importante de ces courbes est la position relative des résultats analytiques et simulés, en effet pour la partie non-saturée de la courbe la valeur analytique est inférieure à la valeur simulée, l'inverse se produisant pour la partie saturée.

Ces 2 phénomènes peuvent d'expliquer de la manière suivante : le modèle utilisé dans le cas non-saturé (système M/M/L) introduit un certain parallélisme quant au séjour des paquets dans le réseau, ce qui tend à minimiser les temps de réponse, par contre dans le cas saturé cette approximation n'est plus utilisée mais l'erreur observée est due à l'hypothèse assimilant la population globale à la population globale moins 1 paquet (cf. 6).

Les 2 autres courbes (Fig. 14 et 15) montrent l'évolution du débit maximal et du temps de réponse maximal - respectivement - en fonction des taux d'arrivée aux autres circuits virtuels. Les erreurs relatives entre les résultats analytiques et la simulation sont très faibles (inférieures à 5 %).

Globalement ces résultats permettent donc d'avoir une très bonne idée du comportement d'un circuit virtuel. Nous obtenons avec une bonne précision les temps de traversée minimal et maximal ainsi que le débit maximal que l'on obtient à saturation du circuit virtuel.

10. Conclusion

Nous avons présenté ici une méthode d'étude des réseaux à commutation de paquets prenant en compte un contrôle de flux. Pour situer cette étude par rapport à des études existantes (en particulier [3] et [4]) nous pouvons dire que nous avons effectué une généralisation de ces deux études ; par rapport au travail de J. Labetoulle et G. Pujolle nous avons considéré des tailles de fenêtres quelconques dans le cas non-saturé et étudié le cas où certains circuits virtuels sont non-saturés et les autres saturés ; enfin l'apport essentiel par rapport à l'étude de M. Reiser (qui était déjà présent dans [3]) est la prise en compte du cas non-saturé qui permet d'obtenir l'évolution d'un circuit virtuel en fonction de sa charge et ceci avant saturation.

Références

[1] CCITT : Recommendation X25. Orange Book, Vol. VIII. Genève 1977

[2] K.C. Sevcik et I. Mitrani, "The Distribution of Queueing Network States at Input and Output Instants". Rapport de Recherche IRIA-LABORIA n° 307, 1978

[3] J. Labetoulle et G. Pujolle, "A Study of Flows Through Virtual Circuits Computer Networks". Computer Networks 5, 1981

[4] M. Reiser, "A Queueing Network Analysis of Computer Communication Networks with Window Flow Control". IEEE Trans. on Com. Vol COM-27, N° 8, Aug. 1979

ANNEXE A

Calcul de f(j)

On rappelle que l'on a :

$$f(j) = 1 + \frac{1}{\rho_j} - \frac{1}{1 - R_j(0)}$$

où $R_j(0)$ est donné par (8).

Nous allons expliciter ici le calcul de f(j) pour des tailles de fenêtre (L_j) égales à 1,2 ou 3.

a) $L_j = 1$

Dans ce cas on a :

$$R_j(0) = 1 + \frac{\rho_j}{1 - \rho_j}$$
$$= 1 - \rho_j$$

D'où

$$f(j) = 1 + \frac{1}{\rho_j} - \frac{1}{1 - (1 - \rho_j)}$$

Soit :

$f(j) = 1$

b) $L_j = 2$

Calcul de $R_j(0)$:

$$R_j(0) = 1 + \rho_j + \frac{\rho_j^2}{2 - \rho_j}$$
$$= \frac{2 - \rho_j}{2 + \rho_j}$$

D'où :

$$f(j) = 1 + \frac{1}{\rho_j} - \frac{1}{1 - \frac{2 - \rho_j}{2 + \rho_j}}$$

$$= 1 + \frac{1}{\rho_j} - \frac{2 + \rho_j}{2\rho_j}$$

Soit :

$$f(j) = \frac{1}{2}$$

c) $L_j = 3$

Calcul de $R_j(0)$:

$$R_j(0) = \left(1 + \rho_j + \frac{\rho_j^2}{2} + \frac{\rho_j^3}{2\ (3 - \rho_j)}\right)^{-1} = \frac{2\ (3 - \rho_j)}{6 + 4\ \rho_j + \rho_j^2}$$

D'où :

$$f(j) = 1 + \frac{1}{\rho_j} - \frac{1}{1 - \frac{2\ (3 - \rho_j)}{6 + 4\ \rho_j + \rho_j^2}}$$

Soit :

$$f(j) = \frac{3}{6 + \rho_j}$$

ANNEXE B

Courbes

Fig. 9 - Temps de reponse moyen du circuit 1 en fonction du taux d'arrivee a ce circuit pour 2 taux d'arrivee aux autres circuits (λ) : 1 et 30 paquets/s.
Taille de fenetre egale a 1

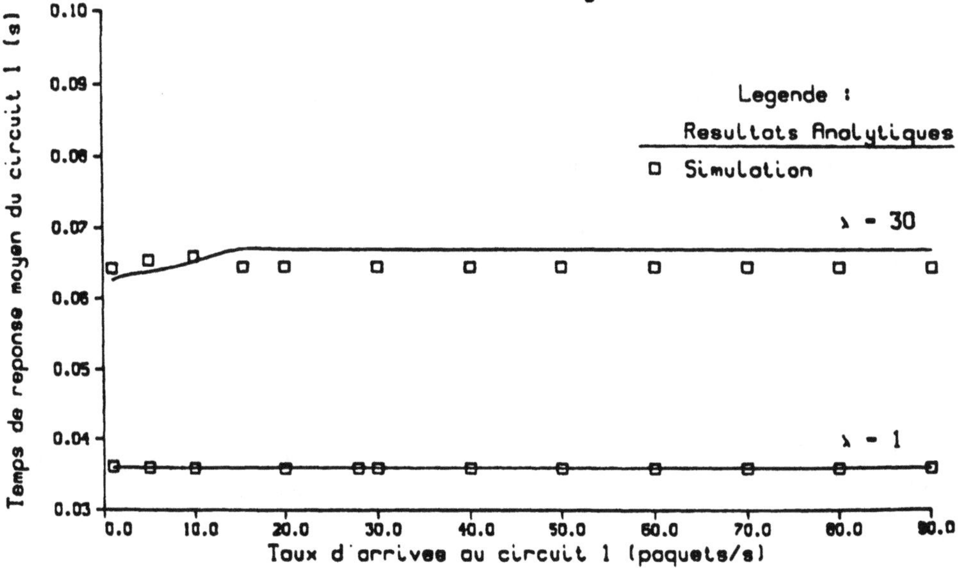

Fig. 10 - Temps de reponse moyen du circuit 1 en fonction du taux d'arrivee a ce circuit pour 2 taux d'arrivee aux autres circuits (λ) : 1 et 30 paquets/s. Taille de fenetre egale a 2

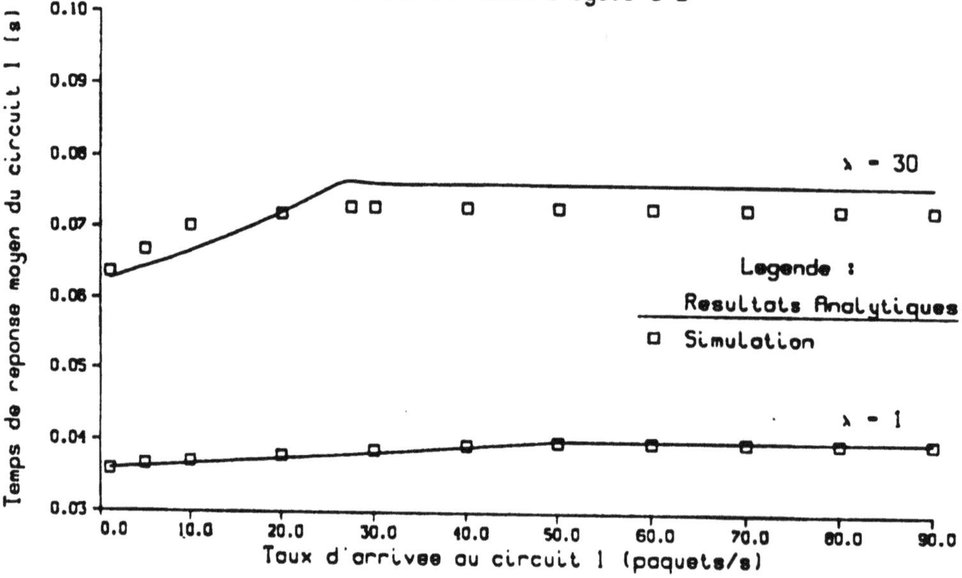

Fig. 11 - Temps de reponse moyen du circuit 1 en fonction du taux d'arrivee a ce circuit pour 2 taux d'arrivee aux autres circuits (λ) : 1 et 30 paquets/s. Taille de fenetre egale a 3

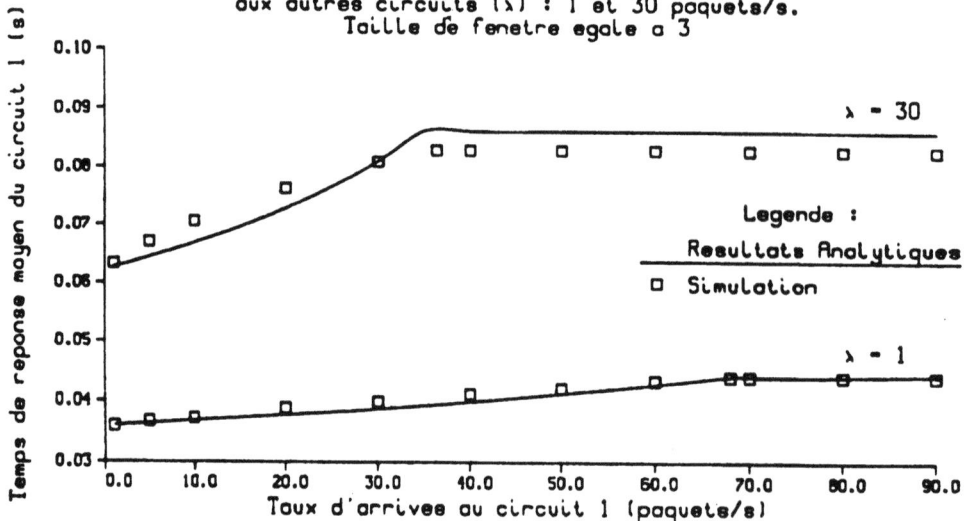

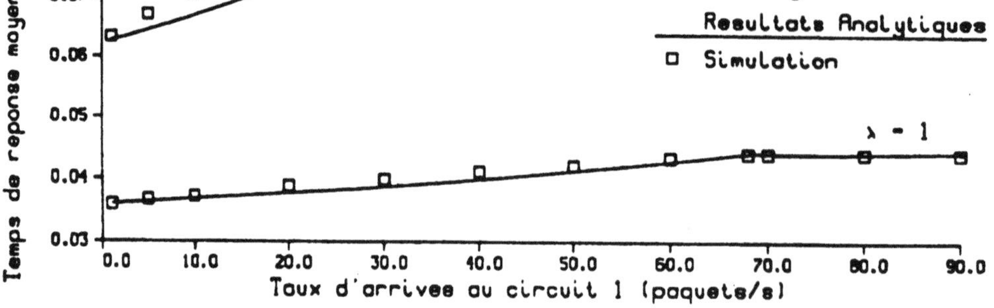

Fig. 12 - Temps de reponse moyen du circuit 1 en fonction du taux d'arrivee a ce circuit pour 2 taux d'arrivee aux autres circuits (λ) : 1 et 30 paquets/s. Taille de fenetre egale a 4

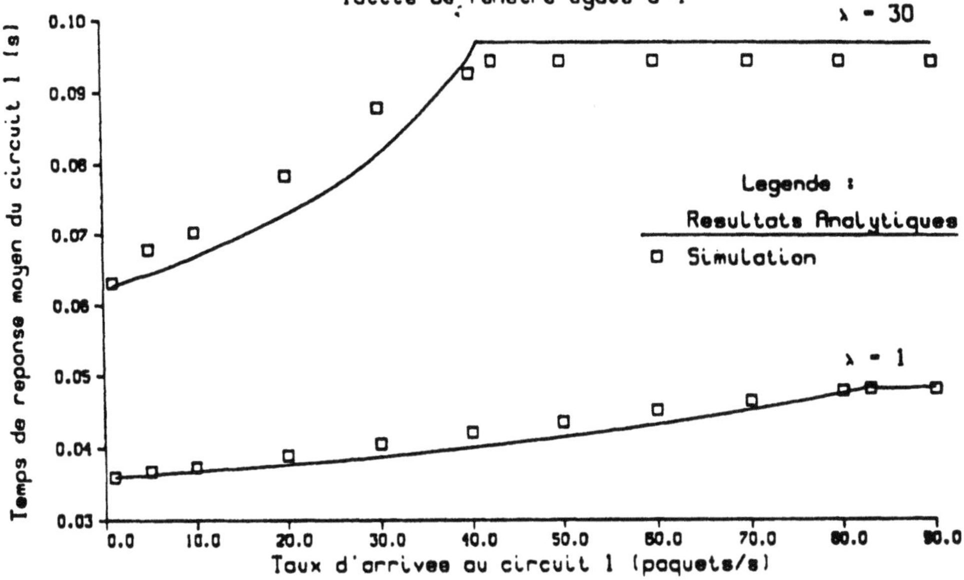

Fig. 13 - Temps de reponse moyen du circuit 1 en fonction du taux d'arrivee a ce circuit pour 2 taux d'arrivee aux autres circuits (λ) : 1 et 30 paquets/s. Taille de fenetre egale a 5

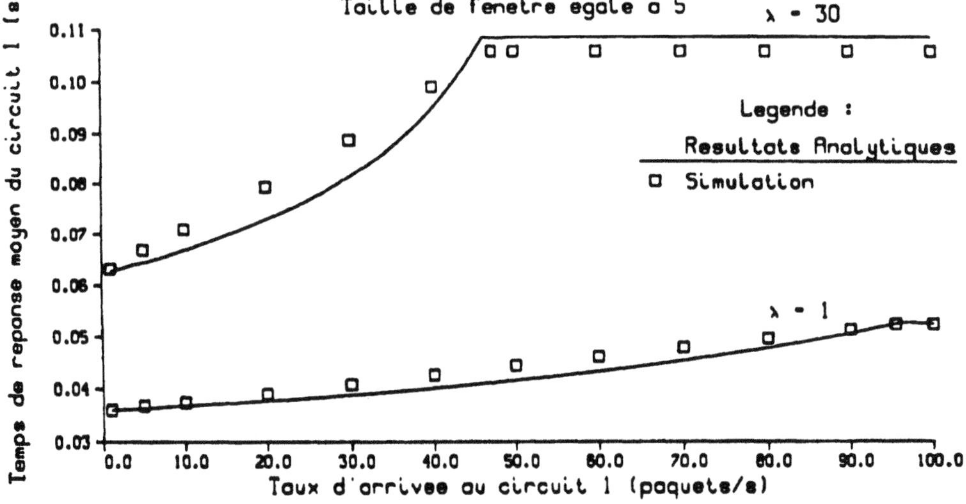

Fig. 14 - Debit maximal moyen du circuit virtuel 1 en fonction du taux d'arrivee aux autres circuits pour des tailles de fenetre (W) de 1 a 5.

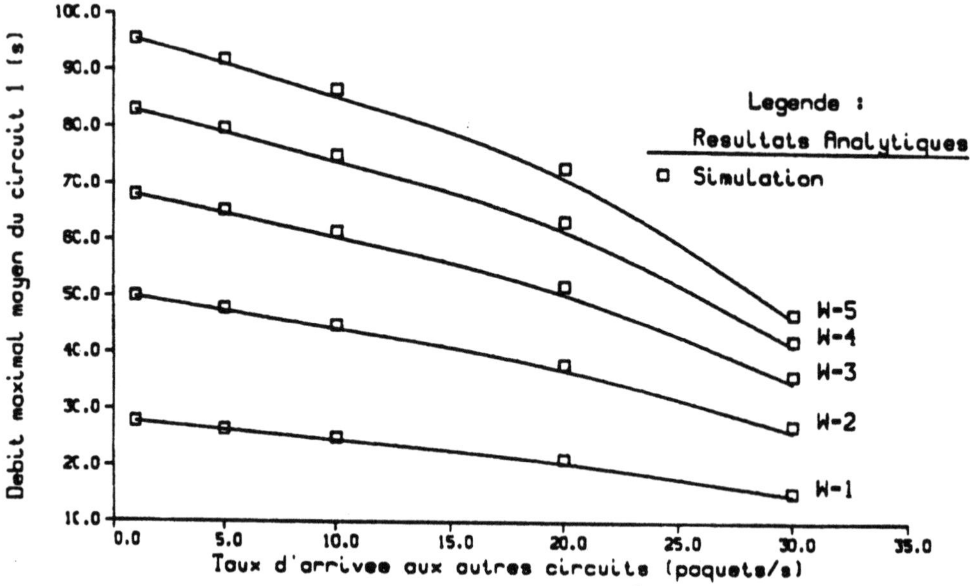

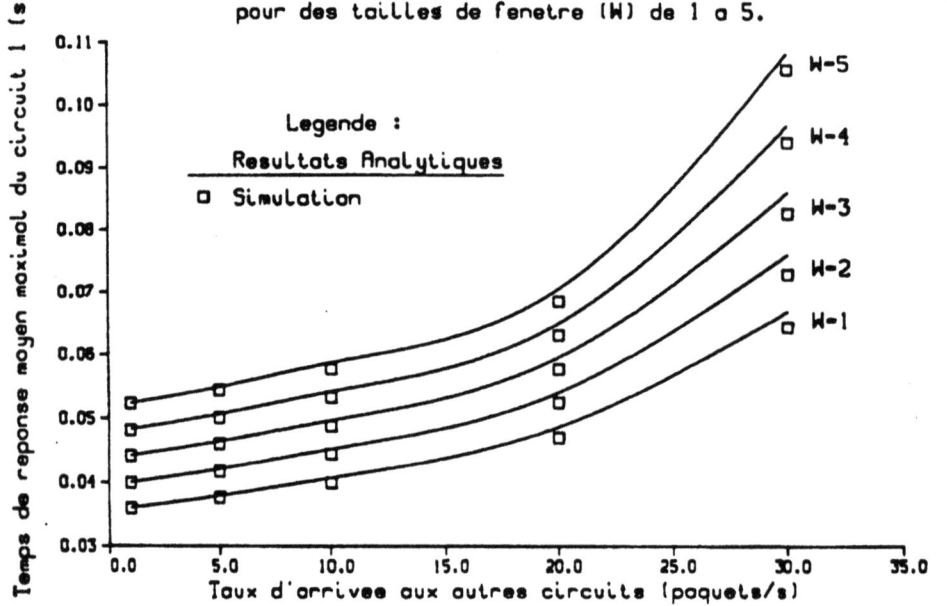

Fig. 15 - Temps de reponse moyen maximal du circuit 1 en fonction du taux d'arrivee aux autres circuits pour des tailles de fenetre (W) de 1 a 5.

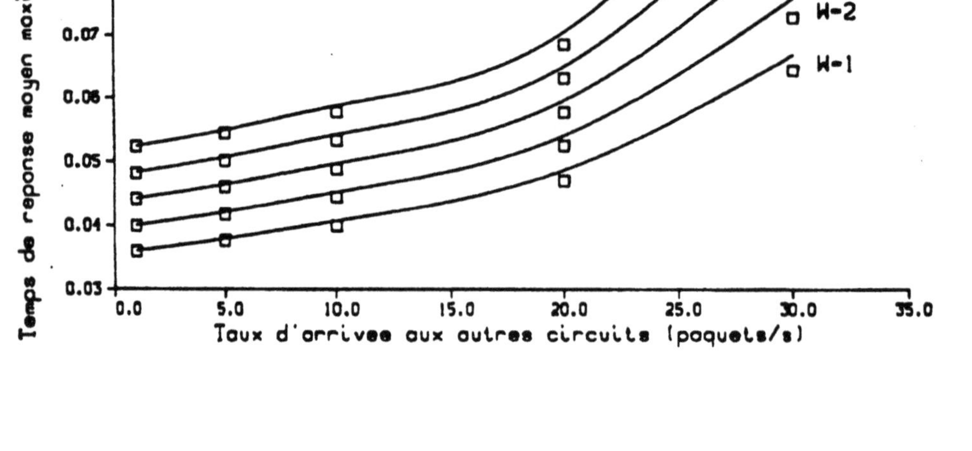

X

QUEUES AND NETWORKS 4

FILES D'ATTENTE ET RESEAUX 4

THE NORMAL APPROXIMATION AND QUEUE CONTROL FOR RESPONSE TIMES IN A PROCESSOR-SHARED COMPUTER SYSTEM MODEL

D. P. Gaver
Naval Postgraduate School

Patricia A. Jacobs
Naval Postgraduate School

Guy Latouche
University Libre Brusselles

ABSTRACT

Represent a time-shared computer system as a group of N terminals, each having submission rate λ and exponential (μ) task durations, with tasks submitted to a central (single) processor. There they are serviced in processor-sharing or time-sliced mode. It is shown that the $R(t)$, the response time conditional on t, the required processing time, becomes approximately normally distributed as t increases. Similar results are derived when N increases.

Variations of the model consider control: an "inside," processor-shared queue services at most c tasks, others queueing first-come first-served "outside." Other possibilities are described and analyzed.

1. Introduction

The abstraction of computer capacity allocation known as processor sharing is an attractive simplification of time slicing, sometimes called round-robin scheduling. The idea is well known (Kleinrock, 1976): given that j jobs or programs are at the execution stage, each receives service equal to one-jth of a time unit per time unit. In other words, if the chance that any single job, processed alone, finishes in $(t,t+h)$ is $\mu h + o(h)$, (exponential-Markov service), then the chance that a particular ("tagged") job in the company of $(j-1)$ others finishes in $(t,t+h)$ is $\mu(h/j) + o(h)$ as $h \to 0$. Processor sharing of the above type tends to be equitable in that it permits short jobs access to processing even if they arrive after, and queue with, longer jobs.

Apparently the first study of delays to arriving and queueing jobs under processor sharing was conducted by Coffman, Muntz, and Trotter (1970). They assumed a steady state M/M/1 system with processor sharing, and were able to determine properties of the response time, R, given the processing time required by the arriving job. Other papers have also appeared.

A recent paper by D. Mitra (1981) analyzes response time, R, under the assumption of a closed system. Idealize the behavior of a system of N terminals and a single computer as a classical machine-repair situation: each thinking terminal (failure-prone machine) applies for computer service at rate λ, and queued or waiting jobs are served at rate μ as long as any jobs are present. Markov assumptions are made throughout, so $X(t)$, the number of jobs at the service stage, is a birth and death process with transition rates

$$
\begin{aligned}
X(t) = j \rightarrow X(t+h) &= j+1 : \quad \lambda_j h + o(h) \\
\rightarrow X(t+h) &= j-1 : \quad \mu_j h + o(h) \\
\rightarrow X(t+h) &= j \quad\;\; : \quad 1-(\lambda_j+\mu_j)h + o(h)
\end{aligned}
\tag{1.1}
$$

and in particular $\lambda_j = \lambda(N-j)$, $\mu_j = \mu$ for $j \geq 1$, otherwise being zero. Let processor sharing govern service effort allocation. In Mitra (1981) the distribution of response time is characterized, and the moments (e.g. mean and variance) are found under interesting conditions, such as that the tagged job requiring t time units of processing arrives to find $j-1$ accompanying jobs; the conditional response time, given only processing requirement t, is given particular attention.

In this paper the previous analysis is generalized and extended. We introduce the idea of processor sharing in an arbitrary birth and death process environment, thus allowing quite general terminal-computer interactions to be represented. In the process, the meaning of "system state at the moment of tagged job arrival" is clarified; see also recent work of Lavenberg and Reiser (1981). Response time characteristics are computed under the assumption that processor-sharing service rates are processor-state-dependent in a more general way than that described earlier; this allows for approximate representation of overhead penalties and also of job scheduling. Other characteristics of tagged job response are also studied, e.g. the accumulated processing work, $W(\tau)$, actually performed on that job by elapsed time $\tau (\tau < t =$ required processing time) following job introduction; note that $W(R) = t$, so the first passage of $W(\tau)$ to t is actually the response time.

Although differential equations may be obtained for transforms of $W(\tau)$ under various initial conditions, and hence, implicitly, for its distribution, the results are far from being explicit and informative. However, central limit theorems for additive functionals of Markov processes, or for cumulative processes, allow the conclusion that the accumulated work accomplished by fixed time τ on a "long" job is approximately normally distributed (Gaussian). This fact in turn allows the conclusion that the response time for a "long" job is also approximately normally distributed. Additionally, the normal approximation may be shown to be valid for our simple model—and probably for others as well—when the number of competing terminals becomes large, i.e. under heavy traffic conditions. The quality of the normal approximations for finite job lengths is currently being assessed by simulation methods.

In the latter part of this paper we describe queue control in a processor-sharing environment. The expedient is to limit the total number of jobs allowed simultaneous processor-shared service at an "inside" queue, with any excess in "first-come, first-served" status in an "outside" queue. Long jobs are also shifted from inside to outside by a sampling mechanism. It is shown that long jobs are favored by a small inside span, c (c being the number simultaneously processor-shared), while short jobs are favored by large c.

2. Mean Response Times

Begin by describing differential equations for the mean response time to be experienced by a tagged, particular, arriving job. Other moments satisfy very similar equations.

(a) Conditioning on Required Time and System State.

Throughout what follows Markovian assumptions are made: service times at the computer are independent and exponential (μ). Generalizations to phase-type distributions are apparently possible.

Let R refer to the response time of a newly arrived job, and

$$m_j(t) = E[R|X(0) = j,\ W(R) = t] \ , \tag{2.1}$$

the conditional expectation of response time, given that the job is initially in the company of j others (arrives to find $j - 1$ present) and requires "work" or processing time equal to t. Let $\lambda_j h$ (resp. $\mu_j h$) for small h identify the infinitesimal generator of the accompanying process, so transition rates are as in (1.1).

Consider the possible system changes in $(0,h)$, and subsequently; the following results occur: $\left(\text{let } \tilde{\mu}_j = (j-1)\frac{r(j)}{j}\right)$:

$$\begin{aligned} m_j(t) &= h + m_j\left(t - \frac{r(j)}{j}\cdot h\right)[1 - (\lambda_j+\tilde{\mu}_j)h] \\ &\quad + \lambda_j h m_{j+1}\left(t - \frac{r(j)}{j}\cdot h\right) + \tilde{\mu}_j h m_{j-1}\left(t - \frac{r(j)}{j}\cdot h\right) + o(h) \ . \end{aligned} \tag{2.2}$$

Allowing $h \to 0$ one finds the differential equations

$$\frac{r(j)}{j}\, m_j{}'(t) = 1 - (\lambda_j + \tilde{\mu}_j)m_j(t) + \lambda_j m_{j+1}(t) + \tilde{\mu}_j m_{j-1}(t) \ . \tag{2.3}$$

This is a standard system of linear differential equations with constant coefficients; initial conditions are $m_j(0) = 0$ for all j.

(b) Conditioning on Required Time.

If one removes the condition that $X(0) = j$ in accordance with the stationary distribution appropriate for an arriving job it follows that the expected response time is <u>linear</u> in the required processing time, t. This holds for quite general birth-and-death process models, and not just for the simple machine-repair setup; see Cohen [1979]. Here is the derivation, in outline.

First, observe that the long-run distribution of $X(0)$, the number of jobs present just after the tagged job enters, is

$$q_j = c\pi_{j-1}\lambda_{j-1} = c\pi_j\mu r(j) \qquad j = 1,2,\ldots,N \ , \tag{2.4}$$

where c is selected so that the q_j's sum to one. Recall that $\pi_j = \pi_0 \frac{\lambda_0\lambda_1\cdots\lambda_{j-1}}{\mu_1\mu_2\cdots\mu_j}$ is the stationary distribution (assumed to exist) of the Markov chain $X(t)$ defined by (1.1) with $\mu_j = \mu r(j)$. This is intuitively apparent, but a formal proof can be based either upon an embedded Markov chain formulation, or upon the theory of additive functionals of a Markov process; see Çinlar ((1975), pp. 269-271). The distribution $\{q_j\}$ has also been given by Kelly, (1979), p. 12.

Use (2.4) to remove the condition that $X(0) = j$; put

$$m(t) = E_{X(0)}E[R|X(0),\ W(R) = t] = \sum_{j=1}^{\infty} q_j m_j(t) \ . \tag{2.5}$$

Then in terms of the differential equations (2.3); after multiplying through by $j/r(j)$ one obtains $(m_0(t) \equiv 0)$

$$m'(t) = \sum_{j=1}^{\infty} \frac{j}{r(j)} q_j + \sum_{j=1}^{\infty} q_j \frac{j}{r(j)} [-(\lambda_j+\tilde{\mu}_j)m_j(t) + \lambda_j m_{j+1}(t) + \tilde{\mu}_j m_{j-1}(t)]$$

$$= \sum_{j=1}^{\infty} \frac{j}{r(j)} q_j \ . \tag{2.6}$$

Thus it follows that the long-run conditional expected response time is linear in the processing time requirement:

$$E[R|W(R) = t] = t \sum_{j=1}^{\infty} \frac{j}{r(j)} q_j = tE\left[\frac{X(0)}{r(X(0))}\right] . \tag{2.7}$$

Apparently no such simple form exists for $\mathrm{Var}[R|W(R) = t]$, although Mitra (1981) has given a formula for a particular case. It will be seen, however, that the above variance is indeed proportional to t if t is large.

3. Total Work Completed on a Tagged Job in a Fixed Time

Turn attention now to $W(\tau)$, the total work expended by the computer on the tagged job by time τ after its arrival, given that the tagged job requires exactly t time-units of work for completion. If when the job arrives there are $X(0) = j$ customers present, then

$$W(\tau) = \int_0^\tau \frac{r(X(u))}{X(u)}\,du, \qquad X(0) = j \geq 1 . \tag{3.1}$$

(a) The Laplace transform of $W(\tau)$.

Here is the derivation of a differential equation for the Laplace transform of $W(\tau)$:

$$\phi_j(s,\tau;t) = E[e^{-sW(\tau)} | X(0) = j,\ R = t] , \quad \text{for} \quad j \geq 1 \tag{3.2}$$

Now argue in a manner analogous to the discussion prior to (2.2) to write a backward equation: for $0 < h < t$

$$\begin{aligned}\phi_j(s,\tau;t) &= (e^{-sr(j)h/j})[1-(\lambda_j+\tilde{\mu}_j)h]\phi_j(s,\tau-h;t)\\ &+ (e^{-sr(j)h/j})\lambda_j h\phi_{j+1}(s,\tau-h;t)\\ &+ (e^{-sr(j)h/j})\tilde{\mu}_j h\phi_{j-1}(s,\tau-h;t) + o(h) ,\end{aligned} \tag{3.3}$$

leading to

$$\frac{d\phi_j}{d\tau} = -(\lambda_j + \tilde{\mu}_j + sr(j)/j)\phi_j + \lambda_j\phi_{j+1} + \tilde{\mu}_j\phi_{j-1} . \tag{3.4}$$

Initial conditions are

$$\phi_j(s,0;t) = E[e^{-sW(0)} | X(0) = j,\ R = t] = 1 , \qquad t > 0 , \tag{3.5}$$

since initially $W(0) = 0$, regardless of the job requirements or the initial environment.

(b) A central limit théorem for $W(\tau)$.

Examination of (3.1) shows that $W(\tau)$ involves sums of contributions to work accumulated while the system inhabits various states during the period $(0,\tau)$. This suggests that, at least for "long" jobs, i.e. such that required processing time $t \to \infty$, one can anticipate a nearly-Normal distribution for $W(\tau)$. An appropriate central limit theorem that establishes this for finite birth-and-death models can be found in Keilson ((1979), p. 121); call this Theorem K. Alternatively, one can make use of the theory of cumulative processes, see Cox ((1962), pp. 99-101); the latter development is adaptable to models more general than the simple birth-and-death process.

In order to apply Theorem K redefine the infinitesimal generator (1.1) to describe the behavior of the accompaniment, $X'(t)$, of the tagged customer; note that the relevant generator is now

$$X'(t) = j \to X'(t+h) = j+1 : \quad \lambda_j' h + o(h)$$

$$\equiv \lambda_{j+1} h + o(h) \tag{3.6}$$

$$\to X'(t+h) = j-1 : \quad \mu_j' h + o(h)$$

$$\equiv \mu_{j+1}\left(\frac{j}{j+1}\right) h + o(h) \tag{3.7}$$

$$\to X'(t+h) = j \quad : \quad 1-(\lambda_j' + \mu_j')h + o(h) \tag{3.8}$$

for $j = 0,1,2,\ldots, \quad N' = N-1$. Then

$$W(\tau) = \int_0^\tau f(X'(u))du \equiv \int_0^\tau \frac{r(X'(u)+1)du}{X'(u)+1} \tag{3.9}$$

and theorem K states that

$$\frac{W(\tau) - \xi\tau}{\sigma\sqrt{\tau}} \overset{(D)}{\Rightarrow} N(0,1) ; \tag{3.10}$$

the constants ξ and σ^2 are such that

$$\xi = \sum_{j=0}^{N'} f(j)\pi_j' \tag{3.11}$$

$$\sigma^2 = 2[f(0),f(1),\ldots,f(N')]\begin{bmatrix} \pi_0' & & 0 \\ & \pi_1' & \\ 0 & & \pi_{N'}' \end{bmatrix} \mathcal{Z} \begin{bmatrix} f(0) \\ f(1) \\ \vdots \\ f(N') \end{bmatrix}, \tag{3.12}$$

where in the present case the definition of $f(\cdot)$ is implicit in (3.9), and $\mathcal{Z}$ is the matrix

$$\mathcal{Z} = \frac{1}{\gamma}\{[I-\mathcal{A}+\mathcal{L}]^{-1} - \mathcal{L}\} \tag{3.13}$$

$\underset{\sim}{I}$ being the identity, and

$$\underset{\sim}{L} = \begin{vmatrix} \pi'_0 & \pi'_1 \cdots \pi'_{N'} \\ \pi'_0 & \pi'_1 \cdots \pi'_{N'} \\ \pi'_0 & \pi'_1 \cdots \pi'_{N'} \end{vmatrix} , \tag{3.14}$$

$\underset{\sim}{L}$-rows are steady-state probabilities for the accompaniment, and $\underset{\sim}{A}$ is defined as follows:

$$A_{0,1} = \lambda'_0/\gamma , \quad A_{0,0} = 1 - \lambda'_0/\gamma$$

$$A_{j,j+1} = \lambda'_j/\gamma , \quad A_{j,j-1} = \mu'_j/\gamma , \quad A_{j,j} = 1 - \nu_j/\gamma \tag{3.15}$$

$$\nu_j = \lambda'_j + \mu'_j ; \quad \gamma = \max_j \nu_j .$$

$$(\nu_0 = \lambda'_0, \ \nu_{N'} = \mu'_{N'})$$

(c) A central limit theorem for response time, $R(t)$.

A graph of $W(\tau)$ vs. τ starts with $W(0) = 0$ and increases in random straight-line segments until $W(\tau) = t$. The value of τ at which this occurs, $\tau(t) \equiv R(t)$, is the first-passage time to t of the work process $\{W(\tau), \tau \geq 0\}$, and is the required response time, so

$$P\{W(\tau)<t\} = P\{R(t)>\tau\} . \tag{3.16}$$

Now invoke the previous theorem (3.10) concerning asymptotic normality of $W(\tau)$ and a standard argument of renewal theory, cf. Karlin and Taylor ((1979), pp. 208-209) to see that if $t = \xi\tau + \sqrt{\sigma^2\tau}x$, then, as $t \to \infty$, $\tau \sim \frac{t}{\xi} - \sqrt{\frac{\sigma^2 t}{\xi^3}}\, x$, from which it follows that

$$\frac{R(t) - \alpha t}{\sqrt{\beta^2 t}} \equiv \frac{R(t) - t/\xi}{\sqrt{\frac{\sigma^2}{\xi^3} t}} \overset{(D)}{\Rightarrow} N(0,1) . \tag{3.17}$$

4. Heavy Traffic Analysis of the Response Time of a Processor-Shared Job

This section investigates the problem of delay of a tagged job requiring t units of processing time when it is accompanied by many others, i.e. is in a heavily loaded system. Restrict attention to the machine repair model in which $\lambda_j = \lambda(N-j)$ and $\mu_j = \mu$, and omit the effect of $r(j)$, i.e. $r(j) \equiv 1$. Let there be N terminals and one processor, with λ^{-1} being the expected terminal think time (exponentially distributed), $\mu = N\mu'$ being the processing rate of arriving jobs; λ and μ' are fixed but N is large and the service rate scaling by N is required in order that queue size be of order N.

Now utilize the fact (Iglehart (1965), and Gaver and Lehoczky (1976)) that if $N(t)$ is the number of jobs at the processing stage at t then $N(t)$ can be approximated by a diffusion process:

$$N(t) = N\,a(t) + \sqrt{N}\,X(t) \tag{4.1}$$

where $a(t)$ is a deterministic function of time and $\{X(t)\}$ is, for the present model, a particular Ornstein-Uhlenbeck process. It turns out that when $N \to \infty$

$$\frac{da(t)}{dt} = \lambda(1-a(t)) - \mu' \tag{4.2}$$

or

$$a(\infty) = 1 - \frac{\mu'}{\lambda} \tag{4.3}$$

which is feasible if $\lambda > \mu'$, i.e. under heavy traffic conditions. Furthermore

$$dX(t) = -\lambda X(t)dt + \sqrt{\lambda(1-a(t)) + \mu'}\,dB(t)\,, \tag{4.4}$$

$\{B(t), t \geq 0\}$ being the standard Wiener process. In the long run,

$$dX(t) = -\lambda X(t)dt + \sqrt{2\mu'}\,dB(t)\,. \tag{4.5}$$

It is in the environment $N(t)$ described by (4.1) that the tagged job enters. It encounters competition for processor-shared service, and so its accumulated work completed by fixed time τ is essentially

$$W(\tau) = \int_0^{\tau} \frac{du}{N(u)}\,. \tag{4.6}$$

Apply the approximation and expand to second order terms in N, the number of terminals, to find

$$W(\tau) \simeq \int_0^\tau \frac{du}{Na(u)\left[1 + \frac{X(u)}{a(u)\sqrt{N}}\right]} \simeq \int_0^\tau \frac{du}{Na(u)} - \int_0^\tau \frac{\sqrt{N}\, X(u)dr}{(Na(u))^2} \tag{4.7}$$

For simplicity, and to enable comparisons with previous results, let $a(r) = a(\infty)$, so the tagged job arrives in the steady state. Expression (4.6) then says that for the approximation advanced here,

$$W(\tau) = \int_0^\tau \frac{du}{E[N(\infty)]} - \frac{\sqrt{N}}{(E[N(\infty)])^2} \int_0^\tau X(u)du \qquad 0 \le \tau \le t \tag{4.8}$$

so the expected amount of work done on the tagged job is nearly $\tau/E[N(\infty)]$, and the actual distribution of total work done is approximately Gaussian (integral of an Ornstein-Uhlenbeck process), where the Gaussian property results from the assumption of many accompanying jobs, and not necessarily because the tagged job is long.

Standard calculations applied to (4.7) show that, as $\tau \to \infty$, $E[\int_0^\tau X(u)du] \simeq 0$ and $\mathrm{Var}[\int_0^\tau X(u)du] \simeq (2\mu'/\lambda^2)$, so the normal approximation to accumulated work $W(\tau)$ has the parameters

$$\xi = \frac{1}{[Na(\infty)]}, \qquad \sigma^2 = \frac{2\mu'}{\lambda^2} \cdot \frac{N}{[Na(\infty)]^4} = \frac{2\mu}{\lambda^2[Na(\infty)]^4} \tag{4.9}$$

from which it follows that the parameters of the normal approximation to $R(t)$ are

$$\alpha = \frac{1}{\xi} = [Na(\infty)] \simeq E[N(\infty)]$$

$$\beta^2 = \frac{2\mu}{\lambda^2}\frac{1}{Na(\infty)} \simeq \frac{2\mu}{\lambda^2}\frac{1}{E[N(\infty)]} = \frac{2}{\lambda}\left\{\frac{N - E[N(\infty)]}{E[N(\infty)]}\right\} . \tag{4.10}$$

These formulas state that if think (demand) rate λ is very large then, since $E[N(\infty)] \to N$, the variance of response time diminishes, while of course expected response time increases like N . This is plausible since in extremely heavy traffic all terminals compete, and the tagged job gets a steady $(1/N)^{\underline{th}}$ of a quantum. For smaller λ the expected response time drops with $E[N(\infty)]$, but response time variance increases.

The above derivations are informative but not rigorous. Semigroup methods of Berman (1979) can be applied to place the results on a mathematically solid basis. Numerical assessment of the results is also of interest.

5. Numerical Comparisons

In this section a brief investigation is reported of the numerical agreement between the very simple formulas from heavy traffic theory for the parameters of the accumulated work distribution and those of direct Markov-chain cumulative process theory origin.

Parameters of Total Work

Examples. Let $\mu = 1$, $N = 25, 50$, with λ varying.

$N = 25$

Rates/λ:	0.01	0.0222...	0.05	0.10	0.15
ξ(M.Ch.)	0.77	0.51	0.16	0.067	.055
(Diffus.)	--	--	0.20	0.067	0.055
σ^2(M.Ch.)	0.44	0.75	0.22	.0045	0.00085
(Diffus.)	--	--	1.28	.0040	0.00079

$N = 50$

Rates/λ:	0.01	0.0222...	0.05	0.10	0.16
ξ(M.Ch.)	0.53	0.13	0.033	0.025	0.023
(Diffus.)	--	0.20	0.033	0.025	0.023
σ^2(M.Ch.)	0.83	0.33	0.0011	0.000081	0.000026
(Diffus.)	--	6.48	0.00099	0.000078	0.000025

The diffusion approximation and Markov chain parameters agree remarkably well when traffic is heavy (large λ), but, as might be feared, diffusion fails miserably for small λ .

6. Queue Control by Service Span and Interruption

In this section we consider queue control. The central processor now has finite service span, c , which may be smaller than the number, N , of terminals. This means that if there are $i \leq c$ jobs in service they are served as before "inside" at a rate $\mu r(i)$, with processor sharing in effect. However, if there are more than c jobs simultaneously requesting service, only c of them are served simultaneously, and at rate $\mu r(c)$, also with processor sharing discipline. The others are queued "outside", with "first-come, first-served" service discipline.

If there are more than c customers requesting service, the customers that are in service "inside" experience independent service interruptions at rate ν. When service is interrupted, each job in service is equally likely to be moved to the end of the queue; thereupon the job at the head of the "outside" queue immediately enters service. Both the imposition of the limited processor sharing, imposed by $c \leq N$, and the interruption process are intended to control queueing by adjusting the relative attention given to short and long jobs.

Markovian assumptions are made throughout, so that $X(t)$, the number of jobs requesting service at time t, is a birth and death process with transition rates given by (1.1).

(a) An Auxiliary Process.

Let R be the response time of a newly arrived tagged job. Since the tagged job may not be served until completion when it first enters service, it is necessary to introduce an auxiliary process $\{Y(t);\ t \geq 0\}$ in order to study R.

In brief summary, if there are i customers (including the tagged job) requesting service at time t and the tagged job is in service, then the state of $Y(t)$ is $(i,0)$. If there are $i > c$ customers requesting service at time t and the tagged job is not in service but is in the jth position in queue, the state of $Y(t)$ is (i,j). One can now describe the possible changes in $Y(t)$ in a time interval of length h; details will be presented elsewhere.

Let

$$m_{(i,j)}(t) = E[R)|Y(0) = (i,j)\ , \quad W(R) = t]\ , \tag{6.1}$$

the conditional expected response time, given that the tagged job is initially in the company of $(i-1)$ others and either it is being served inside (if $j = 0$) or it is jth in the outside queue (if $j > 0$).

Arguments similar to those of Section 2 yield differential equations for $m_{(i,j)}(t)$ which can be solved numerically. A closed-form solution is complicated and uninformative. It is possible to numerically evaluate the mean response time,

$$m(t) = E[R|W(R) = t] = \sum_{i,j} q_{(i,j)} m_{(i,j)}(t)\ ,$$

where $q_{(i,j)}$ is the initial distribution encountered by the tagged job.

The mean response time is not generally linear in t for this model. Note that if $c = N$ then this model is equivalent to that considered in Section 2 and hence as shown is Section 2, $m(t)$ is linear in t for that special case.

(b) An Approximation to Expected Response Time.

A useful approximation to the expected response time for a job requiring t units of work is obtained by the following argument. Assume that the service rate for the tagged job is the same throughout its processing and is equal to the rate that it experiences when it first enters the system. Thus the tagged job requires $t^* = \frac{it}{r(i)}$ units of processing time if it enters when these are $i \leq c$ jobs (including the tagged one) requesting processing. If $i > c$, then $t^* = \frac{ct}{r(c)}$. If $i > c$, then the number of service interruptions during t is Poisson with rate $\frac{\nu}{c}$. Each time service is interrupted, the tagged job spends an expected amount of time $(i-c)\ [\nu + \mu r(c)]^{-1}$ in queue. Thus the expected time spent in queue because of service interuptions is $(i-c)(\frac{\nu}{c})t^*[\nu + \mu r(c)]^{-1}$. If $i > c$, the expected initial wait in queue until the tagged job starts service is $(i-c)[\nu + \mu r(c)]^{-1}$. The resulting approximation to the expected response time is (see (2.4))

$$A = \sum_{i=1}^{c} q(i)\frac{it}{r(i)} + \sum_{i=c+1}^{N} q(i)\frac{ct}{r(c)} + \sum_{i=c+1}^{N} q(i)(i-c)[\nu + \mu r(c)]^{-1}\left[\frac{\nu t}{r(c)} + 1\right]. \tag{6.2}$$

Table 1 gives values for the expected response time, $m(t)$, and the above approximation for various values of λ, μ, ν, c, and t for $r(j) \equiv 1$ and $N = 25$ terminals. The quality of the approximation (6.2) appears to be excellent for all cases considered.

(c) Numerical Implications.

Aspects of the behavior of $m(t)$ to be noted from the table are as follows. If the amount of processing time required, t, is "small", then expected response time is minimized when c is maximized (here $c = 25$); that is, when there is maximal processor sharing and no outside queue. If t is "large", then expected response time is minimized when $c = 1$; that is, when the processor is dedicated solely to the job that is being served, and other jobs queue outside in turn. Note that increasing the rate of service interruptions by changing ν can either increase or decrease the expected response time, depending upon job time requirements.

These behavioral aspects also appear by taking derivatives of the approximate average response time A. In particular,

$$\frac{\partial}{\partial \nu} A \begin{cases} < 0 & \text{if} \quad t < \frac{1}{\mu}, \\ = 0 & \text{if} \quad t = \frac{1}{\mu}, \\ > 0 & \text{if} \quad t > \frac{1}{\mu}. \end{cases}$$

If $r(j) \equiv 1$ $j = 1,\ldots,N$, then A is decreasing in c if $t < \frac{1}{\mu}$; A is increasing in c if $t > \frac{1}{\mu}$, and A is constant in c if $t = \frac{1}{\mu}$.

Finally, arguments similar to those in Section 3 will show that the response time is approximately normally distributed when the required work is large. Again, details will be provided in later work.

The Expected Response Time

λ	μ	ν	t	Actual	Approx.	C	1	3	5	7	9	11	13	15	17	19	21	23	25
0.1	2	0	0.1	X			2.76	2.04	1.47	1.07	0.82	0.70	0.65	0.64	0.63	0.63	0.63	0.63	0.63
					X		2.76	2.05	1.48	1.07	0.83	0.7	0.65	0.64	→	→	→	→	→
0.1	2	1	0.1	X			2.03	1.56	1.18	0.92	0.76	0.68	0.64	0.63	→	→	→	→	→
					X		2.05	1.57	1.19	.92	.76	.68	.64	.63	→	→	→	→	→
0.1	2	1	0.5	X			3.11	3.12	3.14	3.14	3.15	3.16	→	→	→	→	→	→	→
					X		3.16	→	→	→	→	→	→	→	→	→	→	→	→
1	0.5	2.0	0.4	X			10.72	10.64	10.56	10.48	10.40	10.33	10.25	10.17	10.09	10.01	9.94	9.86	9.8
					X		10.72	10.64	10.56	10.48	10.4	10.33	10.25	10.17	10.09	10.01	9.94	9.86	9.8
.01	2	1	0.1	X			0.15	0.11	→	→	→	→	→	→	→	→	→	→	→
					X		0.15	0.11	→	→	→	→	→	→	→	→	→	→	→

Table 1.

APPENDIX

To show (2.7) and (3.10) yield the same value for $E(R|W(\tau) = t]$ for t large for the case $\mu_j = \mu r(j)$.

First:

$$q(j) = c\pi_j\mu_j$$

$$= c\,\frac{\lambda_0 \times \dots \times \lambda_{j-1}}{\mu_1 \times \dots \times \mu_{j-1}}\,.$$

Hence,

$$m \equiv \sum_{j=1}^{\infty} \frac{j}{r(j)}\, q_j$$

$$= c \sum_{j=1}^{\infty} \frac{j}{r(j)} \frac{\lambda_0 \times \dots \times \lambda_{j-1}}{\mu_1 \times \dots \times \mu_{j-1}}$$

Since,

$$\mu_j = \mu r(j)$$

$$m = c\mu \sum_{j=1}^{\infty} j\, \frac{(\lambda_0 \times \dots \times \lambda_{j-1})}{\mu_1 \times \dots \times \mu_j}$$

$$= \frac{c\mu}{\prod_{k=1}^{\infty} \mu_k} \sum_{j=1}^{\infty} j(\lambda_0 \times \dots \times \lambda_{j-1}) \left(\prod_{i=j+1}^{\infty} \mu_i\right)$$

Further,

$$c = \frac{\prod_{k=1}^{\infty} \mu_k}{\sum_{j=1} \lambda_0 \times \dots \times \lambda_{j-1} \prod_{k=j}^{\infty} \mu_k}\,.$$

Hence,

$$m = \mu\, \frac{\sum_{j=1}^{\infty} j(\lambda_0 \times \dots \times \lambda_{j-1}) \left(\prod_{k=j+1}^{\infty} \mu_k\right)}{\sum_{j=1}^{\infty} (\lambda_0 \times \dots \times \lambda_{j-1}) \left(\prod_{k=j}^{\infty} \mu_k\right)} = \frac{\sum_{j=1}^{\infty} j \left(\prod_{i=1}^{j-1} \lambda_i\right) \left(\prod_{k=j+1}^{\infty} \mu_k\right)}{\sum_{j=1}^{\infty} \left(\prod_{i=1}^{j-1} \lambda_i\right) \left(\prod_{k=j}^{\infty} \mu_k\right)}$$

where $\lambda_1 \times \lambda_0 = 1$ by convention.

Second:

$$\pi'(j) = \frac{\lambda_1 \times \dots \times \lambda_j}{\mu_2 \times \dots \times \mu_{j+1}} (j+1)\pi'(0) \qquad j = 0,1,\dots \qquad [\lambda_1 \times \lambda_0 \equiv 1]$$

where

$$\pi'(0) = \frac{\prod_{k=2}^{\infty} \mu_k}{\sum_{j=1}^{\infty} j(\lambda_1 \times \dots \times \lambda_{j-1}) \prod_{k=j+1}^{\infty} \mu_k}$$

$$\xi = \sum_{j=0}^{\infty} \frac{r(j+1)}{j+1} \pi'(j) = \sum_{j=1}^{\infty} \frac{r(j)}{j} \pi'(j-1)$$

$$= \sum_{j=1}^{\infty} r(j) \frac{\lambda_1 \times \dots \times \lambda_{j-1}}{\mu_2 \times \dots \times \mu_j} \pi'(0)$$

$$= \frac{1}{\mu} \sum_{j=1}^{\infty} \frac{\lambda_1 \times \dots \times \lambda_{j-1}}{\mu_2 \times \dots \times \mu_{j-1}} \pi'(0)$$

since

$$\mu_{j+1} = \mu r(j+1) \text{ . Thus}$$

$$\xi = \frac{\pi'(0)}{\left(\prod_{k=2}^{\infty} \mu_k\right)} \frac{1}{\mu} \sum_{j=1}^{\infty} \lambda_1 \times \dots \times \lambda_{j-1} \left(\prod_{k=j}^{\infty} \mu_k\right)$$

$$= \frac{\sum_{j=1}^{\infty} (\lambda_1 \times \dots \times \lambda_{j-1}) \left(\prod_{k=j}^{\infty} \mu_k\right)}{\mu\left(\sum_{j=1}^{\infty} j(\lambda_1 \times \dots \times \lambda_{j-1}) \prod_{k=j+1}^{\infty} \mu_k\right)}$$

A comparison of ξ with m shows that

$$m = \frac{1}{\xi} .$$

REFERENCES

Burman, D. (1979), "An analytic approach to diffusion approximations in queueing". Unpublished Doctoral Dissertation, New York University.

Cinlar, E. (1975), Introduction to Stochastic Processes, Prentice-Hall, Englewood Cliffs, N.J.

Coffman, E. G., Muntz, R. R. and Trotter, H. (1970), "Waiting time distribution for processor-sharing systems," J. Assn. for Comp. Mach., 17, pp. 123-130.

Cox, D. R. (1962), Renewal Theory, Methuen Monograph.

Gaver, D. P., and Lehoczky, J. P. (1976), "Gaussian approximation to service problems: a communications system example," J. Appl. Prob., 13, pp. 768-780.

Gaver, D., Jacobs, P. and Latouche, G. (1981). "Finite birth and death models in randomly changing environments". To appear J. Appl. Prob.

Iglehart, D. L. (1965), "Limiting diffusion approximations for the many-server queue and the repairman problem," J. Appl. Prob., 2, pp.

Karlin, S., and Taylor, H. M. (1975), A First Course in Stochastic Processes, (Second Edition). Academic Press, New York.

Keilson, J. (1979), Markov Chain Models - Rarity and Exponentiality, Springer-Verlag, New York.

Kelly, F. P. (1979), Reversibility and Stochastic Networks, John Wiley and Sons, New York.

Kleinrock, L. (1976), Queueing Systems, Vol. II, Wiley-Interscience.

Lavenberg, S. S. and Reiser, M. (1980), "Stationary state probabilities at arrival instants for closed queueing networks with multiple types of customers," J. Appl. Prob., 17, pp. 1048-1061.

Mitra, D. (1981), "Waiting time distributions from closed queueing network models of shared processor systems," Bell Laboratories Report.

Cohen, J. W. (1979), "The multiple phase service network with generalized processor sharing. Acta Informatica 12 245-284.

TIME DEPENDENT ANALYSIS OF A QUEUEING MODEL BY FORMULATING A BOUNDARY VALUE PROBLEM

by J.P.C. Blanc
Mathematical Centre, Kruislaan 413,
1098 SJ Amsterdam, The Netherlands

Abstract

The analysis of queueing models which can be characterized as a random walk in the first quadrant of the plane often leads to the problem of solving a functional equation for a bivariate generating function. Recently, a method has been developed by which a rather general class of such functional equations related to stationary distributions can be solved with the aid of the theory of boundary value problems, see [1],[3],[4],[5],[6],[7],[10]. In the present study we shall show that the same method can be applied in the analysis of the time dependent behaviour of this class of queueing models. For this discussion a relatively simple model with two types of customers, Poissonian arrival streams, paired services and a general service time distribution will be considered. The generating function of the joint queue length distribution at the nth departure instant will be determined. This function forms the starting point for the analysis of the asymptotic behaviour of the process as $n \to \infty$.

Key words: queueing system, two-dimensional state space, time dependent behaviour, functional equation, boundary value problem

1. Introduction, the model

In [3] the stationary M/G/1 queueing system with alternating service has been studied. The functional equation for the generating function of the joint queue length distribution at departure epochs has been reduced to two Riemann-Hilbert boundary value problems. In the present study it will be shown that this technique of solving functional equations by formulating Riemann-Hilbert problems, or by formulating a related Hilbert problem (see [9] for this terminology) can also be applied in the time dependent analysis of queueing models with a two-dimensional state space. To show this we shall consider a queueing system with two types of customers and paired services, of which the functional equation for the generating function of the joint distribution of the number of type 1 and of type 2 customers left behind in the system at the nth departure instant has the same structure as the functional equation analysed in [3], but is of a simpler form.

This model is as follows. Two types of customers arrive independently at a single service facility. For type j customers the interarrival times are independent random variables with a common negative exponential distribution with mean α_j (j=1,2). An arriving customer who finds the system empty is immediately taken into service; otherwise he joins queue 1 or 2 depending on his type. As soon as a service has been completed, a new service is started if any customer is present. In general a couple of two customers of different type is simultaneously served. If after the completion of a service only customers of one type are present, a customer of this type is individually served. In each queue customers are served in order of their arrival. Successive service times are independent random variables with a common distribution function B(t), for paired services as well as for individual services.

Denote by $\underline{x}_j(n)$, $n=0,1,2,..$, $j=1,2$, the number of type j customers left behind in the system at the nth service completion instant. It is assumed that the process starts for $n=0$ with an empty system (this assumption is not essential, see [1]). It is readily seen that the process $\{(\underline{x}_1(n),\underline{x}_2(n)),n=0,1,..\}$ is an irreducible, aperiodic, discrete time Markov chain with state space $\{0,1,2,..\}\times\{0,1,2,..\}$. In the sequel this Markov chain will be analysed. For this we introduce the generating function: for $|r|<1$, $|p_1|\leqslant 1$, $|p_2|\leqslant 1$,

$$\text{(1)}\qquad \Phi(r;p_1,p_2) := \sum_{n=0}^{\infty} r^n\, E\{p_1^{\underline{x}_1(n)}\, p_2^{\underline{x}_2(n)} \mid \underline{x}_1(0)=0,\ \underline{x}_2(0)=0\}.$$

Further we define

$$\text{(2)}\qquad \frac{1}{\alpha} := \frac{1}{\alpha_1}+\frac{1}{\alpha_2}; \qquad\qquad c_j := \alpha/\alpha_j, \quad j=1,2;$$

$$\text{(3)}\qquad \beta(s) := \int_0^{\infty} e^{-st}\, dB(t), \quad \operatorname{Re} s \geqslant 0;$$

(4) $$\beta_k := \int_0^\infty t^k \, dB(t), \quad k=1,2,..;$$

(5) $$a := \beta_1/\alpha.$$

2. The functional equation

From the definition of the queueing process it follows that for $j=1,2$, the series $\{\underline{x}_j(n), n=0,1,..\}$ satisfies the relations

(6) $$\underline{x}_j(n) = [\underline{x}_j(n-1) - 1]^+ + \underline{\xi}_j(n), \quad n=1,2,..; \qquad \underline{x}_j(0) = 0;$$

here $\underline{\xi}_j(n)$, $n=1,2,..$, stands for the number of type j customers who arrive during the nth service. The generating function of the distribution of $(\underline{\xi}_1(n), \underline{\xi}_2(n))$ is given by: for $|p_1| \leq 1$, $|p_2| \leq 1$,

(7) $$E\left\{p_1^{\underline{\xi}_1(n)} \, p_2^{\underline{\xi}_2(n)}\right\} = \beta\left(\frac{1-c_1p_1-c_2p_2}{\alpha}\right), \quad n=1,2,.. .$$

From the relations (6) the following functional equation for the generating function $\Phi(r;p_1,p_2)$ is deduced by straightforward calculations: for $|r| < 1$, $|p_1| \leq 1$, $|p_2| \leq 1$,

(8) $$\left[p_1p_2 - r\,\beta\left(\frac{1-c_1p_1-c_2p_2}{\alpha}\right)\right] \Phi(r;p_1,p_2) = p_1p_2 + r\,\beta\left(\frac{1-c_1p_1-c_2p_2}{\alpha}\right) \times$$
$$\times [(p_2-1)\,\Phi(r;p_1,0) + (p_1-1)\,\Phi(r;0,p_2) + (p_1-1)(p_2-1)\,\Phi(r;0,0)].$$

Because the generating function $\Phi(r;p_1,p_2)$ is uniquely determined by the relations (6), this functional equation must have at least one solution with the properties of a generating function.

As a first investigation we take $p_2 = 1$ in equation (8). This leads to: for $|r| < 1$, $|p_1| \leq 1$,

(9) $$\left[p_1 - r\,\beta\left(\frac{1-p_1}{\alpha_1}\right)\right] \Phi(r;p_1,1) = p_1 + r\,\beta\left(\frac{1-p_1}{\alpha_1}\right)(p_1-1)\,\Phi(r;0,1).$$

This is a well-known equation from the theory of the M/G/1 queueing system, cf. [2], p.240. Hence, for $|r| < 1$,

(10) $$\Phi(r;0,1) = \frac{1}{1-\mu_1(r)},$$

here $p_1 = \mu_1(r)$ is the unique solution inside the unit circle of the equation

$$(11) \qquad p_1 - r\,\beta\!\left(\frac{1-p_1}{\alpha_1}\right) = 0.$$

An analogous result can be obtained by taking $p_1 = 1$ in equation (8). But for obtaining the complete solution of this equation more powerful techniques are required.

3. Analysis

Throughout this section, r is fixed and real, $0 < r < 1$.
Equation (8) relates the bivariate function $\Phi(r;p_1,p_2)$ to two univariate functions $\Phi(r;p_1,0)$, $\Phi(r;0,p_2)$, and a constant $\Phi(r;0,0)$. A central role in the analysis is played by the *kernel*

$$(12) \qquad p_1p_2 - r\,\beta\!\left(\frac{1-c_1p_1-c_2p_2}{\alpha}\right),$$

because if for a pair (p_1,p_2), $|p_1| \leq 1$, $|p_2| \leq 1$, this kernel vanishes, then the righthand side of equation (8) must also vanish. The existence of such pairs (p_1,p_2) can be shown with Rouché's theorem, cf. [1], p.49. This provides us with a relation between the functions $\Phi(r;p_1,0)$ and $\Phi(r;0,p_2)$, which can be written in the following form:

$$(13) \qquad \frac{\Phi(r;p_1,0)}{1-p_1} + \frac{\Phi(r;0,p_2)}{1-p_2} = \Phi(r;0,0) + \frac{1}{(1-p_1)(1-p_2)},$$

$$(14) \qquad \text{if} \quad p_1p_2 = r\,\beta\!\left(\frac{1-c_1p_1-c_2p_2}{\alpha}\right), \quad |p_1| \leq 1,\ p_1 \neq 1,\ |p_2| \leq 1,\ p_2 \neq 1.$$

Note that the cases $p_1 = 1$ and $p_2 = 1$ have already been discussed in section 2.
From the above functional relation the functions $\Phi(r;p_1,0)$ and $\Phi(r;0,p_2)$ have to be determined. For this purpose we shall first introduce a parameter δ in order to describe the zeros (p_1,p_2) of the kernel (12) as functions of this parameter, cf. [3]. Hence, let

$$(15) \qquad \delta := c_1p_1 + c_2p_2, \qquad w := 2c_1p_1.$$

Substitution of (15) in equation (14) leads to the equation

$$(16) \qquad w^2 - 2\delta w + 4c_1c_2\,r\,\beta\!\left(\frac{1-\delta}{\alpha}\right) = 0.$$

This equation defines a two-valued function $w(r;\delta)$ of δ which is given by

$$(17)\qquad w(r;\delta) = \delta \pm \sqrt{\delta^2 - 4c_1c_2\, r\, \beta\!\left(\frac{1-\delta}{\alpha}\right)}.$$

LEMMA 1. *In the domain* Re $\delta < 1$ *the function* $w(r;\delta)$ *is a two-valued analytic function with exactly two branch points, say* $\delta_1(r)$ *and* $\delta_2(r)$, *which are the roots in the domain* Re $\delta < 1$ *of the equation*

$$(18)\qquad \delta^2 - 4c_1c_2\, r\, \beta\!\left(\frac{1-\delta}{\alpha}\right) = 0.$$

PROOF. Because $\beta(.)$ is the Laplace-Stieltjes transform of a positive random variable, the function $\beta((1-\delta)/\alpha)$ is regular for Re $\delta < 1$ and bounded in absolute value by one for Re $\delta \leqslant 1$. Hence, the function $w(r;\delta)$ is analytic in Re $\delta < 1$, except at points where the discriminant, cf. (18), of equation (16) vanishes. Because $c_1 + c_2 = 1$, cf. (2), we have for Re $\delta = 1$ as well as for $|\delta| \to \infty$, Re $\delta < 1$, the inequalities

$$(19)\qquad |\delta^2| \geqslant 1 > |r| \geqslant \left|4c_1c_2\, r\, \beta\!\left(\frac{1-\delta}{\alpha}\right)\right|.$$

With Rouché's theorem it follows that equation (18) has exactly two zeros in the domain Re $\delta < 1$. □

By considering equation (18) for real δ on $(-\infty,1]$ it is seen that the two roots of this equation in Re $\delta < 1$ are real (since r has been chosen to be real, positive), and that they can be chosen such that

$$(20)\qquad -1 < \delta_1(r) < 0 < \delta_2(r) < 1.$$

Now we can say that equation (13) holds if, cf. (15),

$$(21)\qquad p_1 = \frac{1}{2c_1}\, w(r;\delta), \qquad\qquad p_2 = \frac{1}{2c_2}\,[2\delta - w(r;\delta)],$$

for one of the two branches of the function $w(r;\delta)$, cf. (17), and for δ such that $|p_1| \leqslant 1$ and $|p_2| \leqslant 1$. If p_1 and p_2 are given by (21), then for every δ, Re $\delta \leqslant 1$, and for each branch of the function $w(r;\delta)$ the following inequality follows from equation (16):

$$(22)\qquad |p_1p_2| = \frac{1}{4c_1c_2}\,|w(r;\delta)\,[2\delta - w(r;\delta)]| = \left|r\,\beta\!\left(\frac{1-\delta}{\alpha}\right)\right| < 1.$$

Hence, because either $|p_1| < 1$ or $|p_2| < 1$ for every δ, Re $\delta \leqslant 1$, and for each branch of the analytic function $w(r;\delta)$ if p_1 and p_2 are given by (21), the relation (13) together with (21) can be continued analytically to the domain Re $\delta < 1$ (principle of permanence).

<u>LEMMA 2.</u> *The functions* $\Phi(r;w(r;\delta)/2c_1,0)$ *and* $\Phi(r;0,\ 2\delta-w(r;\delta)\ /2c_2)$ *each possess a continuation as a two-valued analytic function into the domain* Re $\delta < 1$, *with no other branch points than* $\delta_1(r)$ *and* $\delta_2(r)$.

<u>PROOF.</u> The assertion will be proved for the first function, for the second one the proof is similar.
By lemma 1 and the properties of the generating function $\Phi(r;p_1,p_2)$ the function $\Phi(r;w(r;\delta)/2c_1,0)$ is regular for those δ and branches of $w(r;\delta)$ for which $|w(r;\delta)| < 2c_1$. Into a subregion of Re $\delta < 1$ where $|w(r;\delta)| \geq 2c_1$ for one of the branches of $w(r;\delta)$ the function $\Phi(r;w(r;\delta)/2c_1,0)$ can be continued analytically by means of relation (13) together with (21), since by (22) then $|2\delta - w(r;\delta)| < 2c_2$ holds. From lemma 1 it is further clear that the only branch points, which the function $\Phi(r;w(r;\delta)/2c_1,0)$ can have in the domain Re $\delta < 1$, are $\delta_1(r)$ and $\delta_2(r)$. □

Next, relation (13) together with (21) will be considered for δ on the real interval between the branch points $\delta_1(r)$ and $\delta_2(r)$, cf. (20). For $\delta \in [\delta_1(r),\delta_2(r)]$ the discriminant of equation (16) is non-positive, so that, cf. (17), for δ on this interval the two branches of the function $w(r;\delta)$ are complex conjungate and lie on the contour

(23) $$L(r) := \{w;\ |w|^2 = 4c_1c_2\, r\, \beta\Big(\frac{1 - \text{Re } w}{\alpha}\Big),\quad \text{Re } w < 1\}.$$

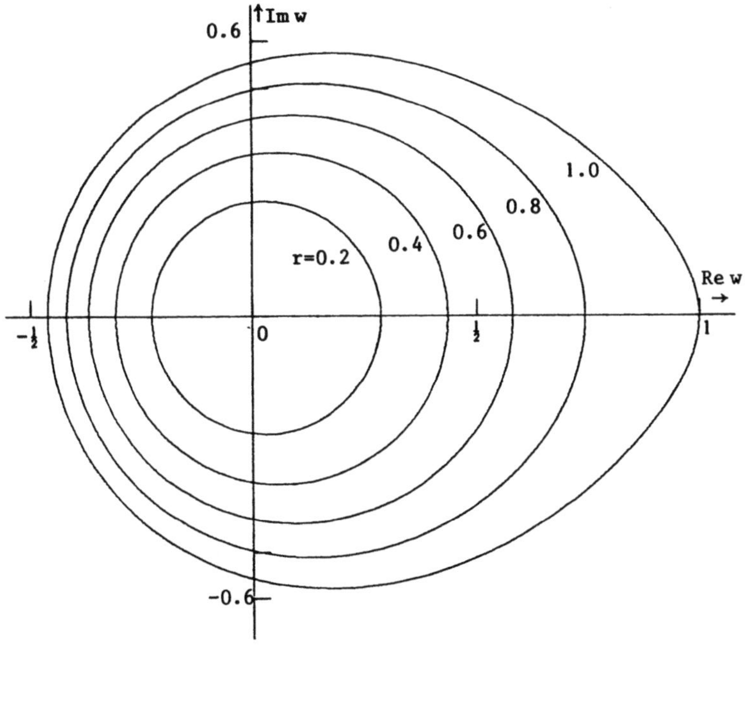

Figure 1

The contour L(r) in the case that $c_1=c_2=\frac{1}{2}$, $\alpha = 1.6$, $\beta(s) = (1+\frac{1}{2}\beta_1 s)^{-2}$, for different values of r

The interior of the contour $L(r)$ will be denoted by $L^+(r)$, its exterior by $L^-(r)$.

LEMMA 3. *For* $w \in L(r)$,

$$(24)\qquad \frac{\Phi(r;w/2c_1,0)}{1-w/2c_1} + \frac{\Phi(r;0,\bar{w}/2c_2)}{1-\bar{w}/2c_2} = \Phi(r;0,0) + \frac{1}{(1-w/2c_1)(1-\bar{w}/2c_2)}.$$

PROOF. From (17) and (23) it is clear that for every $w \in L(r)$ there exists a δ on the interval $[\delta_1(r),\delta_2(r)]$, namely $\delta = \text{Re } w$, such that $w = w(r;\delta)$ for one of the branches of the function $w(r;\delta)$. Moreover, then $2\delta - w(r;\delta) = \bar{w}$. Hence, equation (24) follows from (13) together with (21) by taking $\delta \in [\delta_1(r),\delta_2(r)]$, which is allowed by lemma 2. □

Equation (24) will be the basis for the formulation of a boundary value problem, which will be discussed in section 4. This section will be concluded with the proof that the functions $\Phi(r;w/2c_1,0)$ and $\Phi(r;0,w/2c_2)$ are regular for $w \in L^+(r)$. For simplicity the discussion will be confined to the case $c_2 \leqslant \frac{1}{2} \leqslant c_1$, cf. (2). This is of course no restriction.

First we consider the question whether the point $2c_2$ lies inside, on or outside the contour $L(r)$. Since this contour crosses the real axis only at $w = \delta_1(r)$ and at $w = \delta_2(r)$, cf. lemma 1, we have $2c_2 \in L(r)$ if and only if $2c_2 = \delta_2(r)$, cf. (20). Hence, we insert $\delta = 2c_2$ in equation (18), which leads to the equation (using $c_1 + c_2 = 1$):

$$(25)\qquad c_2 = (1-c_2)\, r\, \beta\!\left(\frac{1-2c_2}{\alpha}\right).$$

This equation inspires us to define the function

$$(26)\qquad R(s) := \frac{s}{1-s}\left[\beta\!\left(\frac{1-2s}{\alpha}\right)\right]^{-1}, \qquad \text{Re } s \leqslant \tfrac{1}{2}.$$

LEMMA 4. *If* $a \leqslant 2$ *then* $R(c_2) < 1$ *for* $c_2 < \frac{1}{2}$ *and* $R(\frac{1}{2}) = 1$. *If* $a > 2$ *then there exists a constant* $c(a)$, $0 < c(a) < \frac{1}{2}$, *such that* $R(c_2) < \frac{1}{2}$ *for* $c_2 < c(a)$ *and* $R(c_2) \geqslant 1$ *for* $c(a) \leqslant c_2 \leqslant \frac{1}{2}$.

PROOF. Rewrite equation (25) as

$$(27)\qquad \frac{s}{1-s} = r\, \beta\!\left(\frac{1-2s}{\alpha}\right), \qquad \text{Re } s \leqslant \tfrac{1}{2}.$$

On the line $\text{Re } s = \frac{1}{2}$ as well as for $|s| \to \infty$, $\text{Re } s < \frac{1}{2}$, the inequality

$$(28)\qquad \left|\frac{s}{1-s}\right| > r \geq r\left|\beta\left(\frac{1-2s}{\alpha}\right)\right|,$$

holds for every r, $0 < r < 1$. Hence, by Rouché's theorem equation (27) has for every r, $0 < r < 1$, exactly one root $s = s_0(r)$ in the region Re $s \leq \frac{1}{2}$. From the properties of the Laplace-Stieltjes transform $\beta(.)$ it is clear that this root $s_0(r)$ must be real and that $0 < s_0(r) < \frac{1}{2}$. Further, $s_0(r)$ is the inverse of the function $R(s)$ for $0 < r < 1$. This implies that the function $R(s)$ increases strictly from zero to one on the interval $0 < s < s_0(1)$, and that $R(s) \geq 1$ for $s_0(1) \leq s \leq \frac{1}{2}$. Whether $s_0(1) < \frac{1}{2}$ or $s_0(1) = \frac{1}{2}$ depends on the derivative of $R(s)$ at $s = \frac{1}{2}$. From (26) we obtain

$$(29)\qquad R'(\tfrac{1}{2}) = 4(1-\tfrac{1}{2}a).$$

Hence, $s_0(1) = \frac{1}{2}$ for $a < 2$, and $s_0(1) < \frac{1}{2}$ for $a > 2$. For $a = 2$ we find by considering higher derivatives of the function $R(s)$ at $s = \frac{1}{2}$ that $s_0(1) = \frac{1}{2}$. By taking $c(a) = s_0(1)$ for $a > 2$ the proof has been completed. □

<u>LEMMA 5.</u> *For* $0 < r < \min\{1, R(c_2)\}$ *we have* $2c_2 \in L^-(r)$. *If* $R(c_2) < 1$ *then* $2c_2 \in L(r)$ *for* $r = R(c_2)$, *and* $2c_2 \in L^+(r)$ *for* $R(c_2) < r < 1$.

<u>PROOF.</u> From (25), (26), and the remark above these formulas we have

$$(30)\qquad 2c_2 \in L(r) \quad\Leftrightarrow\quad 2c_2 = \delta_2(r) \quad\Leftrightarrow\quad r = R(c_2) < 1.$$

From equation (18) it is clear that $\delta_2(r)$ is a continuous function of r for $0<r<1$, and that $\delta_2(r) \downarrow 0$ as $r \downarrow 0$, so that $2c_2 \in L^-(r)$ for $r \downarrow 0$. Further, it is seen from (18) that $\delta_2(r)$ as function of r, $0 < r < 1$, has an inverse, so that it must be a strictly increasing function of r, $0 < r < 1$. With (30) this is sufficient to prove the assertion. □

<u>LEMMA 6.</u> *The functions* $\Phi(r;w/2c_1,0)$ *and* $\Phi(r;0,w/2c_2)$ *are regular in the domain* $L^+(r)$ *and continuous up to the boundary* $L(r)$.

<u>PROOF.</u> Because $\Phi(r;p_1,p_2)$ is a bivariate generating function of a probability distribution in p_1 and in p_2, the functions $\Phi(r;p,0)$ and $\Phi(r;0,p)$ are regular for $|p|<1$ and continuous for $|p| \leq 1$. Further, it follows from the monotonicity of the function $\beta(.)$ and from the fact that Re $w \leq \delta_2(r)$ for $w \in L(r)$, cf. (23) and lemma 1, that for $w \in L(r)$, cf. (23), (18), (20), and therefore also for $w \in L^+(r)$,

$$(31)\qquad |w| \leq 2\sqrt{c_1c_2\, r\, \beta\left(\frac{1-\delta_2(r)}{\alpha}\right)} = \delta_2(r).$$

Hence, the assertions for the function $\Phi(r;w/2c_1,0)$ are obvious since we have chosen $c_1 \geq \frac{1}{2}$, and $\delta_2(r) < 1$ by (20). Also, the assertions for the function $\Phi(r,0,w/2c_2)$ have been proved by the above in the case $2c_2 > \delta_2(r)$, i.e. for $0 < r < \min\{1, R(c_2)\}$,

cf. lemma 5.
Finally, suppose $R(c_2) < 1$, cf. lemma 4, and $R(c_2) \leqslant r < 1$. In this case we use the analytic continuation of the function $\Phi(r;0,[2\delta-w(r;\delta)]/2c_2)$ discussed in lemma 2. By letting δ tend to $\delta_2(r)$ in equation (13) together with (21) it is seen that $\Phi(r;0,\delta_2(r)/2c_2)$ is finite. Because the function $\Phi(r;0,w/2c_2)$ has a power series expansion at $w=0$ with positive coefficients, cf. (1), it follows that this function is regular in the disk $|w| < \delta_2(r)$ and continuous for $|w| \leqslant \delta_2(r)$. With (31) this proves the assertions for $\Phi(r;0,w/2c_2)$ in the present case. □

4. Formulation as a Hilbert boundary value problem

Throughout this section r *is fixed and real,* $0 < r < 1$, *and* $c_2 \leqslant \frac{1}{2} \leqslant c_1$.

As in [3] equation (24) can be reduced to two Riemann-Hilbert problems on the contour $L(r)$. However, in the following we shall give a slightly different approach by formulating a single Hilbert problem, cf. [9], §§34-37. This method is somewhat simpler, and above it has the advantage that it is still applicable in the analysis of a generalization of the present model in which the duration of individual services has not the same distribution as that of paired services, see [1], §IV.2.
As in [3] equation (24) is transformed into a relation on the unit circle by introducing a conformal mapping.

LEMMA 7. *There exists a conformal mapping* $g(r;z)$ *of the unit disk* $|z| < 1$ *onto the domain* $L^+(r)$. *This conformal mapping is uniquely determined by the conditions*

$$(32) \qquad g(r;0) = 0, \qquad g'(r;0) > 0.$$

The conformal mapping $g(r;z)$ *is continuous for* $|z| \leqslant 1$ *and establishes a one-to-one correspondence between this region and* $L^+(r) \cup L(r)$. *Further it satifies the relation*

$$(33) \qquad g(r;\bar{z}) = \overline{g(r;z)}, \qquad |z| \leqslant 1.$$

PROOF. Because $L^+(r)$ is a simply connected domain, cf. (23), the existence of the conformal mapping $g(r;z)$ follows from Riemann's mapping theorem, cf. [8], vol.III, §2, theorem 1.2. The uniqueness theorem for conformal mapping, cf. [8], vol. III, §2, theorem 1.3, implies the uniqueness of $g(r;z)$ given the conditions (32). The assertions for $|z| = 1$ follow from the boundary correspondence theorem, cf. [8], vol.III, §8, theorem 2.24. Finally, relation (33) is a consequence of the property that the real axis is an axis of symmetry of the contour $L(r)$, cf. (23), and of the choice made in (32). □

In the sequel the unit circle will be denoted by C.

THEOREM 1. *If* $2c_2 \in L^-(r)$, *then for* $t \in C$,

$$(34)\qquad \frac{\Phi(r;\frac{1}{2c_1}g(r;t),0)}{1-\frac{g(r;t)}{2c_1}} - \Phi(r;0,0) + \frac{\Phi(r;0,\frac{1}{2c_2}g(r;\frac{1}{t}))}{1-\frac{g(r;1/t)}{2c_2}} = \frac{1}{\left[1-\frac{g(r;t)}{2c_1}\right]\left[1-\frac{g(r;1/t)}{2c_2}\right]},$$

and the functions

$$(35)\qquad \frac{\Phi(r;g(r;t)/2c_1,0)}{1-g(r;t)/2c_1} - \Phi(r;0,0), \qquad \text{and} \quad \frac{\Phi(r;0,g(r;t)/2c_2)}{1-g(r;t)/2c_2},$$

are regular for $|t| < 1$.
This defines a Hilbert boundary value problem on the unit circle.

PROOF. The boundary condition (34) follows from lemma 3 by inserting $w = g(r;t)$, $t \in C$, cf. lemma 7, and by noting that (33) implies:

$$(36)\qquad \overline{g(r;t)} = g(r;\tfrac{1}{t}), \qquad t \in C.$$

Lemma 6 , the regularity of the conformal mapping $g(r;z)$ for $|z| < 1$, and the assumption $2c_2 \in L^-(r)$ imply the regularity for $|t| < 1$ of the functions in (35). According to the definitions in [9], §37, a Hilbert boundary value problem is defined by the relation (34) for the regular functions in (35) if the known function at the righthand side of (34) satisfies a Hölder condition on C, cf. [9], §3. Such a Hölder condition depends on the boundedness of $\frac{\partial}{\partial z} g(r;z)$ in the region $|z| < 1$, which can be proved by using smoothness properties of the contour $L(r)$. For the details of this proof the reader is referred to [1], lemma II.6.2. □

The conformal mapping $g(r;z)$ has an inverse for $|z| \leq 1$, cf. lemma 7. This inverse will be denoted by $g_0(r;w)$, $w \in L^+(r) \cup L(r)$.

THEOREM 2. *If* $2c_2 \in L^-(r)$, *i.e.* $0 < r < \min\{1, R(c_2)\}$, *then the generating function* $\Phi(r;p_1,p_2)$ *is given by: for* $2c_1p_1 \in L^+(r)$, $2c_2p_2 \in L^+(r)$,

$$(37)\qquad \Phi(r;p_1,p_2) = \frac{r(1-p_1)(1-p_2)\beta\left(\frac{1-c_1p_1-c_2p_2}{\alpha}\right)}{p_1p_2 - r\,\beta\left(\frac{1-c_1p_1-c_2p_2}{\alpha}\right)}\Bigg[\frac{p_1p_2}{r(1-p_1)(1-p_2)\beta\left(\frac{1-c_1p_1-c_2p_2}{\alpha}\right)} -$$

$$-\frac{1}{2\pi i}\int_C \frac{1}{\{1-g(r;t)/2c_1\}\{1-g(r;1/t)/2c_2\}}\,\frac{t+g_0(r;2c_1p_1)}{t-g_0(r;2c_1p_1)}\,\frac{dt}{2t} -$$

$$-\frac{1}{2\pi i}\int_C \frac{1}{\{1-g(r;1/t)/2c_1\}\{1-g(r;t)/2c_2\}}\,\frac{t+g_0(r;2c_2p_2)}{t-g_0(r;2c_2p_2)}\,\frac{dt}{2t}\Bigg].$$

PROOF. The Hilbert boundary value problem formulated in theorem 1 is of a simple form, cf. [9], §37. It is easily solved by applying the operator

$$(38) \qquad \frac{1}{2\pi i}\int_C \cdots\cdots \frac{dt}{t-z},$$

on both sides of equation (34), for $|z| < 1$ and for $|z| > 1$. By noting that, cf. (32),

$$(39) \qquad \lim_{z\to\infty} \frac{\Phi(r;0,g(r;1/z)/2c_2)}{1-g(r;1/z)/2c_2} = \Phi(r;0,0),$$

the operation (38) on equation (34) leads with the residu theorem to: for $|z| < 1$,

$$(40) \qquad \frac{\Phi(r;g(r;z)/2c_1,0)}{1-g(r;z)/2c_1} = \frac{1}{2\pi i}\int_C \frac{1}{\{1-g(r;t)/2c_1\}\{1-g(r;1/t)/2c_2\}}\frac{dt}{t-z};$$

and similarly to: for $|z| > 1$,

$$(41) \qquad \Phi(r;0,0) - \frac{\Phi(r;0,g(r;1/z)/2c_2)}{1-g(r;1/z)/2c_2} = \frac{1}{2\pi i}\int_C \frac{1}{\{1-g(r;t)/2c_1\}\{1-g(r;1/t)/2c_2\}}\frac{dt}{t-z}.$$

By taking $z = 0$ in (40) we obtain the unknown constant in (41):

$$(42) \qquad \Phi(r;0,0) = \frac{1}{2\pi i}\int_C \frac{1}{\{1-g(r;t)/2c_1\}\{1-g(r;1/t)/2c_2\}}\frac{dt}{t}.$$

Next, by substituting $z = g_0(r;2c_1p_1)$ in (40) and $1/z = g_0(r;2c_2p_2)$ in (41) together with (42), expressions for the functions $\Phi(r;p_1,0)$ and $\Phi(r;0,p_2)$ are obtained for $2c_1p_1 \in L^+(r)$ and $2c_2p_2 \in L^+(r)$ respectively. Finally, by substituting these expressions for $\Phi(r;p_1,0)$ and for $\Phi(r;0,p_2)$ and (42) for $\Phi(r;0,0)$ in the functional equation (8) the relation (37) is obtained after some simple rearrangements. □

REMARK 1. As it has been noted in section 2, the functional equation (8) must have at least one solution which is a generating function of a joint probability distribution in p_1 and p_2, and which is a generating function of a series with coefficients bounded in absolute value by one in r. With our analysis it has been proved that equation (8) possesses at most one solution with these properties, for $0 < r < \min\{1,R(c_2)\}$. This implies that the righthand side of (37) represents the generating function defined in (1) for $2c_1p_1 \in L^+(r)$, $2c_2p_2 \in L^+(r)$, $0 < r < \min\{1,R(c_2)\}$. Moreover, the expression (37) determines the power series expansion of the function $\Phi(r;p_1,p_2)$ at $r = 0$, $p_1 = 0$, $p_2 = 0$. Hence, by analytic continuation the function $\Phi(r;p_1,p_2)$ has been uniquely determined in theorem 2 for $|r| < 1$, $|p_1| \leq 1$, $|p_2| \leq 1$.

REMARK 2. Explicit formulas for the function $\Phi(r;p_1,p_2)$ for $2c_1p_1 \in L^-(r)$ and/or $2c_2p_2 \in L^-(r)$, $0 < r < \min\{1,R(c_2)\}$, can be obtained by using the analytic continuation of the function $g_0(r;w)$ into $L^-(r)$, and by applying the Plemelj formulas, cf. [9], §17, to the integrals in (37), see [1], theorem II.7.2.

The main interest of the solution of the time dependent distribution of the Markov chain $\{(\underline{x}_1(n),\underline{x}_2(n)),n=0,1,..\}$ is that it forms the basis for the asymptotic analysis of this Markov chain as $n \to \infty$. For this purpose it is important to derive explicit expressions for the function $\Phi(r;p_1,p_2)$ for r in a neighbourhood of one, because e.g. (see [2], p.18 and appendix 1):

$$(43)\qquad \lim_{n\to\infty} \Pr\{\underline{x}_1(n) = 0,\ \underline{x}_2(n) = 0\} = \lim_{r\to 1} (1-r)\ \Phi(r;0,0).$$

However, from lemma 4 and 5 it is seen that theorem 2 does not provide us with such an expression for all values of the parameters α and c_2. Therefore we shall derive below an expression for $\Phi(r;p_1,p_2)$ in the case $R(c_2) < 1$, cf. lemma 4, for $R(c_2) < r < 1$, i.e. for $2c_2 \in L^+(r)$, cf. lemma 5. Hence, suppose $2c_2 \in L^+(r)$. Then theorem 1 is still valid, except that the second function in (35) possesses a single pole in the region $|t| \leqslant 1$ due to a zero of the denominator at the point

$$(44)\qquad z_0(r) := g_0(r;2c_2).$$

With this observation the following result is obtained:

THEOREM 3. *If* $2c_2 \in L^+(r)$, *i.e.* $R(c_2) < r < 1$, *then the generating function* $\Phi(r;p_1,p_2)$ *is given by: for* $2c_1p_1 \in L^+(r)$, $2c_2p_2 \in L^+(r)$,

$$(45)\qquad \Phi(r;p_1,p_2) = \frac{r(1-p_1)(1-p_2)\beta\left(\frac{1-c_1p_1-c_2p_2}{\alpha}\right)}{p_1p_2 - r\,\beta\left(\frac{1-c_1p_1-c_2p_2}{\alpha}\right)} \Bigg[\frac{p_1p_2}{r(1-p_1)(1-p_2)\beta\left(\frac{1-c_1p_1-c_2p_2}{\alpha}\right)} -$$

$$- \frac{1}{2\pi i}\int_C \frac{1}{\{1-g(r;t)/2c_1\}\{1-g(r;1/t)/2c_2\}}\,\frac{t+g_0(r;2c_1p_1)}{t-g_0(r;2c_1p_1)}\,\frac{dt}{2t} -$$

$$- \frac{1}{2\pi i}\int_C \frac{1}{\{1-g(r;1/t)/2c_1\}\{1-g(r;t)/2c_2\}}\,\frac{t+g_0(r;2c_2p_2)}{t-g_0(r;2c_2p_2)}\,\frac{dt}{2t} -$$

$$- \frac{2c_2}{g'(r;z_0(r))}\,\frac{1}{1-\mu_1(r)}\,\frac{1-g_0(r;2c_1p_1)\,g_0(r;2c_2p_2)}{\{1-z_0(r)g_0(r;2c_1p_1)\}\{z_0(r)-g_0(r;2c_2p_2)\}}\Bigg];$$

here

(46) $$g'(r;z_0(r)) := \frac{\partial}{\partial z} g(r;z)\Big|_{z=z_0(r)}.$$

PROOF. By applying - as in theorem 2 - the operator (38) on both sides of equation (34), and by taking into account the pole at $z_0(r)$, cf. (44), the residue theorem leads to: for $|z| < 1$,

(47) $$\frac{\Phi(r;g(r;z)/2c_1,0)}{1-g(r;z)/2c_1} + \frac{2c_2}{z_0(r)g'(r;z_0(r))}\frac{\Phi(r;0,1)}{z_0(r)z-1} =$$
$$= \frac{1}{2\pi i}\int_C \frac{1}{\{1-g(r;t)/2c_1\}\{1-g(r;1/t)/2c_2\}}\frac{dt}{t-z},$$

and for $|z| > 1$,

(48) $$\Phi(r;0,0) + \frac{2c_2}{z_0(r)g'(r;z_0(r))}\frac{\Phi(r;0,1)}{z_0(r)z-1} - \frac{\Phi(r;0,g(r;1/z)/2c_2)}{1-g(r;1/z)/2c_2} =$$
$$= \frac{1}{2\pi i}\int_C \frac{1}{\{1-g(r;t)/2c_1\}\{1-g(r;1/t)/2c_2\}}\frac{dt}{t-z},$$

here $g'(r;z_0(r))$ is given by (46). The constant $\Phi(r;0,1)$ has been determined in (10). By taking $z=0$ in (47) the last unknown constant is obtained:

(49) $$\Phi(r;0,0) = \frac{2c_2}{z_0(r)g'(r;z_0(r))}\frac{1}{1-\mu_1(r)} +$$
$$+ \frac{1}{2\pi i}\int_C \frac{1}{\{1-g(r;t)/2c_1\}\{1-g(r;1/t)/2c_2\}}\frac{dt}{t}.$$

In a similar way as in the proof of theorem 2 the relations (47), (48) and (49) lead to the expression (45). □

REMARK 3. An expression for the function $\Phi(r;p_1,p_2)$ in the case $2c_2 \in L(r)$, i.e. $r = R(c_2) < 1$, can be derived from (37) or (45) by using the continuity of this function, see [1], theorem II.7.5.

5. Concluding remarks

The results of theorem 2 and 3 form the starting point for the analysis of the asymptotic behaviour of the Markov chain $\{(\underline{x}_1(n),\underline{x}_2(n)), n=0,1,..\}$. Especially in the case $c_1 = c_2 = \frac{1}{2}$ this analysis requires an extensive use of theorems on the boundary behaviour of conformal mappings and their derivatives. Here we merely state the main results of this asymptotic analysis, which has been carried through in [1], §II.8. It turns out that the Markov chain $\{(\underline{x}_1(n),\underline{x}_2(n)), n=0,1,..\}$ is transient if

$\max\{c_1,c_2\}a > 1$, that it consists of null states if $\max\{c_1,c_2\}a = 1$, and that it is recurrent if $\max\{c_1,c_2\}a < 1$. In the last case the Markov chain possesses a stationary distribution, and the stationary probability ϕ_0 of the empty state (0,0) is given by (here $c_2 \leq \frac{1}{2} \leq c_1$), cf. (43):

$$(50) \qquad \phi_0 := \lim_{r\to 1} (1-r)\ \Phi(r;0,0) = \frac{2c_2(1-c_1a)}{\lim\limits_{r\to 1} z_0(r)\ g'(r;z_0(r))}.$$

For $c_1a < 1$ the limit in the denominator in (50) is finite and non-vanishing, and it can be numerically evaluated, cf. [3], §6.

The technique of solving functional equations by formulating a Hilbert boundary value problem can also be applied in the analysis of the continuous time parameter queueing process. For the present model this leads to a stationary distribution which differs from that of the imbedded discrete time parameter process, see [1], §III.8.

If the present queueing model is generalized by allowing a service time distribution $B_j(t)$ for individual services of type j customers ($j = 1,2$), which may differ from the service time distribution for paired services, the boundary condition as in (34) becomes more intricate, but it still defines a Hilbert boundary value problem which can be solved with the general method given in [9], §§34-37; see [1], §IV.2.

References

[1] BLANC, J.P.C. (1982) *Application of the Theory of Boundary Value Problems in the Analysis of a Queueing Model with Paired Services*, Mathematical Centre Tracts 153, Amsterdam.

[2] COHEN, J.W. (1982) *The Single Server Queue*, North-Holland Publ. Co., Amsterdam, 2nd ed..

[3] COHEN, J.W. & BOXMA, O.J. (1981) *The M/G/1 queue with alternating service formulated as a Riemann-Hilbert problem*, Performance '81 (ed. F.J. Kylstra), North-Holland Publ. Co., Amsterdam, pp. 181-199.

[4] FAYOLLE, G (1979) *Méthodes Analytiques pour les Files d'Attente Couplées*, Thesis, Univ. de Paris VI, Paris.

[5] FAYOLLE, G. & IASNOGORODSKI, R. (1979) *Two coupled processors: The reduction to a Riemann-Hilbert problem*, Z. Wahrscheinlichkeitstheor. Verw. Geb. 47, pp. 325-351.

[6] FAYOLLE, G., KING, P.J.B. & MITRANI, I. (1982) *The solution of certain two-dimensional Markov models*, Adv. Appl. Probab. 14, pp. 295-308.

[7] IASNOGORODSKI, R. (1979) *Problèmes Frontières dans les Files d'Attente*, Thesis, Univ. de Paris VI, Paris.

[8] MARKUSHEVICH, A.I. (1977) *Theory of Functions of a Complex Variable*, Chelsea Publ. Co., New York.

[9] MUSKHELISHVILI, N.I. (1953) *Singular Integral Equations*, Noordhoff, Groningen.

[10] NAIN, Ph. (1983) *Workload analysis of a two-queue system by formulating a boundary value problem*, these proceedings.

Workload analysis of a two-queue system by formulating a boundary value problem

Philippe Nain
INRIA
Domaine de Voluceau
Rocquencourt
78150 Le Chesnay
FRANCE

Résumé

Nous étudions un nouveau protocole de communication qui réalise l'insertion de messages spéciaux dans un flot de messages réguliers. L'analyse se ramène à l'étude de deux files d'attente et d'un serveur unique. La discipline de service dépend de la charge d'une des files. Nous calculons, à l'état stationnaire, les transformées de Fourier -Stieltjes et Laplace- Stieltjes de la distribution de la charge du système, en résolvant deux équations fonctionnelles.

Abstract

We analyse a new communication protocol which regulates the merging of special messages in a regular flow. The study is carried out via a queueing model consisting of two waiting lines and one single server facility. The server sharing policy depends on the workload of one of the waiting lines. We derive the Fourier -Stieltjes and Laplace- Stieltjes transforms of the joint stationary distribution of the system-workload by solving two functional equations.

Keywords : Markov process ; Functional equations ; Wiener-Hopf factorization ; Algebric curve ; Dirichlet problem.

1. Introduction

Queueing model with state-dependency, connected to the coupling of processors in computer systems, have been extensively studied recently. Analytic methods have been developed by Fayolle and Iasnogorodski allowing the solution of various coupling problems [6], [10], [7], and leading to a fairly general methodology in this field [3], [12], [2] , [1], [8].

The approach involves the solution of functional equations. Generally the unknown functions are the generating functions of the joint stationary distribution of the number of jobs in the system. This, in turn, leads to the resolution of boundary value problems (Dirichlet and Riemann-Hilbert problems).

We use a similar machinery to study a new communication protocol which regulates the insertion of special (priority) messages in a regular flow.

The analysis is carried out via a queueing model consisting of two queues and one single server. The communication protocol can be described as follows : the customers in the special messages queue are served when the workload -the amount of required service times- in the regular messages queue remains below a given threshold. Section I give a more carefull description of the model.

The Fourier -Stieltjes and Laplace- Stieltjes transforms of the joint stationary distribution of the system -workload are obtained, by solving an "exterior" boundary value problem on a circle.

In Section 2, we get a functional equation satisfied by the workload of the system.

Then, an intermediate Wiener-Hopf factorization [Section 3] allows the reduction to an exterior boundary value problem on a circle, which can be solved. Closed forms are obtained [Section 4].

The results are summarized in a theorem. Figures are given, showing the effect of this server sharing policy upon the mean workload in each queue.

1. The model

We consider the following queueing model :

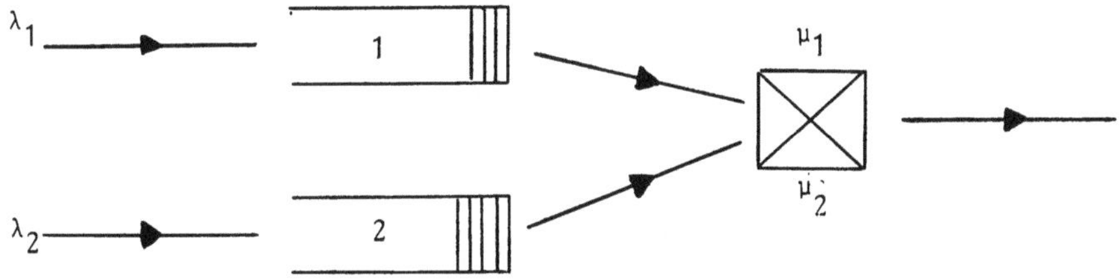

The arrivals form two independent Poisson processes with parameters λ_1, λ_2.

The service times of the customers of queue j ($\stackrel{def}{=}$ customers of type j) form a renewal process with an absolutely continuous but otherwise arbitrary renewal distribution and with mean $1/\mu_j$ (j=1,2). All the processes are assumed to be mutually independent.

The service discipline is first-in-first-out in each queue. The server sharing policy between the two types of customers is the following : when the occupation time of the server -the workload- w.r.t. the customers of type 1 is greater than c [resp. smaller or equal to c] (c is fixed, $c \in \mathbb{R}^+$) a customer of type 1 [resp. type 2] is served. Moreover, customers of each type can be preempted and they will resume their service demand ("preemptive resume priority").

For j = 1,2, we define :

- $\beta_j(s)$ the Laplace-Stieltjes transform (L.S.T.) of the service times distribution of customers of type j, for Re s ≥ 0. Re s denotes the real part of the complex number s.

- $V_j(t)$ the workload of the server w.r.t. queue j at time t (t > 0).

2. The functional equation

The stochastic process $V \stackrel{def}{=} \{ V_1(t), V_2(t), t > 0 \}$ is a Markov process.

Let $\nu^j_{\Delta t}$ be the number of arrivals in queue j in (t, t+ Δt], τ_j be the service time required by the arriving customer in queue j in (t, t+ Δt], (j=1,2) and let us define :

$$H(x,y;t) \stackrel{def}{=} E\{e^{-xV_1(t)-yV_2(t)}\} \quad \text{for Re } x \geq 0,\ \text{Re } y \geq 0,\ t > 0$$

We now establish the sample path equations of the V-process. For $\nu^1_{\Delta t} = 0$, $\nu^2_{\Delta t} = 0$, we have for small Δt :

$$\begin{aligned} V_1(t+\Delta t) &= [V_1(t)-\Delta t]^+ && \text{if } \{V_1(t) > c\} \text{ or } \{V_1(t) \leq c, V_2(t) = 0\} \\ &= V_1(t) && \text{if } \{V_1(t) \leq c, V_2(t) > 0\} \end{aligned}$$

$$\begin{aligned} V_2(t+\Delta t) &= V_2(t) && \text{if } \{V_1(t) > c\} \text{ or } \{V_1(t) \leq c, V_2(t) = 0\} \\ &= V_2(t)-\Delta t && \text{if } \{V_1(t) \leq c, V_2(t) > 0\} \end{aligned}$$

For $\nu^1_{\Delta t} = 0,\ \ \nu^2_{\Delta t} = 1$, we have :

$$
\begin{aligned}
V_1(t+\Delta t) &= V_1(t) - \Delta t && \text{if } \{V_1(t) > c\} \\
&= [V_1(t) - \varepsilon\,\Delta t]^+ && \text{if } \{V_1(t) \le c,\ V_2(t) = 0\} \\
&= V_1(t) && \text{if } \{V_1(t) \le c,\ V_2(t) > 0\}
\end{aligned}
$$

$$
\begin{aligned}
V_2(t+\Delta t) &= V_2(t) + \tau_2 && \text{if } \{V_1(t) > c\} \\
&= \tau_2 - (1-\varepsilon)\Delta t && \text{if } \{V_1(t) \le c,\ V_2(t) = 0\} \\
&= V_2(t) - \Delta t + \tau_2 && \text{if } \{V_1(t) \le c,\ V_2(t) > 0\}
\end{aligned}
$$

For $\nu^1_{\Delta t} = 1,\ \nu^2_{\Delta t} = 0$, we have :

$$
\begin{aligned}
V_1(t+\Delta t) &= V_1(t) - \Delta t + \tau_1 && \text{if } \{V_1(t) > c\} \text{ or } \{0 < V_1(t) \le c,\ V_2(t) = 0\} \\
&= V_1(t) - \varepsilon'\,\Delta t + \tau_1 && \text{if } \{V_1(t) + \tau_1 > c,\ V_1(t) \le c,\ V_2(t) > 0\} \\
&= V_1(t) + \tau_1 && \text{if } \{V_1(t) + \tau_1 \le c,\ V_1(t) \le c,\ V_2(t) > 0\} \\
&= \tau_1 - \varepsilon'\,\Delta t && \text{if } \{V_1(t) = V_2(t) = 0\}
\end{aligned}
$$

$$
\begin{aligned}
V_2(t+\Delta t) &= V_2(t) && \text{if } \{V_1(t) > c\} \text{ or } \{V_1(t) \le c,\ V_2(t) = 0\} \\
&= V_2(t) - \Delta t && \text{if } \{V_1(t) + \tau_1 \le c,\ V_1(t) \le c,\ V_2(t) > 0\} \\
&= V_2(t) - (1-\varepsilon')\Delta t && \text{if } \{V_1(t) + \tau_1 > c,\ V_1(t) \le c,\ V_2(t) > 0\}
\end{aligned}
$$

where ε and ε' are two independent random variables taking their values in $]0,1[$.

From the above relations and the Poissonian arrival assumptions, it readily follows that for Re $x \ge 0$, Re $y \ge 0$:

$$
\begin{aligned}
E\{e^{-xV_1(t+\Delta t)-yV_2(t+\Delta t)} / \nu^1_{\Delta t} = \nu^2_{\Delta t} = 0\}\, P(\nu^1_{\Delta t} = \nu^2_{\Delta t} = 0) &= H(x,y;t) \\
&+ (y-x)\, E\{e^{-xV_1(t)-yV_2(t)} (V_1(t) \le c,\ V_2(t) > 0)\} \\
&- x\, P(V_1(t) = V_2(t) = 0) + o(\Delta t)
\end{aligned}
$$

and

$$E\{e^{-xV_1(t+\Delta t)-yV_2(t+\Delta t)}/(\nu^1_{\Delta t},\nu^2_{\Delta t}) = (k,\ell)\}\, P((\nu^1_{\Delta t},\nu^2_{\Delta t}) = (k,\ell))$$

$$= \lambda_m\, \beta_m(s_m)\, H(x,y;t) + o(\Delta t)$$

where $(k,\ell) \in \{(0,1),(1,0)\}$ and where $m = 1$ if $k = 1$ and $m = 2$ if $k = 0$.

Summing these three relations, dividing by Δt and letting $\Delta t \downarrow 0$, we obtain the following time dependent functional equation for the workload :

$$\frac{\partial}{\partial t} H(x,y\,;t) = (x - \lambda_1\gamma_1(x) - \lambda_2\gamma_2(y))\, H(x,y;t)$$

$$+ (y-x)\, E\{e^{-xV_1(t)-yV_2(t)}(V_1(t) \le c,\ V_2(t) > 0)\} \qquad (1)$$

$$- x\, P(V_1(t) = V_2(t) = 0)$$

where $\gamma_j(s) \overset{\text{def}}{=} 1 - \beta_j(s)$ for $\text{Re } s \ge 0$ and $j = 1,2$.

Now few words about the ergodicity condition of this queueing system. The values of the parameters λ_1, λ_2, μ_1, μ_2 for which the system has a steady state could be derived from the functional analysis of Section 4. (see [7]).

Nevertheless in this particular case a direct treatment is more suitable for sake of clearness.

Since an interrupted customer will resume his service demand, it is easy to see that the duration of a busy period of the whole system (customers of type 1 and 2) is the same as the one of a M/G/1 queue with input parameter $\lambda_1+\lambda_2$ and with L.S.T. of the service times distribution $\frac{\lambda_1}{\lambda_1+\lambda_2}\beta_1(s) + \frac{\lambda_2}{\lambda_1+\lambda_2}\beta_2(s)$ for $\text{Re } s \ge 0$. So the ergodicity condition is then simply $\rho_1 + \rho_2 < 1$ with $\rho_j \overset{\text{def}}{=} \lambda_j/\mu_j$ $(j=1,2)$ [4]. In the following, we assume that this ergodicity condition holds.

Under this assumption, the Markov process $V = \{V_1(t), V_2(t), t > 0\}$ possesses a unique stationary distribution.

Let V_1, V_2 be two random variables with distribution this stationary distribution and let us define $H(x,y) \overset{\text{def}}{=} \lim_{t\to+\infty} H(x,y;t)$.

From equation (1) we then obtain for Re $x \geq 0$, Re $y \geq 0$, the following basic functional equation characterizing the workload of the system at steady state :

$$(x - \lambda_1\gamma_1(x) - \lambda_2\gamma_2(y))\, H(x,y) = (x-y)E\{e^{-xV_1-yV_2}(V_1 \leq c, V_2 > 0)\} + x\, E\{(V_1 = V_2 = 0)\} \tag{2}$$

Some remarks concerning equation (2) :

i) Taking $y = 0$, dividing by x and letting $x \downarrow 0$ in (2), leads to :

$$1 - \rho_1 = E\{(V_1 \leq c, V_2 > 0)\} + P(V_1 = V_2 = 0) \tag{3}$$

In the same way, taking $x = 0$, dividing by y and letting $y \downarrow 0$, leads to :

$$\rho_2 = P(V_1 \leq c, V_2 > 0) \tag{4}$$

From relations (3) and (4), we deduce the expected result :

$$P(V_1 = V_2 = 0) = 1 - \rho_1 - \rho_2 \tag{5}$$

ii) If $c = 0$, the service discipline is the classical preemptive resume priority, where the customers of type 1 hold the higher degree of priority. In this case and for $y = 0$ equation (2) becomes the well known TAKÁCS equation, stating the workload in a M/G/1 queueing system at steady state [14].

iii) If $c \to +\infty$, the same comments to the ones above can be made, changing "type 1" for "type 2" and "y=0" for "x=0".

In what follows, we shall assume that $0 < c < +\infty$ and that the service times of customers of type 1 are exponentially distributed, with mean $1/\mu_1$ (this implies $\gamma_1(s) = \frac{s}{\mu_1 + s}$).

3. The Wiener-Hopf decomposition

We obtain, in this section, a Wiener-Hopf decomposition of the function $E\{e^{-xV_1-yV_2}\}$ in order to solve (section 4) equation (2) for Re $x = 0$, Re $y \geq 0$ [5]. Set :

$$H^-(x,y) \overset{\text{def}}{=} E\{e^{-x(V_1-c)-yV_2}(V_1 \le c)\}$$
$$H^+(x,y) \overset{\text{def}}{=} E\{e^{-x(V_1-c)-yV_2}(V_1 > c)\} \qquad (6)$$

$H^-(x,y)$ [resp $H^+(x,y)$] is analytic in x for Re $x < 0$ [resp. Re $x > 0$] and is continuous in x for Re $x \le 0$ [resp. Re $x \ge 0$].

For Re $x = 0$, Re $y \ge 0$, it follows :

$$H(x,y) = e^{-xc}(H^-(x,y) + H^+(x,y))$$

Let us define for Re $y \ge 0$:

$$\frac{P^-(x,y)}{\mu_1 + x} \overset{\text{def}}{=} (y - \lambda_1\gamma_1(x) - \lambda_2\gamma_2(y))\, H^-(x,y) - x(1-\rho_1-\rho_2)e^{xc} + (x-y)\, E\{e^{-x(V_1-c)}(V_1 < c, V_2 = 0)\} \quad \text{for Re } x \le 0 \quad (*)$$

$$\frac{P^+(x,y)}{\mu_1 + x} \overset{\text{def}}{=} -(x - \lambda_1\gamma_1(x) - \lambda_2\gamma_2(y))\, H^+(x,y) \qquad \text{for Re } x \ge 0$$

Hence from (6), it is seen that P^-[resp. P^+] is regular in x for Re $x < 0$ [resp. Re $x > 0$] and is continuous in x for Re $x \le 0$ [resp. Re $x \ge 0$].

Moreover, for Re $x = 0$, Re $y \ge 0$, we have from (2) and (6) :

$$P^-(x,y) = P^+(x,y)$$

On the other hand, $|P^-(x,y)|$ [resp. $|P^+(x,y)|$] is $O(|x^2|)$ where $|z|$ denotes the modulus of the complex number z. Applying Liouville's theorem it is seen that P defined by :

$$P(x,y) \overset{\text{def}}{=} \begin{cases} P^-(x,y) & \text{if Re } x \le 0 \\ P^+(x,y) & \text{if Re } x \ge 0 \end{cases} \qquad \text{for Re } y \ge 0$$

(*) We obviously have $P(V_1 = c, V_2 = 0) = 0$ since we assumed $c \neq 0$.

and which is therefore regular in x, is a polynomial of degree at most two in x. Dividing P by x^2 and letting $|x| \to +\infty$ with for instance Re x > 0, it is readily seen that the coefficient of x^2 must be zero.

For Re y ≥ 0, we then obtain the following system :

$$(7)\begin{cases} R(x,y)\, H^-(x,y) = f(x) + g(x,y)\, A(x) + \dfrac{x\, B(y) + C(y)}{\mu_1 + x} & \text{for Re } x \le 0 \quad (7.1) \\ S(x,y)\, H^+(x,y) = -\dfrac{x\, B(y) + C(y)}{\mu_1 + x} & \text{for Re } x \ge 0 \quad (7.2) \end{cases}$$

where :

$$A(x) \overset{def}{=} E\{e^{-x(V_1 - c)}(V_1 < c\,,\; V_2 = 0)\}$$

B(y), C(y) are the unknown coefficients in y of the polynomial in x, P(x,y) (8)

$$R(x,y) \overset{def}{=} (y - \lambda_1\gamma_1(x) - \lambda_2\gamma_2(y))$$

$$S(x,y) \overset{def}{=} (x - \lambda_1\gamma_1(x) - \lambda_2\gamma_2(y))$$

$$f(x) \overset{def}{=} x(1 - \rho_1 - \rho_2)e^{cx}$$

$$g(x,y) \overset{def}{=} y - x$$

A central role in the analysis is played by the "kernel" R(x,y) [resp. S(x,y)] because if for a pair (x,y), Re x ≤ 0 [resp. Re x ≥ 0], Re y ≥ 0, this kernel vanishes, then the righthand side of equation (7.1) [resp. (7.2)] must also vanish.

Using this property, we see that it is easy to find a relation between the two unknown coefficients B(y) and C(y), from equation (7.2).

For a fixed y, Re y ≥ 0, the equation S(x,y) = 0 possesses a unique root x = Z(y) in the half-plane Re x ≥ 0 ([4], p. 536).

Moreover, Z(y) is an analytic function for Re y > 0, and is given by :

$$Z(y) = \lambda_2\gamma_2(y) + \lambda_1(1 - E\{e^{-\lambda_2\gamma_2(y)\tilde{P}}\}) \qquad (9)$$

where $\tilde{P}$ stands for the busy period of a M/G/1 queueing system with input parameter λ_1 and L.S.T. of the service times distribution $\beta_1(s)$, Re $s \geq 0$.

We then have :

$$C(y) = -Z(y)\ B(y)\ \forall \text{ Re } y \geq 0 \tag{10}$$

The first equation (7.1) of system (7) becomes for Re $x \leq 0$, Re $y \geq 0$:

$$\boxed{R(x,y).H^-(x,y) = f(x) + g(x,y)\ A(x) + h(x,y)\ B(y)} \tag{11}$$

where :

$$h(x,y) \overset{\text{def}}{=} \frac{x-Z(y)}{\mu_1+x} \tag{12}$$

In the sequel, we examine carefully the algebric curve defined by $T(x,y) \overset{\text{def}}{=} (\mu_1+x)(\mu_2+y)R(x,y) = 0$ in the whole complex plane.

Remarks :

i) Taking $x = 0$ in (7.2), we see that $C(y)$ must be analytic for Re $y > 0$ and continuous for Re $y \geq 0$ since $\gamma_2(y)$ and $H^+(0,y)$ are analytic for Re $y > 0$ and continuous for Re $y \geq 0$. Moreover $C(0) = 0$ since $\gamma_2(0) = 0$.

ii) $Z(y)$ being analytic for Re $y > 0$ and continuous for Re $y \geq 0$ we deduce from i) and equation (10) that $B(y)$ must also be analytic for Re $y > 0$ and continuous for Re $y \geq 0$ ($Z(y) = 0$ for Re $y \geq 0$ iff $y = 0$ and $C(0) = 0$ from i)).

iii) If $\beta_j(s) = \frac{\mu_j}{\mu_j+s}$, Re $s \geq 0$, $j = 1, 2$, we have :

$$Z(y) = \frac{\lambda_2 y + (\mu_2+y)(\lambda_1-\mu_1) + \sqrt{((\lambda_1+\mu_1)(\mu_2+y)+\lambda_2 y)^2 - 4\lambda_1\mu_1(\mu_2+y)^2}}{2(\mu_2 + y)} \tag{13}$$

for Re $y \geq 0$ (cf. [11], p. 215).

iv) Let us assume that $c = 0$. Equation (2) becomes :

$$S(x,y)\ H(x,y) = (x-y)\ E\{e^{-yV_2}(V_1 = 0, V_2 > 0)\} + x(1 - \rho_1 - \rho_2)$$

Since the kernel $S(x,y)$ vanishes whenever $x = Z(y)$ for $\text{Re}\, y \geq 0$ we get :

$$E\{e^{-yV_2}(V_1 = 0,\ V_2 > 0)\} = \frac{(1-\rho_1-\rho_2)\, Z(y)}{y - Z(y)} \quad \text{for } \text{Re}\, y \geq 0$$

and

$$H(x,y) = \frac{(1-\rho_1-\rho_2)y}{S(x,y)} \left(\frac{x-Z(y)}{y-Z(y)}\right) \qquad \text{for } \text{Re}\, x \geq 0,\ \text{Re}\, y \geq 0$$

If we now assume that $c = +\infty$, we find in a similar way that :

$$H(x,y) = \frac{(1-\rho_1-\rho_2)x}{R(x,y)} \left(\frac{y-\tilde{Z}(x)}{x-\tilde{Z}(x)}\right) \qquad \text{for } \text{Re}\, x \geq 0,\ \text{Re}\, y \geq 0$$

where $y = \tilde{Z}(x)$ is the unique root of the equation $R(x,y) = 0$ such that $\text{Re}\, y \geq 0$ for $\text{Re}\, x \geq 0$.

4. Determination of A and B

From now on, we shall assume that $\beta_j(s) = \dfrac{\mu_j}{\mu_j+s}$, $j = 1,2$ (the service times distribution of customers of type j are exponentially distributed, with mean $1/\mu_j$, $j = 1,2$) and $\lambda_1 \neq 0$, $\lambda_2 \neq 0$.

4.1. $T(x,y) = 0$

Solving $T(x,y) = 0$ in y, x being fixed, $x \in \mathbb{C}$, we obtain the algebraic function :

$$Y(x) \stackrel{\text{def}}{=} \frac{\lambda_1 x-(\mu_2-\lambda_2)(\mu_1+x) \pm \sqrt{\Delta(x)}}{2(\mu_1+x)} \tag{14}$$

where :

$$\Delta(x) \stackrel{\text{def}}{=} (\lambda_1 x - (\mu_2-\lambda_2)(\mu_1+x))^2 + 4\lambda_1\mu_2 x(\mu_1+x)$$

and with

$$\sqrt{z} = \sqrt{p}\ e^{i\frac{\Theta}{2}} \quad \text{if } z = pe^{i\Theta},\ p \geq 0,\ -\pi < \Theta \leq \pi$$

The two branches give a two sheeted covering of the complex plane.

Lemma 1

The algebraic function $Y(x)$ defined by $T(x,y) = 0$ has two real branch points x^{**}, x^{*} with $-\mu_1 < x^{**} < x^{*} < 0$. □

Proof

From (14), the branch points of $Y(x)$ are the roots of $\Delta(x)$. Let x^{**} and x^{*} be these roots. It is easy to see that :

$$-\mu_1 < x^{**} \overset{\text{def}}{=} -\mu_1 \frac{(\sqrt{\lambda_2} + \sqrt{\mu_2})^2}{(\sqrt{\lambda_2} + \sqrt{\mu_2})^2 + \lambda_1} < x^{*} \overset{\text{def}}{=} -\mu_1 \frac{(\sqrt{\lambda_2} - \sqrt{\mu_2})^2}{(\sqrt{\lambda_2} - \sqrt{\mu_2})^2 + \lambda_1} < 0$$

□

Lemma 2

For Re $x = 0$, the equation $T(x,y) = 0$ has one root $Y_1(x)$ in the right half-plane and one root $Y_2(x)$ in the left half-plane. □

Proof

For Re $x = 0$, $T(x,y) = 0 \Longleftrightarrow y - \lambda_1\gamma_1(x) - \lambda_2\gamma_2(y) = 0$.

It is well known (cf. [4], p. 536) that, since $\beta_2(.) = 1 - \gamma_2(.)$ is not a lattice distribution, $y - \lambda_1\gamma_1(x) - \lambda_2\gamma_2(y) = 0$ has exactly 2 zeros $Y_1(x)$ and $Y_2(x)$ with Re $Y_1(x) \geq 0$, Re $Y_2(x) < 0$ if Re $x = 0$, $x \neq 0$. If $x = 0$, $Y_1(x) = 0$ and $Y_2(x) = \lambda_2 - \mu_2 < 0$. □

Let us denote $Y_1(x)$ the root of $T(x,y)$ such that Re $Y_1(x) \geq 0$ if Re $x = 0$. We have :

$$Y_1(x) = \frac{\lambda_1 x - (\mu_2-\lambda_2)(\mu_1+x) + \sqrt{\Delta(x)}}{2(\mu_1 + x)} \tag{15}$$

$Y_2(x)$ will denote the other root.

From lemma 1 and (15), we see that the algebraic function defined by $Y_1(x)$ is analytic in the whole complex plane cut along $[x^{**}, x^{*}]$ (the numerator of $Y_1(x)$ vanishes when $x = -\mu_1$).

Lemma 3

Y_1 and Y_2 map the cut $[x^{**}, x^*]$ onto the circle $\mathcal{C}$ with centre $-\mu_2$ and radius $\sqrt{\lambda_2\mu_2}$. □

Proof

For $x \in [x^{**}, x^*]$, $\Delta(x) \leq 0$ and $Y_1(x)$ and $Y_2(x)$ are complex conjugate.

Let $a(x)$ and $b(x)$ be respectively the real part and the imaginary part of $Y_1(x)$.

If $x \in [x^{**}, x^*]$ then :

$$\begin{cases} Y_1(x) + Y_2(x) = \dfrac{\lambda_1 x}{\mu_1 + x} + (\lambda_2 - \mu_2) = 2a(x) \\ Y_1(x)\, Y_2(x) = \dfrac{-\lambda_1\mu_2 x}{\mu_1 + x} = a^2(x) + b^2(x) \end{cases}$$

It is readily seen that $a(x)$ and $b(x)$ are the real solutions of the equation $(a(x) + \mu_2)^2 + b^2(x) = \lambda_2\mu_2$. □

From the assumption $\rho_1 + \rho_2 < 1$, we see that $-\mu_2 + \sqrt{\lambda_2\mu_2} < 0$. Hence $\mathcal{C}$ is entirely contained in the left half-plane $\{ y \in \mathbb{C} \ / \ \mathrm{Re}\, y < 0\}$.

Solving now the equation $T(x,y) = 0$ in x for fixed y, $y \in \mathbb{C}$, we find a unique solution :

$$X(y) \overset{\text{def}}{=} \mu_1 \frac{y(\lambda_2 - \mu_2 - y)}{y^2 + y(\mu_2 - \lambda_2 - \lambda_1) - \lambda_1\mu_2} \tag{16}$$

$X(y)$ has two real poles which are :

$$y^* \overset{\text{def}}{=} \frac{\lambda_1 + \lambda_2 - \mu_2 + \sqrt{(\mu_2 - \lambda_2 - \lambda_1)^2 + 4\lambda_1\mu_2}}{2} > 0$$

$$y^{**} \overset{\text{def}}{=} \frac{\lambda_1 + \lambda_2 - \mu_2 - \sqrt{(\mu_2 - \lambda_2 - \lambda_1)^2 + 4\lambda_1\mu_2}}{2} < 0 \tag{17}$$

It is easily seen that $-\mu_2 < y^{**} < -\mu_2 + \sqrt{\lambda_2\mu_2}$.

Lemma 4

For Re $y = 0$, the unique root $X(y)$ of the equation $T(x,y) = 0$ is located in the left half-plane. □

Proof

For Re $y = 0$, $T(x,y) = 0 \Longleftrightarrow \lambda_1\gamma_1(x) = y - \lambda_2\gamma_2(y)$.

Hence if Re $y = 0$ and $T(x,y) = 0$ then $\mathrm{Re}\,\lambda_1\gamma_1(x) = \mathrm{Re}\,\dfrac{\lambda_1 x}{\mu_1 + x} = -\,\mathrm{Re}\,\lambda_2\gamma_2(y) \le 0$.

The above inequality necessarily entails that Re $x \le 0$. □

Lemma 5

X maps conformally the imaginary axis onto a simple closed curve entirely contained in the left half-plane, symmetrical relative to the real axis and cuting the real axis at points $x = -\mu_1$ and $x = 0$. □

Proof

$X(it) = u(t) + i\,v(t)$ for $t \in \mathbb{R}$ where $u(t)$ and $v(t)$ are real functions. From equation (16) we have :

$$u(t) = -\mu_1 t^2 \frac{t^2 + (\lambda_2 - \mu_2)^2 + \lambda_1\lambda_2}{(t^2 + \lambda_1\mu_2)^2 + t^2(\mu_2 - \lambda_2 - \lambda_1)^2} \le 0$$

and

$$v(t) = \lambda_1\mu_1 t \frac{t^2 - \lambda_2\mu_2 + \mu_2^2}{(t^2 + \lambda_1\mu_2)^2 + t^2(\mu_2 - \lambda_2 - \lambda_1)^2}$$

Lemma is proven when considering the following properties of $u(t)$ and $v(t)$:

i) $u(0) = v(0) = 0$

ii) $\lim\limits_{t \to \mp\infty} u(t) = -\mu_1$ and $\lim\limits_{t \to \mp\infty} v(t) = 0$

iii) $-\mu_1 < u(t) \le 0 \quad \forall\, t \in \mathbb{R}$

iv) $v(t)$ is an odd function. □

Hence, from lemma 1, it is seen that the cut $[x^{**}, x^{*}]$ lies inside $X(\mathrm{Im})$, where Im denotes the imaginary axis. (see Fig. 1).

Lemma 6

$X(Y_i(x)) = x, \forall x \in \mathbb{C}, i = 1,2.$ □

Proof

Let $x \in \mathbb{C}$. Then from the previous considerations the couples $(x,Y_1(x))$, $(x,Y_2(x))$ are solutions of $T(x,y) = 0$. In the same way, the couples $(X(Y_1(x)),Y_1(x))$, $(X(Y_2(x)), Y_2(x))$ are also solutions of $T(x,y) = 0$.

Since $T(x,y)$ possesses a unique root for fixed y, we necessarily have that $X(Y_i(x)) = x$, for $i = 1,2$. □

We introduce the intermediate function $\tilde{B}$ defined by :

$$\tilde{B}(y) \overset{\text{def}}{=} B(y) \,.\, h(X(y),y) \text{ for } \mathrm{Re}\, y \geq 0, \tag{18}$$

4.2 Meromorphic continuation of $\tilde{B}(y)$

The domain within the contour $\mathcal{C}$ is called the interior domain and denoted as $\mathcal{C}^+$, whilst the complementary domain to $\mathcal{C}^+ + \mathcal{C}$ is called the exterior domain and denoted by $\mathcal{C}^-$.

Let $\mathcal{D}$ be defined as $\mathcal{D} \overset{\text{def}}{=} \mathcal{C}^- \cap \{y \in \mathbb{C} \ / \ \mathrm{Re}\, y < 0\}$. In the following Im will denote the imaginary axis of $\mathbb{C}$.

Lemma 7

The algebraic function X maps conformally $\mathcal{D}$ onto the domain situated between the cut $[x^{**}, x^*]$ and the curve X(Im). (see Fig. 1). □

Proof

X is a conformal mapping on $\mathcal{D}$ since X(y) is analytic for $y \in \mathcal{D}$ and since the two roots $-\mu_2 \pm \sqrt{\lambda_2\mu_2}$ of $\frac{\partial}{\partial y}$ X(y) do not belong to $\mathcal{D}$.

The proof of $X(\mathcal{C}) = [x^{**}, x^*]$ is obviously obtained from lemma 3 and lemma 6. □

We can now continued $\tilde{B}$ as a meromorphic function to $\mathcal{C}^-$. This is done in the following lemma.

<u>Lemma 8</u>

The function $\tilde{B}(y)$ which is meromorphic for Re $y \geq 0$ (it possessesa pole at infinity) can be continued as a meromorphic function to $\mathcal{C}^-$ (it also possesses a pole at infinity for Re $y < 0$) and as a continuous function to $\mathcal{C}^- + \mathcal{C}$. □

<u>Proof</u>

From lemma 4 and the continuity of X(y) (in particular) in the neighbourhood of the imaginary axis, there exists a region $\mathcal{R}_y$ in the right half-plane, containing the imaginary axis and such that Re $X(y) \leq 0$, $\forall y \in \mathcal{R}_y$.

Hence, for $y \in \mathcal{R}_y$:

$$f(X(y)) + g(X(y),y)\, A(X(y)) + \tilde{B}(y) = 0 \tag{19}$$

Using lemma 7, equation (19) and the analycity of A in the left half-plane, we do a meromorphic continuation for $\tilde{B}(y)$ for all $y \in \mathcal{C}^-$.

The continuity of $\tilde{B}(y)$ in $\mathcal{C}^- + \mathcal{C}$ follows from the continuity of A in the left half-plane. □

At this point of the study, it is interesting to notice the following fact, in order to explain the introduction of the intermediate function $\tilde{B}$. In the problems of the same type already solved [6] , [10] , the unknown functions (here B) could usually be continued as meromorphic functions to the interior and/or the exterior domain of closed curves (here $\mathcal{C}$). But in this case, the known functions of the right hand of equation (11) cannot all be continued as meromorphic functions inside or outside the circle $\mathcal{C}$.

Indeed, it is readily seen that f(X(y)) has an essential singular point for $y = y^{**} \in \mathcal{C}^+$ and that h(X(y),y) has two real branch points located in $]-\mu_2, 0[$ (see equations (8), (12), (13) and (16)).

4.3. Reduction to a non-homogeneous Dirichlet problem on $\mathscr{C}$

We define :

$$U(y) \overset{\text{def}}{=} \frac{1}{g(X(y),y)} , \qquad V(y) \overset{\text{def}}{=} \frac{f(X(y))}{g(X(y),y)} \tag{20}$$

Before reducing the problem (of finding B(y) for Re y ≥ 0) to the resolution of a boundary value problem, we need the following lemma.

Lemma 9

$$\text{Let } \hat{B}(y) \overset{\text{def}}{=} \tilde{B}(y)\, U(y) - B_1(y) \tag{21}$$

where :

$$B_1(y) \overset{\text{def}}{=} \sum_{j=1}^{2} \frac{r_j}{y-y_j} \tag{22}$$

y_1, y_2 are the two distinct negative real roots of the polynomial

$$P(y) \overset{\text{def}}{=} y^2 + y(\mu_1 - \lambda_1 + \mu_2 - \lambda_2) + \mu_1(\mu_2 - \lambda_2) + \mu_2(\mu_1 - \lambda_1) - \mu_1\mu_2,$$

$$\text{and } r_j \overset{\text{def}}{=} \begin{cases} -\dfrac{f(X(y_j))(y_j-y^*)(y_j-y^{**})}{y_j P'(y_j)} & \text{if } y_j \in \mathscr{D}+\mathscr{C} \\ 0 & \text{otherwise} \end{cases}$$

where P'(y) denotes the derivative of P(y), (j = 1,2).

Then, $\hat{B}(y)$ is analytic in $\mathscr{C}^-$ and continuous in $\mathscr{C}^- + \mathscr{C}$. □

Proof

Let $P(y) \overset{\text{def}}{=} y^2 + y(\mu_1-\lambda_1+\mu_2-\lambda_2) + \mu_1(\mu_2-\lambda_2) + \mu_2(\mu_1-\lambda_2) - \mu_1\mu_2$. From equations (8), (16), (21) we get :

$$U(y) = \frac{(y - y^{**})(y - y^*)}{y\, P(y)}$$

We show in Appendix that the polynomial $P(y)$ always has two distinct negative real roots y_1, y_2. The location of these roots on the negative real axis w.r.t. the circle $\mathscr{C}$ depends on the parameters λ_1, λ_2, μ_1, μ_2.

From lemma 8, it is readily seen that $\frac{\tilde{B}(y)}{y}$ is analytic in $\mathscr{C}^-$ - $(\tilde{B}(0) = 0)$ -and continuous in $\mathscr{C}^- + \mathscr{C}$.

On the other hand, the rational function $y\,U(y)$ is analytic in $\mathscr{C}^- + \mathscr{C}$ except when $y_j \in \mathscr{D} + \mathscr{C}$, where it has a first order pole at this point, $(j = 1,2)$.

Let R_j be the residue of $y\,U(y)$ at $y = y_j$ if $y_j \in \mathscr{D} + \mathscr{C}$.

Obviously, we have :

$$R_j = \frac{(y_j - y^{**})(y_j - y^{*})}{P'(y_j)}, \quad (j = 1,2).$$

Hence the function $\hat{B}(y)$ defined as

$$\hat{B}(y) \stackrel{\text{def}}{=} \tilde{B}(y)\,U(y) - \sum_{j=1}^{2} \frac{r_j}{y-y_j} \quad \text{with} \quad r_j \stackrel{\text{def}}{=} \begin{cases} \dfrac{\tilde{B}(y_j)R_j}{y_j} & \text{if } y_j \in \mathscr{D}+\mathscr{C} \\ 0 & \text{otherwise} \end{cases}$$

is clearly analytic in $\mathscr{C}^-$ and continuous in $\mathscr{C}^- + \mathscr{C}$, $(j = 1,2)$.

Finally, it is seen from equation (19) and the definition of y_j given above that $\tilde{B}(y_j) = -f(X(y_j))$, $(j = 1,2)$. □

We proceed as in [7]. First of all, let us notice from the definition of $A(x)$ (equation (8)) that $\operatorname{Im} A(x) = 0$ for $x \in \mathbb{R}$, where $\operatorname{Im} z$ is the imaginary part of the complex number z.

Hence, in particular, $\operatorname{Im} A(X(y)) = 0$ for $y \in \mathscr{C}$ since $X(\mathscr{C}) = [x^{**}, x^{*}]$.

We now use the meromorphic continuation of $\tilde{B}(y)$ to $\mathscr{C}^- + \mathscr{C}$ (lemma 8).

Multiplying equation (19) by $U(y)$ and taking the imaginary part we get :

$$\operatorname{Im} U(y)B(y) = - \operatorname{Im} V(y) \quad \text{for } y \in \mathscr{C} \tag{23}$$

Using lemma 9 we may rewrite equation (23) as

$$\boxed{\operatorname{Im} \hat{B}(y) = \Psi(y) \qquad \text{for } y \in \mathscr{C}} \tag{24}$$

where :

$$\Psi(y) = - \operatorname{Im} (V(y) + B_1(y)) \tag{25}$$

$\Psi(y)$ is continuous for $y \in \mathscr{C}$, even if $y_j \in \mathscr{C}$ since $-r_j$ is also the residue of $V(y)$ at the point y_j for $y_j \in \mathscr{C}$, $j = 1,2$. The problem is now reduced to the following : find a function $\hat{B}$ analytic in $\mathscr{C}^-$, continuous in $\mathscr{C}^- + \mathscr{C}$, satisfying the boundary condition (24), where $\Psi(y)$ is a known function, continuous on $\mathscr{C}$.

This is an exterior non-homogeneous Dirichlet problem for the circle $\mathscr{C}$. The solution is given by : [9], [13]

$$\hat{B}(y) = - \frac{1}{2\pi} \int_C \Psi(\omega(t)) \frac{t+\omega^{-1}(y)}{t-\omega^{-1}(y)} \frac{dt}{t} + D , \quad y \in \mathscr{C}^- \tag{26}$$

where D is a real constant, and $\omega(y) \overset{\text{def}}{=} \sqrt{\lambda_2\mu_2}\; y - \mu_2$ maps conformally the domain outside the unit circle C onto $\mathscr{C}^-$.

The constant D will be determined in the next section.

4.4. Workload distribution

From lemma 9, equations (12) and (18) we get :

$$B(y) = \frac{(y-X(y))}{X(y)-Z(y)} (\mu_1 + X(y))(\hat{B}(y) + B_1(y)) \qquad \text{for Re } y \geq 0 \tag{27}$$

We must verify that $X(y) - Z(y) \neq 0$ for Re $y \geq 0$, $y \neq 0$. Let us assume that $X(y) = Z(y)$. From the definitions of X and Z this implies that $X(y) = y$ or equivalently $y\, P(y) = 0$. Appendix A shows us that P only vanishes in the left half-plane. Hence $X(y) - Z(y) \neq 0$ for Re $y \geq 0$, $y \neq 0$ and $B(y)$ given by equation (27) is analytic for Re $y > 0$.

This enables us to determine $A(x)$ for Re $x = 0$. From lemma 2, we know that Re $Y_1(x) \geq 0$ for Re $x = 0$ and that $Y_1(x)$ is analytic in the whole complex plane cut along $[x^{**}, x^*]$.

Hence, using equation (27) and lemma 6, we find :

$$A(x) = -\frac{f(x)}{Y_1(x)-x} - \hat{B}(Y_1(x)) - B_1(Y_1(x)) \quad \text{for Re } x = 0 \tag{28}$$

The continuity of A(x) for Re x = 0 follows from the fact that $Y_1(x) - x$ has no root for Re x = 0. Indeed, let us assume that $Y_1(x) = x$. This implies that $x\,P(x) = 0$. We again conclude using Appendix A.

It remains to compute the constant D of equation (26). It is easily seen that $Y_2(x)$ given by

$$Y_2(x) = \frac{\lambda_1 x - (\mu_2-\lambda_2)(\mu_1+x) - \sqrt{\Delta(x)}}{2(\mu_1 + x)} \qquad \text{(see section 4.1)}$$

is positive for $x \in]-\infty, -\mu_1[$.

So, for $x \in]-\infty, -\mu_1[$ we have :

$$A(x) = -\frac{f(x)}{Y_2(x) - x} - \hat{B}(Y_2(x)) - B_1(Y_2(x)) \tag{29}$$

On the other hand, $\lim\limits_{\substack{x\to-\infty \\ x \in \mathbb{R}}} A(x) = 0$. (see equations (8)).

Hence, from equation (29),

$$\hat{B}(y^*) + B_1(y^*) = 0 \tag{30}$$

since $\lim\limits_{\substack{x\to-\infty \\ x \in \mathbb{R}}} Y_2(x) = y^*$ and $\lim\limits_{\substack{x\to-\infty \\ x \in \mathbb{R}}} \frac{f(x)}{Y_2(x) - x} = 0$.

Finally, we obtain from equations (26) and (30) :

$$D = \frac{1}{2\pi} \int_C \Psi(\omega(t)) \frac{t+\omega^{-1}(y^*)}{t-\omega^{-1}(y^*)} \frac{dt}{t} - \sum_{j=1}^{2} \frac{r_j}{y^*-y_j} \tag{31}$$

The results of this study are summarized in the following theorem.

<u>Theorem</u>

We have for Re x = 0, Re y ≥ 0 :

$$H(x,y) = e^{-xc}\left\{\frac{(1-\rho_1-\rho_2)xe^{xc} + (y-x)\,A(x) + (x-Z(y))\,B(y)\,/\,(\mu_1 + x)}{y - \lambda_1\gamma_1(x) - \lambda_2\gamma_2(y)} - \frac{(x - Z(y))\,B(y)}{(\mu_1+x)(x-\lambda_1\gamma_1(x) - \lambda_2\gamma_2(y))}\right\}$$

where A(x) and B(y) are respectively given in equations (28) and (27). □

5. Numerical results

Theorem of section 4 enables us to compute the mean workload in each queue. Formulas are given in Appendix B.

For two values of traffic intensity (weak and heavy) we have plotted (Fig.2,3) the mean workload in each queue versus the control parameter c. It is not surprising to observe that the mean workload in queue 1 [resp. queue 2] increases [resp. decreases] when c increases. As previously noticed, the cases "c = 0" and " c = +∞ " (corresponding to a preemptive resume priority) are symmetrical, namely $E(V_1/c = +\infty) = E(V_2/\ c = 0)$. The third interesting point is the point E for which $E(V_1) = E(V_2)$.

Hence, this server sharing policy can also be seen as a mean to balance the workload of two non identical waiting lines ($\lambda_1 \neq \lambda_2$ and $\mu_1 \neq \mu_2$).

6. Conclusion

We have studied a queueing system consisting of two queues and one single server. The service discipline between the two types of customers depends on the workload of one of the queues (queue 1) and can be summarized as follows : when the workload in queue 1 remains below a given threshold c, a customer of queue 2 is served (if any) ; otherwise a customer of queue 1 is server.

By solving two functional equations, we have been able to derive a closed form for the Fourier-Stieltjes and Laplace-Stieltjes transforms of the joint stationary distribution of the system-workload. Numerical values of the mean workload in each queue have been obtained versus the threshold c.

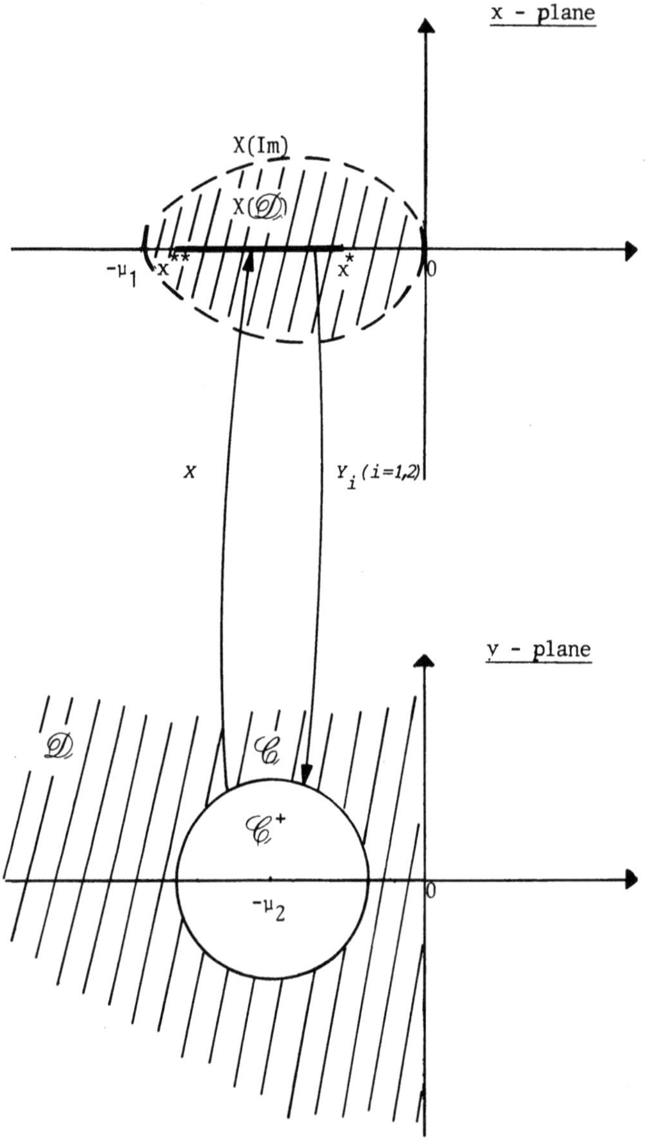

FIGURE 1

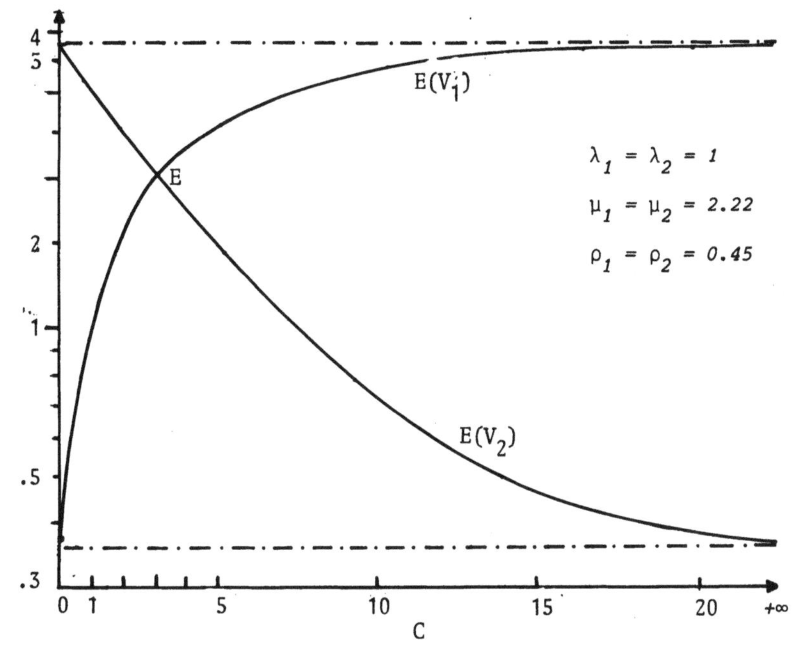

FIGURE 2 : Mean workload for a heavy traffic intensity.

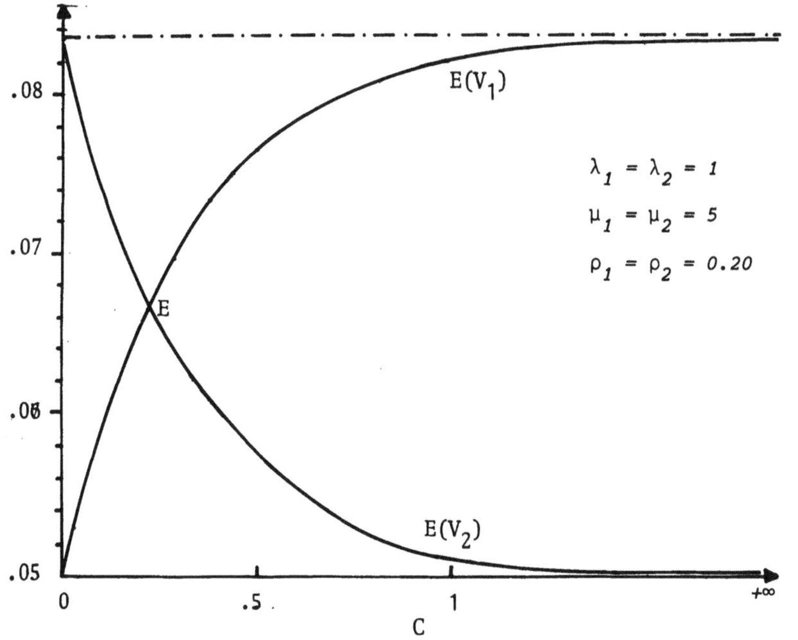

FIGURE 3 : Mean workload for a weak traffic intensity.

APPENDIX A

Lemma

$$P(y) \overset{\text{def}}{=} y^2 + y(\mu_1-\lambda_1+\mu_2-\lambda_2) + \mu_1(\mu_2-\lambda_2) + \mu_2(\mu_1-\lambda_1) - \mu_1\mu_2$$

always has two distinct negative real roots. Moreover, the position of these roots on the negative real axis with respect to the circle $\mathcal{C}$, depends on the values of the parameters λ_1, λ_2, μ_1, μ_2. □

Proof

We assume that $\rho_1 + \rho_2 < 1$ and $\lambda_1 \neq 0$, $\lambda_2 \neq 0$.

Let us define the polynomial T(z) as

$$T(z) \overset{\text{def}}{=} z^2 + z(\alpha^2(1-\rho_1) + 1 - \rho_2) + \alpha^2(1-\rho_1-\rho_2)$$

where :

$$\alpha \overset{\text{def}}{=} \sqrt{\frac{\mu_1}{\mu_2}} \neq 0.$$

We have :

$$P(y) = \mu_2^2 \; T\left(\frac{y}{\mu_2}\right).$$

The discriminant of the polynomial T is :

$$\Delta_T(\alpha) = (\alpha^2(1-\rho_1) + 1 - \rho_2)^2 - 4\alpha^2(1-\rho_1-\rho_2)$$

$$= (\alpha^2(1-\rho_1) + (1-\rho_2) + 2\alpha\sqrt{1-\rho_1-\rho_2})(\alpha^2(1-\rho_1) + (1-\rho_2) - 2\alpha\sqrt{1-\rho_1-\rho_2})$$

Hence :

$$\text{sgn}(\Delta_T(\alpha)) = \text{sgn}(\alpha^2(1-\rho_1) + (1-\rho_2) - 2\alpha\sqrt{1-\rho_1-\rho_2}).$$

Since the discriminant of this polynomial in α is always strictly negative, it follows that $\Delta_T(\alpha) > 0$.

The easiest way to prove the second part of the lemma is to choose particular values for ρ_1, ρ_2, α.

First, let us examine the case where $\rho_1 = \rho_2 = \rho$ and $\alpha = 1$. Then the roots of T are $z_1 = -1$ and $z_2 = -1 + 2\rho$. These roots must be positioned with respect to the interval $I_\rho \overset{\text{def}}{=} [-1 - \sqrt{\rho}\ ;\ -1 + \sqrt{\rho}]$.

- if $\rho < \frac{1}{4}$ then $z_1, z_2 \in I_\rho$ $((z_1, z_2) \in \mathscr{C}^+)$.
- if $\rho = \frac{1}{4}$ then $z_1, z_2 \in I_{1/4}$ and $z_2 = -\frac{1}{2}$ $(z_1 \in \mathscr{C}^+, z_2 \in \mathscr{C})$.
- if $\rho > \frac{1}{4}$ then $z_1 \in I_\rho$ and $z_2 \notin I_\rho$ $(z_1 \in \mathscr{C}^+, z_2 \in \mathscr{C}^-)$.

The case remains where z_1 and z_2 do not belong to I_{ρ_2}. This happens, in particular, if $\rho_1 = \frac{7}{10}$, $\rho_2 = \frac{1}{4}$ and $\alpha = 3$. Then $(z_1, z_2) \in \mathscr{C}^-$. □

APPENDIX B

From the theorem of section 4.4 we have :

$$E\{V_1\} = -\frac{\partial}{\partial x} H(x,0) \Big|_{x=0}$$

$$E\{V_2\} = -\frac{\partial}{\partial y} H(0,y) \Big|_{y=0}$$

A tedious computation , involving first and second order expansions of X, Y_1, Z as well as their derivatives, yields :

$$E\{V_1\} = -\frac{1}{\lambda_1} \left(A(0)(1-\mu_1 c) + \mu_1 A'(0) + cB(0) - (1-\rho_1-\rho_2)\right) - \left(\frac{1+(\mu_1-\lambda_1)c}{(\mu_1-\lambda_1)^2}\right) B(0)$$

$$E\{V_2\} = \frac{\rho_2}{\mu_2(1-\rho_2)^2} A(0) + \frac{1-2\rho_2}{\lambda_2\mu_1(1-\rho_2)^2} B(0)\, Z'(0) + \frac{1}{\mu_1\rho_2(1-\rho_2)} \left(\frac{B(0)Z''(0)}{2} + B'(0)Z'(0)\right)$$

where :

$$Y_1'(0) = \frac{\rho_1}{1-\rho_2} ,$$

$$Z'(0) = \frac{\rho_2}{1-\rho_1} ,$$

$$X'(0) = \frac{1-\rho_2}{\rho_1} ,$$

$$Y_1''(0) = \frac{(\mu_2-\lambda_2)^2 - 3\lambda_1\mu_2 + \lambda_1\lambda_2}{2(\mu_2-\lambda_2)\mu_1^2} + b,$$

$$b \stackrel{\text{def}}{=} \frac{1}{2\mu_1^2(\mu_2-\lambda_2)} \left((\lambda_1+\lambda_2-\mu_2)^2 + 4\lambda_1\mu_2 - \frac{((\mu_2-\lambda_2)^2 + \lambda_1(\lambda_2+\mu_2))(2(\mu_2-\lambda_2)^2 + \lambda_1(\lambda_2+\mu_2))}{(\mu_2-\lambda_2)^2} \right),$$

$Z''(0)$ = the same expression as $Y_1''(0)$, changing respectively $\lambda_1,\lambda_2,\mu_1,\mu_2$ for $\lambda_2,\lambda_1,\mu_2,\mu_1$.

$$X''(0) = 2\mu_1 \frac{(\mu_2-\lambda_2)^2 + \lambda_1\lambda_2}{\lambda_1^2\,\mu_2^2},$$

$$A(0) = \frac{B(0)}{\mu_1-\lambda_1} + 1 - \rho_2 \qquad \text{, (from equation (28))}$$

$$B(0) = -\mu_1(1-\rho_1)\Phi(0) \qquad \text{, (from equation (27))}$$

$$B'(0) = \mu_1\Phi(0)\left(\frac{Z''(0)(1-X'(0)) - X''(0)(1-Z'(0))}{2(X'(0)-Z'(0))^2}\right) + \frac{(X'(0)\Phi(0) + \mu_1\Phi'(0))\,(1-X'(0))}{X'(0) - Z'(0)},$$

$$\Phi(y) \stackrel{\text{def}}{=} \hat{B}(y) + B_1(y),$$

$$\Phi(0) = \frac{1}{2\pi}\int_C \Psi(\omega(t))\left(\frac{t+\omega^{-1}(y^*)}{t-\omega^{-1}(y^*)} - \frac{t+\omega^{-1}(0)}{t-\omega^{-1}(0)}\right)\frac{dt}{t} - \sum_{j=1}^{2}\frac{r_j}{y^*-y_j} - \sum_{j=1}^{2}\frac{r_j}{y_j},$$

$$\Phi'(0) = -\frac{\sqrt{\lambda_2\mu_2}}{\pi}\int_C \frac{\Psi(\omega(t))}{(t\sqrt{\lambda_2\mu_2} - \mu_2)^2}\,dt - \sum_{j=1}^{2}\frac{r_j}{y_j^2},$$

$$A'(0) = \left(\frac{1-\rho_1-\rho_2}{1-Y_1'(0)}\right)\left(\frac{Y''(0)}{2(1-Y_1'(0))} + c\right) - Y_1'(0)\,\Phi'(0).$$

I should like to thank G. Fayolle for suggesting this research topic and for many helpful discussions during the course of this work, and also to express my appreciation to F. Baccelli and Ph. Robert for their help.

REFERENCES

[1] BLANC, J.P.C. : Application of the theory of boundary value problems in the analysis of a queueing model with paired services. Thesis, University of Utrecht, Utrecht, (1982).

[2] BACCELLI, F. and FAYOLLE, G. : Two dimensional diffusion processes with boundary and jumps. Application to coupled queues. Rapport de Recherche INRIA, N° 160, (1982).

[3] BOXMA, O.J. and COHEN, J.W. : The M/G/1 queue with alternating service formaluted as a Riemann-Hilbert problem. Perfórmance 81, F.J. Kylstra Ed. North-Holland Publishing Company, (1981).

[4] COHEN, J.W. : The single server. Amsterdam, North-Holland Publishing Company, (1969).

[5] COHEN, J.W. : On the M/G/2 queueing model. Stoch. Proc. Appl. 12, 3, (1982).

[6] FAYOLLE, G. : Méthodes analytiques pour les files d'attente couplées. Thèse, Université Paris VI, (1979).

[7] FAYOLLE, G. and IASNOGORODSKI, R. : Two coupled processors : the reduction to a Riemann-Hilbert próblem. Wahrscheinlichkeitsth 47, 325-351, (1979).

[8] FAYOLLE, G. and KING, P.J.B. and MITRANI, I. : The solution of certain two-dimensional Markov models. Adv. Appl. Prob. 14, 295-308, (1982).

[9] GHAKOV, F.D. : Boundary value problems. Pergamon Press, Oxford, (1966).

[10] IASNOGORODSKI, R. : Problèmes-frontières dans les files d'attente. Thèse, Université Paris VI, (1979).

[11] KLEINROCK, L. : Queueing systems. Vol. 1, Academic Press, (1975).

[12] MIKOU, N. : Modèles de réseaux de files d'attente avec pannes. Thèse, Université Paris XI, Orsay, (1981).

[13] MUSKHELISHVILI, N.I. : Singular integral equations. Groningen, Holland-Moscow, P. Noordhoff, (1946).

[14] TAKÁCS, L. : Introduction to the theory of queues. Oxford Univ. Press, (1952).

XI

MODELS OF STORAGE PROCESSES

MODELES D'ALLOCATION DE RESSOURCES

THE SPATIAL REQUIREMENT OF AN M/G/1 QUEUE, OR: HOW TO DESIGN FOR BUFFER SPACE

Bhaskar Sengupta
Bell Laboratories (HO4M303)

Holmdel, NJ 07733, U.S.A.

ABSTRACT

Consider an M/G/1 queue where each customer (independently of others) has a space requirement which may depend on its service time. We find the steady state distribution of the total space required by all customers present in the system. An application of the model is in the field of message switching where the memory space required by a message is (usually) a monotonic function of the service time. By using the processor sharing version of the M/G/1 queue, another application may be in the design of storage space required for time sharing computers.

1. Introduction

A common problem faced by designers of computer or communicating systems is to decide on the amount of memory space that should be made available. For example, a message switch receiving messages of varying sizes must be engineered to protect against buffer overflow. In this example, the buffers occupied by a message and the time to switch the message are monotonic functions of the message length. We discuss the problem of finding the steady state distribution of the total buffers used by the messages present in the system. As another example, consider the arrival of jobs of varying sizes at a time shared computer where the space required by a job may depend on the service time. The problem is to characterize the distribution of the space required by all jobs present in the system.

We model the problem as an M/G/1 queue first with first come first served discipline and then the processor sharing discipline. The arrival rate is assumed to be λ and the service time density is $f(t)$ (assumed to exist) with a mean of $1/\mu$. We assume that the space required by a customer is independent of that of other customers. The space is occupied at the instant he arrives and is released entirely at the instant he completes service. We allow the space required by a customer to depend on his service time in the sense that there is a known probability distribution function $g(y,t)dt = P(Y\leq y,\ t<T\leq t+dt)$ where Y is the space required by a customer and T is its service time. We are concerned with finding the steady state distribution of the total space required by all customers present in the queueing system.

This problem is interesting because there is a dependence between the space required by the waiting customers and the space required by the customer in service. For instance, in our message switching example, a long message length will usually imply large buffer usage and a long service time. Consequently, one may expect to see a larger number of waiting customers behind a long message in service. One contribution of this paper is to explicitly characterize this dependence between the space needed for the waiting customers and that by the customer in service.

In sections 2 through 5, we solve the problem of the M/G/1 queue with first come first served discipline. In Section 6 we discuss the processor sharing version of the same queue. In section 7, we discuss an application of the method to message switching.

2. Space Required by the Waiting Customers

One method of solving this problem is based on the treatment of the M/G/1 queue by Kosten [1] using the supplementary variable technique. Kosten defines $q_r(x)dx$ to be the (steady state) probability that the server is busy, there are r customers waiting (excluding the one in service) and that the customer in service has a past duration (or age) of service time between x and x + dx. Let N_q be the number of customers waiting and X the past duration of the customer in service. Then (see pages 58, 59 of ref. 1)

$$q_r(x)dx = P(\text{Server busy},\ N_q=r,\ x<X\leq x+dx).$$

If $Q(\alpha,x)$ is the generating function of $q_r(x)$, then

$$Q(\alpha,x) = K(\alpha)(1-F(x))e^{-\lambda(1-\alpha)x} \tag{1}$$

where $F(x) = P(T \le x) = \int_0^x f(u)\,du$,

$$K(\alpha) = \frac{(1-\alpha)\lambda(1-\rho)}{J(\alpha)-\alpha},$$

$$\rho = \lambda/\mu$$

and $$J(\alpha) = \int_0^\infty e^{-\lambda(1-\alpha)u} f(u)\,du.$$

Let the distribution function of the space required by a customer be $d(y)$. Then

$$d(y) = P(Y \le y) = \int_0^\infty g(y,t)\,dt$$

Let $D(\alpha)$ be the Laplace-Stieltjes transform of $d(y)$. Let B_q be the space occupied by the waiting customers and

$$r(y,x)\,dx = P(\text{Server busy}, B_q \le y, x < X \le x+dx).$$

Further, let $R(\alpha,x)$ be the Laplace-Stieltjes transform of $r(y,x)$. Let Y_i be the space occupied by the ith waiting customer. Then $B_q = \sum_{i=1}^{N_q} Y_i$, the Y_i's are iid and N_q is independent of Y_i. Thus

$$R(\alpha,x) = Q(D(\alpha), x) \tag{2}$$

and the transform of the space occupied by waiting customers is

$$\int_0^\infty R(\alpha,x)\,dx.$$

Further, $EB_q = EY\,EN_q$ and $\text{Var } B_q = EN_q \text{Var } Y + (EY)^2 \text{Var } N_q$.

3. Space Required by the Customer in Service

Let $\phi(y|x)$ be the probability that the space required by the customer in service (B_s) is less than or equal to y given that the past duration is x. Then

$$\phi(y|x) = \int_x^\infty g(y,t)\,dt/(1-F(x)) \tag{3}$$

and let $\Phi(\alpha|x)$ be the corresponding Laplace-Stieltjes transform. From the fact that the density function of X is $\mu(1-F(x))$ and that $B_s = 0$ whenever the server is idle, we have

$$EB_s = \rho \lim_{\alpha\to 0} \int_0^\infty -\Phi'(\alpha|x)\mu(1-F(x))\,dx = \lambda E(YT). \tag{4}$$

It is possible to arrive at this result heuristically by using Little's law on the server. The arrival rate of buffers is λEY. The average time spent by a buffer with the server is $E(YT)/EY$. The product yields EB_s.

Further,

$$EB_s^2 = \rho \lim_{\alpha\to 0} \int_0^\infty \Phi''(\alpha|x)\mu(1-F(x))\,dx = \lambda E(Y^2T). \tag{5}$$

4. Total Space Required by All Customers

Let $B(=B_s+B_q)$ be the total space required by all customers. Then

$$P(B\le y) = P(\text{Server idle}) + \int_0^\infty P(\text{Server busy, } B_s+B_q\le y|X=x)\mu(1-F(x))\,dx$$

because the density function of the past duration is $\mu(1-F(x))$. But, given the past duration X, B_s and B_q are independent random variables. This is true because N_q (and consequently B_q) at any time depends on the customer in service only through his past duration. Viewed in another way, suppose at some time we know the past duration of the customer service, then the distribution of the number of customers waiting at this time will not be influenced by the knowledge of the time at which this customer will complete his service. This is equivalent to saying that B_s and B_q are independent given X. If $B(\alpha)$ is the Laplace-Stieltjes transform of the distribution function of B, we have

$$B(\alpha) = 1 - \rho + \int_0^\infty \Phi(\alpha|x)R(\alpha,x)dx. \tag{6}$$

By using (1), (2) and (3), it is possible to show that (see Appendix I)

$$B(\alpha) = 1 - \rho + \frac{K(D(\alpha))}{\lambda(1-D(\alpha))}[D(\alpha) - G(\alpha, \lambda(1 - D(\alpha)))] \tag{7}$$

where

$$G(\alpha,\sigma) = \int_0^\infty\int_0^\infty e^{-\alpha y}e^{-\sigma t}d_y g(y,t)dt.$$

By taking the first derivative of (6), we obtain

$$E(B) = \lambda E(YT) + E(Y)E(N_q)$$

where the first term is EB_s (from (4)) and the second is $E\,B_q$. Further, from the second derivative of (6) we have (see Appendix II),

$$E(B^2) = E(B_q^2) + \lambda E(Y^2T) + 2\lambda E(N_q)E(Y)E(YT) + \lambda^2 E(Y)E(YT^2). \tag{8}$$

From this, we can easily obtain the correlation coefficient γ between B_s and B_q,

$$\gamma = \frac{\lambda^2 EYE(YT^2)}{2\sqrt{Var\ B_s Var\ B_q}}.$$

5. Three Special Cases

Consider first, the special case where each customer brings a spatial requirement of 1 unit with probability one, i.e., $g(y,t) = 0$ for $y < 1$ and $g(y,t) = f(t)$ if $y \geq 1$ for all t. It is then easy to show that $B(\alpha)$ reduces to the generating function of number of customers in the system, which is the familiar Pollaczek-Khintchine equation.

Next, consider the case where the space occupied by a customer is independent of the service time. In this case, $D(\alpha) = \Phi(\alpha|x)$ for all x and from (6),

$$B(\alpha) = 1 - \rho + D(\alpha) \int_0^\infty R(\alpha,x)\,dx.$$

This result bears out the fact that the space required by the customer in service is independent of that for the waiting customers, conditional on the server being busy. Further if $N(\alpha)$ is the generating function of the number of customers in the system, it is possible to show that $B(\alpha) = N(D(\alpha))$ which is what we expect.

Finally, we show that (6) can be used to calculate the distribution of the unfinished work or the virtual waiting time for the M/G/1 queue. In this case, we will assume that each customer brings a spatial requirement which is exactly equal to the service time. However, the spatial requirement of the customer in service depletes steadily as he receives service and is given by

$$P(Y \leq y|X=x) = \phi(y|x) = P(T \leq x+y|T>x) = (F(x+y) - F(x))/(1 - F(x)).$$

Further, $g(y,t) = 0$ if $y < t$ and $g(y,t) = f(t)$ if $y \geq t$ for all t. Thus, $D(\alpha) = \tilde{f}(\alpha)$ which is the Laplace Transform of $f(t)$. By using (1), (2) and (6), it is now possible to show that (see Appendix III)

$$B(\alpha) = \frac{(1-\rho)\alpha}{\alpha-\lambda(1 - \tilde{f}(\alpha))} \qquad (9)$$

which is the Laplace-Stieltjes transform of the unfinished work in the system.

6. The Processor Sharing Discipline

The basic idea in finding the distribution of the space required for the M/G/1 processor sharing queue is exactly the same as before. We look at the joint distribution of the queue size and the attained service of each customer in the system. Let X_i be the attained service of the ith customer (without regard to any ordering requirement within the customers) and N the number of customers. Then let

$$P_n(x_1,\dots,x_n) = \lim_{\Delta\to 0} \frac{1}{\Delta^n} P(N=n,\ x_i<X_i\le x_i+\Delta;\ i=1,\dots,n).$$

Sakata, et al., [2] have shown that

$$P_n(x_1,\dots,x_n) = \rho^n(1-\rho) \prod_{i=1}^{n} \bar{f}(x_i), \qquad 0<x_i<\infty \tag{10}$$

where $\bar{f}(x)$ is the density of the backward recurrence time of the service distribution. Then

$$P(B\le y) = \sum_{n=0}^{\infty} \int_0^{\infty}\cdots\int_0^{\infty} P(B\le y\,|\,N=n,\ X_i=x_i;\ i-1,\dots,n)P_n(x_1,\dots,x_n)dx_1\dots dx_n.$$

Given the attained service of each customer, their spatial requirements are independent. From (10), we have

$$B(\alpha) = \sum_{n=0}^{\infty} \rho^n(1-\rho) \prod_{i=1}^{n} \int_0^{\infty} \Phi(\alpha|x_i)\bar{f}(x_i)dx_i = \frac{1-\rho}{1-\rho \displaystyle\int_0^{\infty} \Phi(\alpha|x)\bar{f}(x)dx}$$

After routine manipulations, this reduces to

$$B(\alpha) = \frac{1-\rho}{1-\lambda E(Te^{-\alpha Y})}.$$

Further, $EB = \lambda E(YT)/(1-\rho)$

and $\quad EB^2 = \lambda E(Y^2T)/(1-\rho) + 2(EB)^2.$

By the following methods of Section 5, it is possible to derive the results for the three special cases where the answers are known.

7. An application to a message switch

In this section, we discuss issues relating to an application of the model in designing the buffer size of a message switch.

A message being switched joins three distinct queues in tandem. The first one allows the message to enter the message switch through an incoming line. The second performs the message switching function and, if necessary, character conversion. The third queue enables the message to leave the message switch on an outgoing line. We will refer to these queues as the incoming queue, the CPU queue and the outgoing queue.

During the waiting time of the incoming queue, the message is resident in a terminal or host and does not occupy any buffers of the message switch. During the service time of this queue, the buffers of the message switch are gradually filled as the message is received. After being completely received, the message remains in the buffers of the message switch until it is transmitted on an outgoing line.

Thus for the purpose of buffer usage, all messages on the CPU queue and the outgoing queue occupy buffers to the full extent needed for the message. For the incoming queue, only the message being received (served) occupies buffers. Further, since the buffers are filled gradually, the actual buffer usage of a message in service in the incoming queue is less than it would be after it has fully entered the switch.

The queueing schematic for the message switch is shown in Figure 1. This figure assumes that there are four full duplex lines. The flow of messages and buffer occupany is also shown in the figure.

The method of calculating the distribution of the buffers occupied in the message switch is based on two approximating assumptions. First, we assume that messages arrive at each queue according to a Poisson process. Second, the queues operate independently of one another. Thus, the switch is viewed as a collection of independent M/G/1 queues. For each of the queues, we first calculate the mean and variance of buffers occupied by the methods of the earlier sections. The mean and variance of buffers occupied in the switch are calculated by adding the same quantities for the individual queues. Finally, we fit a Gamma distribution to the calculated values of the mean and variance to address questions about the distribution of buffer occupancy.

We add that we have actually used the above method to calculate the buffer occupancy of a message switch. In spite of the approximating assumptions, we were able to obtain results that agreed very well with computer simulations. A comparison of the results of simulation and analysis are shown in Figure 2. The analytical approximation underestimates the simulation result slightly. One possible reason may be that the assumptions of Poisson input to independent queues is not quite valid.

We end this section with a discussion on the use of the method of earlier sections. For the incoming queue, only the message in service occupies buffers and that too partially. We assume that the buffers occupied by the message is uniformly distributed between zero and B_s as long as the queue is nonempty. Further, the buffers required by a message is a monotonically increasing step function of the service (transmission) time. In particular, $Y = i$ whenever $a_i < T \leq a_{i+1}$; $i = 1,2,\ldots$ Also, for $a_i < t < a_{i+1}$,

$$g(y,t) = \begin{cases} f(t) & \text{if } y \geq i \\ 0 & \text{if } y < i. \end{cases}$$

Using these and (4) and (5), we can find the first and second moments of the buffers occupied by the message in the incoming queue.

For the CPU queue, the service time was found to be deterministic and thus the random variables Y and T are independent. This is merely a special case of the results of Section 4. For the outgoing queues, the relationship between Y and T is the same as that for incoming queues. Using this fact and the results of Section 4, one can easily compute the first and second moments of buffer occupancy.

Finally, we state that not all lines of the message switch were full duplex. Some of the lines were polled half duplex where the message switch had nonpreemptive priority over the adjacent terminal or host in gaining control over the line. This would imply that there is a single server serving the outgoing and incoming parts of a line where the outgoing messages have priority over incoming messages. In this situation, we calculated the moments of buffer occupancy separately for the message in service and the waiting messages for the high priority queue by using results from the M/G/1 priority queues. Then the mean buffer occupancy was calculated by adding the individual means. The variance was calculated by assuming that the correlation coefficient was equal to one. This method, naturally, gives an upper bound on the variance. The buffer occupancy for the low priority queue was calculated in the same manner as the buffer occupancy for the incoming queue.

8. Conclusion

This paper was concerned with finding the distribution of the spatial requirements of all customers in an M/G/1 queue. At issue was the dependence between B_s and B_q. This dependence was resolved by claiming that the random variables are independent, conditional on the age of the customer in service. In principle, this procedure will work in other single server queueing situations where $R(\alpha,x)$ is known or can be calculated. An additional difficulty may be in carrying out the integration in (5). A worthwhile extension of the results may be for the case where the queueing system has a finite space available.

Acknowledgment

I am grateful to Dave Jagerman and Robert Morris for useful discussions.

References

[1] L. Kosten, "Stochastic Theory of Service Systems," Pergamon Press, 1973.

[2] M. Sakata, S. Noguchi, J. Oizumi, "Analysis of a Processor Shared Queuing Model for Time Sharing Systems," Proceedings of the Second Hawaii International Conference on Systems Sciences, 625-528, January, 1969.

Appendix I

Equation (7) may be derived as follows:

$$\int_0^\infty \Phi(\alpha|x)R(\alpha,x)dx = \int_0^\infty \int_0^\infty e^{-\alpha y} d_y \phi(y|x) Q(D(\alpha),x)dx$$

$$= \int_0^\infty \int_0^\infty e^{-\alpha y} \int_x^\infty d_y g(y,t)dt \cdot K(D(\alpha))\exp\{-\lambda(1 - D(\alpha))x\}dx$$

$$= K(D(\alpha)) \int_0^\infty \int_0^\infty e^{-\alpha y} d_y g(y,t) \int_0^t \exp\{-\lambda(1 - D(\alpha))x\}dx\ dt$$

$$= \frac{K(D(\alpha))}{\lambda(1-D(\alpha))} \int_0^\infty \int_0^\infty e^{-\alpha y} d_y g(y,t)(1 - \exp\{-\lambda(1 - D(\alpha))t\})dt$$

$$= \frac{K(D(\alpha))}{\lambda(1-D(\alpha))} [D(\alpha) - G(\alpha,\lambda(1-D(a)))] .$$

Appendix II

To derive (8), we first note that

$$B'' = \int \Phi''R + \int \Phi R'' + 2\int \Phi'R'. \tag{11}$$

Since $\Phi(0|x) = 1$, we have

$$\int_0^\infty \Phi(0|x)R''(0,x)\,dx = EB_q^2. \tag{12}$$

Also, $R(0,x) = \lambda(1 - F(x))$ gives us from (5)

$$\int_0^\infty \Phi''(0|x)R(0,x)\,dx = EB_s^2. \tag{13}$$

Finally,

$$\Phi'(0|x) = \int_x^\infty \int_0^\infty y\,d_y g(y,t)\,dt/(1 - F(x)) \tag{14}$$

and from (1) and (2),

$$R'(0,x) = (1 - F(x))\exp\{-\lambda(1 - D(0))x\}(K'(D(0))D'(0)$$

$$+ K(D(0))D'(0)\lambda x)$$

But

$$D(0) = 1$$

$$D'(0) = EY$$

$$K(1) = \frac{\lambda(1-\rho)}{1-J'(1)} = \lambda$$

If we let $N(\alpha)$ be the generating function of the number in the system, it is easy to see that

$$K(\alpha)J(\alpha) = \lambda N(\alpha)$$

By taking a derivative we obtain

$$K'(1) = \lambda N'(1) - \lambda\rho = \lambda EN_q.$$

Thus

$$R'(0,x) = (1 - F(x))(\lambda EN_q EY + \lambda^2 xEY) \tag{15}$$

From (14) and (15) we obtain

$$\int_0^\infty \Phi'(0|x)R\ (0,x)dx = \int_0^\infty \int_0^\infty y d_y g(y,t) \int_0^t \lambda EY(EN_q+\lambda x)dx\ dt$$

$$= \lambda EY \int_0^\infty \int_0^\infty y\left(tEN_q + \frac{\lambda t^2}{2}\right)d_y g(y,t)dt$$

$$= \lambda EYEN_q E(YT) + \frac{\lambda^2}{2} EYE(YT^2).$$

And equation (8) is now a result of (11), (12), (13) and the above.

Appendix III

We derive equation (9) as follows:

$$D(\alpha) = \tilde{f}(\alpha)$$

$$J(\alpha) = \tilde{f}(\lambda(1-\alpha))$$

$$R(\alpha,x) = \frac{\lambda(1-\rho)(1-\tilde{f}(\alpha))}{\tilde{f}(\lambda(1-\tilde{f}(\alpha)))-\tilde{f}(\alpha)} \cdot ((1 - F(x))\exp\{- (1-\tilde{f}(\alpha))x\}$$

$$\Phi(\alpha|x) = \int_0^\infty e^{-\alpha y} d_y \phi(y|x) = \int_0^\infty e^{-\alpha y} f(x+y)dy/(1 - F(x))$$

and

$$\int_0^\infty \Phi(\alpha|x)R(\alpha,x)$$

$$= \frac{\lambda(1-\rho)(1-\tilde{f}(\alpha))}{\tilde{f}(\lambda(1-\tilde{f}(\alpha)))-\tilde{f}(\alpha)} \int_0^\infty \exp\{(\alpha-\lambda(1-\tilde{f}(\alpha)))x\} \int_x^\infty e^{-\alpha z} f(z)dz\ dx.$$

By an interchange of the order of integration, this expression becomes

$$\frac{\lambda(1-\rho)(1-\tilde{f}(\alpha))}{\alpha-\lambda(1-\tilde{f}(\alpha))},$$

from which it is easy to see that $B(\alpha)$ is the Laplace-Stieltjes transform of the unfinished work in the system.

FIGURE 1

QUEUEING SCHEMATIC FOR A MESSAGE SWITCH

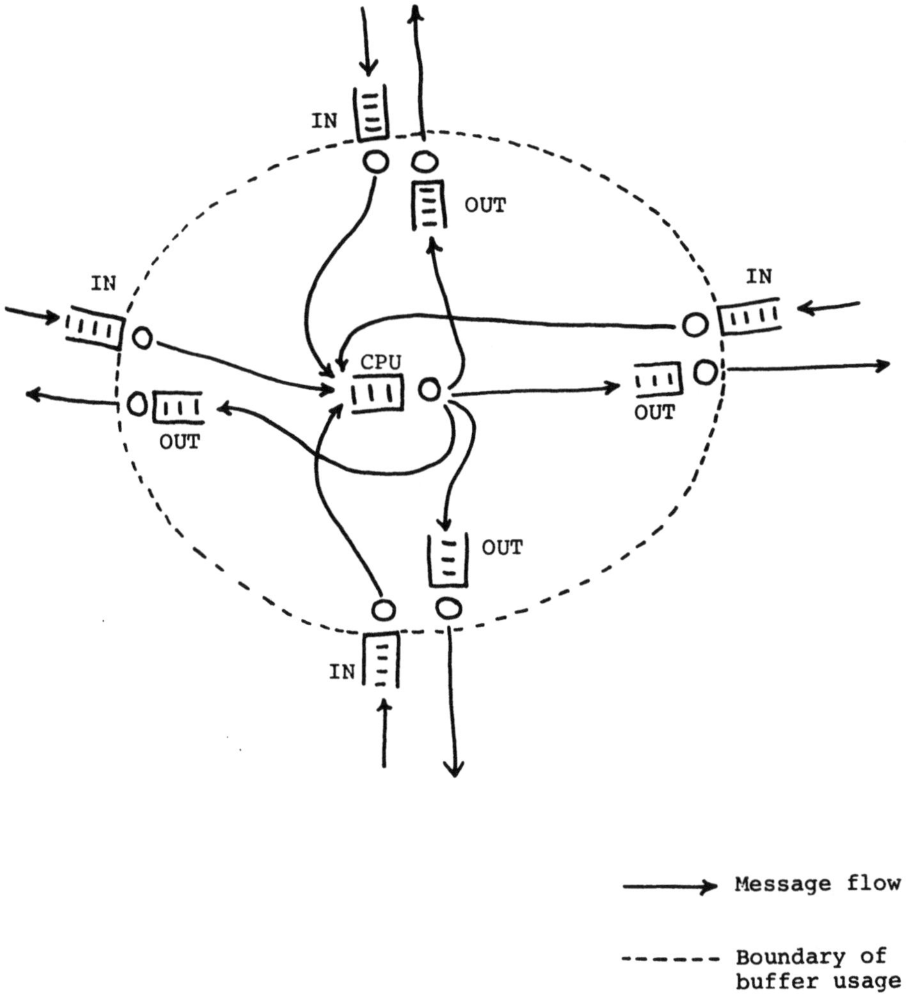

FIGURE 2

COMPARISON OF SIMULATION VS. ANALYTICAL

BUFFER USAGE DISTRIBUTION

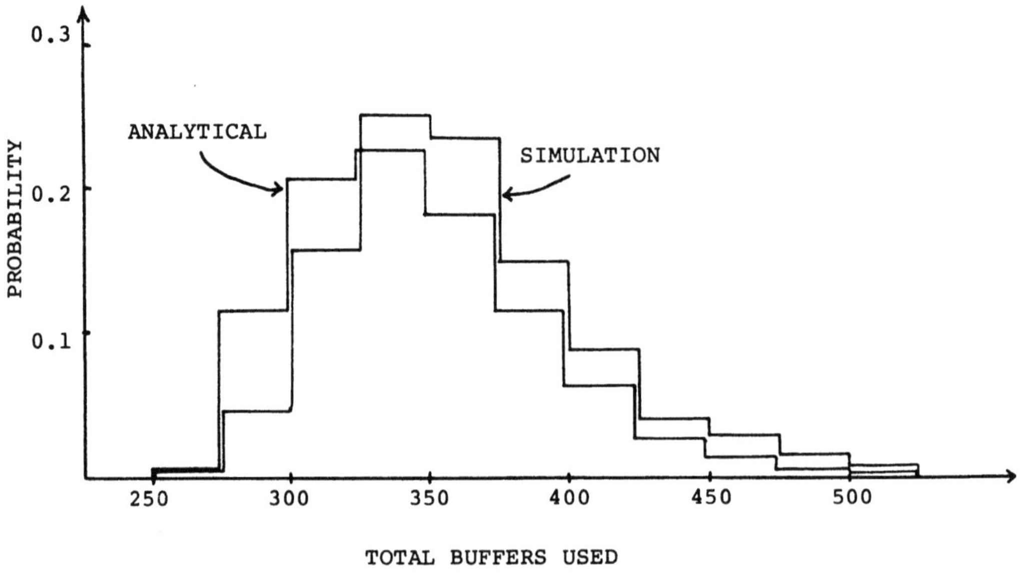

	SIMULATION	ANALYTICAL
MEAN	361	349
STD. DEV.	43.3	44.4
95th PERCENTILE	445	425
99th PERCENTILE	490	470

XII

COMPUTER SYSTEMS EVALUATION 2

EVALUATION DE SYSTEMES INFORMATIQUES 2

ANALYSIS OF PROTOCOLS OF MULTIPLE ACCESS

by

Anthony Ephremides
Professor of Electrical Engineering
University of Maryland
College Park, Maryland 20742

ABSTRACT

We review the process of establishing a multiple access protocol, and focus on its analytical performance evaluation. In cases where due to either instability or queue interaction the analysis is hindered, we suggest approximate approaches. In other cases traditional point process techniques seem to work. We focus on two case studies. The first concerns the long-outstanding problem of delay analysis in random access protocols with queueing. The second concerns performance evaluation in access protocols without feedback.

1. INTRODUCTION

It is well-known that many random access protocols are unstable and, even those that aren't or those that can be stabilized, involve complex queue interation that make exact analytical performance evaluations almost impossible. Among the various approaches to this problem [1-13] we present in section 2, an approximate analysis that suggests how classical techniques may be used for a straight-forward intuitive approximation. The proposed approach leads to exact analysis in cases where the queue interation is somewhat controlled.

These issues are present in the usual case of multiple access with channel feedback. It is the feedback feature (that is, the ability to control future individual transmission policies on the basis of what occured in previous transmission slots on the channel), which couples the different users and creates analytical difficulties. However when there is no feedback (as in the cases of no acknowledgements and/or strong local interference) the users operate in an open-loop fashion and the difficulties disappear. Standard point process techniques can then be used as illustrated in Section 3.

2. APPROXIMATE ANALYSIS OF A RANDOM ACCESS MODEL

Consider M users each of which receives packets according to an independent Bernoulli process with parameter σ. Packets can be locally buffered (queued). The buffers have unlimited capacity. Each user attempts transmission of a packet with probability one, if a packet is there and if a collision has not been experienced. If a collision has been experienced transmission takes place in the next slot with probability p (the user is then called <u>blocked</u>). The user persists until that packet is successfully transmitted. When that happens he becomes unblocked. When unblocked the user can be <u>idle</u>, if there is no packet in his buffer, and <u>active</u> otherwise.

The precise modeling of this system is straight forward by means of a M-dimensional discrete random walk with appropriate boundaries. It is of course these boundaries that cause complications and make the analysis extremely difficult or impossible.

There have been several approximate approaches to this problem. Here we outline one.

Consider the state of each terminal to be adequately characterized by the size of the buffer contents and by the indication whether it is blocked or unblocked. Thus the state is denoted by (i,j), where i is zero if the terminal is blocked and one if it is unblocked (idle or active), and where j=0,1,2,... indicates the number of packets in the buffer.

Let $\pi_{ij} \triangleq$ steady state probability of the state (i,j). Let

$$G_i(z) \Delta = \sum_{n=0}^{\pi} \pi_{i,n} z^{-n} \quad , \quad k=0,1 \qquad (1)$$

Further, define

$r \triangleq Pr$ [successful transmission/ user is blocked]

$q_1 \triangleq Pr$ [collision/user is idle]

$q_2 \triangleq Pr$ [collision/user is active]

Assuming that these quantities suffice, we may characterize the transition probability of the user Markov chain as it appears in Fig. 1.

We can also obtain that

$$\pi_{1,0} = \frac{r(1-\sigma)-\sigma q_2}{r(1-\sigma)-\sigma(q_2-q_1)} \qquad (2)$$

$$G_0(1) = \frac{(1-\sigma)\sigma q_1}{r(1-\sigma)-\sigma(q_2-q_1)} \qquad (3)$$

$$G_1(1) = \sigma+(1-\sigma)\frac{r(1-\sigma)-\sigma q_2}{r(1-\sigma)-\sigma(q_2-q_1)} \qquad (4)$$

$$\pi_{1,1} = \frac{\sigma}{1-\sigma}\pi_{0,0} \qquad (5)$$

$$\pi_{0,0} = \frac{\sigma q_1}{\sigma(1-q_2) + r(1-\sigma)}\pi_{1,0} \qquad (6)$$

$$Q = \frac{\sigma^2(1-\sigma)q_1}{[r(1-\sigma)-\sigma_2][r(1-\sigma)-\sigma(q_2-q_1)]} \qquad (7)$$

These expressions are derived by straight forward analysis of the user Markov chain. The quantity Q represents the average queue size.

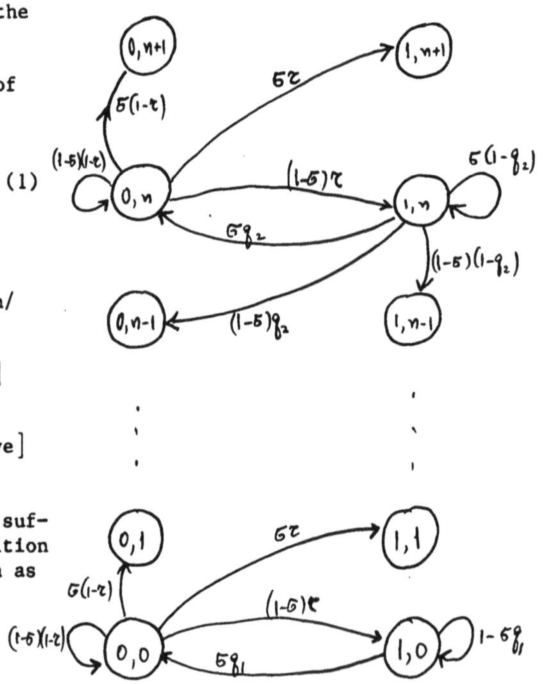

Figure 1

Let us now consider another Markov chain that describes the state of the system. This is described by the number of terminals that are in each of the three state, blocked (B), active (A), and idle (I). Let

$n \triangleq$ number of blocked terminals

$n_1 \triangleq$ number of active terminals

then

$M-(n+n_1) \triangleq$ number of idle terminals

It is important to observe that n_1 can either be 0 or 1. The reason is the following: If there were two or more active terminals at the beginning of a slot, then there should have been two or more successful transmissions during the preceeding slot (an impossibility), since an active terminal attempts transmission with probability one and a terminal becomes active only if it transmits successfully a "new" or an "old" packet in the previous slot. Thus, the state of the system is described by the pair (n_1, n), where $n=0,1,\ldots,M$ if $n_1=0$, and $n=0,1,\ldots M-1$ if $n_1=1$.

The Markov chain is illustrated in Fig. 2 where the transition probabilities can be calculated and are given in the Appendix.

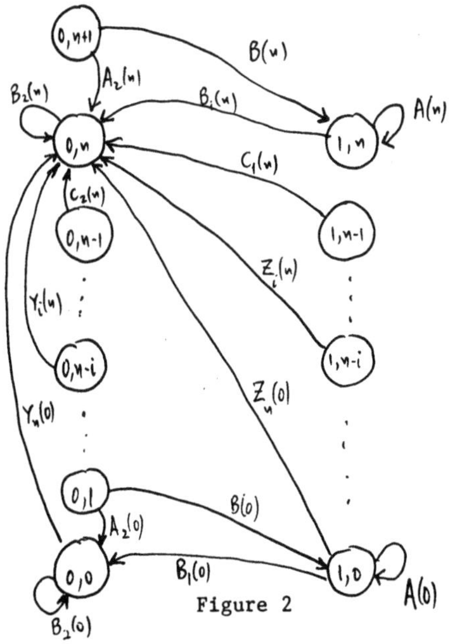

Figure 2

We denote by $P_{n_1,n}$ the steady state probability that the system is in state (n_1,n).

The two chains considered are coupled in the following way.

Note first that

$$G_0(1) = \Pr[\text{a user is blocked}] = \sum_{j=0}^{\infty} \pi_{0j}$$

$$G_1(1) = \Pr[\text{a user is unblocked}] = \sum_{j=0}^{\infty} \pi_{1j}$$

$$\pi_{10} = \Pr[\text{a user is idle}]$$

$$G_1(1)-\pi_{10} = \Pr[\text{a user is active}]$$

and $G_0(1) + G_1(1) = 1$.

By conditioning on the state of the system we obtain:

$$r = \frac{p \sum_{n=1}^{M} P_{0,n}(1-p)^{n-1}(1-\sigma)^{M-n}}{1-P_{0,0}-P_{1,0}} \tag{8}$$

where the expression in the denominator represents the probability that there is at least one blocked user, p in the numerator represents the probability that the blocked user under consideration attempts transmission, $(1-p)^{n-1}$ the probability that no other blocked user transmits and $(1-\sigma)^{M-n}$ the probability that no idle user receives a packet arrival (and, therefore, transmits). Finally $P_{0,n}$ is the probability of state (0,n). Similarly,

$$q_2 = 1-\frac{\sum_{n=0}^{M-1} (1-\sigma)^{M-n-1}(1-p)^n P_{1,n}}{\sum_{n=0}^{M-1} P_{1,n}} \tag{9}$$

Similarly,

$$q_1 = 1-\frac{\sum_{n=0}^{M-1} (1-\sigma)^{M-n-1}(1-p)^n P_{0,n}}{1-P_{0,M}-P_{1,M-1}} \tag{10}$$

Note that, if we don't assume an arrival to the idle terminal under consideration, the probability of successful transmission is given by

$$h = \sigma(1-q_1)$$

Eqs. (2) through (7) provide the values of the quantities that were needed in order to solve the system equations given in Appendix A. However, these quantities are not expressed in terms of only the system parameters M,p,σ, but, in addition, in terms of the success and collision probabilities r,q_1,q_2. These last probabilities were calculated above in terms of the system parameters <u>and</u> the system probabilities $p_{n_1,n}$. Thus, in total, we have a set of simultaneous coupled non-linear equation in r,q_1,q_2 and in $\{p_{n_1,n}\}$. Solving these equations requires computational methods. Assuming the solution is obtained, it is possible to calculate the average waiting time.

To obtain D, the total average delay per packet, we need to obtain W and S, the waiting and service times. As far as W is concerned, it can be computed from Little's formula as

$$W = \frac{Q}{\sigma}$$

From Eq. (7) and Little's result we obtain

$$W = \frac{\sigma(1-\sigma)q_1}{[r(1-\sigma) - \sigma q_2][r(1-\sigma) - \sigma(q_2-q_1)]} \tag{11}$$

For the calculation of the other component of the delay, the service time S, we observe that "holding" times between attempted transmissions are geometrically distributed with parameter p, since we assumed that p is the probability of transmission in any given slot given that the terminal is blocked. Thus, the average "holding time" is equal to p^{-1}. Since r is the average probability of success for a blocked packet and r/p is the probability of the same event given that a transmission is attempted by the blocked terminal we immediately obtain that the conditional average transmission time, given that a first collision occurred when the terminal was either idle or active, is

$$\frac{1}{p\,\frac{r}{p}} = \frac{1}{r}$$

The probability of the conditioning event, namely, that there was a first collision for a terminal that was either idle or active, is given by

$$\Pr\left[\text{terminal is active}\right]\cdot q_2 + \Pr\left[\text{terminal is idle}\right]\cdot q_1$$

$$= \frac{[G_1(1) - \pi_{10}]}{G_1(1)} q_2 + \frac{\pi_{10}}{G_1(1)} q_1$$

Thus

$$S = \frac{q_2 G_1(1) + \pi_{10}(q_1-q_2)}{rG_1(1)} \tag{12}$$

and then

$$D = S + W$$

It is possible to solve these equations numerically and to also optimize with respect to p. Detailed results are supplied in [14]. A recently obtained simulation of the system [15] shows a reasonable agreement with the computed results for certain regions of throughput values, but some disagreement in other regions.

It is important to note that there are no other approximations that we are aware of that produce a better match.

The same modeling approach fares much better in case of controlled interaction between the users. For example in case of a reservation-based access scheme such as the ones described in [16] we can use a similar analysis to the one described here and obtain precise and accurate results. More details of this can be found in [16].

3. ANALYSIS OF THE NO-FEEDBACK CASE

The usual basic assumption in the study of most multiple access protocols is the existence of perfect ternary feed-back. This is to say that every user is able to determine with some delay (which is often assumed to be negligible) whether during a slot there were zero, one, or more than one packets transmitted (this is accomplished by monitoring the channel) and, furthermore, he is able to determine whether his transmissions are successfully received (either via this ternary feedback when noise is negligible or via acknowledgement message mechanisms).

Acknowledgement messages are generally important because they can be used for flow control in addition to their error control function implied so far. Also they are often indispensable in order for the useful ternary feedback assumption to be valid; that is, monitoring of the channel by the transmitting node is in practice insufficient to determine whether successful reception took place.

In several applications acknowledgement messages are not permissible or are not reliable. For example some nodes are passive receivers in order to conceal their location. Other nodes may be very near interfering sources and thus can serve as "transmit-only" nodes. In such cases it is important to devise other schemes for securing the delivery of radio messages. Conceptually it is easiest to start with the assumption that there can be no reliable feedback at all. That is, there are no acknowledgements and there is no channel monitoring. This is the case we want to examine here.

In [17] an information theoretic approach to the problem establishes the feasibility of achieving the maximum possible throughput rate permitted by the basic multiple access protocol used provided there is enough redundancy at the packet level and sufficient tolerance of delay. We do not simply mean use of forward error correction techniques but rather randomized retransmissions of packets or, more generally, transmissions of redundant packets.

Here we focus on specific schemes that employ the above philosophy and are easy to implement, and we seek to establish their performance. We consider three policies. The first one assumes random access, with each packet being retransmitted randomly, again and again, until a fixed period of time W from the time of its initial generation elapses. The second policy is identical to the first one except that the retransmissions continue randomly until a fixed number M of them is completed. Assuming the rescheduling times are independent and exponentially distributed with parameter λ and setting $\lambda W=M$ we can consider an obviously superior third policy that requires the retransmissions to continue either until M are completed or until W seconds pass, whichever occurs first. We want to calculate three quantities for each policy: 1) the channel traffic induced by the policy, 2) the throughput achieved, and 3) the probability of successful delivery of a packet. Let us assume that there are K users, each generating <u>new</u> packets according to a Poisson process of rate λ. The retransmissions continue independently using the same parameter λ.

1) <u>Channel Traffic</u>

a) <u>First Policy</u>: Let us look at one of the K users. Let N_t denote the number of new packets generated by him in the time interval $(0,t]$ and let g_t denote the total number of packet transmissions or retransmissions in $(0,t]$. Then

$$E[dg_t/\text{past until } t] = P[dg_t = 1/\text{past until } t] =$$

$$= \lambda dt + (N_t - N_{t-W})\lambda dt$$

$$E[dg_t] = \lambda dt + \lambda^2 W dt$$

$$g = \frac{[Edg_t]}{dt} = \lambda + \lambda^2 W$$

and

$$G = Kg = K\lambda(1+\lambda W)$$

while the input is only $K\lambda$.

b) <u>Second Policy</u>: Here we need the notion of the number of "active" packets at time t. Let us denote it by N_t'. In the first policy N_t' was simply $N_t - N_{t-W}$. Here it is

$$N_t' = N_t - k$$

where k is defined by

$$\sum_{i=1}^{k} \tau_i \quad t - \frac{M}{\lambda} < \sum_{i=1}^{k+1} \tau_i$$

where τ_i is the ith interarrival time between new packets of the user. Thus

$$E[dg_t/\text{past of } N_t \text{ and of } g_t] =$$

$$P[dg_t = 1/\text{pasts of } N_t \text{ and } g_t] =$$

$$= \lambda dt + N_t'\lambda dt = \lambda dt + (N_t - k)\lambda dt$$

Then

$$E[dg_t/\text{past of } N_t \text{ only}] = N_t \frac{t - \frac{M}{\lambda}}{t}$$

(this is due to the fact that for given N_t the arrival points in $(0,t]$ are uniformly distributed [18]). Finally

$$E[dg_t] = (\lambda + M\lambda)dt$$

and

$$g = \lambda(1+M) \text{ or } G = K\lambda(1+M)$$

c) <u>Third Policy</u>: Choose W and M such that $M = \lambda W$.

Let N_t' = number of packets "born" within the last W seconds.

N_t'' = number of new packets born within the W window that have not completed M retransmissions.

Again

$$E[dg_t/\text{past}] = \lambda dt + N_t''\lambda dt$$

and

$$E[dg_t/\text{past of } N_t \text{ only}] =$$

$$\lambda dt + E[N_t''/\text{past of } N_t]\lambda dt$$

$$= \lambda dt + (N_t' - k)\lambda dt$$

where k = number of packets for which $t - t_i$ $E[X]$ and where t_i is the arrival instant of the (i^{th}) packet and X is the "time-to-death" of an active packet (a r.v. equal to the sum of M i.i.d. exponential variables truncated at the value W). We find that

$$E[X] = Wq_W - \frac{M}{\lambda}\frac{(\lambda W)^M}{M!} e^{-\lambda W} + \frac{M}{\lambda}(1-q_W)$$

where

$$q_W = \int_W^\infty e^{-\lambda w} \frac{(\lambda x)^{M-1}}{(M-1)!} \lambda dx =$$

$$P[\text{less than M attempts in W seconds}] =$$

$$e^{-\lambda W}\left[1+\lambda W + \frac{(\lambda W)^2}{2} + \ldots + \frac{(\lambda W)^{M-1}}{(M-1)!}\right]$$

Thus

$$E[dg_t/N_t] = \lambda dt + N_t \frac{E[x]}{t} \lambda dt$$

and

$$E[dg_t] = \lambda dt + \lambda^2 E[x]dt$$

Combining the above equations we finally obtain

$$G = K\lambda(1+M(1-\frac{M^M}{M!}e^{-M}))$$

2) <u>Throughput</u>

We need an approximation due to the analytical complexity of the precise expressions for point processes with time-varying rates. Namely we assume the packet duration ε to be small relative to $\frac{1}{\lambda}$ (a perfectly reasonable assumption for a large number of users and in order for the self-interference due to overlapping transmissions by the same user to be small). We take ε to be one. Thus λ must be very small (besides, $K\lambda$ must be 1 anyway).

a) <u>First Policy</u>. Given the past history of all transmissions the probability of a successful transmission (or retransmission) by one user at time t is given by

$$(\lambda + \lambda(N_t - N_{t-W}))dt. \quad P[\text{no other user initiates a transmission within } (t-1),\ t+1]].$$

The transmission processes of all users are independent but the probability of zero transmissions in an interval for a process with time-varying rate is given by

$$e^{-\int_{t-1}^{t+1} \lambda_\tau \, d\tau} = e^{-\int_{t-1}^{t+1} (N_\tau - N_{\tau-W})\lambda d\tau}$$

Because of our assumption, $N_\tau - N_{\tau-W}$ is approximately constant for $\tau\varepsilon(t-1,t+1]$. Thus, if by s we denote the throughput for one user, we have

$$sdt = \lambda(1+\lambda W)dt[e^{-2\lambda}E(e^{-\lambda 2(N_t - N_{t-W})}]^{K-1}$$

or

$$s = \lambda(1+\lambda W)[e^{-2\lambda}e^{-\lambda W(1-e^{-2\lambda})}]^{K-1}$$

and for small λ

$$1-e^{-2\lambda} \approx 2\lambda$$

thus

$$s = \underbrace{\lambda(1+\lambda W)}_{g} \; e^{-2\lambda\overbrace{(1+\lambda W)}^{g}(K-1)}$$

For slotted random access we can get rid of the 2 factor throughout. Let

$$S = s.K \text{ and } G = s.k. \quad \text{Thus}$$

$$S = Ge^{-G\frac{K-1}{K}}$$

which yields

$$S_{max} = \frac{K}{K-1} e^{-1} \text{ for } G_{max} = \frac{K}{K-1}$$

and optimum window size

$$W_0 = \frac{\frac{1}{(K-1)\lambda} - 1}{\lambda}$$

for any $\lambda \frac{1}{K-1}$. Since a successful transmission may be one of a previously successfully transmitted packet we must adjust the value of throughput to its "real" level by dividing by $1+\lambda W$. Thus

$$S_{real} = K\lambda e^{-(K-1)\lambda(1+\lambda W)}$$

which is always (for any λ) inferior to the level achieved by plain random access ($W=0$ and $S=K\lambda e^{-(K-1)\lambda}$ as opposed to $K\lambda e^{-1}$). The explanation of the apparent paradox is that the "real" throughput is measured as rate of "true" successes in a given time period, while this policy involves possible successes later on, after, sometimes considerable, retransmission delay. We shall see that the probability of success of a <u>given</u> packet increases with our policy. Thus we will confirm the conceptual result from information theory that guarantees successful deliveries up to a maximum rate but with increasing delay.

b) <u>Second Policy</u>. The approach is exactly as before. Because of our approximation we can reduce the result to

$$sdt = \lambda(1+M)dt \; e^{-\lambda(1+M)(K-1)}$$

and

$$S = Ge^{-G\frac{K-1}{K}}$$

with

$$S_{max} = \frac{K}{K-1} e^{-1} \text{ for } G_{max} = \frac{K}{K-1} \text{ and}$$

$$M_0 = \frac{1}{(K-1)\lambda} - 1$$

Again, the determination of the real throughput requires an adjustment and yields

$$S_{real} = K\lambda e^{-\lambda(1+M)(K-1)}$$

which is again worse than plain random access without retransmissions. Of course the same explanation applies here as before.

(c) <u>Third Policy</u>. For $\lambda W = M$ the expressions for S and S_{real} are exactly as before except that G is given by the appropriate expression corresponding to the third policy.

3) Success Probability

Obviously retransmission methods must enhance the chances of the successful transmission of a given packet if overloading of the channel is avoided.

We can confirm this expectation by analyzing the proposed policies.

a) First Policy. First note that with plain random access (W=0) the success probability is $e^{-\lambda(K-1)}$. For W 0 we calculate it by conditioning on the number x of retransmissions afforded within the fixed window W. Obviously x is a Possion random variable.

$$P[\text{success}] = 1-P[\text{all collisions}] = 1 - \sum_{x=0}^{\infty} P[x \text{ retransmissions}].$$

$$P\left[\begin{matrix}\text{collision in all x and in} \\ \text{original transmission}\end{matrix}\right] =$$

$$1 - \sum_{x=0}^{\infty} \frac{(\lambda W)^x}{x!} e^{-\lambda W} \cdot \rho^{x+1}$$

where

$$\rho = \Pr[\text{collision in single try}] = 1 - e^{-\lambda(1+\lambda W)(K-1)}$$

After performing the summation we find

$$P[\text{success}] = 1 - \rho e^{-\lambda W(1-\rho)}$$

We can easily determine that this expression yields a better value (for $\lambda \ \frac{1}{K-1}$) than the one for W = 0.

b) Second Policy. A similar calculation as before yields

$$P[\text{success}] = 1 - (1-e^{-\lambda(1+M)(K-1)})^{M+1}$$

which, again, is greater than $e^{-\lambda(K-1)}$

c) Third Policy. The actual calculation here is much more complex than before. However if M=λW we can compare the first and the second policies. This amounts to comparing

$$(1 - e^{-\lambda(K-1)(M+1)})^M \text{ vs } e^{-Me^{-\lambda(K-1)(M+1)}}$$

$$\text{or} \quad (1-x^M)^M \text{ vs } e^{-Mx^M}$$

for $x = e^{-\lambda(K-1)}$. If λ is very small (as assumed), x is close to one and then we can verify that the quantity on the left is less than the one on the right and thus the second policy is preferable. Of course detailed study of these curves can establish what happens for higher values of λ and whether there is a crossing of the two curves as M increases.

Of course it is of interest to explore other retransmission policies as well and to incorporate in their study the effect of some form of limited feedback as well as that of other phenomena such as the capture phenomenon.

4. CONCLUSION

We reviewed some approximate techniques for the analysis of interacting queueing systems. We focused on one technique that shows a limited success that is typical of all such approximations.

Then we reviewed the case where, due to the lack of coupling between queueing systems, analysis is successful. In both cases we believe there is an illustration of the factors that are involved in the conception of a protocol of multiple access, its analysis and its eventual validation.

APPENDIX

Consider first the following Lemma.

Lemma

the conditional probability that the

buffer is empty given that the user is blocked

= the conditional probability that the buffer size exceeds one given that the user is Active

Proof:

The left hand side of the above equals to $\frac{\pi_{00}}{G_0(1)}$ and the R.H.S. equals to $\frac{\pi_{11}}{G_1(1)-\pi_{10}}$. By using Eqs. [2-6], the Lemma is proved.

Now let

$$g_i(n) \triangleq \Pr\,[\text{i out of n blocked users attempt transmission}]$$

$$= \binom{n}{i} p^i(1-p)^{n-i}$$

$$q_j(n+n_1) \triangleq \Pr[\text{j out of } M-(n+n_1) \text{ idle users attempt transmission}]$$

$$= \binom{M-(n+n_1)}{j} \sigma^j(1-\sigma)^{M-(n+n_1)-j}$$

and $E \triangleq \frac{\pi_{00}}{G_0(1)} = \frac{\pi_{11}}{G_1(1)-\pi_{10}}$

Referring to Fig. 2 the transition probabilities to (0,n), 0 n M are as follows:

$B_1(n)$ = Pr [only the active terminal (with buffer contents = 1) transmits]

$= q_0(n+1)g_0(n)E$

$C_1(n)$ = Pr [the active terminal and one or more blocked ones transmit]

$= q_0(n)[1-g_0(n-1)] \qquad 2<n$

$Z_i(n)$ = Pr [the active terminal and packets arrive at (and are, therefore, transmitted by) i-1 idle terminals] =

$q_{i-1}(n-i+1) \qquad 2<i<n, \quad 2<n$

$A_2(n)$ = Pr [only one blocked terminal (with zero contents in the buffer) transmits]

$= q_0(n+1)g_1(n+1)\ E$

$B_2(n)$ = Pr [either none or at least two blocked terminals transmit or no blocked but one idle transmit]

$= q_0(n)[1-g_1(n)] + q_1(n)g_0(n)$

$C(n)$ = Pr [one idle terminal and at least one blocked terminal transmit]

$= q_1(n-1)[1-g_0(n-1)]$

$Y_i(n)$ = Pr [packets arrive at i (1) idle terminals]

$= q_i(n-i) \qquad 2<i<n, \quad 2<n$

We now determine the transition probabilities to (1,n), 0<n

$B(n)$ = Pr [only the blocked terminal (with buffer contents 0) transmits]

$= q_0(n+1)g_1(n+1)[1-E]$

$A(n)$ = Pr [only the active terminal (with buffer contents 1) transmits]

$= q_0(n+1)g_0(n)[1-E]$

Finally, we look at the end states:

$B_1(0) = q_0(1)E$

$A_2(0) = q_0(1)g_1(1)E$

$B_2(0) = q_0(0) + q_1(0)$

To (0,0): $B(0) = q_0(1)g_1(1)[1-E]$

$A(0) = q_1(1)g_0(0)[1-E]$

The equations for this chain are:

$$P_{1,0} = A(0)P_{1,0} + B(0)P_{0,1}$$

$$P_{1,n} = A(n)P_{1,n} + B(n)P_{0,n+1}$$

$$1 \leq n \leq M-1$$

$$P_{0,n} = A_2(n)P_{0,n+1} + B_2(n)P_{0,n} + C_2(n)P_{0,n-1} + \sum_{i=2}^{n} Y_i(n)P_{0,n-i} + B_1(n)P_{1,n} + C_1(n)P_{1,n-1} + \sum_{i=2}^{n} Z_i(n)P_{1,n-i}$$

$$2 \leq n \leq M-1$$

$$P_{0,1} = A_2(1)P_{0,2} + B_2(1)P_{0,1} + B_1(1)P_{1,1}$$

$$P_{0,0} = A_2(0)P_{0,1} + B_2(0)P_{0,0} + B_1(0)P_{1,0}$$

We designate these equations as the system equations.

REFERENCES

1. Abramson, N., "The ALOHA System - Another Alternative for Computer Communications". AFIPS Conference Proceedings, 1970 Fall Joint Computer Conference, Vol. 37, 281-285.

2. Capetanakis, J.I., "Tree Algorithms for Packet Broadcast Channels", IEEE Trans. on IT, Vol. 25, No. 5, Sept. 1979, pp. 505-515.

3. Fayolle, G., E. Gelenbe and J. Labetoulle, "Stability of Optimal Control of the Packet Switching Broadcast Channel", JACM, Vol. 24, No. 3, July 1977, pp. 375-386.

4. Kamal, S.S., and S.A. Mahmoud, "A Study of User's Buffer Variations in Random Access Satellite Channels", IEEE Trans. on Communications, Vol. 27, No. 6, June 1979, pp. 857-868.

5. Kleinrock, L., and F.A. Tobagi, "Packet Switching in Radio Channels: Part I - Carrier Sense Multiple-Access Modes and Their Throughput-Delay Characteristics," IEEE Transactions on Communications, Vol. COM-23, 1400-1416, 1975.

6. Tobagi, F., "Multiaccess Protocols in Packet Communication Systems", IEEE Trans. on Communications, Vol. 28, pp. 468-488, April 1980.

7. Wieselthier, J.E. and Ephremides, A., "A New Class of Protocols for Multiple Access in Satellite Networks", IEEE Transactions on Automatic Control, October 1980.

8. Lam, S.S., "Packet Broadcast Networks - A Performance Analysis of the R-ALOHA Protocol", IEEE Transactions on Computers, Vol. C-29, No. 7, July 1980.

9. Balagangadhar and R.L. Pickholtz, "Analysis of a Reservation Multiple Access Technique for Data Transmission Via Satellites", IEEE Trans. on Communications, Vol. COM-27, No. 10, October 1979.

10. Fayolle, G. and Iasnogorodski, "Two Coupled Processors: The Reduction to a Riemann-Hilbert Problem" Wahrscheinlichkeits Theorie, Springer-Verlag, 1979.

11. Leibowitz, M.A., "An Approximate Method for Treating a Class of Multiqueue Problems," IBM Journal of Research and Development, Vol. 5, No. 3, 204-209, July 1981.

12. Hashida, O. and K. Ohara, "Line Accomodation Capacity of a Communication Control Unit", Review of the Electrical Communication Laboratories, Vol. 20, No. 3-4, March-April, 1972.

13. Hashida, O., "Analysis of Multiqueue", Review of the Electrical Comm. Lab., Nippon Telegraph and Telephone Public Corporation, Vol. 20, No. 3-4, March-April, 1972.

14. Saadawi, T.N. and A. Ephremides, "Analysis, stability and Optimization of Slotted ALOHA with a Finite Number of Buffered Users", IEEE Trans. on AC, June, 1981.

15. J. Massey, R. Rueppel, J. Ruprecht, Private Communication, September 1982.

16. Saadawi, T.N. and A. Ephremides, "Analysis of the Tree Algorithm with a Finite Number of Buffered Users", Proceedings ICC, June, 1981.

17. D. Cohn, "Redundant Packets in Multiple Access Systems", International Symposium on Information Theory, les Arcs, France, June, 1982.

18. D. Snyder, Random Point Processes, Wiley, 1975.

PERFORMANCE ANALYSIS OF A LINK LEVEL PROTOCOL FOR PACKET SWITCHING NETWORKS COMBINING TWO DIFFERENT RETRANSMISSION STRATEGIES

G. Wieber

Institut für Netzwerk- und Signaltheorie
Fachbereich Regelungs- und Datentechnik
Technische Hochschule Darmstadt
Germany

ABSTRACT

In this paper, a link level protocol combining two basic protocols is analyzed through a product-form queueing network. The basic protocols differ in the retransmission scheme implemented to correct transmission errors due to the loss of transmitted packets or returning acknowledgements. One strategy allows retransmission of packets independent of their original sequence, whereas the other one maintains the packet order. Both alternatives are analyzed through exact or approximate Markov queueing systems. It turns out, that retransmission of erroneously transmitted packets can be interpreted as either an errorfree transmission over a link of reduced capacity or a transmission of packets arriving with an increased arrival rate. The compound model combines the two basic retransmission schemes within one system. Its analysis yields explicit results for the average transmission time and the probability distribution of the number of packets in the station. Both results depend on important design or system parameters as timeout delay, mean transmission rate, and error probability. A numerical evaluation comparing the different schemes is presented showing the influence of the parameters on the system performance. Some remarks on buffer design criteria are also included.

Key words: packet switching network, link level, retransmission scheme, queueing system, product-form queueing network, packet sequence, timeout

1. INTRODUCTION

In packet switching networks transmission errors between neighboring nodes are corrected by the retransmission of copies of the packets assumed to be lost. For that reason acknowledgements are exchanged either confirming the correct reception of sent packets or informing the transmitter of detected errors (optional).

There are two basic strategies one may think of. In the first one the sequence order of packets is not taken into account, whereas in the second one it is strictly maintained. According to the network performance both schemes exhibit one inherent advantage as well as one disadvantage. In strategy 1 only the lost frames have to be retransmitted, whereas in scheme 2 the additional retransmission of all packets, sent after the first one not acknowledged in time, is required. Therefore the packet sequence is maintained in the latter case, whereas it is generally lost in the first [1].

Thus it might be of considerable interest to study a link protocol which combines the advantages and avoids the disadvantages. Single packet messages should be allowed to travel through the network independently, whereas long user messages, entering the network as consecutive packets, should traverse the subnet in a way that preserves their original sequence. In order to meet different traffic patterns, none of the basic protocols should be implemented solely, but a combination of both.

It is the aim of this paper to present Markovian queueing network models for two retransmission strategies and to combine them in a compound model. Based on the system parameters analytic expressions for the mean transmission time and the probability distribution of the number of packets in a station are derived. In section 2, queueing networks for the two basic protocols are presented and analyzed. The transmission queues are represented by queueing systems of the M/M/1 and M/G/1 type, respectively.

The M/G/1 system of model 2 is gained by introducing virtual transmission times [2]. It is then approximated by a M/M/1 system with identical mean service time. The approximate model is interpreted as a feedback system with increased arrival and service rate. In section 3, a combination of the two former models is derived yielding the queue length distribution and average transmission time. In section 4, finally, some numerical results are presented showing the influence of some of the parameters on the system performance. The results give an idea of how the transmission speed may be influenced if individual packets are allowed to travel through the network independently.

In the subsequent sections the following notation will be used:

λ, λ_i, $\tilde{\lambda}$, $\tilde{\lambda}_i$	mean arrival rate of information packets, $[\lambda] = \frac{\text{packets}}{\text{sec}}$ (The subscripts refer to the strategy applied.)
μ, μ_i, $\tilde{\mu}$	mean service rate, $[\mu] = \frac{\text{packets}}{\text{sec}}$
ρ	utilization factor, $\rho = \frac{\lambda}{\mu}$
p, p_i, p_i'	transmission error probability
$P(n)$, $P(N)$	probability of finding n customers in the send queue or N customers in the transmitter, respectively
$\overline{N}$	mean number of customers in the transmitter
T_0	constant timeout time ($M/D/\infty$ system)
T_a	random variable of the service time in the $M/G/\infty$ acknowledgement system
$\overline{T}_a$	mean acknowledgement time
$B(s)$, $S(s)$, $H(s)$	generating functions of service time, transmitting time and holding time density, respectively.

In order to arrive at mathematically tractable models, some assumptions are necessary:

Interarrival and transmission times are exponentially distributed. Errors occur independently of each other and do not depend on the packet length. Erroneous packets may be repeated indefinitely. All queues are assumed to be stable ($\rho < 1$). No explicit window mechanism is taken into account limiting the number of unacknowledged packets. The modulus of the packet sequence numbers is chosen large enough, so that a possible suspension of transmissions caused by a lack of available numbers can be neglected. All models are described from node A's point of view.

2. MODELING AND ANALYSIS OF TWO RETRANSMISSION STRATEGIES

In this section, two different protocols are described. Two neighboring nodes are connected through a full duplex channel. Because of the symmetry of the given connection, it suffices to describe the models from node A's point of view.

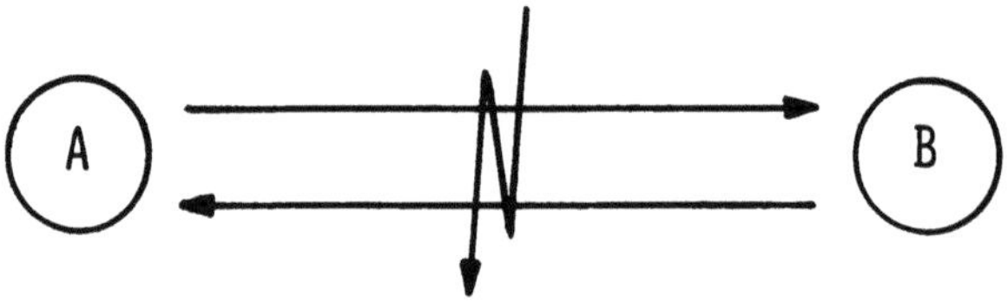

Fig.1 Erroneous transmission channel

Packets are assumed to arrive at the transmission facility according to a Poisson process with mean arrival rate λ. They are put into the send queue waiting for transmission. In order to cope with transmission errors, copies of the sent packets are kept in the transmitting station. Retransmission will take place after an error has been realized. Successfully received packets are acknowledged by the receiver via returning frames. If, after expiry of a certain amount of time, a packet has not been acknowledged, the transmitter will initiate its retransmission.

In the first model presented, packets are corrected individually, whereas in model 2 the packet sequence is maintained.

Model 1

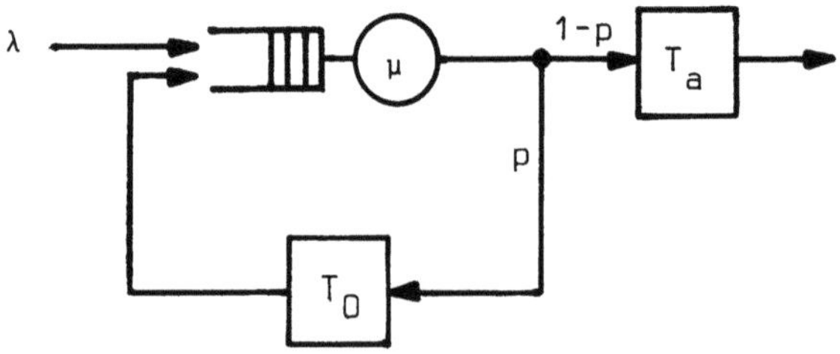

Fig.2 Non-sequential packet retransmission scheme

In the 3 station BCMP [3] queueing network of fig. 2, the M/M/1 system represents the send queue where packets are waiting to be transmitted, transmission rate μ. If with probability 1-p a packet is transmitted correctly, a positive acknowledgement will be received after time T_a (M/G/∞ system). The buffer containing a copy of the acknowledged packet will then be released. If, on the other hand, an error occures (probability p), the packet under consideration will move into the timeout station (M/D/∞) representing the time until the transmitter will know about the loss. After T_0 the timeout expires. The packet returns to the sending system to be transmitted again [4,5]. Markov queueing network analysis yields the following well known results. (See appendix I for a derivation of eq. (2.1).)

$$P(N)=(1-\frac{\tilde{\lambda}}{\mu})e^{-\tilde{\lambda}(pT_0+(1-p)\overline{T}_a)} \cdot (\frac{\tilde{\lambda}}{\mu})^N \cdot \sum_{k=0}^{N} \frac{[\mu(pT_0+(1-p)\overline{T}_a)]^k}{k!} \tag{2.1}$$

$\tilde{\lambda}$ is defined as follows

$$\tilde{\lambda} = \frac{\lambda}{1-p} \tag{2.2}$$

The mean number of packets yields

$$\overline{N} = \frac{\tilde{\lambda}}{\mu-\tilde{\lambda}} + p\tilde{\lambda}T_0 + \lambda\overline{T}_a \tag{2.3}$$

Assuming small error probabilities the distribution of the transmission time for a successful transmission and the transmitter holding time can approximately be computed through generating functions.

Transmission time

$$S(s) = (1-p)S_s(s) + p(1-p)S_s^2(s)T_0(s) + \dots$$

$$= \frac{(1-p)S_s(s)}{1-pS_s(s)T_0(s)} = \frac{\mu(1-p)-\lambda}{s+[\mu(1-p)-\lambda]\cdot\frac{1-p\cdot e^{-sT_0}}{1-p}} \tag{2.4}$$

Mean Transmission time

$$\overline{T} = \frac{1}{\mu(1-p)-\lambda} + \frac{p}{1-p} T_0 \tag{2.5}$$

Variance

$$\sigma_t^2 = \overline{T}^2 + \frac{p}{1-p} T_0^2 \tag{2.6}$$

Equivalently, the following relations hold:

Holding time

$$H(s) = S(s) \cdot B_a(s) \tag{2.7}$$

Mean holding time

$$\overline{H} = \overline{T} + \overline{T}_a \tag{2.8}$$

Variance

$$\sigma_h^2 = \sigma_t^2 + \sigma_a^2 \tag{2.9}$$

It should be mentioned that the blocking of new-arriving packets due to a lack of free buffers can easily be taken into account by means of a closed queueing network. The analysis of a model limited to M buffers is very similar to that presented here. For a calculation see Appendix II.

Model 2

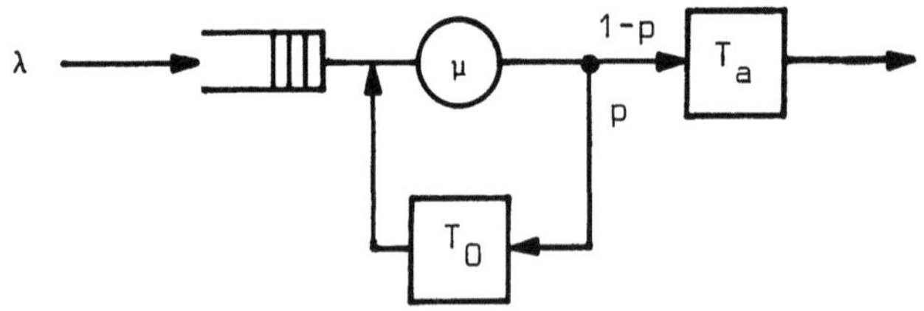

Fig.3 Packet retransmission scheme preserving the packet order

Since in this scheme the order of packets shall be maintained, packets sent in the sequel of an erroneous frame have to be retransmitted, too. By introducing virtual transmission times, an equivalent server for the sender queueing system can be derived.

Necessary retransmissions of subsequent packets are interpreted as an additional portion of the transmission time of the first erroneously transmitted packet. Thus, the service time distribution of the transmitter queueing system can be represented by a Cox-like station [6].

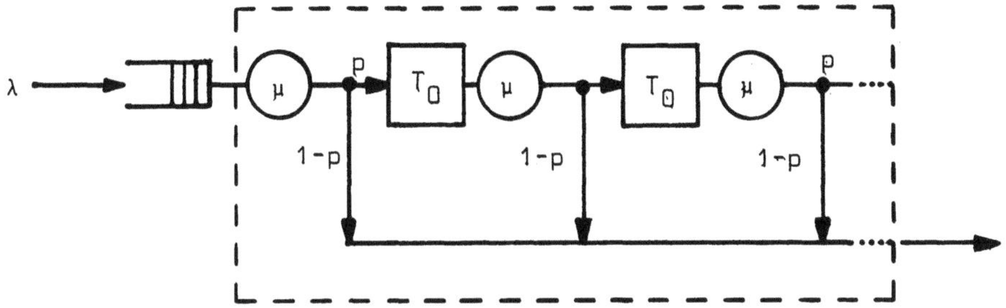

Fig.4 Cox-like representation of virtual transmission times

The analysis yields (M/G/1 system, [7]):

Laplace-transform of the service time density

$$B_s(s) = (1-p)B(s) + p(1-p)B(s)^2 T_0(s) + \ldots$$

$$B_s(s) = \frac{\mu(1-p)}{s+\mu(1-p\cdot e^{-sT_0})} \quad (2.10)$$

Mean service time

$$\bar{T}_s = \frac{1 + p\mu T_0}{\mu(1-p)} \quad (2.11)$$

Variance

$$\sigma_s^2 = \bar{T}_s^2 + \frac{p}{1-p} T_0^2 \quad (2.12)$$

Squared coefficient of variation (SCV)

$$c_s^2 = 1 + p \frac{(1-p)\mu^2 T_0^2}{(1+p\mu T_0)^2} \qquad (2.13)$$

Apparently, the error probability p leads to a reduction of the average transmission rate. For a SCV not much greater than 1, i.e.

$$p \frac{(1-p)\mu^2 T_0^2}{(1+p\mu T_0)^2} << 1$$

the service time is assumed to be approximately exponentially distributed with

$$\tilde{\mu} = \mu \cdot \frac{1-p}{1+p\mu T_0} \qquad (2.14)$$

The generating function for the service time density yields

$$B_s(s) = \frac{\tilde{\mu}}{s+\tilde{\mu}} \qquad (2.15)$$

To describe the behavior of model 2 more accurately, the resulting M/M/1 sending system is replaced by an equivalent feedback station.

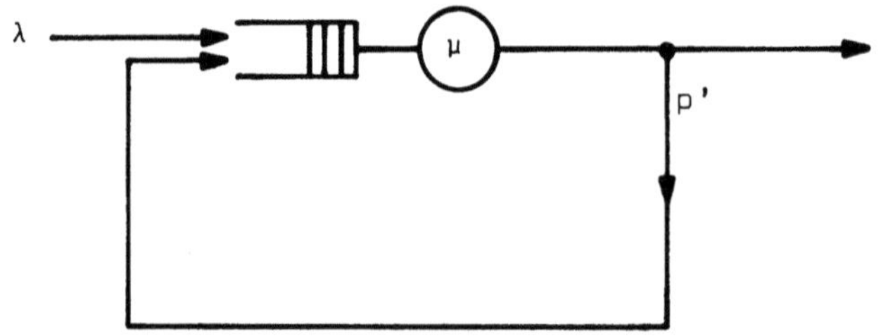

Fig.5 M/M/1 feedback system representation of model 2

The number of necessary repetitions is now determined by a new and greater error probability. p' may be computed easily, if equal utilization factors for the original and the feedback system are required. For

$$\frac{\lambda}{\tilde{\mu}} = \frac{\lambda}{\mu(1-p')}$$

p' yields

$$p' = p \cdot \frac{1+\mu T_0}{1+p\mu T_0} \tag{2.16}$$

The probability distribution of the queue length and the generating functions of the transmission and holding time density may then be computed as follows ($\tilde{\lambda} = \frac{\lambda}{1-p'}$):

Number in system (c.f. (2.1))

$$P(N) = (1-\frac{\tilde{\lambda}}{\mu})e^{-\lambda \bar{T}_a} \cdot (\frac{\tilde{\lambda}}{\mu})^N \cdot \sum_{k=0}^{N} \frac{(\mu(1-p')\bar{T}_a)^k}{k!} \tag{2.17}$$

Transmission time

$$S(s) = \frac{\mu(1-p')-\lambda}{s+\mu(1-p')-\lambda} \tag{2.18}$$

Holding time

$$H(s) = S(s) \cdot B_a(s) \tag{2.19}$$

Despite of different retransmission strategies, the two models finally exhibit a similar structure. In both cases transmission errors lead to an additional number of transmissions, which obviously reduce the total service capacity of the network.

3. PERFORMANCE ANALYSIS OF THE COMPOUND MODEL

In this section a protocol is analyzed, which, at the same time, permits the packets to be transmitted either according to strategy one or according to strategy two. The model combining the two retransmission schemes is called compound model.

Model 3

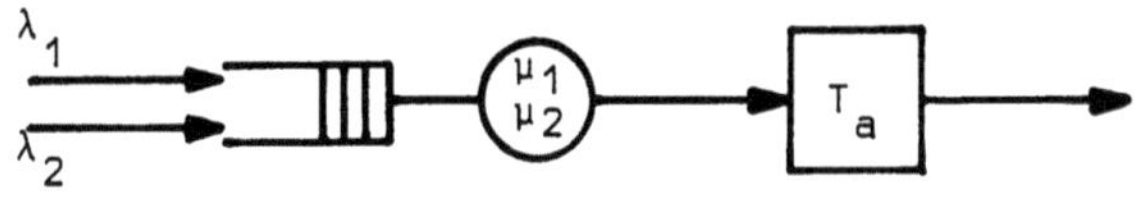

Fig.6 Compound model

The two packet streams are assumed to be independent and to behave according to the particular basic model. Packets of type 1 are retransmitted individually, whereas packets of type 2 maintain their original order. Assuming the same exponentially distributed length for both types of packets, model 3 also becomes a Markovian queueing network, this time, however, with two classes of customers.

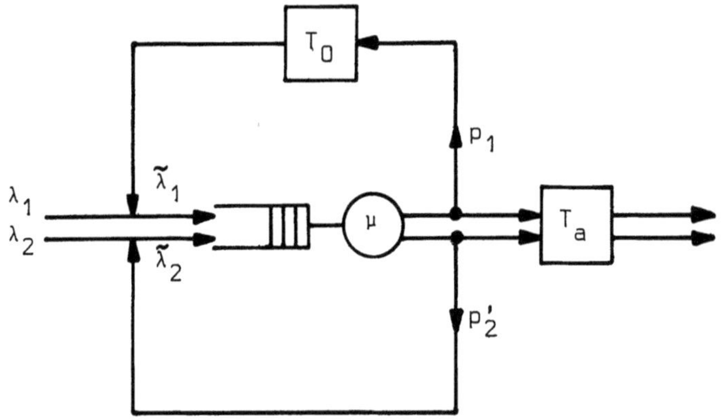

Fig.7 Queueing network representation of model 3

Before the model can be analyzed, an expression for p_2' has to be evaluated. Because of the interference of type 1 packets, p_2' is different from p' derived for model 2. In case of model 2, the term $\frac{1-p}{1+p\mu T_0}$ represented the blocking of new transmissions during a timeout interval, c.f. eq. (2.16). This is no longer true in model 3, since type 1 packets may leave the sender correctly, although a type 2 packet is presently in the timeout system. The factor, which represents the increase of transmissions, can be computed as follows. Every time, a type 2 packet waits for retransmission in the timeout system, on the average $\tilde{\lambda}_1 T_0$ customers of type 1 will leave the send queue consuming an average amount of virtual transmission time of length $\frac{\tilde{\lambda}_1 T_0}{\mu}$. This time, used for transmissions of type 1 packets, therefore has to be subtracted from T_0.

Thus, with an actual service rate

$$\mu_2 = \mu \frac{1-p_2}{1+p_2(\mu-\tilde{\lambda}_1)T_0} \tag{3.1}$$

the virtual error probability p_2' yields

$$p_2' = p_2 \frac{1+(\mu-\tilde{\lambda}_1)T_0}{1+p_2(\mu-\tilde{\lambda}_1)T_0} \tag{3.2}$$

The queueing network is now analyzed easily. Letting

$$\tilde{\lambda}_1 = \frac{\lambda_1}{1-p_1}, \quad \tilde{\lambda}_2 = \frac{\lambda_2}{1-p_2'},$$

$$\tilde{\lambda} = \tilde{\lambda}_1 + \tilde{\lambda}_2, \text{ and } \lambda = \lambda_1 + \lambda_2$$

the following results are obtained.

Number of customers in the system (c.f. appendix I)

$$P(N) = (1-\frac{\tilde{\lambda}}{\mu}) \cdot e^{-(\lambda \overline{T}_a + \frac{p_1\lambda_1}{1-p_1}T_0)} \cdot (\frac{\tilde{\lambda}}{\mu})^N \cdot \sum_{k=0}^{N} \frac{[\frac{\mu}{\tilde{\lambda}}(\lambda \overline{T}_a + \frac{p_1\lambda_1}{1-p_1}T_0)]^k}{k!} \tag{3.3}$$

Mean number of customers

$$\overline{N} = \frac{\tilde{\lambda}}{\mu - \tilde{\lambda}} + \lambda \overline{T}_a + \frac{p_1 \lambda_1}{1 - p_1} \cdot T_0 \qquad (3.4)$$

Mean holding time

$$\overline{H} = \frac{\overline{N}}{\lambda} = \frac{1}{\lambda} \cdot \frac{\tilde{\lambda}}{\mu - \tilde{\lambda}} + \overline{T}_a + \frac{p_1}{1 - p_1} \cdot \frac{\lambda_1}{\lambda} \cdot T_0 \qquad (3.5)$$

Except for the virtual transmission time of type 2 packets, which was approximated by an exponential distribution, the given expressions are exact. Thus, a mathematically tractable model has been derived, which permits two different retransmission schemes to be combined within one protocol.

4. RESULTS

The compound model of section 3 comprises the features of the three models presented. Therefore, only the results obtained from model 3 are discussed in this section. The most fundamental equations are the mean number of packets in the system (with and without the acknowledgement station) and the mean holding time. Assuming equal error probability for both retransmission schemes, $p_1 = p_2 = p$, and introducing a new service rate,

$$\tilde{\mu} = \mu \cdot \frac{\lambda}{\tilde{\lambda}}$$

equations (3.4) and (3.5) may be written as follows:

$$\overline{N} = \frac{\lambda}{\tilde{\mu}-\lambda} + \frac{p}{1-p}\lambda_1 T_0 + \lambda\overline{T}_a \tag{4.1}$$

$$\overline{H} = \frac{1}{\tilde{\mu}-\lambda} + \frac{p}{1-p}\cdot\frac{\lambda_1}{\lambda}T_0 + \overline{T}_a \tag{4.2}$$

For $q = \frac{\lambda_1}{\lambda}$, $\tilde{\mu}$ yields

$$\tilde{\mu} = \mu \cdot \frac{1-p}{1+p\mu T_0(1-q)(1-q\cdot\frac{p}{1-p})} \tag{4.3}$$

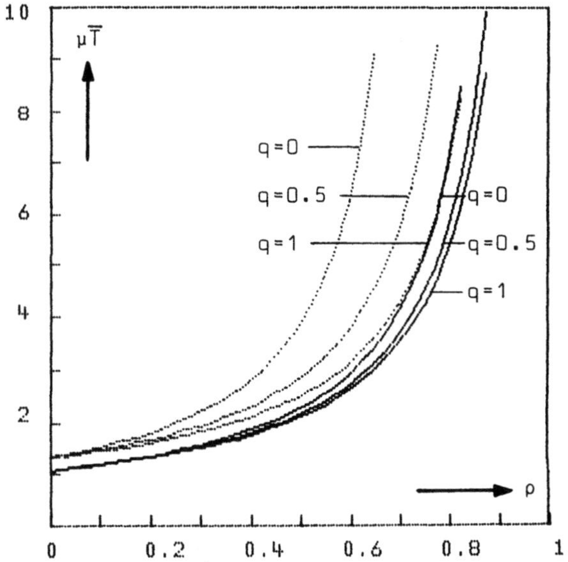

Fig.8 Mean transmission time versus utilization factor (μT_0=5; p=0.01 —— ; p=0.05 ······)

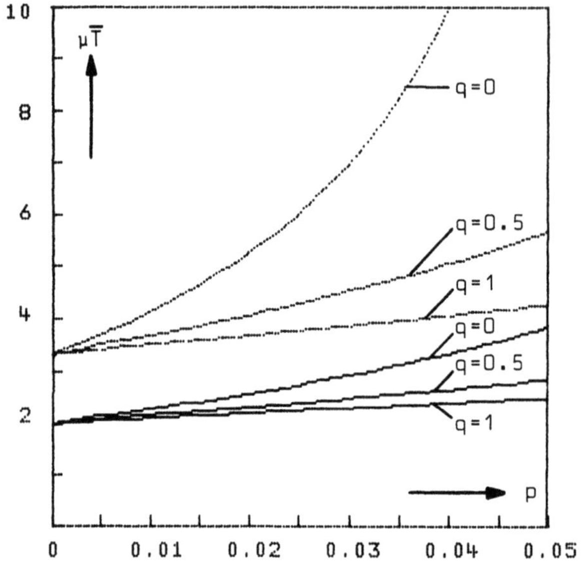

Fig.9 Mean transmission time versus error probability ($\mu T_0=5$; $\rho=0.5$ —; $\rho=0.7$ —)

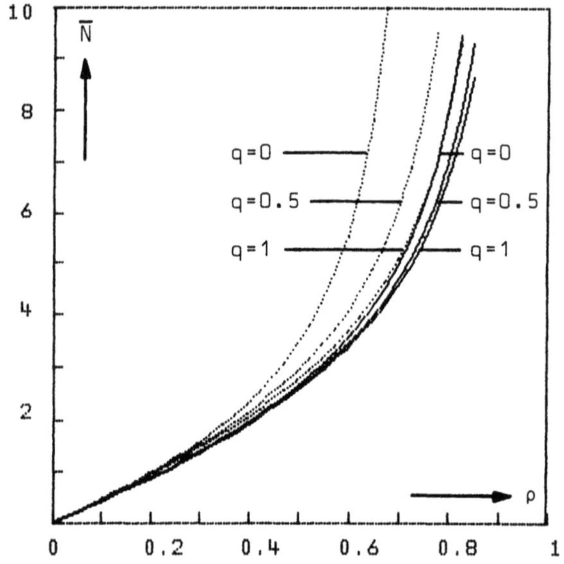

Fig.10 Mean number of packets versus utilization factor ($\mu T_0=5$; $\mu \overline{T}_a=3$; $p=0.01$ —; $p=0.05$ ········)

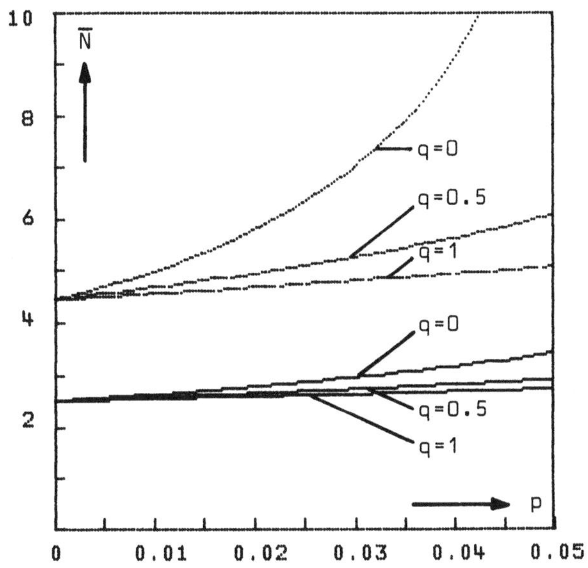

Fig.11 Mean number of packets versus error probability (μT_0=5; $\mu \bar{T}_a$=3; ρ=0.5 ——; ρ=0.7 ········)

For the choice of the parameters T_0 and p, see appendices III und IV.

The main conclusions are as follows:

The retransmission scheme influences the packet transmission time and the buffer utilization of the transmitter. Individual retransmission always causes less delay and leads to a smaller average number of packets in a node. However, if the system under consideration exhibits only small error probabilities and is moderately loaded, the differences are small, so that either scheme can be chosen with small consequence on the system performance.

The decision, which scheme to implement, therefore can be based on different criteria. For instance, in a network with a great portion of 'Virtual Circuit' traffic model 2 should be preferred, because possible recording at the destination node can thus be avoided. Moreover, in a case like this, each scheme can be approximately modeled by any of the three models presented. Therefore the model best adapted to mathematical analysis may be chosen.

5. CONCLUSIONS

Two link transmission protocols are combined in order to permit packets either to be sent in sequence or independently of each other. Based on a mathematically tractable model explicit results are given for the buffer occupation in the sending node and the mean transmission time. The results show the influence of the most important system parameters - timeout delay, packet loss probability, and packet type ratio - on the average system performance. It is also shown, that for lightly loaded systems and small error probabilities any model may be replaced by any other one without changing the performance essentially.

ACKNOWLEDGEMENT

I would like to thank Prof. Spaniol and Prof. Tzschach for many helpful discussions and their constructive suggestions.

REFERENCES:

[1] D. Davies, D. Barber, W. Price and C. Solomonides,
Computer networks and their protocols
(John Wiley Interscience), 1979

[2] W. Bux, K. Kümmerle, H.L. Truong
Balanced HDLC procedures: A performance analysis
IBM Report, Oct. 1979

[3] F. Baskett, K. Chandy, R. Muntz and F. Palacios
Open, closed and mixed networks of queues with different classes of customers
J. ACM 22 (1975)

[4] M. Buttò, G. Colombo, A. Tonietti
Packet network performance analysis
CSELT Rapporti tecnici, Vol. IX, N. 1, Feb. 1981

[5] H. Besier, P. Heuer, G. Kettler, H. Willie
Verkehrstheoretische Untersuchungen von elektronischen Datenvermittlungssystemen
Mitteilungen aus dem Forschungsinstitut der Deutschen Bundespost, Heft 5, 1980

[6] H. Kobayashi
Modeling and Analysis
(Addison-Wesley), 1978, p. 190

[7] G. Pujolle, O. Spaniol
Modeling and Evaluation of Several Internal Network Services
Performance Evaluation, Vol. 1, No. 1, Jan. 1981

APPENDIX

I. Derivation of the probability distribution P(N) of model 1, c.f. eq. (2.1).

n_s, n_t, n_a represent the number of packets in the sending system, the timeout system, and the acknowledgement system, respectively.

$$P(N) = \sum_{\substack{\forall\, n_s, n_t, n_a \\ (n_s+n_t+n_a=N)}} P(n_s) \cdot P(n_t) \cdot P(n_a) \tag{I.1}$$

$$= \sum \left(1-\frac{\lambda}{\mu(1-p)}\right)\left(\frac{\lambda}{\mu(1-p)}\right)^{n_s} \cdot e^{-\frac{p\lambda}{1-p}T_0} \cdot \frac{\left(\frac{p\lambda}{1-p}T_0\right)^{n_t}}{n_t!} \cdot$$

$$\cdot\, e^{-\lambda\overline{T}_a} \cdot \frac{(\lambda\overline{T}_a)^{n_a}}{n_a!}$$

$$= \left(1-\frac{\tilde{\lambda}}{\mu}\right) e^{-\tilde{\lambda}(pT_0+(1-p)\overline{T}_a)} \cdot \sum_{\substack{\forall\, n_s, k \\ (n_s+k=N)}} \left(\frac{\tilde{\lambda}}{\mu}\right)^{n_s} \cdot$$

$$\cdot \sum_{n_t=0}^{k} \frac{(p\tilde{\lambda}T_0)^{n_t}}{n_t!} \cdot \frac{(\lambda\overline{T}_a)^{k-n_t}}{(k-n_t)!}$$

$$= \left(1-\frac{\tilde{\lambda}}{\mu}\right) e^{-\tilde{\lambda}(pT_0+(1-p)\overline{T}_a)} \cdot \sum_{k=0}^{N} \left(\frac{\tilde{\lambda}}{\mu}\right)^{N-k} \cdot \frac{[\tilde{\lambda}(pT_0+(1-p)\overline{T}_a)]^k}{k!}$$

$$= \left(1-\frac{\tilde{\lambda}}{\mu}\right) e^{-\tilde{\lambda}(pT_0+(1-p)\overline{T}_a)} \cdot \left(\frac{\tilde{\lambda}}{\mu}\right)^{N} \cdot \sum_{k=0}^{N} \frac{[\mu(pT_0+(1-p)\overline{T}_a)]^k}{k!} \tag{I.2}$$

The probability distribution of model 3 (Eq. (3.3)) is derived equivalently.

II. Calculation of the blocking probability p_B

For the sake of simplicity the calculation of p_B is limited to model 1, but it can easily be extended to the other models.

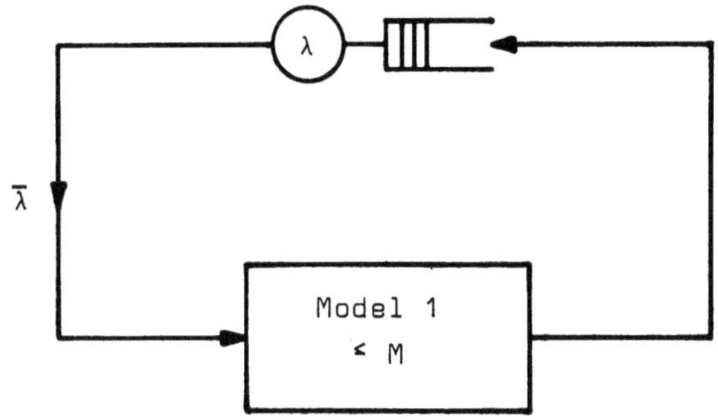

Fig.12 Closed queueing network representation of model 1

Starting with the probability of finding n_s, n_t and n_a packets in the transmitter and n_n customers in the arrival queue

$$p(n_s,n_t,n_a,n_n) = \frac{1}{G}\cdot\left(\frac{1}{\tilde{\mu}}\right)^{n_s}\cdot\left(\frac{1}{\lambda}\right)^{n_n}\cdot\frac{\left(\frac{p}{1-p}T_0\right)^{n_t}}{n_t!}\cdot\frac{\bar{T}_a^{\,n_a}}{n_a!} \qquad \text{(II.1)}$$

where

$$G = \frac{S_M(\lambda T) - \left(\frac{\lambda}{\tilde{\mu}}\right)^{M+1} S_M(\tilde{\mu}T)}{\left(1-\frac{\lambda}{\tilde{\mu}}\right)\lambda^M} \qquad \text{(II.2)}$$

with

$$\tilde{\mu}=\mu(1-p),\quad T = \frac{p}{1-p}T_0 + \bar{T}_a,\quad \text{and}\quad S_M(x) = \sum_{i=0}^{M}\frac{x^i}{i!}$$

the probability of finding N packets in the transmitter can be calculated:

$$P(N) = \frac{\left(1-\frac{\lambda}{\tilde{\mu}}\right)\left(\frac{\lambda}{\tilde{\mu}}\right)^N S_N(\tilde{\mu}T)}{S_M(\lambda T) - \left(\frac{\lambda}{\tilde{\mu}}\right)^{M+1} S_M(\tilde{\mu}T)} \qquad \text{(II.3)}$$

From this result the blocking probability is easily derived:

$$p_B = P(M) = \frac{(1-\frac{\lambda}{\tilde{\mu}})(\frac{\lambda}{\tilde{\mu}})^M}{\frac{S_M(\lambda T)}{S_M(\tilde{\mu}T)} - (\frac{\lambda}{\tilde{\mu}})^{M+1}} \qquad \text{(II.4)}$$

Since the blocking of arriving packets shall not degrade the performance of the transmission facility drastically, p_B should not exceed 1%. The minimum buffer size M necessary to meet this requirement is shown in the following table:

$\tilde{\mu}T$ \ $\frac{\lambda}{\tilde{\mu}}$	0,3	0,5	0,7
2,5	5	7...8	12
3,5	5...6	8	12...13
4,5	6	9	13

III. Estimation of the timeout interval T_0

a) Lower bound

The timer is started after the last bit of the sent packet has left the transmitter. After a propagation delay T_{prop} (negligible on terrestial links) the packet arrives at the receiving node, where it consumes a certain amount of processing time T_{proc}, until it is accepted. The sender can be informed of the receipt not before the packet presently occupying the transmitter has completely left the node. The minimum possible time between sending and receiving an acknowledgement therefore can be estimated

$$T_0 \geq 2\,(T_{prop} + T_{proc}) + \frac{2}{\mu}$$

yielding

$$T_{0_{min}} \approx \frac{3}{\mu} \qquad \text{(III.1)}$$

b) Upper bound

There is no explicit upper bound for T_0 but apparent practical considerations require its limitation. Copies of all packets not yet acknowledged have to be stored in the transmitter. An increasing timeout delay therefore generally requires an increasing amount of buffers. This is particularly true for the transmission scheme which preserves the packet order. Here, for great values of T_0 the mean transmission time exceeds all reasonable values (see e.g. eq. (4.3)). Following these arguments, T_0 should be chosen as small as possible. But, in order to avoid a retransmission just because the related acknowledgement does not arrive in time, T_0 must be great enough to permit even very long acknowledging packets to reach the transmitter before the timeout expires. A reasonable value therefore seems to be

$$T_0 = \frac{5}{\mu} \qquad \text{(III.2)}$$

IV. Exemplified estimation of the error probability p

Generally, p is combined of two parts, the first one being the error-prone link causing transmission errors, and the second one being the loss probability due to a lack of available buffers at the receiver.

Since the first part depends on the bit error rate p_b across the link, packets of different lengths will generally have different error probabilities:

$$p_E(n) = 1 - (1-p_b)^n \qquad \text{(IV.1)}$$

$p_E(n)$ is the error probability of a packet of length n, if independent bit errors are assumed. Since it is necessary for the calculation to have a constant packet error probability, p_E is approximated by a fixed value. Due to the exponential distribution of the packet length, 99% of the packets will be shorter than L bits, if the following relation holds ($\eta := \frac{\mu}{C}$, C:= link capacity):

$$0.99 = 1 - e^{-\eta L} \qquad \text{(IV.2)}$$

For an average packet length of for instance 10^3 bits, 99% of the packets will contain less than about 5000 bits. The packet error probability therefore may be upper bounded by

$$p_E = 1 - (1-p_b)^L = p_b L \qquad \text{(IV.3)}$$

If, for instance, the bit error probability amounts to 10^{-6} and if $L = 5000$, p_E approximately yields

$$p_E \approx 0.005 \qquad \text{(IV.4)}$$

The second component adding to the error probability is the lack of free buffers at the receiving node. A node is assumed to have a common pool of K buffers which are shared among all incoming and outgoing links according to a predefined policy. In order to make efficient use of the buffers, they should be overallocated, c.f. [7] for similar reasoning. That is why a packet occasionally will not find any vacant buffer on its arrival. The receiver ignores this packet, which is therefore lost and has to be retransmitted. Despite of the fact, that the loss or blocking probability p_B generally depends on the utilization of a node, in this paper it is assumed to be constant. In order to guarantee a reasonable performance, p_B should lie in the region of only a few per cent, even for high utilization factors. Following appendix II, a mean buffer size of about 9 buffers allocated to each link will cause an average loss probablility in the region of 1%.

The total error probability can therefore be approximately estimated:

$$p = p_E + p_B = 0.01 \;...\; 0.02 \qquad \text{(IV.5)}$$

XIII

QUEUES AND NETWORKS 5

FILES D'ATTENTE ET RESEAUX 5

An Approximate Analysis on Controlled Tandem Queues

Kunio Goto, Yutaka Takahashi, and Toshiharu Hasegawa

Department of Applied Mathematics and Physics
Faculty of Engineering
Kyoto University
Kyoto, Japan

An approximation method for M/M type tandem queueing systems, in which the number of customers in each stage is restricted to a finite level and also the total number of customers in the system is limited, is presented. It is tested through several examples in comparison with exact results or simulation. Moreover interesting features in allocation of stages are also shown.

1. Introduction

Recently, queueing networks have been increasing in their importances to estimate stochastic behaviors of complex systems such as computer communication networks and operating systems. In these complex systems, because of interdependencies between facilities, analyses which involve whole system are needed. Concerning the systems where the queues are unrestricted, the solvable classes are widely extended by Jackson [5], Baskett, Chandy et al. [1]. In those classes, where local balance is satisfied, equilibrium state probability can be represented by a well-known product form. Though, in actual system, resources are restricted, then queueing network with blocking must be used to model them mathematically. However, blocking causes interdependencies between queues, and as a result local balance is not satisfied. Therefore, simulation results or numerical calculations if they are available are used so far. But simulation is expensive and not robust, and numerical calculation is limited by cpu time or available memory space of a computer.

So, in this paper, we suggest an approximation method for M/M type tandem queueing system in which the number of customers in each stage is restricted to a finite level and also the total number of customers in the system is limited. This constraint of our model is aimed to describe the actual systems that have finite capacities, for example, a switching node which shares a common buffer pool, a packet communication line in which the number of packets on a line is limited (for instance, window control in ARPA and CYCLADE), a production line which has multiprocessors and shares a common storage, etc.

In the case of no limitation of the total number of customers in the system, an efficient numerical calculation for 2-stage model is shown in [10] . Because of the limitation of the total number of customers in the system, numerical calculation becomes more complex. In our method, basically the whole system is replaced with a waiting room and an exponential server with its rate depending on the number of customers in the system. Using this method, we can evaluate the characteristic quantities. It is found that almost same throughputs and shorter delays are obtained by limiting the total number of customers in the system, comparing with totally unlimited case, and in some of 2-stage cases, same throughput, delay, and probability distribution of the number of customers in the system are derived by exchanging the order of servers.

Section 2 is devoted to introduce our model, and Section 3 explains our approximation method. In Section 4, the estimations are compared with exact solutions and simulation results. Also, some cosiderations about the behavior of the system are in Section 4. And Conclusion is available in Section 5.

2. Model

A n-stage tandem queueing system is considered (Fig.1), where it is assumed that interarrival time and service time at each stage are exponentially distributed with parameters λ and μ_k (k=1,...n,), respectively, and the number of customers in each stage is restricted to a finite level (less than or equal to N_k), further the total number of customers in all stages is limited (less than or equal to N). So, the constraint concerning queue length causes loss of arrivals or blocking phenomena.

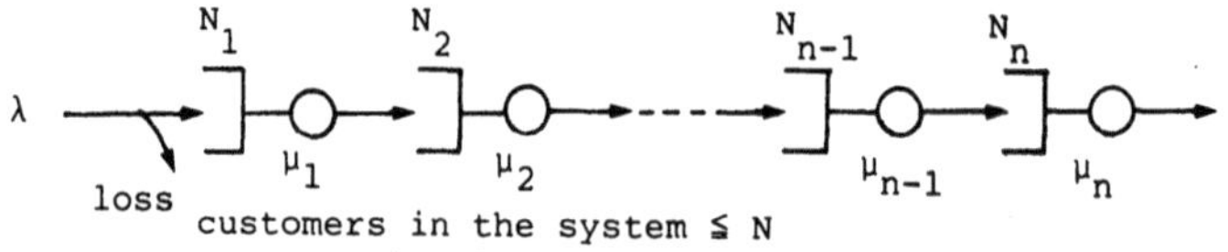

Fig. 1

Blocking between stages k-1 and k occurs when stage k is full of N_k customers. In this case, the server of stage k-1 stops processing until stage k releases a customer (i.e. the service to the customer at the head of stage k has been finished.), and then begins to offer its exponential service. When the sum of the number of customers in all stages becomes N or the first stage is full of N_1 customers, further arrivals are lost.

If $\sum_k N_k \leq N$, this model is a traditional tandem queueing systems with finite waiting rooms, studied in some papers [5], [6], [7]. But if $\sum_k N_k > N$,this is the case we are interested in, this model is used to describe those actual systems which have finite capacities.

3. Approximation Method

There are many difficulties in analyzing queueing systems with blocking. It's mainly because local balance isn't satisfied. If local balance is satisfied, equilibrium state probability can be represented by a well-known product form [1]. Then we must solve the global balance equation. Though, the global balance equation cannot be solved analytically, and getting its numerical results is limited in nature by cpu time or memory size. So, some approximate analyses are needed.

In such a queueing system shown in Fig.1, we are interested mainly in its mean throughput and delay. An approximate analysis to get their estimations using composite queues is suggested. In this method, the whole system is replaced with a waiting room and an exponential server with its rate depending on the number of customers in the system. As for locally balanced system (in the case that $N_k \geq N$ for all k in our model), it is well known that the method gives exact results for the estimations of queue length, delay, and throughput [2]. In our model except for some cases, because of blocking and loss phenomena, this method is only an approximation but a powerful tool for estimating characteristic quantities.

The approximation method is described below.

Replacing with Composite Queue

A queueing system in Fig.1 is considered. We reduce the n-stage tandem queue to a composite queue (Fig.7). At first the last two stages (stage n-1 and stage n) are reduced to one by Step 1. By repeating Step 2, all the stages are reduced to one. And finally mean throughput and delay are estimated in Step 3. (in 2-stage case, only Step 1 and Step 3 are executed).

Step 1

The last two stages in the system of Fig.1 are replaced with a composite queue which has the exponential server (Fig.2) whose service rate $\mu^*_{n-1}(i)$ is assumed to be the throughput of the closed queueing system shown in Fig.3, where i is the number of customers in stages n-1 and n (i.e. the number of customers in the closed queueing system of Fig.3).

$\mu^*_{n-1}(i)$ is obtained as follows.

$$N^*_{n-1} = \min(N, N_{n-1}+N_n)$$

for $i = 0,1,\ldots,N^*_{n-1}$

$$jmin = \max(0, i-N_{n-1}),\quad jmax = \min(N_n, i)$$

$$\begin{cases} P_c(j+1) = \dfrac{\mu_{n-1}}{\mu_n} P_c(j), \quad (j = jmin,\ldots,jmax-1) \\ \sum_{j=jmin}^{jmax} P_c(j) = 1 \end{cases}$$

$$\mu^*_{n-1}(i) = \sum_{j=\max(1,jmin)}^{jmax} \mu_n P_c(j) \quad .$$

where N^*_{n-1} means the maximum number of customers in stages n-1 and n, and $P_c(j)$ denotes the probability that the number of customers in stage 2 in Fig.3 is j.

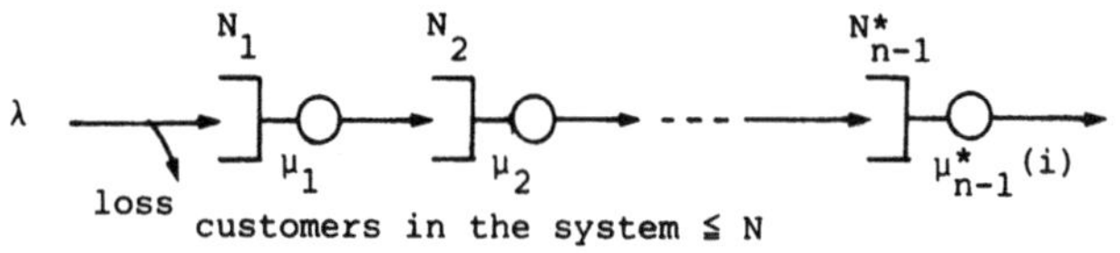

Fig. 2

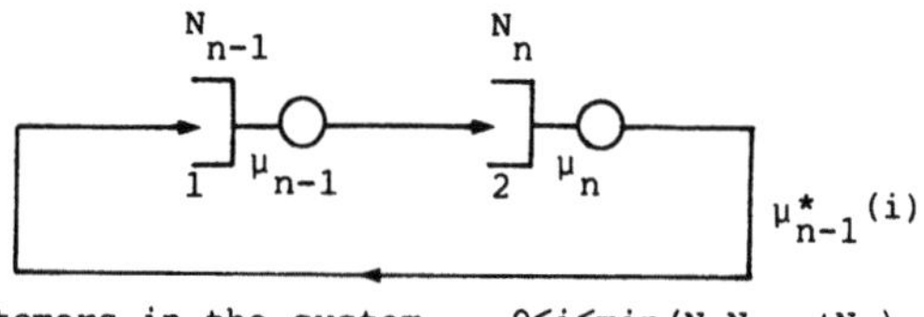

Fig. 3

Step 2

When the last k stages have been reduced to one (Fig.4), we aggregate the last two stages to one in the same way as Step 1. But the blocking between stages n-k and n-k+1 in Fig.4 actually occurs between stages n-k and n-k+1 in the original system of Fig.1. So, this actual blocking must be taken into consideration to reduce the last two stages in Fig.4.

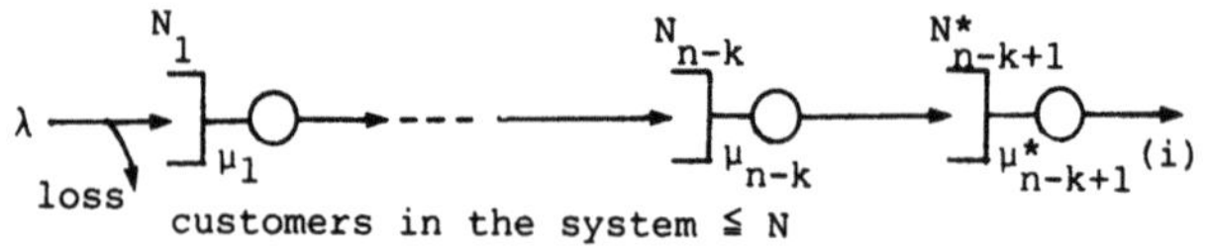

Fig. 4

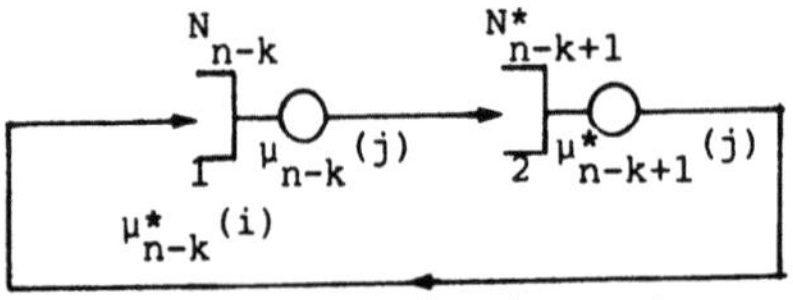

Fig. 5

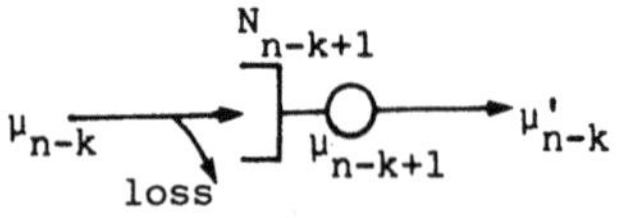

Fig. 6

As in Step 1, a closed queueing system is considered (Fig.5), though the rate μ_{n-k} is replaced with $\mu_{n-k}(j)$ where j is the number of customers in stage n-k+1 in the system of Fig.4.

$\mu_{n-k}(j)$ is defined as follows

$$\mu_{n-k}(j) = \begin{cases} \mu_{n-k} & (0 \leqq j < N_{n-k+1}) \\ \mu'_{n-k} & (N_{n-k+1} \leqq j \leqq N) \end{cases}$$

where μ'_{n-k} is the throughput of the system of Fig.6, and obtained by following

$$\begin{cases} P'(j+1) = \dfrac{\mu_{n-k}}{\mu_{n-k+1}} P'(j) & (0 \leqq j \leqq N_{n-k+1} - 1) \\ \sum_{j=0}^{N_{n-k+1}} P'(j) = 1 & \end{cases}$$

$$\mu'_{n-k} = (1 - P'(N_{n-k+1}))\, \mu_{n-k} \quad ,$$

where P'(j) is the probability that the number of customers in the system of Fig.6 is j. Then the service rate of the composite queue of stages n-k and n-k+1 in Fig.4 is derived in similar way as in Step 1. This Step 2 should be repeated until the whole queueing system is reduced into the composite queue of Fig.7.

Step 3

As in Step 2 taking account of the loss of arrivals in the first stage (actually, the loss is caused by two conditions, one is that the whole system is full of N customers, which is always taken into consideration while reducing, another is that the first stage is full of N_1 customers), let the arrival rate to the composite queue of Fig.7 be $\lambda^*(i)$, which is defined as follows

if $N_1 < N$

$$\lambda^*(i) = \begin{cases} \lambda & (i < N_1) \\ \lambda & (i \geqq N_1) \end{cases}$$

otherwise

$$\lambda^*(i) = \lambda$$

where λ' is the throughput of the system of Fig.8, which is obtained as follows

$$\begin{cases} P'(j+1) = \dfrac{\lambda}{\mu_1} P'(j) & (0 \leqq j \leqq N_1 - 1) \\ \sum_{j=0}^{N_1} P'(j) = 1 & \end{cases}$$

$$\lambda' = (1 - P'(N_1))\, \lambda \quad .$$

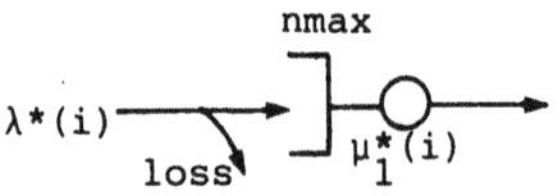

Fig. 7

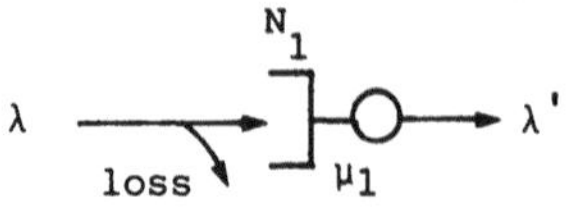

Fig. 8

Then calculating the following M/M/1 type balance equations, we obtain the eqilibrium state probabilities of the number of customers in the system, P(i), and mean throughput and delay.

$$\begin{cases} P(i+1) = \dfrac{\lambda^*(i)}{\mu_1^*(i)} P(i) & (0 \leq i \leq nmax - 1) \\ \sum_{i=0}^{nmax} P(i) = 1 & \end{cases}$$

where $nmax = \min(N, \sum_{k=1}^{n} N_k)$.

Then mean throughput $T(\lambda)$ is obtained as follows

$$T(\lambda) = \begin{cases} (1-P(nmax))(1-P'(N_1))\lambda & (N_1 < nmax) \\ (1-P(nmax))\lambda & (N_1 \geq nmax) \end{cases} .$$

Mean delay $D(\lambda)$ is

$$D(\lambda) = \frac{\sum_{i=0}^{nmax} i \cdot P(i)}{T(\lambda)} .$$

4. Numerical Results

Our approximation method stated in Section 3 is extensively evaluated in comparison with exact solutions, if possible, and with simulation results. Concerning 2-stage and 3-stage cases, the global balance equations are directry solved. However, the more the number of different states increases, the more difficult and expensive it becomes explosively to obtain the exact solutions because solving simultaneous equations consumes much memory space and computational time. Therefore, as for the system which has more than 4 stages or more than 500 different states, our method is compared with simulation results. These simulation results are shown in [15].

4.1 Comparison with Exact Solutions and Simulation Results

To compare the estimations by the approximations with exact solutions or simulation results, there are a huge number of the combinations of parameters. So we classify them from two points of view. One is whether the blocking in each stage occurs or not, and the loss of arrivals is only caused when the first stage is full or also caused when the system is full of N customers. Another is the combination of service rates.

In this sense the representative parameters of 2-stage, 3-stage, 4-stage, and 6-stage model are chosen as shown in Figs.9,11,13,15, and 17, respectively. The solid lines and symbols in Figs. from 9 to 14 denote the approximations and the exact solutions, respectively, and in Figs. from 15 to 18 represent the approximations and the simulation results, respectively.

In 2 and 3-stage cases, comparing with the exact solutions, in most cases, mean throughput, delay, and the probability distribution of the number of customers in the system are well estimated. Especially in the case that local balance is satisfied (cases 1 and 2 in Fig.9, cases 1 and 2 in Fig.13), these approximations are precisely same as the exact solutions (see Figs.9,10,13, and 14), since the approximation method in these cases is equivalent to the parametric analysis described in [2]. In the paper, it is shown that Norton's theorem (of electric network) holds in certain classes of queueing network that obey local balance. And also in cases 7 and 8 concerning 2-stage (Figs.11 and 12), and other cases in which $N_1 \geq N$, $N_2 \leq N$, and $N_1+N_2 \geq N$, where local balance is no longer satisfied, those approximations are precisely same as the exact results.

In 4 and 6-stage cases, as illustrated in Figs.15,16,17, and 18, these approximations are compared with the simulation results shown in [15]. All the approximations are close to the simulation results. Also in 4 and 6-stage cases, the approximations are exact when local balance is satisfied, because of the same reason in 2 and 3-stage cases.

This approximation method in nature will be of good performance if the interarrival time distribution at each stage is nearly exponential, since these approximations are obtained by repeatedly reducing the system to a composite queue with an exponential server. Indeed, it is recognized through some examples as shown in Table 1.

However, on the condition that each stage has a small waiting room (say 2 or 3), this method becomes a little less useful than otherwise. It may be because interarrival time distribution at each stage becomes far from exponential because of frequent occurence of blocking (see Table 1).

Concerning the computational time, all the computations were done by FACOM M-200 system, and it took about 10 seconds to get an exact solution in the case of about 350 states, for a set of parameters (for example, 3-stage, $N_1=N_2=N_3=N=10$, $\lambda=1.0$, $\mu_1=1.5$, $\mu_2=1.0$, $\mu_3=0.5$). And the computational time become longer nearly in proportion to the square of the number of the states. And the simulation needed about 0.7 second for a set of parameters with 10000 transactions, which has

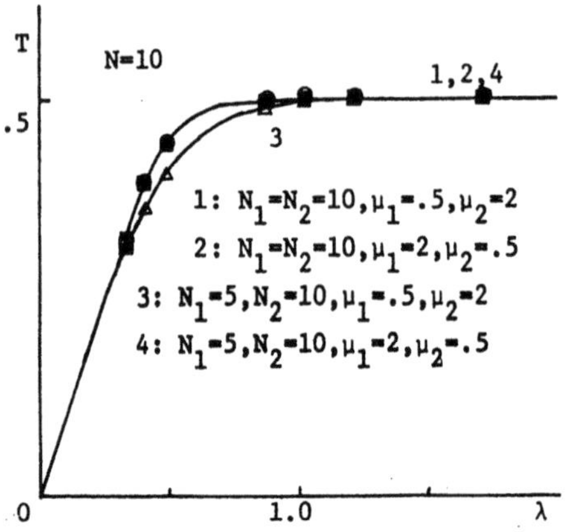

Fig. 9. Throughput v.s. arrival rate in 2-stage model.

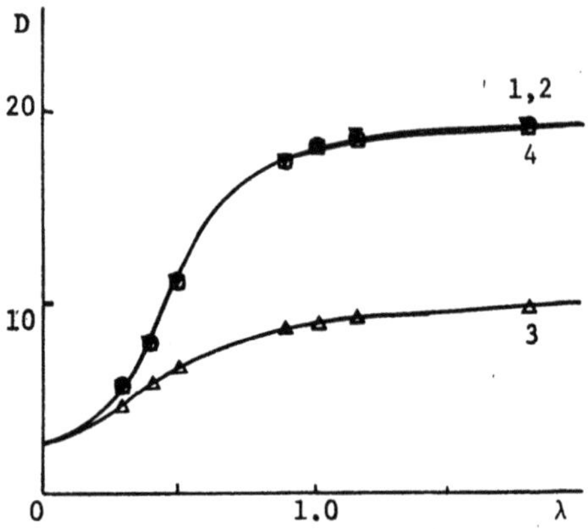

Fig. 10. Delay v.s. arrival rate in 2-stage model.

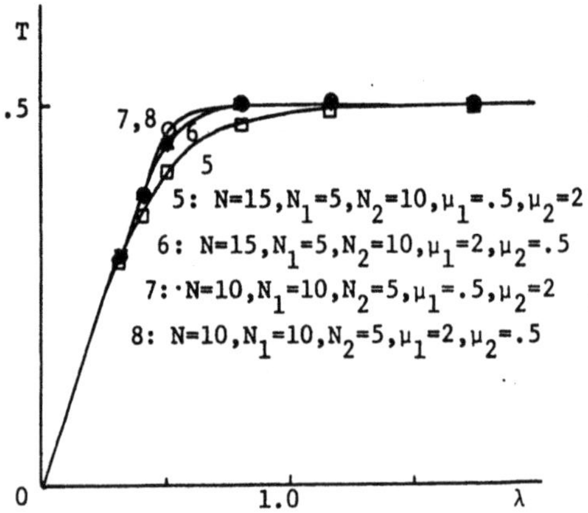

Fig. 11. Throughput v.s. arrival rate in 2-stage model.

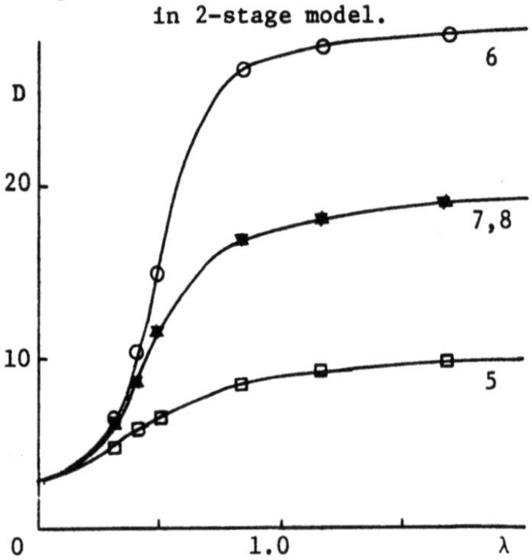

Fig. 12. Delay v.s. arrival rate in 2-stage model.

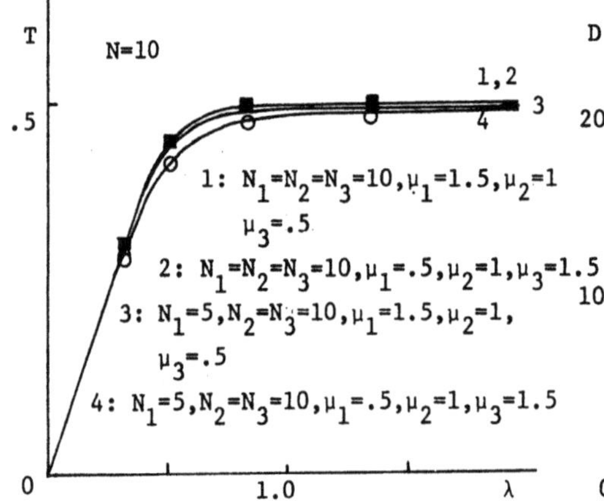

Fig. 13. Throughput v.s. arrival rate in 3-stage model.

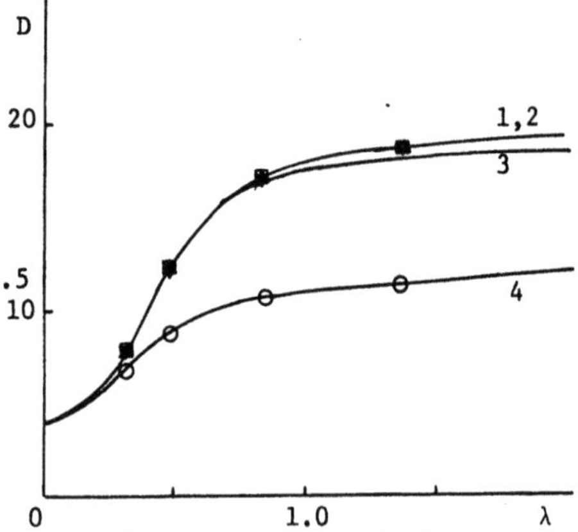

Fig. 14. Delay v.s. arrival rate in 3-stage model.

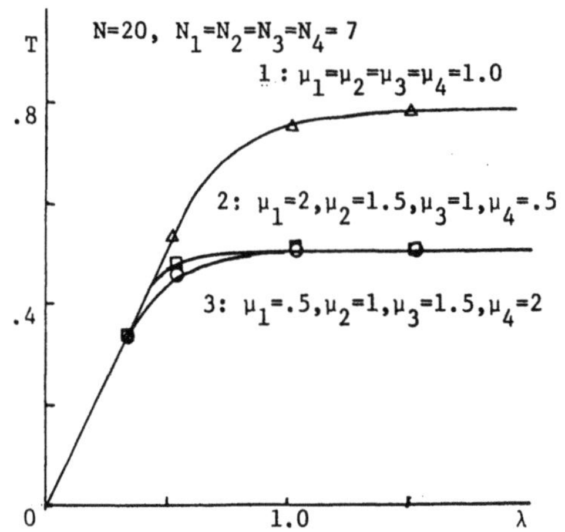

Fig. 15. Throughput v.s. arrival rate in 4-stage model.

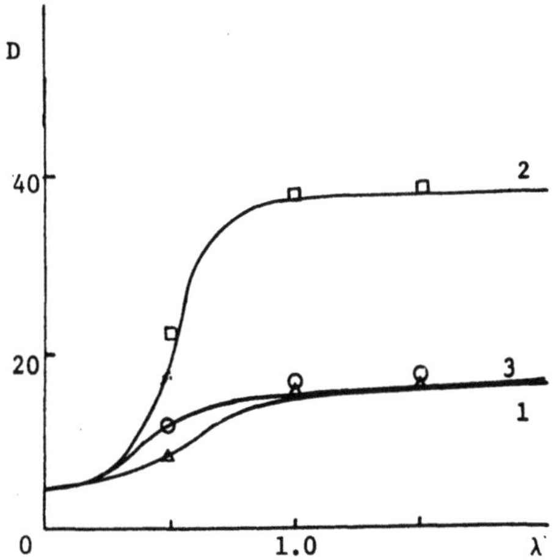

Fig. 16. Delay v.s. arrival rate in 4-stage model.

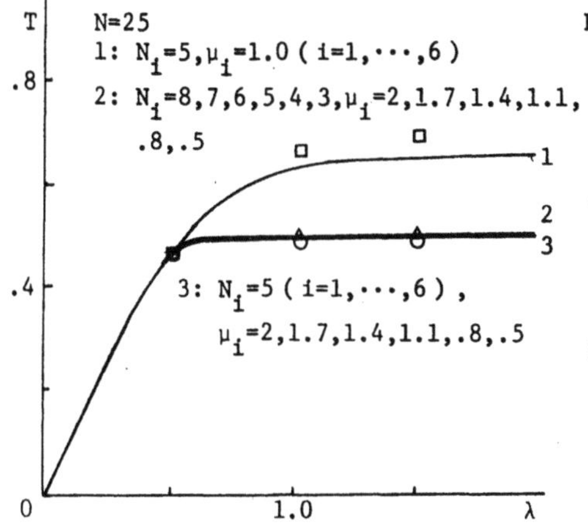

Fig. 17. Throughput v.s. arrival rate in 6-stage model.

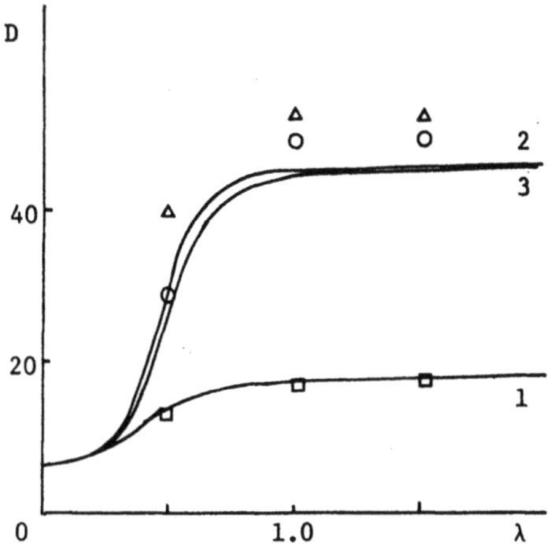

Fig. 18. Delay v.s. arrival rate in 6-stage model.

Table 1. Characteristic quantities at each stage.

λ=1.0,Te,Ts,Ta,De,Ds,Da: throughput, delay (exact, simulation, approximation).
Cvi: coefficient of variation of interarrival time at each stage
(d: departure) obtained from simulation.

2-stage $N=10,N_1=10,N_2=5,$ $\mu_1=.5,\mu_2=2$ (case 7 in Fig.11)	3-stage $N=10,N_1=5,N_2=N_3=10,\mu_1=1.5,\mu_2=1,\mu_3=.5$ (case 3 in Fig.13)
Te=.499 De=18.0 Cv1=1.00 Ts=.497 Ds=18.2 Cv2=1.01 Ta=.499 Da=18.0 Cvd=1.01	Te=.498 De=18.1 Cv1=1.00 Cv2=.99 Ts=.500 Ds=18.1 Cv3=.99 Cvd=.99 Ta=.491 Da=17.6
2-stage $N=3,N_1=3,N_2=2,$ $\mu_1=\mu_2=1$	4-stage $N=20,N_i=7,\mu_i=1$ (i=1,...,4) (case 1 in Fig.15)
Te=.582 De=3.54 Cv1=.78 Ts=.593 Ds=3.43 Cv2=.70 Ta=.571 Da=3.58 Cvd=.76	Ts=.797 Ds=17.1 Cv1=.95 Cv2=.94 Cv3=.94 Ta=.752 Da=14.8 Cv4=.94 Cvd=.96
	6-stage $N=25,N_i=5,\mu_i=1$ (i=1,...,6) (case 1 in Fig.17)
	Ts=.717 Ds=21.0 Cv1=.95 Cv2=.91 Cv3=.90 Ta=.626 Da=17.6 Cv4=Cv5=.90 Cv6=.91 Cvd=.97

no concern with the number of the states. In this approximation method, it is no need to repeat all the steps for each value of λ, so only Step 3 should be done, it took only 50 milliseconds to get the estimations for 12 given sets of parameters in 3-stage case, even in 6-stage case only about 60 milliseconds were needed.

4.2 Considerations

We examined these tandem queueing systems for more than 300 sets of parameters (n, N, N_k, μ_k, $k=1,\ldots,n$), and noticed some natures of them.

One of them is the exchangeability of the order of servers in some cases. In the locally balanced cases ($N_k \geqq N$ for all k, for example 2-stage cases 1,2, 3-stage cases 1,2, as mentioned before), the probability distribution of the number of customers in the system, mean throughput and delay are exactly same in case of any ordering of service rates μ_k ($k=1,\ldots,n$; n: the number of stages), which can be proved by the B.C.M.P.'s thorem [1]. Further if not locally balanced, only in the 2-stage cases where $N_1 \geqq N$, $N_1+N_2 \geqq N$ (for example, 2-stage cases 7,8 in Fig.9), same exchangeability of μ_1 and μ_2 can be conjectured, but it has not been proved. And in above cases, the estimations by the approximation are precisely identical to the exact solutions. These exchangeabilities has been checked by more than 50 examples.

Other features of these systems indicated by the estimations from the approximation are as follows. Concerning throughput, at light or heavy load, it is almost same by exchanging the order of servers, though at moderate load, in most cases, higher throughput is given in the case that poorer servers are situated at backward stages and the greater part of the total of waiting rooms are allocated to quick servers(see Figs.9,11,13, and 15). It is because the probabibity that no customer is present at each stage is less than that of inverse case. In this case the loss is not mainly caused by the backlog in the whole system, but in a few of forward stages. Therefore, once a customer is admitted to join the first queue, it can traverse the system with less delay.

And by the limitation of the total number of customers in the system, mean delay can be reduced holding almost same throughput (see Figs.9,10,11, and 12). This phenomenon is explained as the below. At light or medium load, loss scarcely occurs, but when the throughput saturates at heavy load, the arriving customers find the system full of N customers and are lost. So,the total waiting line in the system is limited to N, and the total delay never exceeds N divided by the throughput.

Using this method, the nearly optimal order of the servers, or allocation of waiting space can be derived in response to various purposes.

5. Conclusion

In this paper, we suggest an efficient approximation method for M/M type tandem queueing system in which the number of customers in each stage is restricted to a finite level and also the total number of customers in all stages is limited. In this method basically the whole system is replaced with a waiting room and an exponential server with its rate depending on the number of customers in the system. Compared with the exact solutions, the estimations from this method (mean throughput, delay and probability distribution of the number of customers in the system) show rather good results, and it takes only a little computation time to get them. Using this method, we examined the characteristics of those queueing systems, and it is found that almost same throughput and shorter delay are obtained by limiting the total number of customers in the system, comparing with totally unlimited case, and that in some 2-stage cases exchanging the order of servers has no effect on throughput, delay and probability distribution of the number of customers in the system. The latter nature should be proved in further reseach. And also we considered the improvement of performance achieved by adequate allocation of the waiting room of each stage and ordering of the servers, and some tendencies about them are obtained. Further work on extending the analysis to other classes of the service time distribution and to more complex networks is in progress.

References

[1] F.Baskett, K.M.Chandy, R.R.Muntz, and F.G.Palacios, "Open, Closed and Mixed Networks of Queues with Different Classes of Customers," Journal of ACM, Vol.22, pp.248-260 (1975).

[2] K.M.Chandy, U.Herzog, and L.Woo, "Parametric Analysis of Queueing Networks," IBM J. Res. & Develop., pp.36-42 (1975).

[3] M.Gerla and L.Kleinrock, "Flow Control : A Comparative Survey," IEEE Trans. on Commun., Vol.COM-28, pp.248-260 (1975).

[4] W.J.Gordon and G.F.Newell, "Closed Queueing Systems with Exponential Service," Opns. Res., Vol.15, pp.254-265 (1967).

[5] F.S.Hillier and R.W.Boling, "The Effect of Some Design Factors on the Efficiency of Production Lines with Variable Operation Times," J. Indust. Engineering 7, pp.651-658 (1966).

[6] F.S.Hillier and R.W.Boling, "Finite Queues in Series with Exponential or Erlang Service Times - a Numerical Approach," Opns. Res., Vol.15, pp.286-303 (1967).

[7] G.C.Hunt, "Sequential Arrays of Waiting Lines," Opns. Res., Vol.4, pp.674-683 (1956).

[8] J.R.Jackson, "Networks of Waiting Lines," Opns. Res., Vol.5, pp.518-521 (1957).

[9] L.Kaufman, B.Gopinath, and E.F.Wunderlich, "Analysis of Packet Network Congestion Control Using Sparse Matrix Algorithms," IEEE trans. on Commun., Vol.COM-29, pp.453-465 (1981).

[10] K.Kawashima and Y.Harada, "An Efficient Numerical Calculation for Two-stage Queueing Models with Finite Waiting Room and Blocking," Trans. IECEJ, Vol.J64-B, pp.769-776 (1981).
[11] L.Kleinrock, "Queueing Systems Vol.1: Theory," John Wiley & Sons, 1976.
[12] A.Kurinckx and G.Pujolle, "Analytic Methods for Multiprocessor System Modeling," Performance of Computer systems M.Arato, A.butrimenco (eds.), North-Holland Publishing Co., 1979.
[13] F.R.Moore, "Computational Model of a Closed Queueing Network with Exponential Servers, " IBM J. Res. & Develop., pp.567-572 (1972).
[14] M.C.Pennoti and M.Schwartz, "Congestion Control in Store and Forward Tandem Links," IEEE Trans. on Commun., Vol.COM-23, pp.1434-1443 (1975).
[15] K.Sato, "A Simulation on a Certain Tandem Queueing System (in Japanese)," Graduate Thesis, Kyoto University, 1982.
[16] Y.Takahashi, "Queueing Network Theory (in Japanese)," Systems and Control, Vol.22, pp.731-737 (1978).
[17] E.F.Wunderlich, L.Kaufman, and B.Gopinath, "The Control of Store and Forward Congestion in Packet Switched Networks," ICCC, pp.851-856 (1980).

TWO IDENTICAL COMMUNICATION CHANNELS IN SERIES WITH A FINITE INTERMEDIATE BUFFER AND OVERFLOW

O.J. Boxma
Mathematical Institute
University of Utrecht

June 1982

Abstract

The queueing analysis of store and forward data communication networks is complicated by the fact that messages preserve their length as they traverse the system: in the queueing model service times of a message at successive queues (channels) are dependent. The present study considers the case of two communication channels in series with identical capacities and with a finite intermediate buffer. That part of a message for which there is no room in the buffer is lost. The resulting queueing model of two queues in series with identical service requirements at both queues is extensively analysed. Results include the distribution of the response time in the second queue (time in buffer plus transmission time in second channel) and of the total amount of work in the second queue. A comparison is made with the analogous queueing model with independent service requirements.

Keywords: Communication network; Message switching; Queueing theory; Independence Assumption; Finite buffer; Loss system; Overflow probability; Response time; Workload.

1. INTRODUCTION

Store and forward data communication networks can be modelled as networks of queues, with communication channels identified as servers, communication centre buffers as waiting rooms and messages as customers, service requirement being equal to message length divided by channel capacity. This is a natural and general representation of the actual operation of such communication systems, but it leads to complicated queueing problems.

In the analysis of queueing networks the service processes at the various queues are almost always assumed to be independent, while waiting rooms (buffers) are generally assumed to have infinite capacity. Both assumptions are seldom realistic in the case of actual communication networks.

Let us first consider the independence of the service processes at the various queues. Messages maintain their length as they traverse the network, and therefore service times of a customer (message) at two consecutive service facilities are strongly related (in fact identical, if the two channel capacities are the same), thus causing in the model the arrival times and service times at a queue to be strongly related. Kleinrock [13] considers a general message-switching communication network with channel capacity $C^{(i)}$ of the i-th channel. He assumes that the external arrival processes at the various centres are independent Poisson processes, and that message lengths are independent negative exponentially distributed random variables. For this general model an exact analysis appeared to be impossible. This has led Kleinrock to the introduction of the very important "Independence Assumption": Each time a message is completely received at a centre, a new length is sampled for this message according to the negative exponential distribution. The resulting model is a well-known exponential queueing network model which can be analysed completely (see e.g. Jackson [10]).

The Independence Assumption clearly does noet correspond to the actual situation in any real communication net, but Kleinrock offers evidence of the fact that the model does rather accurately describe the behaviour of the message delay in many real networks. His argument is, that in networks with considerable mixing of the traffic streams the correlation between service times and interarrival times at one and the same service station almost disappears. If there is not much mixing, in particular if the network is a tandem connection, then the Independence Assumption cannot be expected to yield very accurate results. Simulation results in [13] and [14] illustrate these statements.

In [1,2] it was shown to be possible to give an exact analysis of the simplest message-switching tandem connection: two queues in series with a Poisson arrival process at the first queue, identical service times of a customer at both queues and infinite waiting rooms. The service time distribution was a general one; specification to the negative exponential distribution yielded much quantitative insight into the accuracy of the Independence Assumption, and its limitations (unrealistic results in the heavy traffic case).

For other previous work on networks of queues with dependent service times we refer to Calo [7], who gives ordering relations for the successive waiting times in a tandem connection of single server queues in which the successive service times experienced by any particular customer are scaled versions of the same random variable, and to Kelly [11], who allows dependent service times in networks of symmetric queues.

Let us next consider the infinite waiting room assumption, usually introduced in the analysis of queueing networks. In any "real" service system, buffer capacities are finite, and blocking or loss occurs. Very few exact results for networks with blocking or loss are known. Some approximate methods for the analysis of such (usually exponential) queueing networks have been developed; we mention Boxma and Konheim [5] (blocking) and Bronshtein and Gertsbakh [6] (loss).

In a recent study Kelly [12] considers a message-switching network consisting of n channels in series with equal channel capacities and with finite buffer capacity B in front of each channel. If a buffer is full blocking occurs. He is able to take both phenomena of identical service times and finite buffer capacity into account in his study of the asymptotic behaviour of the throughput of this system as a function of n and B.

In the present study we analyse a message-switching tandem connection with finite intermediate buffer and loss (overflow). The queueing model consists of two queues Q_1,Q_2 in series, with an infinite buffer in front of Q_1 and a finite buffer with capacity C (bits) in front of Q_2. If a message, arriving at this buffer, cannot be completely stored then the excess part is lost. Apart from this the model is identical to the one studied in [1,2]. The model can be analysed completely, and a comparison with the case of independent service times at both queues is possible.

The following observations have been made in [1,2] for the case of identical service times at both queues and *infinite* buffer sizes (the discussion is concentrated on the response time distribution at the second queue).

In low traffic the model behaves almost exactly like the corresponding model with independence of the service times at both queues (Kleinrock's Independence Assumption hence yields accurate results in the low traffic situation). However, in the heavy traffic case a completely different situation occurs. Now the assumption of identical service times appears to have a strong regularizing effect on the response times in Q_2; its mean and variance are much smaller than those in the model with independent service times (see Figure 5 in Section 3 of the present paper).

Obviously the assumption of finite buffer size in itself also has a regularizing effect. These effects will be studied in some more detail for the model with negative exponentially distributed service times. It will be seen that the additional effect of the finite buffer size is small, when service times are identical: under the assumption of finite buffer size the difference between the model with independent service times and that with identical service times is less pronounced. A number of graphs and a table serve to illustrate the separate and combined influence of finite buffer size and identical service times.

The organization of the paper is as follows. Section 2 contains a detailed description of the queueing model. The distribution of the response times at the second service facility (waiting time plus service time) is studied in Section 3. This section also considers the above-mentioned investigation of the effects of finite buffer size and of identical service times on response times. Section 4 is devoted to an analysis of the distribution of the amount of work in the second service facility at an arbitrary epoch. We also obtain the overflow probability and the fraction of work lost.

Finally a remark about related models of queues in series with identical service times and a finite capacity. First consider the case that a message which can not be completely stored is lost (not just the excess part). The property of identical service times implies that only those messages are lost which have a length greater than C; an analysis of the resulting model should not present unsurmountable difficulties.

The situation is much more complicated when the number of messages, and not the amount of work, determines whether loss (or blocking) occurs. The nice properties resulting from the assumption of identical service times are lost, and it seems unlikely that detailed analytic results can be obtained (apart from special cases like: no waiting room).

2. DESCRIPTION OF THE QUEUEING MODEL

We consider a model of two single server queues Q_1, Q_2 in series. Q_1 has an infinite storage capacity, but Q_2 has a storage capacity $C < \infty$. Customers enter the tandem system individually at Q_1. After completion of his service at Q_1 a customer enters immediately Q_2; however, if his arrival at Q_2 causes the total amount of work present at Q_2 to exceed C, then the excess amount is lost (the excess part of the message is lost: overflow). Customers are served individually and at both counters the service discipline is first come-first served.

Let $\underline{t}_n$ denote the arrival epoch of the n-th customer C_n at the system, with $\underline{t}_1 \stackrel{\text{def}}{=} 0$; it will always be assumed that C_1 meets an empty tandem system. The arrival process at Q_1 is assumed to be a Poisson process with rate α^{-1}, i.e. the interarrival times $\underline{t}_n - \underline{t}_{n-1}$ are independent and negative exponentially distributed with mean α.

Denote by $\underline{\tau}_n$ the service time of C_n at Q_1, $n=1,2,\ldots$. The stochastic variables (s.v.) $\underline{\tau}_n$ are independent of the arrival process, and form a sequence of independent identically distributed positive s.v. with distribution

$$B(t) \stackrel{\text{def}}{=} \Pr\{\underline{\tau}_n < t\}, \qquad t > 0; \qquad (2.1)$$

We define

$$\beta \stackrel{\text{def}}{=} \int_0^\infty t \, dB(t) < \infty,$$
$$\beta(\rho) \stackrel{\text{def}}{=} \int_0^\infty e^{-\rho t} \, dB(t), \qquad \text{Re } \rho \geqslant 0. \qquad (2.2)$$

The second special feature of the model, apart from the overflow property, is that the service requirement of C_n at Q_2 is also equal to $\underline{\tau}_n$, $n=1,2,\ldots$, i.e. a customer has identical service requirements at both queues (but part of the service request of a customer at Q_2 may be lost, as described above).

We further define for n=1,2,..., t > 0:

$\underline{r}_n$ = departure epoch of C_n from Q_1;

$\underline{z}_n$ = number of customers in Q_1 immediately after $\underline{r}_n$;

$\underline{\upsilon}_n$ = number of arrivals at Q_1 during the service of C_n in Q_1;

$\underline{\delta}_{n+1}$ = time interval that Q_1 is empty during $(\underline{r}_n, \underline{r}_{n+1}]$;

$\underline{s}_n^{(2)}$ = sojourn time (response time; waiting plus actual service time) of C_n in Q_2;

$\underline{x}_t$ = number of customers in Q_1 at t; (2.3)

$\underline{\eta}_t$ = past service time of the customer in service in Q_1 at t ($\underline{\eta}_t \overset{def}{=} 0$ if Q_1 is empty at t);

$\underline{v}_t^{(2)}$ = amount of work in Q_2 at t;

(A) = 1 if A occurs,
= 0 if A does not occur. (indicator function)

3. THE RESPONSE TIME DISTRIBUTION IN THE SECOND QUEUE

It can be easily seen that the stochastic process $\{(\underline{z}_n, \underline{s}_n^{(2)}), n=1,2,...\}$ (cf. (2.3)) forms a two-dimensional imbedded Markov chain; in particular we have the following recurrence relations for the queue length in Q_1 just after a departure epoch and for the response time in Q_2, for n=1,2,... (cf. Fig. 1):

$$\underline{z}_{n+1} = \max(0\ ,\ \underline{z}_n - 1) + \underline{\upsilon}_{n+1}, \tag{3.1}$$

$$\begin{aligned} \underline{s}_{n+1}^{(2)} &= \min(C\ ,\ \max(\underline{s}_n^{(2)}\ ,\ \underline{\tau}_{n+1})) && \text{if } \underline{z}_n > 0, \\ &= \min(C\ ,\ \max(\underline{s}_n^{(2)} - \underline{\delta}_{n+1}, \underline{\tau}_{n+1})) && \text{if } \underline{z}_n = 0; \end{aligned} \tag{3.2}$$

$$\underline{s}_1^{(2)} = \underline{\tau}_1.$$

A closer examination of (3.2) already reveals that a message is distorted at Q_2 (overflow occurs) iff its corresponding service request is larger than C:

$$\Pr\{\text{overflow}\} = 1 - B(C+). \tag{3.3}$$

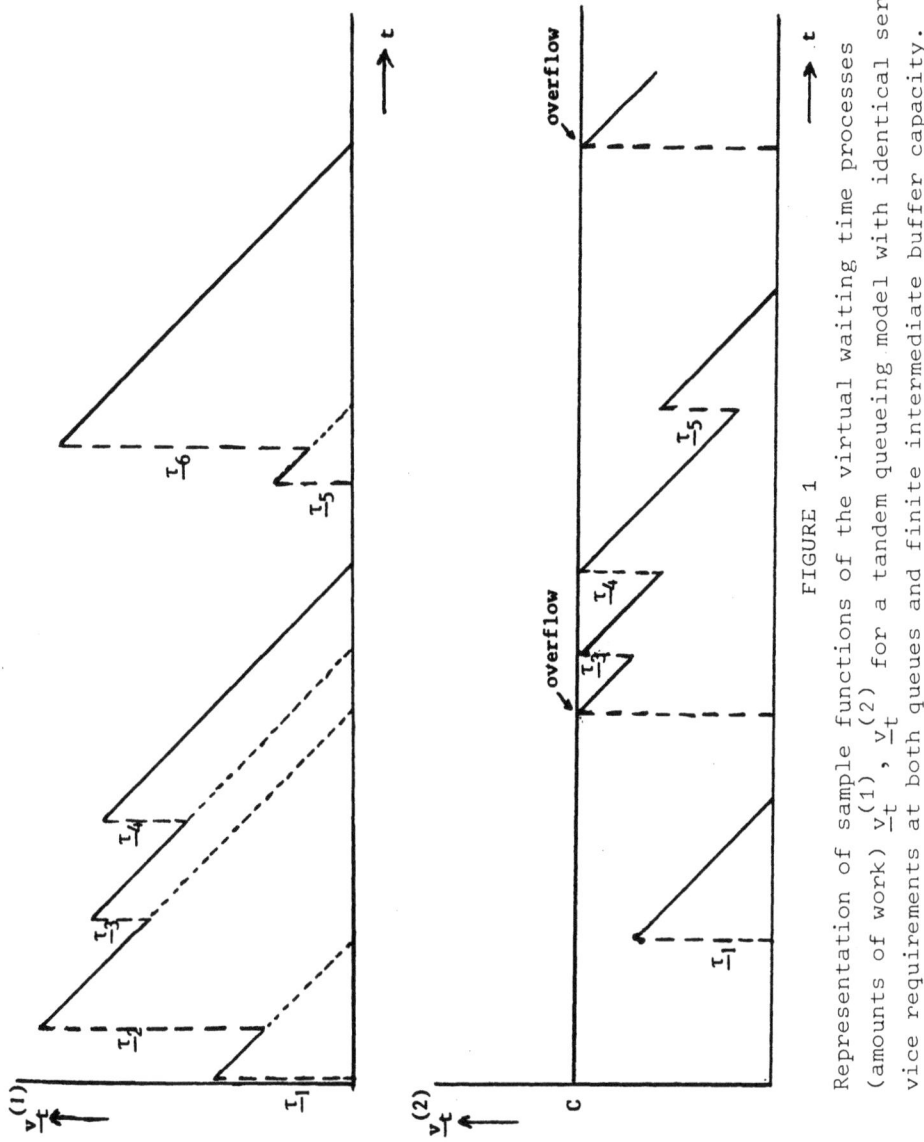

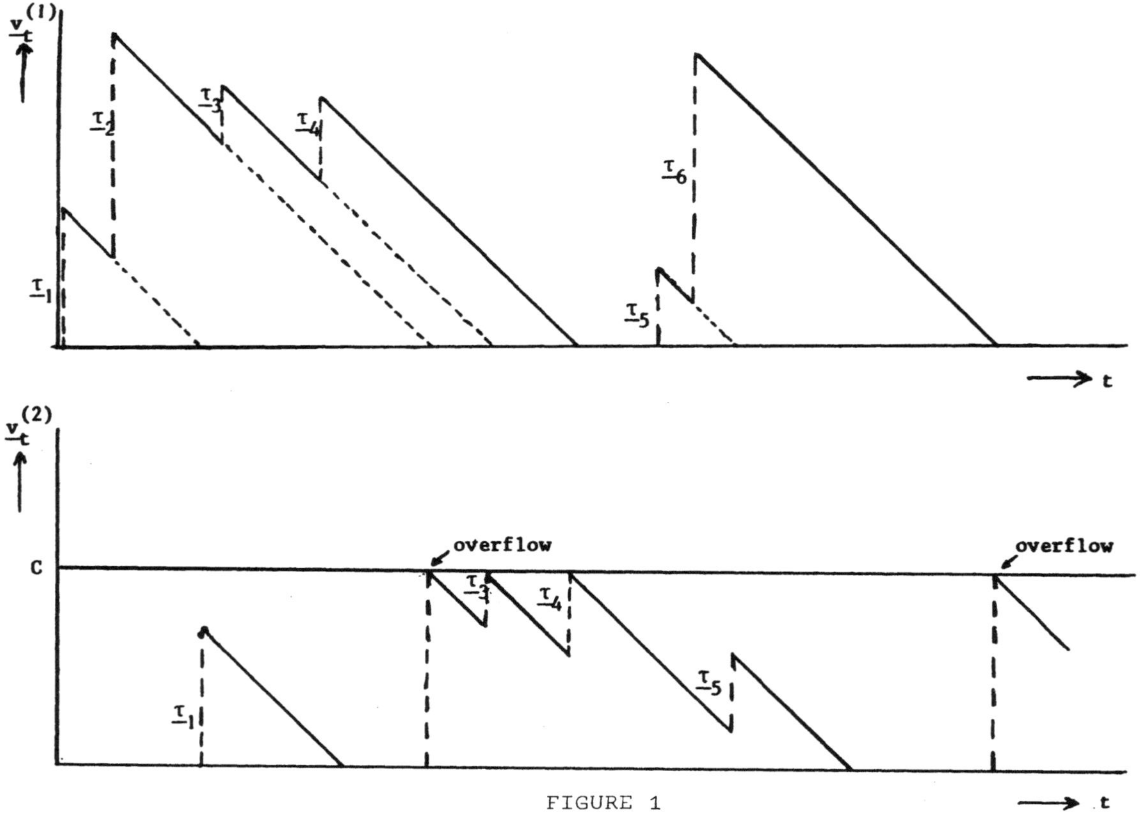

FIGURE 1

Representation of sample functions of the virtual waiting time processes (amounts of work) $\underline{v}_t^{(1)}$, $\underline{v}_t^{(2)}$ for a tandem queueing model with identical service requirements at both queues and finite intermediate buffer capacity.

From (3.1) and (3.2), for n=1,2,...:

$$E[r^{\underline{z}_{n+1}}(\underline{s}_{n+1}^{(2)} < w)]$$

$$= \frac{1}{r} E[r^{\underline{z}_n + \underline{\upsilon}_{n+1}}(\max(\underline{s}_n^{(2)}, \underline{\tau}_{n+1}) < w)] -$$

$$- \frac{1}{r} E[r^{\underline{\upsilon}_{n+1}}(\max(\underline{s}_n^{(2)}, \underline{\tau}_{n+1}) < w, \underline{z}_n = 0)] +$$

$$+ \quad E[r^{\underline{\upsilon}_{n+1}}(\max(\underline{s}_n^{(2)} - \underline{\delta}_{n+1}, \underline{\tau}_{n+1}) < w, \underline{z}_n = 0)], \; 0 < w \leqslant C;$$

$$E[r^{\underline{z}_{n+1}}(\underline{s}_{n+1}^{(2)} = C)] \tag{3.4}$$

$$= \frac{1}{r} E[r^{\underline{z}_n + \underline{\upsilon}_{n+1}}(\max(\underline{s}_n^{(2)}, \underline{\tau}_{n+1}) \geqslant C)] -$$

$$- \frac{1}{r} E[r^{\underline{\upsilon}_{n+1}}(\max(\underline{s}_n^{(2)}, \underline{\tau}_{n+1}) \geqslant C, \underline{z}_n = 0)] +$$

$$+ E[r^{\underline{\upsilon}_{n+1}}(\max(\underline{s}_n^{(2)} - \underline{\delta}_{n+1}, \underline{\tau}_{n+1}) \geqslant C, \underline{z}_n = 0)]$$

$$= \frac{1}{r} E[r^{\underline{z}_n + \underline{\upsilon}_{n+1}}(\underline{s}_n^{(2)} = C)] + \frac{1}{r} E[r^{\underline{z}_n + \underline{\upsilon}_{n+1}}(\underline{s}_n^{(2)} < C, \underline{\tau}_{n+1} \geqslant C)] -$$

$$- \frac{1}{r} E[r^{\underline{\upsilon}_{n+1}}(\underline{s}_n^{(2)} = C, \underline{z}_n = 0)] -$$

$$- \frac{1}{r} E[r^{\underline{\upsilon}_{n+1}}(\underline{s}_n^{(2)} < C, \underline{z}_n = 0, \underline{\tau}_{n+1} \geqslant C)] +$$

$$+ E[r^{\underline{\upsilon}_{n+1}}(\underline{\tau}_{n+1} \geqslant C, \underline{z}_n = 0)].$$

Introduce for n=1,2,..., $|r| \leqslant 1$, $w > 0$:

$$K_n(r,w) \stackrel{\text{def}}{=} E[r^{\underline{z}_n}(\underline{s}_n^{(2)} < w)], \tag{3.5}$$

$$H_n(w) \stackrel{\text{def}}{=} \Pr\{\underline{z}_n = 0, \underline{s}_n^{(2)} < w\}. \tag{3.6}$$

Observing that $\underline{\tau}_{n+1}$ and $\underline{\upsilon}_{n+1}$ are both independent of $\underline{z}_n$ and $\underline{s}_n^{(2)}$, and realizing that $\underline{\delta}_{n+1}$ (when positive) is negative exponentially distributed with mean α, no matter what the values of $\underline{\tau}_{n+1}$ and $\underline{s}_n^{(2)}$ are, it follows from (3.3) - (3.6) (and (2.1)) for n=1,2,..., $|r| \leqslant 1$,

$$K_{n+1}(r,w) = \frac{1}{r} [\int_0^w e^{-(1-r)t/\alpha} \, dB(t)] \, K_n(r,w) -$$

$$- \frac{1}{r} [\int_0^w e^{-(1-r)t/\alpha} \, dB(t)] \, H_n(w) +$$

$$+ \left[\int_0^w e^{-(1-r)t/\alpha} \, dB(t) \right] \left[\int_{g=0}^{C-w} \frac{1}{\alpha} e^{-g/\alpha} H_n(w+g) \, dg + \right.$$

$$\left. + \int_{g=C-w}^{\infty} \frac{1}{\alpha} e^{-g/\alpha} \Pr\{\underline{z}_n = 0\} dg \right],$$

$$0 < w \leq C; \qquad (3.7)$$

$$E[r^{\underline{z}_{n+1}}] - K_{n+1}(r,C) = \frac{1}{r} \beta\left(\frac{1-r}{\alpha}\right) [E[r^{\underline{z}_n}] - K_n(r,C)] +$$

$$+ \frac{1}{r} \left[\int_C^{\infty} e^{-(1-r)t/\alpha} \, dB(t) \right] K_n(r,C) -$$

$$- \frac{1}{r} \beta\left(\frac{1-r}{\alpha}\right) [\Pr\{\underline{z}_n = 0\} - H_n(C)] -$$

$$- \frac{1}{r} \left[\int_C^{\infty} e^{-(1-r)t/\alpha} \, dB(t) \right] H_n(C) +$$

$$+ \left[\int_C^{\infty} e^{-(1-r)t/\alpha} \, dB(t) \right] \Pr\{\underline{z}_n = 0\}. \qquad (3.8)$$

In [1,2] the analysis of the model with $C = \infty$ is pursued by studying the generating functions of $K_n(r,w)$ and $H_n(w)$. A similar approach is possible here. However, to make the analysis more transparent, we have decided to restrict ourself here to the stationary case. We assume that

$$a \overset{\text{def}}{=} \beta/\alpha < 1. \qquad (3.9)$$

This is the ergodicity condition for the M/G/1 queue Q_1; it can be verified that Q_2 is ergodic, irrespective of the value of a, due to the overflow property. The following limits now exist: for $|r| \leq 1$, $w > 0$,

$$\lim_{n\to\infty} K_n(r,w) \overset{\text{def}}{=} K(r,w) = E[r^{\underline{z}_\infty}(\underline{s}_\infty^{(2)} < w)], \qquad (3.10)$$

$$\lim_{n\to\infty} H_n(w) \overset{\text{def}}{=} H(w) = \Pr\{\underline{z}_\infty = 0, \underline{s}_\infty^{(2)} < w\}; \qquad (3.11)$$

$\underline{z}_\infty$ ($\underline{s}_\infty^{(2)}$) denotes a s.v. with distribution the limiting distribution of $\underline{z}_1, \underline{z}_2, \ldots$ ($\underline{s}_1^{(2)}, \underline{s}_2^{(2)}, \ldots$).
From (3.7), (3.8), (3.10) and (3.11) for $a < 1$,

$$K(r,w) = \left[1 - \frac{1}{r} \int_0^w e^{-(1-r)t/\alpha} \, dB(t)\right]^{-1} \frac{1}{r} \int_0^w e^{-(1-r)t/\alpha} \, dB(t) .$$

$$.\left[- H(w) + r\, e^{w/\alpha} \int_w^C \frac{1}{\alpha} e^{-y/\alpha} H(y) \, dy + r\, e^{-(C-w)/\alpha} \Pr\{\underline{z}_\infty = 0\} \right],$$

$$|r| \leq 1,\ 0 < w \leq C, \qquad (3.12)$$

$$K(r,C) = [1 - \frac{1}{r}\int_0^C e^{-(1-r)t/\alpha}\, dB(t)]^{-1}.$$

$$.[(1 - \frac{1}{r}\,\beta(\frac{1-r}{\alpha}))\, E[r^{\underline{z}_\infty}] - \frac{1}{r}\,[\int_0^C e^{-(1-r)t/\alpha}\, dB(t)]\, H(C) +$$

$$+ \Pr\{\underline{z}_\infty = 0\}\, (\frac{1}{r}\,\beta(\frac{1-r}{\alpha}) - \int_C^\infty e^{-(1-r)t/\alpha}\, dB(t))], \qquad |r| \leqslant 1. \quad (3.13)$$

$K(r,w)$ and $H(w)$, $0 < w \leqslant C$, must be determined from (3.12), (3.13) and the fact that $K(r,w)$, $0 < w \leqslant C$ is an analytic function of r in the region $|r| \leqslant 1$. Rouché's theorem implies that the denominator of the righthand side of (3.12) has exactly one zero $r = m(w)$ in the region $|r| < 1$; similarly $r = m(C)$ for the righthand side of (3.13). In [3] it is proved that we can write, denoting by $\underline{m}$ the supremum of the service times of the customers in a busy cycle of Q_1,

$$m(w) = \Pr\{\underline{m} < w\}, \qquad w > 0. \qquad (3.14)$$

See [3] and [4] for detailed results concerning this distribution. Here we only mention that

$$1 - B(w) \leqslant 1 - m(w) \leqslant (1 - B(w))/(1 - a), \qquad w > 0. \qquad (3.15)$$

We now prove:

<u>Theorem 3.1</u>

$$K(r,w) = E[r^{\underline{z}_\infty}(\underline{s}_\infty^{(2)} < w)] = [1 - \frac{1}{r}\int_0^w e^{-(1-r)t/\alpha} dB(t)]^{-1}.$$

$$.\frac{1}{r}\int_0^w e^{-(1-r)t/\alpha}\, dB(t)\, .\, H(w)(\frac{r}{m(w)} - 1), \quad |r| \leqslant 1,\ 0 < w \leqslant C, \qquad (3.16)$$

$$H(w) = \Pr\{\underline{z}_\infty = 0,\ \underline{s}_\infty^{(2)} < w\}$$

$$= (1 - a)\, m(w) \exp[-\frac{1}{\alpha}\int_w^C (1 - m(u))\, du], \quad 0 < w \leqslant C. \quad (3.17)$$

<u>Proof</u>

(3.12), the analyticity argument for $K(r,w)$ mentioned below (3.13), and (3.14) lead to the following integral equation: for $0 < w \leqslant C$,

$$0 = - H(w) + m(w)\, e^{w/\alpha} \int_w^C \frac{1}{\alpha}\, e^{-y/\alpha}\, H(y)\, dy + m(w)\, e^{-(C-w)/\alpha}(1 - a), \qquad (3.18)$$

(note that from M/G/1 theory $\Pr\{\underline{z}_\infty = 0\} = 1 - a$).

Putting

$$f(w) = H(w)/m(w), \qquad 0 < w \leqslant C, \qquad (3.19)$$

it follows that for $0 < w \leqslant C$,

$$f(w) = e^{w/\alpha} \int_w^C \frac{1}{\alpha} e^{-y/\alpha} f(y)\, m(y)\, dy + e^{-(C-w)/\alpha}(1-a).$$

Differentiation yields after a simple calculation,

$$\frac{df(w)}{dw} = \frac{1}{\alpha}(1 - m(w))\, f(w), \qquad 0 < w \leqslant C;$$

hence

$$f(w) = K \exp[-\frac{1}{\alpha} \int_w^C (1 - m(u))\, du], \qquad 0 < w \leqslant C,$$

with

$$f(C) = H(C)/m(C) = K.$$

So

$$H(w) = m(w)\, \frac{H(C)}{m(C)} \exp[-\frac{1}{\alpha} \int_w^C (1 - m(u))du], \quad 0 < w \leqslant C. \qquad (3.20)$$

$m(C)$ is known, but $H(C)$ has yet to be determined. The analyticity argument for $K(r,C)$ mentioned below (3.13) implies that the numerator of the righthand side of (3.13) should be zero for $r = m(C)$. Further, from M/G/1 theory (cf. Cohen [8], p. 238),

$$E[r^{\underline{z}_\infty}] = (1 - a)(1 - r)\, \frac{\beta(\frac{1-r}{\alpha})}{\beta(\frac{1-r}{\alpha}) - r}, \qquad |r| \leqslant 1. \qquad (3.21)$$

Combination of these two facts finally implies:

$$H(C) = (1 - a)\, m(C). \qquad (3.22)$$

(3.17) now follows from (3.20) and (3.22), and (3.16) follows from (3.12) and (3.18). Note that in the present approach (with $H(\infty)$ and $K(r,\infty)$ known) the analysis can be somewhat simplified: (3.22) immediately follows from (3.18). □

Putting $r = 1$ in (3.16) yields

$$S^{(2)}(w) \overset{\text{def}}{=} \Pr\{\underline{s}_\infty^{(2)} < w\} = \frac{B(w)}{1 - B(w)}\, H(w)\, \frac{1 - m(w)}{m(w)}$$

$$= \frac{B(w)}{1 - B(w)}\, (1 - m(w))(1 - a) \exp[-\frac{1}{\alpha} \int_w^C (1 - m(u))\, du],$$

$$0 < w \leqslant C. \qquad (3.23)$$

In particular,

$$\Pr\{\underline{s}_{\infty}^{(2)} = C\} = 1 - \Pr\{\underline{s}_{\infty}^{(2)} < C\} = 1 - \frac{B(C)}{1-B(C)}(1-m(C))(1-a). \tag{3.24}$$

Note that this probability is larger than the probability of overflow, $1 - B(C)$ (cf. (3.3) and (3.15)); indeed there is a positive probability that a waiting time plus a complete service time in Q_2 exactly add up to C (cf. Fig. 1).

Remark 3.1

In [1,2] we found for $C = \infty$, $w > 0$:

$$\Pr\{\underline{s}_{\infty}^{(2)} < w\} = \frac{B(w)}{1-B(w)}(1-m(w))(1-a)\exp[-\frac{1}{\alpha}\int_{w}^{\infty}(1-m(u))du],$$

$$\Pr\{\underline{z}_{\infty} = 0, \underline{s}_{\infty}^{(2)} < w\} = (1-a)\, m(w)\exp[-\frac{1}{\alpha}\int_{w}^{\infty}(1-m(u))\,du].$$

Note that $\Pr\{\underline{s}_{\infty}^{(2)} < w\}$ and $\Pr\{\underline{z}_{\infty} = 0, \underline{s}_{\infty}^{(2)} < w\}$ are, for fixed w, non-increasing in C (see (3.17), (3.23)).

Remark 3.2

In [1,3]

$$G(w) \overset{\text{def}}{=} \frac{B(w)}{1-B(w)}(1-m(w))(1-a), \qquad w > 0, \tag{3.25}$$

is shown to be the limiting distribution for $n \to \infty$ of the s.v.

$\underline{g}_n \overset{\text{def}}{=}$ the supremum of the service times in Q_1 of the n-th customer C_n and of the customers who have arrived before C_n and belong to the same busy cycle of Q_1 as C_n.

Furthermore,

$$Y(w) \overset{\text{def}}{=} \exp[-\frac{1}{\alpha}\int_{w}^{C}(1-m(u))\,du], \qquad 0 < w \leq C,$$

is also a proper probability distribution (in fact it can be shown to be the limiting distribution for $k \to \infty$ of the amount of work in Q_2 at the start of the k-th busy cycle of Q_1). An interpretation of

$$\Pr\{\underline{s}_{\infty}^{(2)} < w\} = G(w)\,Y(w), \qquad 0 < w \leq C, \tag{3.26}$$

is now apparent from Fig. 1.

Remark 3.3

It is interesting to observe that in the present model one can derive an explicit expression for the response time distribution in Q_2

FIGURE 2

Comparison of the response time distributions $S(w)$ (----) and $S^{(2)}(w)$ (——) for $C = 2$, with $B(w) = 1 - e^{-w/0.75}$.

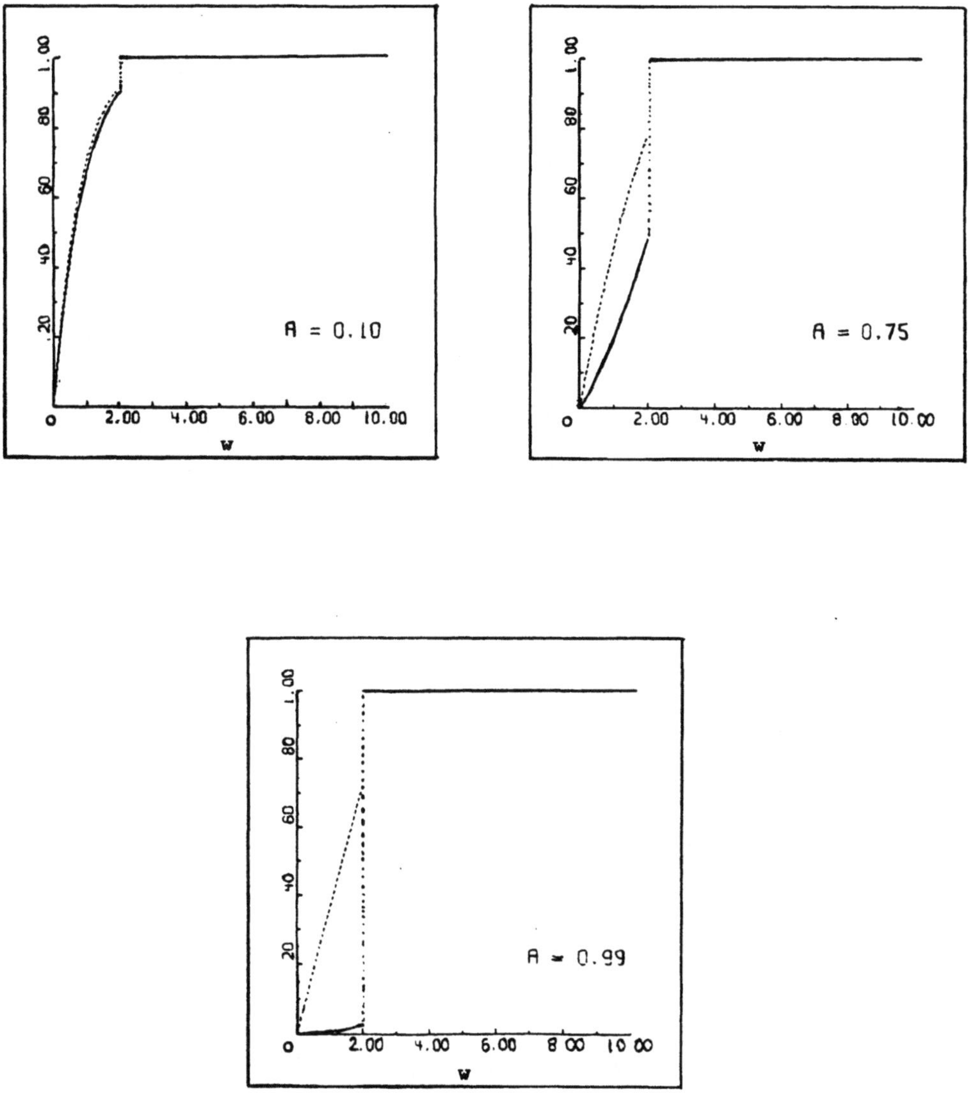

FIGURE 3

Comparison of the response time distributions S(w) (----) and $S^{(2)}(w)$ (——) for C = 4, with $B(w) = 1 - e^{-w/0.75}$.

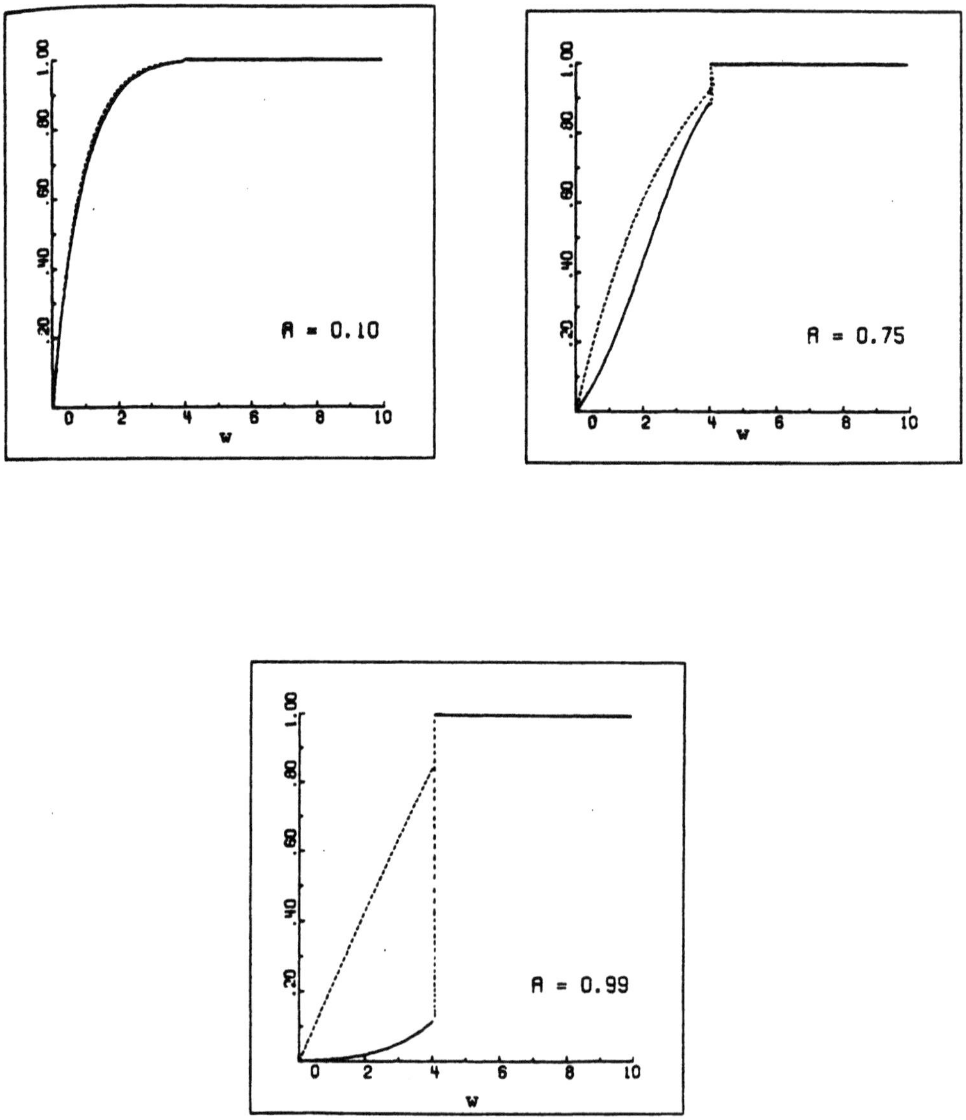

FIGURE 4

Comparison of the response time distributions S(w) (----) and $S^{(2)}(w)$ (——) for C = 8, with $B(w) = 1 - e^{-w/0.75}$.

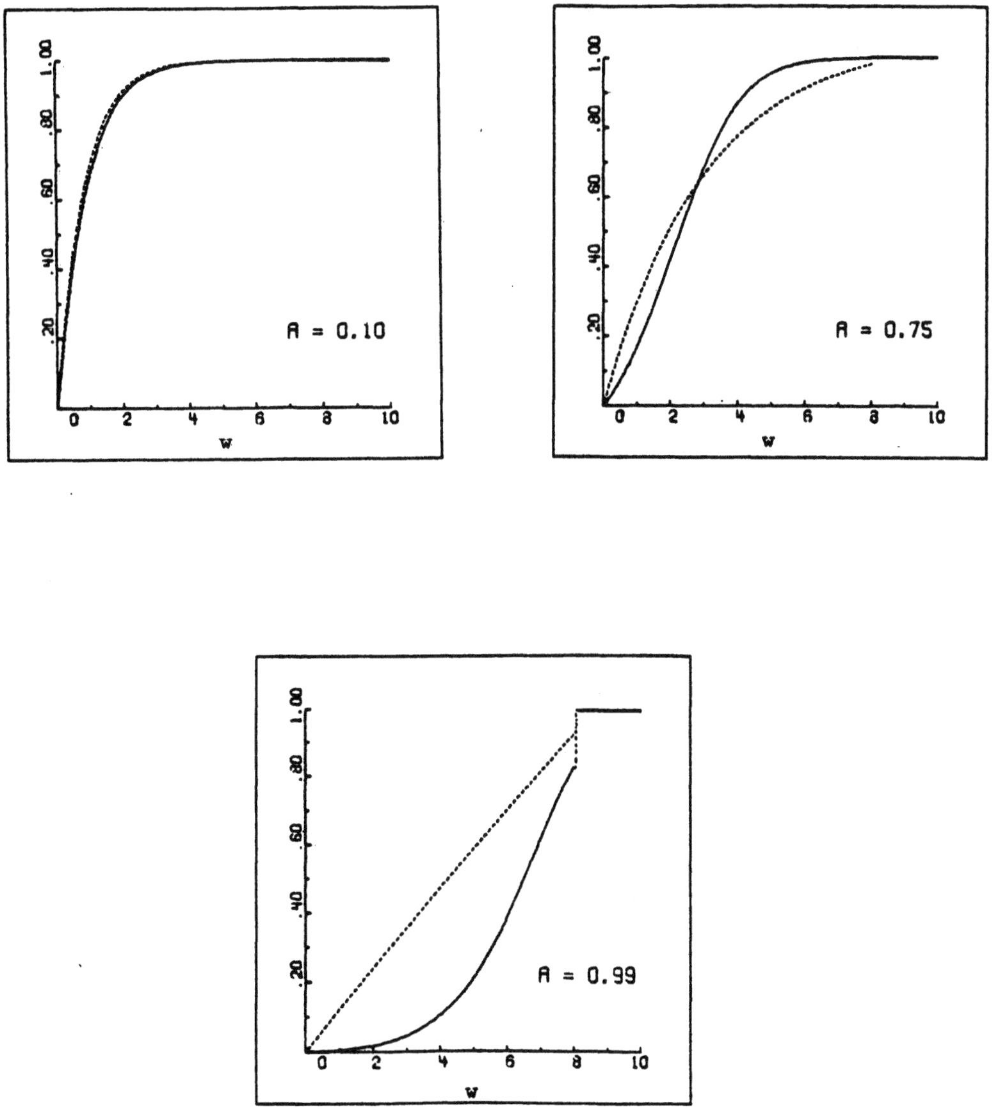

FIGURE 5

Comparison of the response time distributions $S(w)$ (----) and $S^{(2)}(w)$ (——) for $C = \infty$, with $B(w) = 1 - e^{-w/0.75}$.

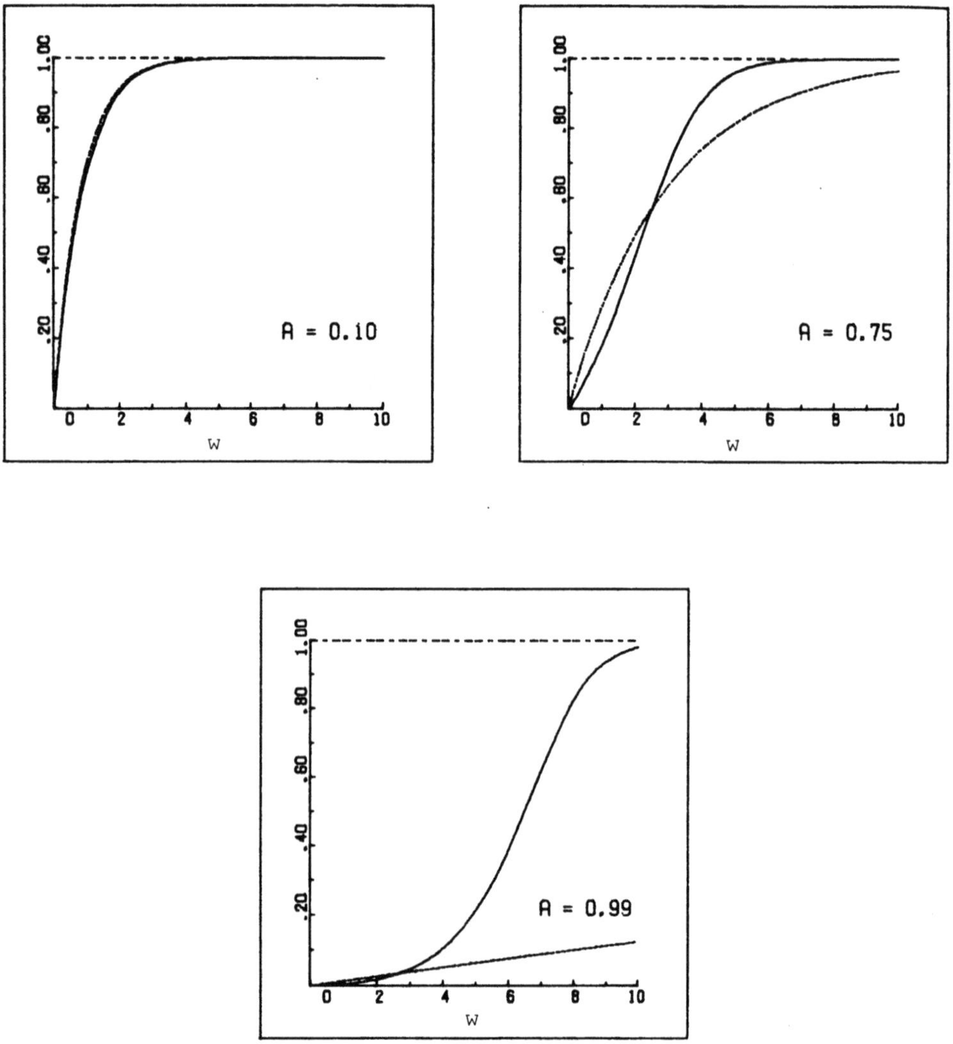

TABLE 1

Comparison of mean response times at Q_2 in the case of independent service times ($E\{\underline{s}_\infty\}$) and in the case of identical service times ($E\{\underline{s}_\infty^{(2)}\}$), with $B(w) = 1 - \exp(-w/0.75)$.

C	a	$E\{\underline{s}_\infty\}$	$E\{\underline{s}_\infty^{(2)}\}$
2	0.10	0.746	0.794
2	0.75	1.121	1.571
2	0.99	1.267	1.981
4	0.10	0.824	0.878
4	0.75	1.768	2.260
4	0.99	2.294	3.873
8	0.10	0.833	0.886
8	0.75	2.505	2.397
8	0.99	4.261	6.218
∞	0.10	0.833	0.886
∞	0.75	3.000	2.398
∞	0.99	75.000	6.382

- apart from the fact that m(w) cannot in general be determined explicitly from its defining equation,

$$m(w) = \int_0^w e^{-(1-m(w))t/\alpha}\, dB(t).$$

This equation is easily solved numerically in an iterative fashion.

It is interesting to compare the behaviour of the response time distribution $S^{(2)}(w)$ with that of the response time distribution $S(w)$ in Q_2 in the case of independent service times at both queues (with the same distribution B(.)); in the last case Q_2 behaves like a ./G/1 queue with finite capacity C. In figures 2-5 and in Table 1 comparison results are presented for the case of a negative exponential service time distribution, for $C = 2,4,8,\infty$ and for the light, medium and heavy traffic situation; the response time distribution $S(w)$ is given by (cf. Cohen [8], p. 546)

$$S(w) = [1 - e^{-(1-a)w/\beta}] / [1 - a\, e^{-(1-a)C/\beta}], \qquad 0 < w \leqslant C.$$

Observe that the assumption of identical service times has a strong regularizing effect in the heavy traffic situation as can be clearly seen in the case of infinite buffer size: the assumption of finite buffer size has a somewhat similar effect.

4. THE DISTRIBUTION OF THE AMOUNT OF WORK IN THE SECOND QUEUE

In order to derive the distribution of the amount of work in Q_2 (the "virtual waiting time"), we observe that the process (cf. (2.3)) $\{(\underline{x}_t,\underline{\eta}_t,\underline{v}_t^{(2)}),\ t \in [0,\infty)\}$ is a vector-valued Markov process. $\underline{\eta}_t$ is a so-called supplementary variable, and $\underline{v}_t^{(2)}$ is the amount of work in Q_2 a time t. Define for $\eta,v,t > 0$,

$$R_0(v,t) = \Pr\{\underline{x}_t = 0,\ \underline{v}_t^{(2)} < v \mid \underline{x}_0 = 0,\ \underline{v}_0^{(2)} = 0\}, \qquad (4.1)$$

$$R_j(\eta,v,t)\, d\eta = \Pr\{\underline{x}_t = j, \eta \leqslant \underline{\eta}_t < \eta + d\eta, \underline{v}_t^{(2)} < v \mid \underline{x}_0 = 0, \underline{v}_0^{(2)} = 0\},$$

$$j = 1,2,\ldots\ . \qquad (4.2)$$

We shall derive a set of relations by considering the process $\{(\underline{x}_t,\underline{\eta}_t,\underline{v}_t^{(2)}),\ t \in [0,\infty)\}$ during a small time interval $(t,t+\Delta t)$, $\Delta t > 0$. Our approach will be similar to the one of Cohen [8] for the M/G/1 queue, and to the one in [1] for the model with $C = \infty$. For those t for which $B(t) < 1$ we can write:

$$R_0(v,t+\Delta t) = [1-\frac{\Delta t}{\alpha}+o(\Delta t)]\,[R_0(v+\Delta t,t) +$$

$$+ \Pr\{\underline{x}_t = 1, \underline{\eta}_{t+\Delta t} = 0, \underline{v}_t^{(2)} + \underline{\tau}_n < v+\theta\,\Delta t \mid \underline{x}_0 = 0, \underline{v}_0^{(2)} = 0\}] + o(\Delta t),$$

$$0 < \theta < 1,\ t > 0,\ 0 < v < C; \quad (4.3)$$

$$R_0(C+,t+\Delta t) = [1-\frac{\Delta t}{\alpha}+o(\Delta t)]\,[R_0(C+,t) +$$

$$+ \Pr\{\underline{x}_t = 1, \underline{\eta}_{t+\Delta t} = 0 \mid \underline{x}_0 = 0, \underline{v}_0^{(2)} = 0\}] + o(\Delta t),$$

$$t > 0. \quad (4.4)$$

(Note that obviously $R_0(C,t) = R_0(C+,t) = R_0(\infty,t)$).
If $B(x) < 1$ then $(B(x+y) - B(x))/(1 - B(x))$ is the conditional probability that a service has a duration less than $x + y$, whenever it has a duration of at least x. Choose t large enough, i.e. $t > C$; we can now rewrite (4.3), (4.4), with $0 < \Delta\eta < \Delta t$:

$$R_0(v,t+\Delta t) = [1-\frac{\Delta t}{\alpha}+o(\Delta t)]\,[R_0(v+\Delta t,t) +$$

$$+ \int_{\eta=\Delta\eta}^{v} R_1(\eta-\Delta\eta, v-\eta, t)\,\frac{B(\eta-\Delta\eta+\Delta t) - B(\eta-\Delta\eta)}{1 - B(\eta-\Delta\eta)}\,d\eta] + o(\Delta t),$$

$$t > C,\ 0 < v < C, \quad (4.5)$$

$$R_0(C+,t+\Delta t) = [1-\frac{\Delta t}{\alpha} + o(\Delta t)]\,[R_0(C+,t) +$$

$$+ \int_{\eta=\Delta\eta}^{C} R_1(\eta-\Delta\eta, C-\eta, t)\,\frac{B(\eta-\Delta\eta+\Delta t) - B(\eta-\Delta\eta)}{1 - B(\eta-\Delta\eta)}\,d\eta +$$

$$+ \int_{\eta=C}^{t} R_1(\eta-\Delta\eta, C+, t)\,\frac{B(\eta-\Delta\eta+\Delta t) - B(\eta-\Delta\eta)}{1 - B(\eta-\Delta\eta)}\,d\eta] + o(\Delta t),\ t > C. \quad (4.6)$$

Next suppose $\underline{\eta}_t > 0$, $\Delta t < \underline{\eta}_t$; then for $\eta, t > 0$, $0 < v < C$:

$$R_1(\eta,v,t)\,d\eta = [1-\frac{\Delta t}{\alpha} + o(\Delta t)]\,\frac{1 - B(\eta)}{1 - B(\eta-\Delta t)}\,R_1(\eta-\Delta t, v+\Delta t, t-\Delta t)\,d\eta + o(\Delta t), \quad (4.7)$$

$$R_j(\eta,v,t)\,d\eta = [1-\frac{\Delta t}{\alpha} + o(\Delta t)]\,\frac{1 - B(\eta)}{1 - B(\eta-\Delta t)}\,R_j(\eta-\Delta t, v+\Delta t, t-\Delta t)\,d\eta +$$

$$+ [\frac{\Delta t}{\alpha} + o(\Delta t)]\,R_{j-1}(\eta-\Delta t, v+\Delta t, t-\Delta t)\,d\eta + o(\Delta t), \quad (4.8)$$

$$j=2,3,\ldots\ .$$

Finally consider the case $\underline{\eta}_t = 0+$, i.e. just before t a new service has been started. It is then impossible to choose Δt so small that $\Delta t < \underline{\eta}_t$. We can write for $t > C$, $0 < v < C$, with θ a number such that $0 < \theta < 1$:

$$R_1(\theta\Delta t,v,t)\,\Delta t = [1-\frac{\Delta t}{\alpha}+o(\Delta t)].$$

$$.[\int_{\eta=\Delta\eta}^{v+\theta\Delta t} R_2(\eta-\Delta\eta,v-\eta,t-\Delta t)\,\frac{B(\eta-\Delta\eta+\Delta t)-B(\eta-\Delta\eta)}{1-B(\eta-\Delta\eta)}\,d\eta]\,+$$

$$+\,[\frac{\Delta t}{\alpha}+o(\Delta t)]\;R_0(v+\Delta t,t-\Delta t)+o(\Delta t), \qquad (4.9)$$

$$R_1(\theta\Delta t,C+,t)\,\Delta t = [1-\frac{\Delta t}{\alpha}+o(\Delta t)].$$

$$.[\int_{\eta=\Delta\eta}^{C} R_2(\eta-\Delta\eta,C-\eta,t-\Delta t)\,\frac{B(\eta-\Delta\eta+\Delta t)-B(\eta-\Delta\eta)}{1-B(\eta-\Delta\eta)}\,d\eta\,+$$

$$+\int_{\eta=C}^{t} R_2(\eta-\Delta\eta,C+,t-\Delta t)\,\frac{B(\eta-\Delta\eta+\Delta t)-B(\eta-\Delta\eta)}{1-B(\eta-\Delta\eta)}\,d\eta]\,+$$

$$+\,[\frac{\Delta t}{\alpha}+o(\Delta t)]\;R_0(C+,t-\Delta t)+o(\Delta t), \qquad (4.10)$$

$$R_j(\theta\Delta t,v,t)\,\Delta t = [1-\frac{\Delta t}{\alpha}+o(\Delta t)].$$

$$.[\int_{\eta=\Delta\eta}^{v+\theta\Delta t} R_{j+1}(\eta-\Delta\eta,v-\eta,t-\Delta t)\,\frac{B(\eta-\Delta\eta+\Delta t)-B(\eta-\Delta\eta)}{1-B(\eta-\Delta\eta)}\,d\eta]+o(\Delta t),$$

$$j=2,3,\ldots, \qquad (4.11)$$

$$R_j(\theta\Delta t,C+,t)\,\Delta t = [1-\frac{\Delta t}{\alpha}+o(\Delta t)].$$

$$.[\int_{\eta=\Delta\eta}^{t} R_{j+1}(\eta-\Delta\eta,C-\eta,t-\Delta t)\,\frac{B(\eta-\Delta\eta+\Delta t)-B(\eta-\Delta\eta)}{1-B(\eta-\Delta\eta)}\,d\eta\,+$$

$$+\int_{\eta=C}^{t} R_{j+1}(\eta-\Delta\eta,C+,t-\Delta t)\,\frac{B(\eta-\Delta\eta+\Delta t)-B(\eta-\Delta\eta)}{1-B(\eta-\Delta\eta)}\,d\eta]+o(\Delta t),$$

$$j=2,3,\ldots\,. \qquad (4.12)$$

Next we divide the relations (4.5) - (4.12) by Δt and proceed to the limit $\Delta t \to 0$. We shall assume that the relevant limits exist (note that it was already tacitly assumed in (4.2) that the joint distribution of $\underline{x}_t$, $\underline{\eta}_t$ and $\underline{v}_t^{(2)}$ possesses a density w.r.t. $\underline{\eta}_t$). See also the discussion in [1]. It follows from (4.5) - (4.12) that for $\eta > 0$, $t > C$, $0 < v < C$, $B(t) \lesssim 1$,

$$\frac{\partial}{\partial t}R_0(v,t) = \frac{\partial}{\partial v}R_0(v,t) - \frac{1}{\alpha}R_0(v,t) + \int_{\eta=0}^{v} R_1(\eta,v-\eta,t)\frac{dB(\eta)}{1-B(\eta)}, \qquad (4.13)$$

$$\frac{\partial}{\partial t}R_0(C+,t) = -\frac{1}{\alpha}R_0(C+,t) + \int_{\eta=0}^{C} R_1(\eta,C-\eta,t)\,\frac{dB(\eta)}{1-B(\eta)}\,+$$

$$+\int_{\eta=C}^{t} R_1(\eta,C+,t)\,\frac{dB(\eta)}{1-B(\eta)}, \qquad (4.14)$$

$$\frac{\partial}{\partial\eta}R_1(\eta,v,t) - \frac{\partial}{\partial v}R_1(\eta,v,t) + \frac{\partial}{\partial t}R_1(\eta,v,t) = -\{\frac{1}{\alpha}+\frac{B'(\eta)}{1-B(\eta)}\}\,\times$$

$$\times\ R_1(\eta,v,t), \tag{4.15}$$

$$\frac{\partial}{\partial\eta} R_j(\eta,v,t) - \frac{\partial}{\partial v} R_j(\eta,v,t) + \frac{\partial}{\partial t} R_j(\eta,v,t) = -\{\frac{1}{\alpha} + \frac{B'(\eta)}{1-B(\eta)}\}\cdot$$

$$\cdot\ R_j(\eta,v,t) + \frac{1}{\alpha} R_{j-1}(\eta,v,t), \qquad j=2,3,\ldots . \tag{4.16}$$

$$R_1(0+,v,t) = \frac{1}{\alpha} R_0(v,t) + \int_{\eta=0}^{v} R_2(\eta,v-\eta,t) \frac{dB(\eta)}{1-B(\eta)}, \tag{4.17}$$

$$R_1(0+,C+,t) = \frac{1}{\alpha} R_0(C+,t) + \int_{\eta=0}^{C} R_2(\eta,C-\eta,t) \frac{dB(\eta)}{1-B(\eta)} +$$

$$+ \int_{\eta=C}^{t} R_2(\eta,C+,t) \frac{dB(\eta)}{1-B(\eta)}, \tag{4.18}$$

$$R_j(0+,v,t) = \int_{\eta=0}^{v} R_{j+1}(\eta,v-\eta,t) \frac{dB(\eta)}{1-B(\eta)}, \tag{4.19}$$

$$R_j(0+,C+,t) = \int_{\eta=0}^{C} R_{j+1}(\eta,C-\eta,t) \frac{dB(\eta)}{1-B(\eta)} +$$

$$+ \int_{\eta=C}^{t} R_{j+1}(\eta,C+,t)\frac{dB(\eta)}{1-B(\eta)}, \qquad j=2,3,\ldots . \tag{4.20}$$

In order to avoid unnecessarily lengthy calculations we assume, like we did in Section 3, that $a = \beta/\alpha < 1$. In [1] it is shown for the case $C = \infty$ that the functions $R_0(v,t)$ and $R_j(\eta,v,t)$, $j=1,2,\ldots,$ have for $t \to \infty$ a stationary limiting distribution if $a < 1$. The same statement holds for $C < \infty$, but we shall omit a proof here. We also make the assumption that the support S of B(.) is $[0,\infty)$. It can however readily be shown that this assumption is not relevant, and that the forthcoming results are also valid for the case that S is bounded.

Now introduce the following notation for $\eta, v > 0$, $j=1,2,\ldots$:

$$\lim_{t\to\infty} \Pr\{\underline{x}_t = 0, \underline{v}_t^{(2)} < v \mid \underline{x}_0 = 0, \underline{v}_0^{(2)} = 0\} \overset{def}{=}$$

$$\Pr\{\underline{x}_\infty = 0, \underline{v}_\infty^{(2)} < v\} \overset{def}{=} R_0(v), \tag{4.21}$$

$$\lim_{t\to\infty} \Pr\{\underline{x}_t = j, \eta \leqslant \underline{\eta}_t < \eta+d\eta, \underline{v}_t^{(2)} < v \mid \underline{x}_0 = 0, \underline{v}_0^{(2)} = 0\} \overset{def}{=}$$

$$\Pr\{\underline{x}_\infty = j, \eta \leqslant \underline{\eta}_\infty < \eta+d\eta, \underline{v}_\infty^{(2)} < v\} \overset{def}{=} R_j(\eta,v)\, d\eta, \tag{4.22}$$

$$Q_0(v) \overset{def}{=} R_0(v), \qquad Q_j(\eta,v) \overset{def}{=} \frac{R_j(\eta,v)}{1-B(\eta)} . \tag{4.23}$$

Letting $t \to \infty$ in (4.13) - (4.20) it follows from (4.21) - (4.23) for $\eta > 0$, $0 < v < C$:

$$0 = \frac{d}{dv} Q_0(v) - \frac{1}{\alpha} Q_0(v) + \int_{\eta=0}^{v} Q_1(\eta,v-\eta)\, dB(\eta), \tag{4.24}$$

$$0 = -\frac{1}{\alpha} Q_0(C+) + \int_{\eta=0}^{C} Q_1(\eta,C-\eta)dB(\eta) + \int_{\eta=C}^{\infty} Q_1(\eta,C+)dB(\eta), \tag{4.25}$$

$$\frac{\partial}{\partial\eta} Q_1(\eta,v) - \frac{\partial}{\partial v} Q_1(\eta,v) = -\frac{1}{\alpha} Q_1(\eta,v), \tag{4.26}$$

$$\frac{\partial}{\partial\eta} Q_j(\eta,v) - \frac{\partial}{\partial v} Q_j(\eta,v) = -\frac{1}{\alpha} Q_j(\eta,v) + \frac{1}{\alpha} Q_{j-1}(\eta,v), \tag{4.27}$$

$j=2,3,\ldots;$

$$Q_1(0+,v) = \frac{1}{\alpha} Q_0(v) + \int_{\eta=0}^{v} Q_2(\eta,v-\eta)\, dB(\eta), \tag{4.28}$$

$$Q_1(0+,C+) = \frac{1}{\alpha} Q_0(C+) + \int_{\eta=0}^{C} Q_2(\eta,C-\eta)\, dB(\eta) + \int_{\eta=C}^{\infty} Q_2(\eta,C+)\, dB(\eta), \tag{4.29}$$

$$Q_j(0+,v) = \int_{\eta=0}^{v} Q_{j+1}(\eta,v-\eta)\, dB(\eta), \tag{4.30}$$

$$Q_j(0+,C+) = \int_{\eta=0}^{C} Q_{j+1}(\eta,C-\eta)\, dB(\eta) + \int_{\eta=C}^{\infty} Q_{j+1}(\eta,C+)dB(\eta), \tag{4.31}$$

$j=2,3,\ldots\ .$

Defining

$$G(r,\eta,v) \overset{\text{def}}{=} \sum_{j=1}^{\infty} r^j\, Q_j(\eta,v), \qquad |r| \leqslant 1,\ \eta,v > 0, \tag{4.32}$$

it follows from (4.26), (4.27), (4.32) for $|r| \leqslant 1$, $\eta > 0$, $0 < v < C$:

$$\frac{\partial}{\partial\eta} G(r,\eta,v) - \frac{\partial}{\partial v} G(r,\eta,v) = -\frac{1-r}{\alpha} G(r,\eta,v). \tag{4.33}$$

The characteristic equations for this partial differential equation are (see e.g. Courant and Hilbert [9]):

$$d\eta = -\,dv = -\frac{\alpha}{1-r}\frac{dG(r,\eta,v)}{G(r,\eta,v)}\ .$$

The general solution of (4.33) reads:

$$G(r,\eta,v) = F(r,v+\eta)\, e^{-(1-r)\eta/\alpha}, \quad |r| \leqslant 1,\ \eta > 0,\ 0 < v < C, \tag{4.34}$$

where $F(.,.)$ is a function to be determined from (4.24), (4.25), (4.28) - (4.32). It follows from (4.28), (4.30), (4.32) for $|r| \leqslant 1$, $0 < v < C$,

$$G(r,0+,v) = \frac{1}{r} \int_{\eta=0}^{v} G(r,\eta,v-\eta)\, dB(\eta) - \int_{\eta=0}^{v} Q_1(\eta,v-\eta)\, dB(\eta) + \frac{r}{\alpha} Q_0(v); \tag{4.35}$$

this relation leads in combination with (4.24) and (4.34) to

$$F(r,v) = [1 - \frac{1}{r} \int_{\eta=0}^{v} e^{-(1-r)\eta/\alpha}\, dB(\eta)]^{-1}\, [-(\frac{1-r}{\alpha})Q_0(v) + \frac{d}{dv}Q_0(v)],$$

$$|r| \leqslant 1,\ 0 < v < C. \tag{4.36}$$

In a similar way (4.25), (4.29), (4.31), (4.32) and (4.34) yield:

$$F(r,C+) = [\,1 - \frac{1}{r} \int_{\eta=0}^{C} e^{-(1-r)\eta/\alpha}\, dB(\eta)]^{-1}.$$

$$.\ [-(\frac{1-r}{\alpha})\, Q_0(C+) + \frac{1}{r} \int_{\eta=C}^{\infty} G(r,\eta,C+)\, dB(\eta)],\ |r| \leqslant 1. \tag{4.37}$$

The functions $F(r,v)$ and $Q_0(v)$, $0 < v < C$, should be determined from the following conditions:

(i) they should satisfy (4.36).

(ii) $F(r,v)$ is according to its definition - and the definition of $G(r,\eta,v)$ - for fixed v, $v > 0$, an analytic function of r in the region $|r| \leqslant 1$.

Similar statements for $F(r,C+)$ and $Q_0(C+)$, combined with a norming condition (summing of probabilities to one for $v \to \infty$) lead to the determination of these two unknown terms. However, it is easier to observe that $Q_0(C+)$ and $F(r,C+)$ (and $G(r,\eta,C+)$) in fact only relate to queue Q_1, an M/G/1 queue. Hence from M/G/1 theory (see Cohen [8], p. 336, or [1], p. 142),

$$Q_0(C+) = \Pr\{\underline{x}_\infty = 0\} = 1 - a, \tag{4.38}$$

$$F(r,C+) = \sum_{j=1}^{\infty} r^j \frac{\partial}{\partial\eta} \Pr\{\underline{x}_\infty = j, \underline{\eta}_\infty < \eta\}\Big|_{\eta=0+}$$

$$= \frac{1-a}{\alpha}\, \frac{1-r}{\beta(\frac{1-r}{\alpha}) - r}\, r, \qquad |r| \leqslant 1, \tag{4.39}$$

$$G(r,\eta,C+) = \frac{1-a}{\alpha}\, \frac{1-r}{\beta(\frac{1-r}{\alpha}) - r}\, r\, e^{-(1-r)\eta/\alpha}, \qquad |r| \leqslant 1,\ \eta > 0. \tag{4.40}$$

Substitution of (4.38) - (4.40) in (4.37) leads to an identity.

We now determine $F(r,v)$ and $Q_0(v)$, $0 < v < C$. From (4.36) and the fact that $F(r,v)$ is an analytic function of r in the region $|r| \leqslant 1$, for fixed $v > 0$, it follows that

$$\frac{d}{dv} Q_0(v) = \frac{1 - m(v)}{\alpha} Q_0(v), \qquad 0 < v < C,$$

hence

$$Q_0(v) = A \exp[-\frac{1}{\alpha} \int_v^C (1 - m(u))\, du], \qquad 0 < v < C.$$

From (4.38) A = 1-a, so that

$$Q_0(v) = (1-a) \exp[-\frac{1}{\alpha} \int_v^C (1 - m(u))\, du],\ 0 < v < C. \qquad (4.41)$$

Finally,

$$F(r,v) = [1 - \frac{1}{r} \int_0^v e^{-(1-r)\eta/\alpha}\, dB(\eta)]^{-1}.$$
$$. \frac{1-a}{\alpha}(r - m(v)) \exp[-\frac{1}{\alpha} \int_v^C (1 - m(u))\, du],\ |r| \leq 1,\ 0 < v < C. \qquad (4.42)$$

It follows from (4.39) and (4.42) that F(r,v) has a discontinuity in v = C. This is not surprising, since F(r,v) is related to arrival epochs at Q_2, and at such epochs there is a positive probability that the amount of work in Q_2 jumps to C. Similarly G(r,η,v) has a discontinuity in v+η = C; for if at an epoch t_1, $\underline{v}^{(2)}_{t_1} = C$ then it is possible that $\underline{v}^{(2)}_{t_1} = C - \eta$, $\underline{\eta}_{t_1} = \eta$. In fact combination of (4.34), (4.40) and (4.42) shows that

$$G(r,\eta,v) = [1 - \frac{1}{r} \int_0^{v+\eta} e^{-(1-r)q/\alpha}\, dB(q)]^{-1} \frac{1-a}{\alpha} (r - m(v+\eta)).$$
$$. \exp[-\frac{1}{\alpha} \int_{v+\eta}^C (1 - m(u))\, du]\, e^{-(1-r)\eta/\alpha},$$
$$|r| \leq 1, \qquad 0 < v+\eta \leq C,$$
$$= \frac{1-a}{\alpha} \frac{1-r}{\beta(\frac{1-r}{\alpha}) - r}\, r\, e^{-(1-r)\eta/\alpha},\ |r| \leq 1,\ v+\eta > C. \qquad (4.43)$$

<u>Theorem 4.1</u>

$$\Pr\{\underline{v}^{(2)}_\infty < v\} = (1-a) \exp[-\frac{1}{\alpha} \int_v^C (1 - m(u))\, du] +$$
$$+ \int_v^C \frac{1-a}{\alpha} \frac{1 - m(z)}{1 - B(z)} \exp[-\frac{1}{\alpha} \int_z^C (1 - m(u))\, du]\, (1 - B(z-v))\, dz +$$
$$+ \frac{1}{\alpha} \int_{C-v}^\infty (1 - B(u))\, du, \qquad 0 < v \leq C, \qquad (4.44)$$

$$\Pr\{\underline{v}^{(2)}_\infty = 0\} = 1 - a + \frac{1}{\alpha} \int_C^\infty (1 - B(u))\, du. \qquad (4.45)$$

<u>Proof</u>

From (4.43),

$$G(1,\eta,v) = \frac{1-a}{\alpha}\,\frac{1-m(v+\eta)}{1-B(v+\eta)}\,\exp[-\frac{1}{\alpha}\int_{v+\eta}^{C}(1-m(u))\,du],$$

$$0 < v+\eta \leqslant C, \quad (4.46)$$

$$= \frac{1}{\alpha}, \qquad v+\eta > C.$$

Furthermore,

$$\Pr\{\underline{v}_\infty^{(2)} < v\} = \Pr\{\underline{x}_\infty = 0, \underline{v}_\infty^{(2)} < v\} + \Pr\{\underline{x}_\infty \geqslant 1, \underline{v}_\infty^{(2)} < v\}$$

$$= Q_0(v) + \int_0^\infty (1-B(\eta))\,G(1,\eta,v)\,d\eta. \quad (4.47)$$

(4.44) now follows from (4.41), (4.46) and (4.47); (4.45) is an immediate consequence of (4.44). □

Remark 4.1

$C \to \infty$ yields the known results for the infinite buffer case (see e.g. [2], p. 642).

Remark 4.2

(3.3) implies that the expected loss per message in Q_2 equals

$$(1-B(C))\,.\,E[(\underline{\tau}_n - C)|\underline{\tau}_n > C] = \int_C^\infty (u-C)\,dB(u) = \int_C^\infty (1-B(u))\,du.$$

Hence the mean amount of work, admitted at Q_2 per unit of time, equals

$$\frac{1}{\alpha}[\beta - \int_C^\infty (1-B(u))\,du].$$

This result is in agreement with (4.45).

REFERENCES

1. Boxma, O.J. Analysis of Models for Tandem Queues (Ph.D. Thesis, University of Utrecht, Utrecht, 1977).

2. Boxma, O.J. On a tandem queueing model with identical service times at both counters I,II, Adv. Appl. Prob. 11 (1979) 616-644-659.

3. Boxma, O.J. On the longest service time in a busy period of the M/G/1 queue, Stoch. Proc. Appl. 8 (1978) 93-100.

4. Boxma, O.J. The longest service time in a busy period,

Zeitschr. für Op. Research 24 (1980) 235-242.

5. Boxma, O.J. & Konheim, A.G. — Approximate analysis of exponential queueing systems with blocking, Acta Informatica 15 (1981) 19-66.

6. Bronshtein, O. & Gertsbakh, I. — An open exponential queueing network with losses and limited waiting spaces: a method of approximate analysis, Rep. Ben Gurion Univ. of the Negev, 1981.

7. Calo, S.B. — Message delays in repeated-service tandem connections, IEEE Trans. on Communications Vol. COM-29 (1981) 670-678.

8. Cohen, J.W. — The Single Server Queue (2nd ed.; North-Holland Publ. Cy., Amsterdam, 1982).

9. Courant, R. & Hilbert, D. — Methoden der Mathematischen Physik Bd. 2 (Springer, Berlin, 1937).

10. Jackson, J.R. — Networks of waiting lines, Op. Research 5 (1957) 518-521.

11. Kelly, F.P. — Reversibility and Stochastic Networks (Wiley, New York, 1979).

12. Kelly, F.P. — The throughput of a series of buffers, Adv. Appl. Prob. 14 (1982) 633-653.

13. Kleinrock, L. — Communication Nets; Stochastic Message Flow and Delay (Mc Graw Hill, New York, 1964).

14. Mitchell, C.R. Paulson, A.S. & Beswick, C.A. — The effect of correlated exponential service times on single server tandem queues, Naval Res. Logist. Quart. 24 (1977) 95-112.

XIV

QUEUES AND NETWORKS 6

FILES D'ATTENTE ET RESEAUX 6

THE CYCLE TIME DISTRIBUTION OF CYCLIC TWO-STAGE QUEUES WITH A NON-EXPONENTIAL SERVER

H. Daduna
Technische Universität Berlin
Fachbereich Mathematik
Strasse des 17. Juni 135
D-1000 Berlin 12

Summary:

We compute a recursion scheme for the (conditional) cycle time in a closed two-stage cycle where one server is exponential while the second has arbitrary service time distribution.

Keywords:

Cycle time, cyclic queue, GORDON-NEWELL networks, moments of cycle times.

1. Introduction:

The fundamental model of a multiprogrammed computer system which consists of a central processor unit (CPU) and a data transmission unit (DTU) is a two-node closed cyclic queueing system with N customers cycling around (Fig. 1). The CPU is represented as a FCFS exponential server, while the DTU (representing any device that supplies data storage, e.g. disk, drum, tape) is modelled by an FCFS server with general service times. The N customers cycling around represent the different programs subject to the multiprogrammed service mode.

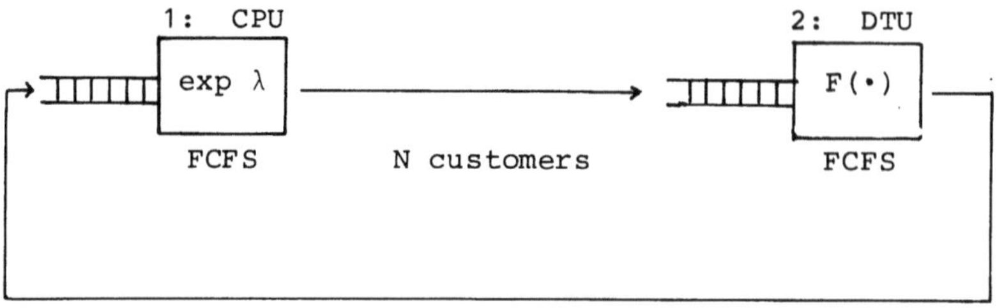

Fig. 1

This model has found considerable interest in the literature (for a review see KLEINROCK [5], p. 233) - firstly because of its importance for system design and secondly because it is the simplest GORDON-NEWELL netowrk of queues if the DTU service times are exponentially distributed.

A survey on GORDON-NEWELL networks may be found in LEMOINE [7]: they are closed networks of queues with FCFS nodes and exponential service times. LEMOINE describes the ways in which these networks are generalized to include non-exponential service times and non-FCFS service disciplines and remarks that up to now there remains as an open problem how to deal with networks like that of GORDON-NEWELL [3] (i.e., with FCFS service discipline), but with arbitrarily distributed service times. The difficulties that arise depend on the fact that such systems are not locally balanced and the equilibrium distributions are no longer of product form. That one can not hope to find simple steady state distributions was shown by LAVENBERG [6] who showed that the equilibrium queue length distribution at the DTU in the system of Fig. 1 is the same as that of an M/G/1 finite capacity queue with N-1 waiting places; the steady-state for this system was computed by KEILSON [4]. Using again the equivalence between our cyclic tandem and the M/G/1 finite capacity queueing system, REISER and KOBAYASHI [10] gave a recurison scheme to compute the equilibrium distribution of the system of Fig. 1. (For some comments on the note of REISER and KOBAYASHI see ALLEN [0].) Using LITTLE's formula this gives the expected cycle time.

Although a general solution of the (generalized) GORDON-NEWELL systems seems to be not possible it should be worth to attack at least special cases hoping to approach in that way the solution of more complicated systems.

Back to our cyclic queueing system: from a customers point of view the most important performance measure of the system might be the distribution of the cycle time, i.e. the time between entering the queue at the CPU and leaving the DTU next (see CHOW [2], who derived the cycle time for a pure exponential system).

Under the assumption of general DTU-time distribution the cycle time distribution is not known. Only recently BOXMA and DONK [1] gave an approximation for the Laplace-Stieltjes transform (LST) of the joint distribution of the sojourn times in both queues, yielding an approximation of the cycle time LST.

In his note [6] LAVENBERG derived an expression for the recursive (with respect to the number N of cycling customers) computation of the LST of the steady-state queueing time at the non-exponential server. The same can easily be obtained for the steady-state queueing time at the exponential server.

In this note we prove a simple recursion scheme for the computation of the LST of the steady-state cycle time in the system of Fig. 1. In fact we do a little more: We compute the conditional cycle time LST if the number of customers waiting at the CPU and DTU are given when the cycle starts. This conditional cycle time LST does not depend on the assumption that the system is in steady-state. So we have - if the systems state distribution is known - computed the non-equilibrium cycle time distribution and - using LAVENBERG's result [6] - the equilibrium cycle time distribution.

The expressions of our recursion scheme can be differentiated readily in order to obtain moments of any order (if they exist) of the equilibrium and non-equilibrium cycle time.

Acknowledgement: I want to thank Prof. Dr. R. Schassberger, TU Berlin, and Dipl.-Ing. Dietmar Voss, Hewlett Packard, for helpful discussions on the subject of this paper.

2. The model:

(See Fig. 1). We consider a two-stage (closed) cyclic queue where N customers are cycling. Service times at server 1 are exponentially distributed with parameter $\lambda > 0$, service times at server 2 have general distribution function $F(\cdot)$ and Laplace-Stieltjes transform (LST) $\phi(s)$; $s \geq 0$, all service times are mutually independent. We assume $F(0+) < 1$.

We consider the system at departure points $(\tau_i : i=1,2,\ldots)$ of customers from server 2. Then the number $n(\tau_i+)$, $i=1,2,\ldots$, of customers at server 1 just after that departure moments is a discrete time Markov chain which is irreducible and aperiodic and has a unique steady state $(\pi_N(k),\ k=1,2,\ldots,N)$.

Suppose that Y and S_n are independent random variables with distribution functions

$$F(\cdot),\ \Gamma_{\lambda,n}(t) = 1 - \sum_{k=0}^{n-1} e^{-\lambda t} \frac{(\lambda t)^k}{k!}\quad,\ \text{resp.,}$$

and define

$$\alpha_n = \int_{[0,\infty)} \Big(1-F(t)\Big) \frac{\lambda^n}{(n-1)!}\, t^{n-1} e^{-\lambda t} dt,\ n=1,2,\ldots$$

and

$$\alpha_0 = 1,$$

$$\left(\text{ i.e. for } n \geq 1 \quad \alpha_n = P(S_n < Y) = \int_{[0,\infty)} F(dt)\ \Gamma_{\lambda,n}(t)\right),$$

then the nonzero transition probabilities of the Markov chain $n(\cdot)$ are given by

$$p_{jk} = \begin{cases} \alpha_l - \alpha_{l+1} & \text{if } k = j+1-l \quad 0 \leq l \leq j-1 \\ \alpha_j & \text{if } k = 1 \end{cases}$$

for $1 \leq j \leq N-1$, and $p_{Nk} = p_{N-1,k}$ if $1 \leq k \leq N$.

If we define further

$$a_0 = 1,$$

$$a_1 = \alpha_1(1-\alpha_1)^{-1},$$

$$a_k = \left(\sum_{l=1}^{k-1} a_l \alpha_{k+1-l} + \alpha_k\right)(1-\alpha_1)^{-1}\quad,\ k \geq 2\ ,$$

$$A_k = \sum_{i=0}^{k} a_i,\ k \geq 0\ ,$$

then we have for the steady state of $\Big(n(\tau_i+) : i=1,2,\ldots\Big)$

$$\pi_N(n) = a_{N-1} A_{N-1}^{-1}\quad,\quad n = 1,\ldots,N.$$

It is shown easily that the following recursion holds:

$$\pi_N(n) = \pi_{N-1}(n-1)\cdot A_{N-2} \cdot A_{N-1}^{-1}\ ,\ 2 \leq n \leq N.$$

(See LAVENBERG [6]).

3. The cycle time:

We consider a typical test customer C at the moment when he leaves server 2 and the conditional LST $f_N(n)(s)$, $s \geq 0$, of the time until he leaves server 2 next time, under the condition that at the starting point of that cycle there are $n \in \{1,\dots,N\}$ customers at server 1 including C.

From the definition we have

$$(1) \quad f_N(N)(s) = \frac{\lambda}{\lambda+s} \; f_N(N-1)(s).$$

For $1 \leq n < N$ we compute $f_N(n)(s)$ by a recursion scheme:
If the test customer C has joined the queue at server 1 (and finds there n-1 customers before him) then we consider the system next time when the first of the following two events shall happen:

- the service time Y of the customer just in service at server 2 ends;
- the test customer C leaves server 1. Let S_n be his sojourn time at server 1.

The distribution of the time until the first of these events shall happen is the same as that of a random variable $T = \min(Y,S_n)$, where S_n is independent of Y and has distribution function $\Gamma_{\lambda,n}(\cdot)$.

Let N_T be the number of customers which are served at server 1 in the time of length T between the departure of C from server 2 and the moment we look at the system next time.

Further let for $t \geq 0$ $\phi_t(s)$, $s \geq 0$, be the LST of a random variable which has distribution function

$$P(Y \leq s+t \mid Y > t), \quad s \geq 0,$$

where Y has LST $\phi(s)$.

Taking $Y, S_n, N_{(\cdot)}$ as given above and observing that any customer who leaves server 2 after C has started his cycle, shall not influence that cycle time of C any longer, we have

$$\begin{aligned} f_N(n)(s) &= \\ &= \int_{[o,\infty)} \Gamma_{\lambda,n}(dt)\, P(Y>t)\, e^{-st} \phi_t(s)\, \phi^{N-1}(s) \; + \\ &+ \int_{[o,\infty)} F(dt) \sum_{k=o}^{n-1} P(N_t=k, S_n \geq t)\, f_{N-1}(n-k)(s)\, e^{-st}, \end{aligned}$$

$$1 \leq n \leq N-1.$$

In this formula S_n is a random variable which represents the time between the moment at which C's cycle starts and the time of his departure from server 1. With this interpretation we have for k=0,1,..,n-1

$$(N_t = k) \subseteq (S_n \geq t).$$

Using this and

$$\phi_t(s) = \frac{e^{ts}}{1-F(t)} \int_{(t,\infty)} F(dx)e^{-sx}$$

we obtain

(2) $f_N(n)(s) =$

$$= \phi^{N-1}(s) \int_{(0,\infty)} F(dx)e^{-sx}\Gamma_{\lambda,n}(x) +$$

$$+ \sum_{k=0}^{n-1} f_{N-1}(n-k)(s) \int_{[0,\infty)} F(dt)e^{-t(s+\lambda)} \frac{(\lambda t)^k}{k!},$$

$$1 \leq n \leq N-1.$$

Together with the initial condition

(3) $f_1(1)(s) = \frac{\lambda}{\lambda+s} \cdot \phi(s)$,

this proves our

1. Proposition: The conditional LST of the cycle time in the two stage closed network of Fig. 1 where one server is exponential and the second server has general service time distribution and all service times are independent is given by (1),(2),(3) in a recursive scheme.

An immediate consequence of the proposition is

2. Corollary:

a) Let X_i, i = 1,2,..., be the time which the customer who caused the i-th jump in the system needs for his cycle which began at τ_i+. Suppose $n(\tau_i+)$ has distribution $(p_i(n): n=1,\dots,N)$. Then the LST $f_{N,X_i}(s)$, $s \geq 0$, of X_i is given by (1),(2),(3) and

(4) $f_{N,X_i}(s) = \sum_{n=1}^{N} p_i(n)f_N(n)(s)$.

b) The steady-state cycle time LST $f_N(s)$, $s \geq 0$, in the system of Fig. 1 is given by (1),(2),(3) and

$$f_N(s) = \sum_{n=1}^{N} \pi_N(n)f_N(n)(s) .$$

3. Remarks:

a) If server 2 has exponential service times with parameter $\mu > 0$, then (1),(2) and (3) turn to

(5) $f_N(n)(s) =$

$$= \left(\frac{\mu}{\mu+s}\right)^N\left(\frac{\lambda}{\lambda+\mu+s}\right)^n + \sum_{k=0}^{n-1} f_{N-1}(n-k)(s) \left(\frac{\lambda}{\lambda+\mu+s}\right)^k \frac{\mu}{\lambda+\mu+s}, \quad 1 \leq n \leq N-1,$$

(6) $f_N(N)(s) = \frac{\lambda}{\lambda+s} f_N(N-1)(s)$, and

(7) $f_1(1)(s) = \frac{\lambda}{\lambda+s} \frac{\mu}{\mu+s}$,

which by virtue of (4) again proves the result of CHOW [2], who computed the LST of the steady-state cycle time in a two-stage exponential system:

$$f_N(s) = \frac{\lambda^N \mu^N}{\lambda^N - \mu^N} \left(\left(\frac{1}{\mu+s}\right)^N - \left(\frac{1}{\lambda+s}\right)^N \right), \quad N \geq 1.$$

b) But the formulas (5), (6), (7) give even more: They determine the solution of the first-entrance-equations for a two-stage cycle which were derived in [9] in order to generalize CHOW's result, and which were not solved there.

c) The class of finite mixtures of Erlang distribution, i.e. of distributions of the form

$$F(t) = \sum_{i=o}^{m} q_i \Gamma_{\mu,i}(t) \quad , \quad q_i \in [0,1], \quad \sum_{i=o}^{m} q_i = 1, \ n > 0, \ m \in \mathbb{N} ,$$

is dense in the class of all probability distributions on the positive half-axis. (See SCHASSBERGER [8], p. 31). With respect to our proposition this class is very easy to work with, since the integrals in (2) reduce to finite polynomials in s:

$$f_1(1)(s) = \frac{\lambda}{\lambda+s} \left(\sum_{i=o}^{m} q_i \left(\frac{\mu}{\mu+s}\right)^i \right)$$

$$f_N(n)(s) =$$

$$= \left(\sum_{i=o}^{m} q_i \left(\frac{\mu}{\mu+s}\right)^i \right)^{N-1} \left(\sum_{i=1}^{m} q_i \left\{ \left(\frac{\mu}{\mu+s}\right)^i - \sum_{k=o}^{n-1} \frac{\mu^i \lambda^k}{(\lambda+\mu+s)^{i+k}} \binom{i+k-1}{i-1} \right\} \right) +$$

$$+ \sum_{k=o}^{n-1} f_{N-1}(n-k)(s) \left(q_o \delta_{ok} + \sum_{i=1}^{m} q_i \frac{\mu^i \lambda^k}{(\lambda+\mu+s)^{i+k}} \binom{i+k-1}{i-1} \right),$$

$$n \leq N-1,$$

where δ_{lr} is the Kronecker delta.

d) By differentiation we obtain from (1), (2), (3) the moments (if they exist) of conditional cycle times. Evaluation of the r-th differential of (2) at point 0+ yields for the r-th moment

$$(-1)^r f_N^{(r)}(n)(o+) =$$

$$(8) \quad = (-1)^r \sum_{l=o}^{r} \binom{r}{l} \left(\phi^{N-1}(o+)\right)^{(l)} \int_{(o,\infty)} F(dx)\,\Gamma_{\lambda,n}(x)\,(-x)^{r-l} +$$

$$+ \sum_{k=o}^{n-1} (-1)^r \sum_{l=o}^{r} \binom{r}{l} f_{N-1}^{(l)}(n-k)(o+) \int_{[o,\infty)} F(dt)\,\frac{(\lambda t)^k}{k!}\,e^{-\lambda t}(-t)^{r-l}$$

if $1 \le n < N$.

If we write $\mu^{(h)}$ for the h-th moment of $F(\cdot)$, then using the definition of the α_k, $k = 0,1,\ldots$, we get for the conditional moments

$$(-1)^r f_N^{(r)}(n)(o+) =$$

$$= (-1)^r \sum_{l=o}^{r} (-1)^{r-l} \binom{r}{l} \left(\phi^{N-1}(o+)\right)^{(l)} \Big[\mu^{(r-l)} - \sum_{k=o}^{n-1} \frac{(r-l+k)!}{k!} \cdot \lambda^{l-r} \cdot$$

$$(9) \quad \cdot \left(\alpha_{r-l+k} - \alpha_{r-l+k+1}\right)\Big] +$$

$$+ \sum_{k=o}^{n-1} \sum_{l=o}^{r} (-1)^l \binom{r}{l} f_{N-1}^{(l)}(n-k)(o+)\left(\alpha_{k+r-l} - \alpha_{k+r-l+1}\right) \cdot \frac{(r-l+k)!}{k!} \cdot \lambda^{l-r}$$

for $1 \le n < N$.

One should notice that for $F(t) = \sum_{i=o}^{m} q_i \Gamma_{\lambda i}(t)$ the α_k, $k = 0,1,\ldots$, may be computed explicitly in simple terms. The same holds if $F(\cdot)$ is a hyperexponential distribution or a discrete distribution.

e) Unconditioning of (8) on n yields the expected cycle time **W** which was obtained by REISER and KOBAYASHI [10] and ALLEN [0] via LITTLE's formula as

$$\mathbf{W} = \rho \cdot \mu ,$$

where $\rho = 1 - p_N(N)$ is the utilization of server 2. ($p_N(N)$ is the steady state probability of finding all customers in server 2.) The problems arising in determining the seemingly simple expression for **W** stem from the fact that $p_N(N)$ is to compute via an N-fold differentiation.

For the case of two-stage hyperexponential and two-stage Erlang distribution REISER and KOBAYASHI reduced that problem to the solution of certain quadratic equations.

f) We checked our result on expected cycle times - obtained from (8) - by comparing it with the results on W by ALLEN and REISER, KOBAYASHI.

4. Numerical results:

From corollary 2.b) and from (9) we get simple recursive expressions for the moments of cycle times in equilibrium.

For different levels of multiprogramming (N=1,2,...,7) we compared the effects of different service time distributions at server 2 on the mean and variance of the cycle time distribution using

i) two-stage hyperexponential distribution

$$H_2(\alpha_1,\alpha_2;\mu_1,\mu_2)(t) =$$
$$= 1-\alpha_1 e^{-\mu_1 t} - \alpha_2 e^{-\mu_2 t}, \quad t \geq 0,$$
$$0 \leq \alpha_i \leq 1, \; i = 1,2$$
$$0 < \mu_i \quad, \; i = 1,2$$
$$\alpha_1 + \alpha_2 = 1 \;;$$

ii) exponential distribution

$$\exp(\mu)(t) = 1-e^{-\mu t} \quad t \geq 0, \; 0 < \mu;$$

iii) Erlang two-stage distribution

$$\Gamma_{2\mu,2}(t) = 1-e^{-2\mu t} - e^{-2\mu t} \cdot 2\mu t, \; t \geq 0, \; 0 < \mu;$$

iv) constant service time distribution

$$D(b)(t) = \begin{cases} 1 & t \geq b \\ 0 & 0 \leq t < b \end{cases} \qquad t \geq 0, \; 0 < b.$$

Firstly we assumed the service time distribution $F(\cdot)$ at server 2 to have mean 1 and varied the parameter λ of the exponential service time at server 1. (For the hyperexponential distribution we used parameters

$$\alpha_1 = \frac{1}{2} - \sqrt{\frac{1}{6}}, \; \alpha_2 = \frac{1}{2} + \sqrt{\frac{1}{6}}, \; \mu_1 = 1-\sqrt{\frac{2}{3}}, \; \mu_2 = 1+\sqrt{\frac{2}{3}}.)$$

Under this assumptions the means of the cycle times behave very regular: they increase almost linear in N; and for fixed N they increase always in order

$D(1)(\cdot)$, $\Gamma_{2,2}(\cdot)$, $\exp(1)(\cdot)$, $H_2(\alpha_1,\alpha_2;\mu_1,\mu_2)(\cdot)$, as expected.

The same behaviour show the variances of cycle times under the above conditions, if the parameter λ of the service times at server 1 is set to be 0,375; 0,75; 0,1.

For $\lambda = 1,5$ we have a fundamental change for deterministic service times (see Fig. 2): the variances of the cycle times decrease. This is easily interpreted: for $\lambda \gg \mu$ and large N server 2 is a bottleneck. Therefore with a great probability almost all customers stay at server 2 when the test customer C arrives at server 1. This implies, that with great probability all customers present at server 1 at the start of C's cycle and the test customer himself will be served at node 1 before server 2 becomes empty next time after C's arrival at node 1. But variation of cycle times under deterministic service time distribution at server 2 stems only from service at node 1.

Now deterministic service times can be approximated by Erlang service times with increasing number of phases. So one should expect the same behaviour of cycle time variances which was found for deterministic service times at node 2 if one has service time distribution $\Gamma_{m\mu,m}(t)$ for great m. But this is not the case (see Fig. 3, 4). In the case of Fig. 3 the variance of the service times at node 2 is not small enough; while in the case of Fig. 4 the interpretation is as follows: the effect of the great variance of server-1 service times vanishes if N increases (as it does in the deterministic case); but from a certain level of multiprogramming $\tilde{N}$ the increase of variance at server 2 exceeds the loss of variance at server 1.

Variance of cycle time versus service time distribution.

Server 1: exp(1.5)

Server 2:

⊙ $H_2(\frac{1}{2} - 6^{-1/2}, \frac{1}{2} + 6^{-1/2}; 1-(\frac{2}{3})^{1/2}, 1+(\frac{2}{3})^{1/2})$

•exp(1)

+$\Gamma_{2,2}$

×D(1)

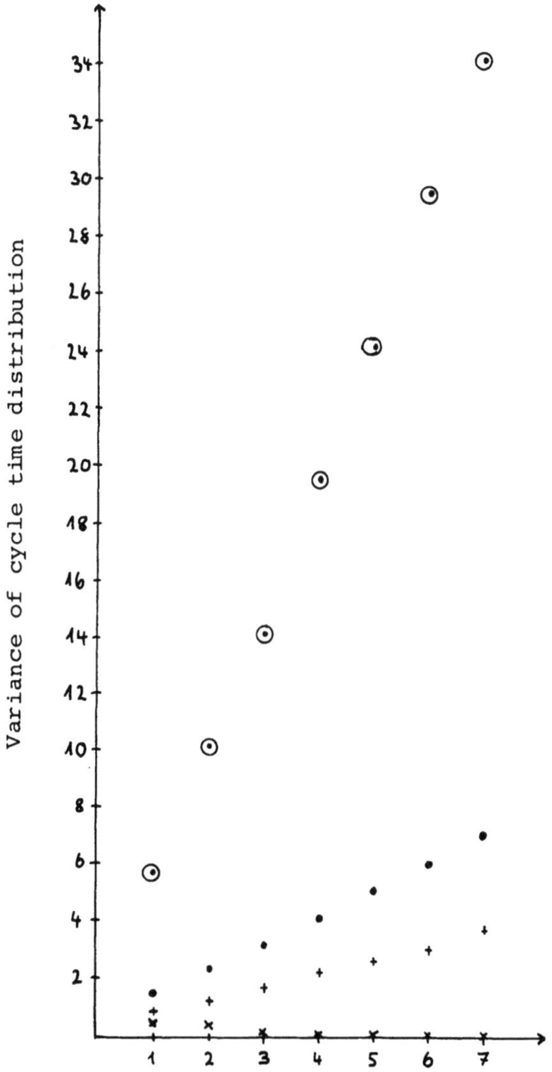

Number of customers in system (level of multiprogramming)

Fig. 2

Approximation of deterministic service time distribution by Erlang distributions

Server 1:
exp(1)
Server 2:
$\Gamma_{2.5,5}$

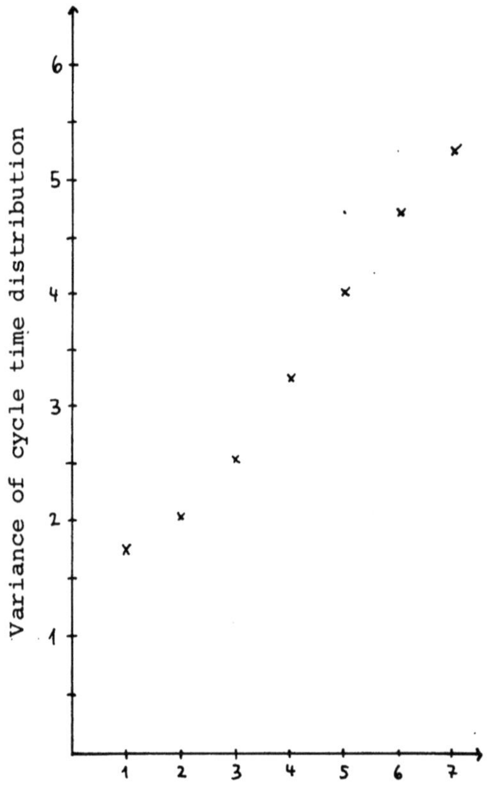

Server 1:
exp(1)
Server 2:
$\Gamma_{25,50}$

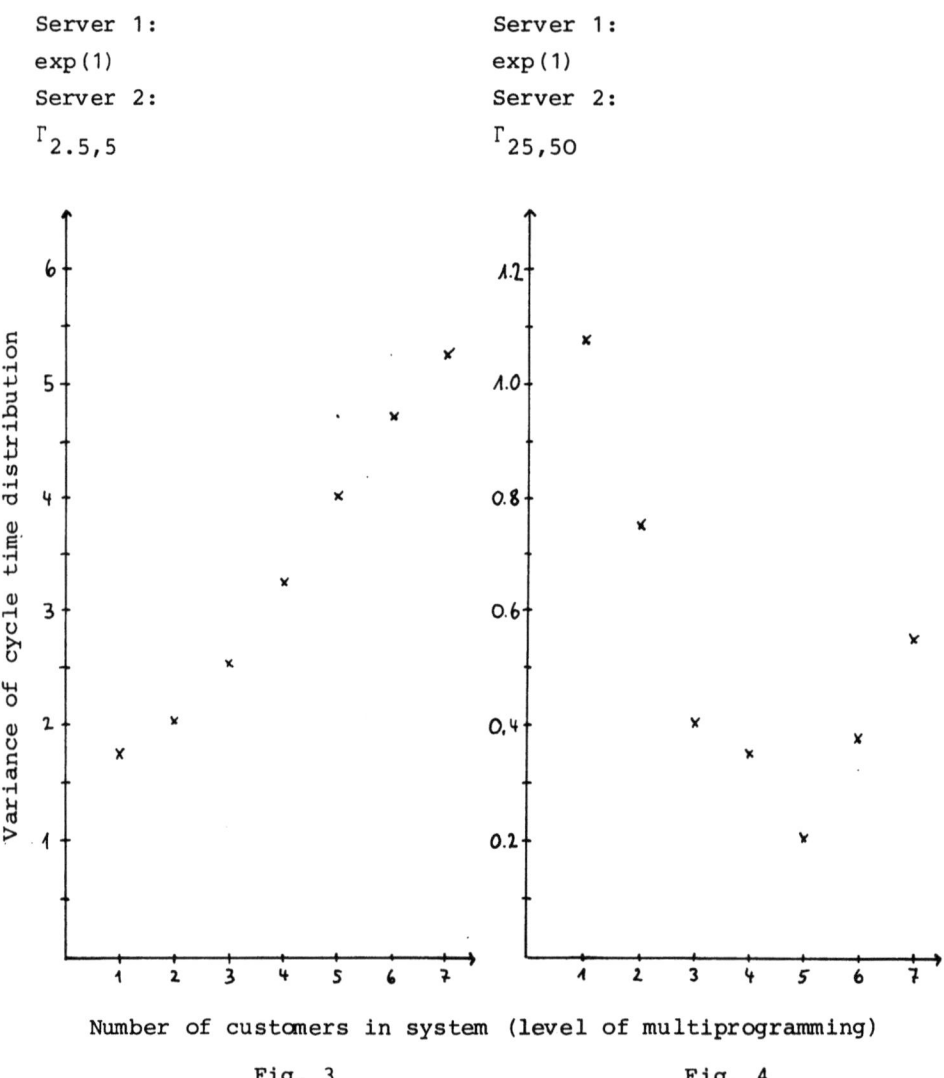

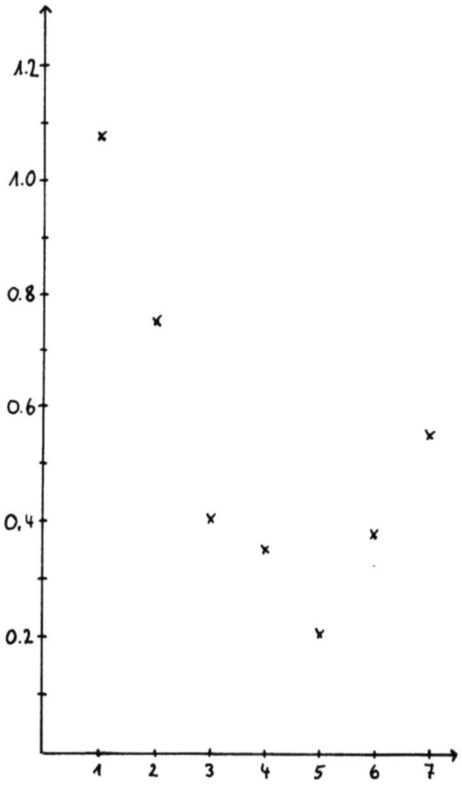

Number of customers in system (level of multiprogramming)

Fig. 3

Fig. 4

References:

[0] ALLEN,A.O.: Probability, statistics and queueing theory, Academic Press, 1978

[1] BOXMA,O.J.; DONK,P.: On response time and cycle time distributionsin a two-stage cyclic queue, Preprint Nr. 203, University of Utrecht, 1981

[2] CHOW,W.M.: The cycle time distribution of exponential cyclic queues, J.Ass.Comp.Mach., 27(2), 1980, 281-286

[3] GORDON,W.T.; NEWELL,G.V.: Closed queueing systems with exponential servers, Operations Res., 15, 1967, 252-265

[4] KEILSON,J.: The ergodic queue length distribution for queueing systems with finite capacity, J. Roy.Stat.Soc.,Ser.B, 28, 1966, 190-201

[5] KLEINROCK,L.: Queueing systems, Vol.2, WILEY, 1976

[6] LAVENBERG,S.S.: The steady-state queueing time distribution for the M/G/1-finite capacity queue, Management Sci. 21, (5), 1975, 501-506

[7] LEMOINE,A.J.: Networks of queues - A survey of equilibrium analysis, Management Sci., 24,(4), 1977, 464-481

[8] SCHASSBERGER,R.: Warteschlangen, SPRINGER, 1973

[9] SCHASSBERGER,R.; DADUNA,H.: The time for a round-trip in a cycle of exponential queues, J.Ass.Comp.Mach. (30), 146-150, 1983

[10] REISER,M.; KOBAYASHI,H.: The effects of service time distributions on system performance, Inform. Process. 74, North-Holland Publ. Company, 1974

Lecture Notes in Control and Information Sciences

Edited by A. V. Balakrishnan and M. Thoma

Vol. 22: Optimization Techniques
Proceedings of the 9th IFIP Conference on Optimization Techniques,
Warsaw, September 4–8, 1979
Part 1
Edited by K. Iracki, K. Malanowski, S. Walukiewicz
XVI, 569 pages. 1980

Vol. 23: Optimization Techniques
Proceedings of the 9th IFIP Conference on Optimization Techniques,
Warsaw, September 4-8, 1979
Part 2
Edited by K. Iracki, K. Malanowski, S. Walukiewicz
XV, 621 pages. 1980

Vol. 24: Methods and Applications in Adaptive Control
Proceedings of an International Symposium
Bochum, 1980
Edited by H. Unbehauen
VI, 309 pages. 1980

Vol. 25: Stochastic Differential Systems – Filtering and Control
Proceedings of the IFIP-WG7/1 Working Conference
Vilnius, Lithuania, USSR, Aug. 28 – Sept. 2, 1978
Edited by B. Grigelionis
X, 362 pages. 1980

Vol. 26: D. L. Iglehart, G. S. Shedler
Regenerative Simulation of Response Times in Networks of Queues
XII, 204 pages. 1980

Vol. 27: D. H. Jacobson, D. H. Martin, M. Pachter, T. Geveci
Extensions of Linear-Quadratic Control Theory
XI, 288 pages. 1980

Vol. 28: Analysis and Optimization of Systems
Proceedings of the Fourth International Conference on Analysis and Optimization of Systems
Versailles, December 16–19, 1980
Edited by A. Bensoussan and J. L. Lions
XIV, 999 pages. 1980

Vol. 29: M. Vidyasagar,
Input-Output Analysis of Large-Scale Interconnected Systems –
Decomposition, Well-Posedness and Stability
VI, 221 pages. 1981

Vol. 30: Optimization and Optimal Control
Proceedings of a Conference Held at Oberwolfach, March 16–22, 1980
Edited by A. Auslender, W. Oettli, and J. Stoer
VIII, 254 pages. 1981

Vol. 31: Berc Rustem
Projection Methods in Constrained Optimisation and Applications to Optimal Policy Decisions
XV, 315 pages. 1981

Vol. 32: Tsuyoshi Matsuo,
Realization Theory of Continuous-Time Dynamical Systems
VI, 329 pages, 1981

Vol. 33: Peter Dransfield
Hydraulic Control Systems –
Design and Analysis of Their Dynamics
VII, 227 pages, 1981

Vol. 34: H.W. Knobloch
Higher Order Necessary Conditions in Optimal Control Theory
V, 173 pages, 1981

Vol. 35: Global Modelling
Proceedings of the IFIP-WG 7/1 Working Conference Dubrovnik, Yugoslavia,
Sept. 1–5, 1980
Edited by S. Krčevinac
VIII, 232 pages, 1981

Vol. 36: Stochastic Differential Systems
Proceedings of the 3rd IFIP-WG 7/1 Working Conference
Visegrád, Hungary, Sept. 15–20, 1980
Edited by M. Arató, D. Vermes, A.V. Balakrishnan
VI, 238 pages, 1981

Vol. 37: Rüdiger Schmidt
Advances in Nonlinear Parameter Optimization
VI, 159 pages, 1982

Vol. 38: System Modeling and Optimization
Proceedings of the 10th IFIP Conference
New York City, USA, Aug. 31 – Sept. 4, 1981
Edited by R.F. Drenick and F. Kozin
XI, 894 pages. 1982

Vol. 39: Feedback Control of Linear and Nonlinear Systems
Proceedings of the Joint Workshop on Feedback and Synthesis of Linear and Nonlinear Systems
Bielefeld/Rom
XIII, 284 pages. 1982

Vol. 40: Y.S. Hung, A.G.J. MacFarlane
Multivariable Feedback:
A Quasi-Classical Approach
X, 182 pages. 1982

Vol. 41: M. Gössel
Nonlinear Time-Discrete Systems –
A General Approach by Nonlinear Superposition
VIII, 112 pages. 1982

Vol. 42: Advances in Filtering and Optimal Stochastic Control
Proceedings of the IFIP-WG 7/1 Working Conference
Cocoyoc, Mexico, February 1–6, 1982
VIII, 391 pages. 1982

Lecture Notes in Control and Information Sciences

Edited by A. V. Balakrishnan and M. Thoma

Vol. 43: Stochastic Differential Systems
Proceedings of the 2nd Bad Honnef Conference
of the SFB 72 of the DFG at the University of Bonn
June 28 – July 2, 1982
Edited by M. Kohlmann and N. Christopeit
XII, 377 pages. 1982.

Vol. 44: Analysis and Optimization of Systems
Proceedings of the Fifth International
Conference on Analysis and Optimization of Systems
Versailles. December 14–17, 1982
Edited by A. Bensoussan and J. L. Lions
XV, 987 pages, 1982

Vol. 45: M. Arató
Linear Stochastic Systems
with Constant Coefficients
A Statistical Approach
IX, 309 pages. 1982

Vol. 46: Time-Scale Modeling of Dynamic Networks
with Applications to Power Systems
Edited by J. H. Chow
X, 218 pages. 1982

Vol. 47: P. A. Ioannou, P. V. Kokotovic
Adaptive Systems with Reduced Models
V, 162 pages. 1983,

Vol. 48: Yaakov Yavin
Feedback Strategies for Partially
Observable Stochastic Systems
VI, 233 pages, 1983

Vol. 49: Theory and Application of Random Fields
Proceedings of the IFIP-WG 7/1
Working Conference
held under the joint auspices of the
Indian Statistical Institute
Bangalore, India, January 1982
Edited by G. Kallianpur
VI. 290 pages. 1983

Vol. 50: M. Papageorgiou
Applications of Automatic Control Concepts
to Traffic Flow Modeling and Control
IX, 186 pages. 1983

Vol. 51: Z. Nahorski, H.F. Ravn, R.V.V. Vidal
Optimization of Discrete Time Systems
The Upper Boundary Approach
V, 137 pages 1983

Vol. 52: A. L. Dontchev
Perturbations, Approximations and Sensitivity Analysis
of Optimal Control Systems
IV, 158 pages. 1983

Vol. 53: Liu Chen Hui
General Decoupling Theory of Multivariable
Process Control Systems
XI, 474 pages. 1983

Vol. 54: Control Theory for Distributed
Parameter Systems and Applications
Edited by F. Kappel, K. Kunisch,
W. Schappacher
VII, 245 pages. 1983.

Vol. 55: Ganti Prasada Rao
Piecewise Constant Orthogonal Functions
and Their Application to Systems and Control
VII, 254 pages. 1983.

Vol. 56: Dines Chandra Saha, Ganti Prasada R
Identification of Continuous
Dynamical Systems
The Poisson Moment Functional
(PMF) Approach
IX, 158 pages. 1983.

Vol. 57: T. Söderström, P. G. Stoica
Instrumental Variable Methods
for System Identification
VII, 243 pages. 1983.

Vol. 58: Mathematical Theory of
Networks and Systems
Proceedings of the MTNS-83 International
Symposium
Beer Sheva, Israel, June 20–24, 1983
Edited by P. A. Fuhrmann
X, 906 pages. 1984

Vol. 59: System Modelling and Optimization
Proceedings of the 11th IFIP Conference
Copenhagen, Denmark, July 25-29, 1983
Edited by P. Thoft-Christensen
IX, 892 pages. 1984

Vol. 60: Modelling and Performance
Evaluation Methodology
Proceedings of the International Seminar
Paris, France, January 24–26, 1983
Edited by F. Bacelli and G. Fayolle
VII, 655 pages. 1984

MIX
Papier aus verantwortungsvollen Quellen
Paper from responsible sources
FSC® C105338

If you have any concerns about our products,
you can contact us on
ProductSafety@springernature.com

In case Publisher is established outside the EU,
the EU authorized representative is:
Springer Nature Customer Service Center GmbH
Europaplatz 3, 69115 Heidelberg, Germany

Printed by Libri Plureos GmbH
in Hamburg, Germany